Advances in crop modelling for a sustainable agriculture

It is widely recognised that agriculture is a significant contributor to global warming and climate change. Agriculture needs to reduce its environmental impact and adapt to current climate change whilst still feeding a growing population, i.e. become more 'climate-smart'. Burleigh Dodds Science Publishing is playing its part in achieving this by bringing together key research on making the production of the world's most important crops and livestock products more sustainable.

Based on extensive research, our publications specifically target the challenge of climate-smart agriculture. In this way we are using 'smart publishing' to help achieve climate-smart agriculture.

Burleigh Dodds Science Publishing is an independent and innovative publisher delivering high quality customer-focused agricultural science content in both print and online formats for the academic and research communities. Our aim is to build a foundation of knowledge on which researchers can build to meet the challenge of climate-smart agriculture.

For more information about Burleigh Dodds Science Publishing simply call us on +44 (0) 1223 839365, email info@bdspublishing.com or alternatively please visit our website at www.bdspublishing.com.

Related titles:

Precision agriculture for sustainability
Print (ISBN 978-1-78676-204-7); Online (ISBN 978-1-78676-206-1, 978-1-78676-207-8)

Robotics and automation for improving agriculture
Print (ISBN 978-1-78676-272-6); Online (ISBN 978-1-78676-274-0, 978-1-78676-275-7)

Assessing the environmental impact of agriculture
Print (ISBN 978-1-78676-228-3); Online (ISBN 978-1-78676-230-6, 978-1-78676-231-3)

Chapters are available individually from our online bookshop:
https://shop.bdspublishing.com

climate•SMART•publishing
IN AGRICULTURAL SCIENCE

BURLEIGH DODDS SERIES IN AGRICULTURAL SCIENCE
NUMBER 75

Advances in crop modelling for a sustainable agriculture

Edited by Emeritus Professor Kenneth Boote,
University of Florida, USA

burleigh dodds
SCIENCE PUBLISHING

Published by Burleigh Dodds Science Publishing Limited
82 High Street, Sawston, Cambridge CB22 3HJ, UK
www.bdspublishing.com

Burleigh Dodds Science Publishing, 1518 Walnut Street, Suite 900, Philadelphia, PA 19102-3406, USA

First published 2020 by Burleigh Dodds Science Publishing Limited

Library of Congress Control Number: 2019951738

British Library Cataloguing in Publication Data
A catalogue record for this book is available from the British Library

ISBN 978-1-78676-240-5 (Print)
ISBN 978-1-78676-243-6 (PDF)
ISBN 978-1-78676-242-9 (ePub)
ISSN 2059-6936 (print)
ISSN 2059-6944 (online)

DOI 10.19103/AS.2019.0061

Typeset by Deanta Global Publishing Services, Dublin, Ireland

Contents

Part 2 Developing whole farm system, landscape and regional models

Series list

Title	Series number
Achieving sustainable cultivation of maize - Vol 1 From improved varieties to local applications Edited by: Dr Dave Watson, CGIAR Maize Research Program Manager, CIMMYT, Mexico	001
Achieving sustainable cultivation of maize - Vol 2 Cultivation techniques, pest and disease control Edited by: Dr Dave Watson, CGIAR Maize Research Program Manager, CIMMYT, Mexico	002
Achieving sustainable cultivation of rice - Vol 1 Breeding for higher yield and quality Edited by: Prof. Takuji Sasaki, Tokyo University of Agriculture, Japan	003
Achieving sustainable cultivation of rice - Vol 2 Cultivation, pest and disease management Edited by: Prof. Takuji Sasaki, Tokyo University of Agriculture, Japan	004
Achieving sustainable cultivation of wheat - Vol 1 Breeding, quality traits, pests and diseases Edited by: Prof. Peter Langridge, The University of Adelaide, Australia	005
Achieving sustainable cultivation of wheat - Vol 2 Cultivation techniques Edited by: Prof. Peter Langridge, The University of Adelaide, Australia	006
Achieving sustainable cultivation of tomatoes Edited by: Dr Autar Mattoo, USDA-ARS, USA & Prof. Avtar Handa, Purdue University, USA	007
Achieving sustainable production of milk - Vol 1 Milk composition, genetics and breeding Edited by: Dr Nico van Belzen, International Dairy Federation (IDF), Belgium	008
Achieving sustainable production of milk - Vol 2 Safety, quality and sustainability Edited by: Dr Nico van Belzen, International Dairy Federation (IDF), Belgium	009
Achieving sustainable production of milk - Vol 3 Dairy herd management and welfare Edited by: Prof. John Webster, University of Bristol, UK	010
Ensuring safety and quality in the production of beef - Vol 1 Safety Edited by: Prof. Gary Acuff, Texas A&M University, USA & Prof. James Dickson, Iowa State University, USA	011
Ensuring safety and quality in the production of beef - Vol 2 Quality Edited by: Prof. Michael Dikeman, Kansas State University, USA	012
Achieving sustainable production of poultry meat - Vol 1 Safety, quality and sustainability Edited by: Prof. Steven C. Ricke, University of Arkansas, USA	013
Achieving sustainable production of poultry meat - Vol 2 Breeding and nutrition Edited by: Prof. Todd Applegate, University of Georgia, USA	014

Introduction

This collection summarises key advances in crop modelling, with a focus on developing the next generation of crop and whole-farm models to improve decision making and support for farmers. Chapters in Part 1 review advances in modelling individual components of agricultural systems, such as plant responses to environmental conditions, crop growth stage prediction, nutrient and water cycling as well as pest/disease dynamics.

Building on topics previously discussed in Part 1, Part 2 addresses the challenges of combining modular sub-systems into whole farm system, landscape and regional models. Chapters cover topics such as integration of rotations and livestock, as well as landscape models such as agroecological zone (AEZ) models. Chapters also review the performance of specific models such as APSIM and DSSAT and the challenges of developing decision support systems (DSS) linked with such models. The final part of the book reviews wider issues in improving model reliability such as data sharing and the supply of real-time data, as well as crop model inter-comparison.

Part 1 Modelling sub-systems

Chapter 1 discusses advances and improvements in modelling plant processes. It provides an overview of the advances made historically and today in modelling key plant processes. It focuses on the processes associated with carbon acquisition and allocation, such as photosynthesis, stomatal conductance, respiration and assimilate partitioning, and provides a section on modelling phenology, the timing of plant developmental events.

Moving on to Chapter 2, advances in functional-structural plant (FSP) modelling of plants and crops are discussed. Crop models have been instrumental in predicting yields in wide ranges of current and future environmental conditions. However, they encounter problems in representing spatial heterogeneity of a plant stand and the associated plant responses under competition from other plants including weeds and intercrops. The FSP models simulate the effects of specific plant traits, management choices that influence plant architecture, and lighting regimes such as those in greenhouses. For such purposes, functional-structural plant (FSP) models have been developed, which simulate individual plants that interact with each other in 3D, with the changes in plant architecture feeding back on the distribution of environmental drivers that make them grow and develop (light, water, nutrients). With this in mind, the authors outline the purposes of FSP models, the components they need to have to serve these purposes and provide an account of recent applications of FSP models.

Chapter 3 reviews improving the modelling of nutrient cycles in crop cultivation. Agricultural productivity depends on crops receiving adequate amounts of essential nutrients from the soil and supplied fertilizers and manures. Through the biogeochemical cycling of nutrients, fertile soils supply the following essential nutrients to plants: nitrogen, phosphorus, potassium, sulfur, calcium, magnesium, iron, zinc, copper, manganese, boron, molybdenum, cobalt, and nickel. Deficiency of any of these nutrients results in lower productivity. The chapter focuses primarily on the nutrient cycle modelling for nitrogen, phosphorus and potassium and their relationships to soil organic matter decomposition. Given the importance of nitrogen in global food production and its impact on the environment, the major emphasis of the chapter is on nitrogen modelling.

The subject of Chapter 4 is improving the modelling of water cycles in crop cultivation. Adequate representation of the water cycle in cropping systems remains as one of the major concerns for crop model developers as it is involved in many critical processes such as the determination of the energy balance and crop temperature, the determination of crop productivity, and the movement of nutrients in the soil and their uptake. The chapter begins by discussing crop water availability and its linkage to weather, soil and management at point, field and watershed scales. It then reviews the determination of the potential and actual crop evapotranspiration and soil water availability in crop models. The chapter also discusses modelling crop water uptake and the integration of soil water balance components in crop models. Finally, the chapter reviews crop water use and atmospheric carbon dioxide concentration effects on transpiration and concludes by highlighting the main limitations of modelling water cycles that are currently observed.

The final chapter of Part 1, Chapter 5, discusses improving crop pest/disease modelling. Dissemination of crop pests and diseases occurs through both natural and anthropogenic processes, facilitated by the increasing interconnectedness of the global food chain. Food security depends on our ability to manage crop pests and diseases effectively. The chapter offers a brief overview of approaches to coupling of models in order to better predict and manage crop pests and diseases. The chapter explains the PEST subroutine in the CROPGRO family of models, and shows how a strategy of keeping the crop model separate from disease/pest models, and coupling through Message Passing Interface (MPI) functionality, can be effective. It also provides a section on insect pest and disease management, specifically focusing on the CROPGRO-soybean simulation model. Finally, the chapter looks ahead to future research trends in this area.

Part 2 Developing whole farm system, landscape and regional models

Part 2 begins with Chapter 6 which reviews whole farm system models in practice. The challenges of contemporary agriculture are causing farmers to seek models directed at whole-farm goals, which requires mathematically-based models grounded in the latest scientific understanding of animal growth and reproduction, crop physiology, soil and environmental sciences, as well as economics and human relations. Such models are often invaluable tools in the process of undertaking nuanced, multifaceted management decisions at the farm system level. The chapter explores model selection and, through a series of case studies, whole farm modelling in both smallholder systems and single enterprise farm businesses. Finally, the chapter looks ahead to future research trends in this area.

Moving on to Chapter 7, the chapter describes the DSSAT crop modelling ecosystem. It begins by providing the history of the DSSAT model development and goes on to discuss the creation of multiple crop models and tools for data preparation, analysis and applications. The chapter also provides an overview of the structure and science of the crop simulation models included in DSSAT. The tools and utilities for weather, soil, and experimental data entry are also discussed, and a section on crop model calibration and evaluation is included. Finally, the chapter reviews the range of applications and concludes by discussing the future of the DSSAT crop modelling ecosystem.

The next chapter of Part 2 reviews modelling crop rotations and capturing short- and long-term feedbacks for sustainability and soil health. Crop rotation is a critical management strategy available to farmers to sustain soil fertility, combat pests, and improve crop yields. Because of the complex feedbacks occurring in the soil-plant-atmosphere continuum, legacy effects and their interactions with climate are generally not well understood, and their ultimate influence on yields and environmental quality is difficult to predict. Crop simulation models offer a way of parsing through this complexity. Chapter 8 illustrates how crop models account for the interactions between soil, cultivars, management, and climate, on crops grown in various rotations, and their effects on yield and environmental outcomes (e.g. nitrate leaching, soil organic carbon sequestration) under current and future climate scenarios by presenting two different case studies and modelling approaches. The chapter concludes by summarizing the advantages and disadvantages of the modelling approaches and offers some recommendations for how to best capture legacy effects.

Chapter 9 examines the integration of livestock production into whole-farm system models of mixed crop-livestock systems. Crop-livestock farming systems integrate various biophysical components with farm decision-making. Whole-farm system models help to understand the complexity arising from this integration and are useful tools to explore effects of interventions and to design improved systems. After describing the generic characteristics of crop-livestock systems, the chapter gives an account of the separate component models, from which whole-farm models are usually assembled. Most attention goes to the wide diversity in pasture, livestock and manure models, while approaches to represent farm management in models are also discussed. The interactions between farm components strongly determine model functioning and performance, and often present modelling challenges to capture them. With a focus on grazing, manure dynamics and animal mobility, the chapter assesses the key processes governing these interactions and reviews how models represent them. The representation of feed quality in crop model outputs and the management of time scales and carry-over effects is also discussed. Finally, the chapter concludes by providing two examples of common whole-farm model applications from contrasting environments.

Moving on to Chapter 10, the chapter examines the integration of economic simulation models with whole-farm system models for ex ante technology impact assessment. The key analytical challenge to developing agricultural system models is to predict the behaviour of agricultural systems in the multiple dimensions that are relevant to their economic, environmental and social sustainability, under recent historically observed conditions and when they are modified to improve economic performance. The chapter discusses ex ante technology impact assessment, defined as forward-looking efforts at technology design and evaluation. Supported by a case study, the chapter reviews methods to combine process-based and economic production models to overcome some of the limitations of each approach. Finally, the chapter looks ahead to future research trends in this area and provides detailed guidance on further reading.

Chapter 11 addresses developing climate-based decision support systems (DSS) from agricultural systems models. Decision support systems (DSSs) in production agriculture are computer software programs that make use of models and other information to make site-specific recommendations for farm management-related activities. Major advances have been made in developing DSS for agriculture, particularly with respect to incorporating climate forecasts into farmers' decision-making process. The chapter describes DSSs in agriculture, focusing on the importance of a participatory approach in DSS development. The chapter offers examples of climate-based DSS for crop and land management, pest and disease management, and livestock (dairy) management. The chapter reflects on lessons learned in the process of developing climate-based DSS. Finally, the chapter looks ahead to future research trends in this area.

The subject of Chapter 12 is the use of landscape models to support sustainable intensification of agroecological systems. Intensification of agriculture to meet the growing demand of food and renewable resources is contrasted with a decrease in species diversity, pollution in water bodies and a decline in freshwater resources. To protect our resources, the natural functioning of ecosystems and the diversity of landscapes, agricultural land use must reduce its resource use and its impact on biodiversity and ecosystems. Addressing both challenges simultaneously requires a perspective that takes a more detailed look into the heterogeneity of climate and soils and their role in agroecosystem functioning, the interactions between flora and fauna, the interdependencies with the green infrastructure at landscape scale and the role of humans and how they organise themselves in this system. In this chapter the authors reflect on the different social, economic and natural features of agricultural landscapes that govern important processes in agro-ecosystems and discuss current approaches for modelling ecosystem services using a landscape approach.

The concluding chapter of Part 2 discusses agroecosystem models for delivering ecosystem services. Agroecosystems are complex representations of the interactions occurring among multiple biological systems at the field and landscape scale. Models simulating these interactions across time and space are not very mature and expanding our current knowledge on crop, pest, and soils models provides a framework for how we use agroecological models to quantify a range of ecosystem services. Chapter 13 examines the need for an agroecosystem approach, outlining the importance of plants, soils, pests, diseases, weeds and pollinators as part of the agroecosystem. Finally, the chapter looks ahead to future research trends in this area.

Part 3 Modelling issues

The first chapter, Chapter 14, examines data for developing, testing and applying crop and farm models. It reviews the kinds of data available to crop and farm modellers. The chapter highlights practices concerning storage, exchanging and combining data from various sources, examining the socio-economic and ethical implications. It also looks at the role of companies in data flows and generating recommendations, and presents examples of successful use of agricultural data, both for science and agricultural practice. The chapter concludes by looking ahead to future research trends in the area.

Moving on to Chapter 15, this chapter examines how to deal with uncertainty in crop models. There is increasing awareness in crop modelling of the importance of uncertainty. The modeller needs uncertainty information to prioritize improvements, while the user needs uncertainty information to make informed decisions. The chapter introduces the concept of model uncertainty, considering such factors as model structure, inputs and parameters. The chapter then looks at ways to reduce uncertainty in crop modelling and looks ahead to future trends in the area. Finally, the chapter provides detailed guidance on further reading on the subject.

Chapter 16 focuses on crop simulation model inter-comparison and improvement. Crop modelling is an integrated discipline of agricultural sciences that contributes to agricultural research, including the areas of agronomic management, precision agriculture, environmental impacts and externalities, input efficiency, managing climate variability and seasonal forecasting, breeding, climate change impacts and adaptations, and agricultural policy. More recently, the crop modelling community has begun to exchange models, model routines, and data in coordinated and protocol-based crop model inter-comparison and improvement projects. The chapter summarises crop model inter-comparison studies and their contribution to sustainable agriculture and discuss gaps and trends in crop simulation model inter-comparisons and improvements. It covers crop model inter-comparison, crop model uncertainty, crop model improvement, and crop model application, respectively. A case study from the Agricultural Model Inter-comparison and Improvement Project for Wheat (AgMIP-Wheat) is included. The chapter concludes with a review of future trends in research.

The final chapter of the book examines the future of crop modelling for sustainable agriculture. Crop growth simulation models have become increasingly important as accepted tools with uses in education, research, crop management, industry, and government policy decisions related to improving and sustaining agricultural production of food, feed, fibre, and ecosystem services. The chapter begins with a brief review of history of crop modeling that has led to excellent collaborations among the world's crop modeling community. It discusses present and future potential uses of crop models in education, academic research, plant breeding, strategic crop management, in-season advisory tools, famine risk tools, climate change impact assessment, ecosystem services, sustaining long-term soil fertility, and integration with livestock and economic models. The chapter concludes with a look to the future with a review of what is new in crop models, the processes that need improvement, along with suggestions of new features needed

Part 1

Modelling sub-systems

Chapter 1

Advances and improvements in modeling plant processes

Soo-Hyung Kim and Jennifer Hsiao, University of Washington, USA; and Hannah Kinmonth-Schultz, University of Kansas, USA

1 Introduction

The growth of a crop is coordinated by multiple physiological processes. Process-based crop models (PBCMs) simulate growth, development, and performance of crop plants by modeling their underlying physiological processes, and the coordination and integration of these processes at the whole-plant and canopy scales. The physiological processes incorporated into PBCMs can include photosynthesis, transpiration, respiration, organ development, and assimilate transport. Most PBCMs are explanatory models by nature as they integrate sub-models that represent the mechanisms and processes at lower levels of biological organization into the whole-crop level responses (Boote et al., 2013; Marcelis et al., 1998; Thornley and Johnson, 1990). For example, in some maize PBCMs the dynamics of total leaf area is simulated by modeling growth and senescence of individual leaves (Kim et al., 2012; Lizaso et al., 2003). A key strength of PBCMs is their ability to mechanistically integrate multiple physiological processes and their interactions into crop performance with respect to environmental, management, and genetic factors. Much progress has been made in our understanding of individual physiological processes; yet, current understanding of some processes may be more advanced than others. This discrepancy is reflected in the degree to which individual physiological

http://dx.doi.org/10.19103/AS.2019.0061.01

mechanisms are implemented into PBCMs - some more thorough and reductionistic (e.g. photosynthesis and floral development) than others (e.g. sink regulations). Using a modular crop model structure (Acock and Reddy, 1997), the sub-models for different processes may be updated periodically to reflect, and include new biological knowledge, thus requiring minimal effort for revamping and restructuring the entire model.

Modeling the processes involved in carbon acquisition and allocation is fundamental in PBCMs. *Carbon dynamics* are determined by the rate of CO_2 assimilation through photosynthesis, release of CO_2 and energy production through respiration, and allocation of assimilated carbon to internal and external sinks (e.g. growing organs and symbionts). Many models and modeling approaches with varying degrees of mechanism and reductionism exist for plant carbon fixation (e.g. Farquhar et al., 1980; Monteith, 1977; Zhu et al., 2013). On the other hand, models for how the carbon assimilates are distributed within plants are less common. Modeling *phenology* is another top priority in PBCMs because accurate prediction of the timing of crop development can dictate the model's ability to simulate subsequent growth and other physiological processes as well as final biomass and yield. Phenology models are also critical for making timely management decisions in season. Crop phenology models have evolved from simple models relating observations of growing-season temperatures to specific plant processes such as the timing of anthesis (Lehenbauer, 1914), to models that routinely incorporate multiple layers of plant organization, specifically, organ initiation, expansion and senescence (Wang et al., 2017).

This chapter provides an overview of the advances made historically and today in modeling key plant processes with a focus on the processes associated with carbon acquisition and allocation (photosynthesis, stomatal conductance, respiration, and assimilate partitioning), and the timing of plant developmental events (phenology). Areas for further improvement are also addressed.

2 Modeling photosynthesis and stomatal conductance for carbon acquisition

Photosynthesis is a key physiological process that determines plant growth, performance, and crop yield. It responds to various environmental factors such as light, temperature, humidity, atmospheric CO_2 concentrations, soil water and nitrogen availability, and is arguably one of the most studied physiological processes within plants. Stomatal conductance controls the exchange of carbon and water that occurs through the simultaneous process of photosynthesis and transpiration. Accurate representations of photosynthesis and stomatal conductance within models are critical in linking plant water relations with carbon gain, and can be considered as one of the critical foundations in crop modeling. Here, we introduce several key modeling approaches developed to describe these processes.

2.1 Models of photosynthesis

2.1.1 Brief history

Some of the earliest attempts to model photosynthesis focused on describing the light interception and distribution within the canopy. Leaf-level photosynthetic rates were modeled as a function of the intercepted light level at different canopy layers, commonly through a rectangular hyperbola function (de Wit, 1965; Duncan et al., 1967; Monsi and Saeki, 1953; Monteith, 1965). Subsequently, several field observations demonstrated biomass accumulation to be proportional to intercepted radiation (Monteith, 1972; Warren Wilson, 1967). Specifically, Monteith (1977) presented how dry-matter production of various economically important crops in Great Britain linearly correlated with the intercepted radiation under ideal growing conditions, leading to the theory that carbon gain in crops through photosynthesis can be estimated based on the amount of sunlight intercepted during the growing season and on the efficiency with which it is converted into biomass. This concept is commonly known as radiation use efficiency (RUE) and has been adopted into many crop simulation models due to its elegance and simplicity (White et al., 2011).

In 1980, Farquhar and colleagues published a biochemical model for C_3 plants that attempted to incorporate and simplify the expanding understanding of mechanisms behind photosynthesis (The FvCB model; Farquhar et al., 1980). The FvCB model was further modified by Harley and Sharkey (1991), and was later expanded for C_4 photosynthesis by von Caemmerer and Furbank (1999). Despite being published nearly four decades ago, the FvCB model is still one of the most frequently used methods for studying photosynthetic response to different environmental perturbations (e.g. Farquhar et al., 2001; Field, 2002; Rogers et al., 2017; Sellers et al, 1997).

To date, a majority of the crop models for staple crops (e.g. maize, rice, wheat, potato) use RUE as the primary method for modeling biomass production, but an increasing number of models have been adopting or transferring to photosynthesis-based approaches at the leaf or canopy level (Asseng et al., 2013; Bassu et al., 2014; Fleisher et al., 2017; Li et al., 2015).

2.1.2 A mechanistic modeling approach for photosynthesis

The FvCB model captures the biochemical demand for CO_2 assimilation at the chloroplast that is scalable to a leaf. The leaf net photosynthetic rate (A) is determined by describing the rate of three key processes within the Calvin cycle:

(1) the rate of carboxylation, which is catalyzed by Rubisco (A_c), (2) the rate of RuBP regeneration, which is controlled by the electron transport rate within the light-dependent reaction of photosynthesis (A_j), and (3) the rate of synthesizing photosynthetic end products (A_p), mainly, converting triose phosphates into sucrose and starch, also known as triose phosphate utilization (TPU; Fig. 1). Since the three processes (i.e., A_c, A_j, A_p) are continuous components within the Calvin cycle, the most rate-limiting process eventually determines the net photosynthetic rate (Eqn. 1):

$$A = min\left(A_c, A_j, A_p\right) \tag{1}$$

The rate of each process responds to enzymatic properties as well as environmental variables such as temperature, light levels, and CO_2 and oxygen concentrations. Here, we briefly describe how the FvCB model individually describes A_c, A_j, and A_p.

A_c represents photosynthetic rates limited by the carboxylation capacity of Rubisco and is calculated based on Rubisco kinetics (Eqn. 2):

$$A_c = V_{c\,max}\left(\frac{C_c - \Gamma^*}{C_c + K_c\left(1 + O / K_o\right)}\right) - R_d \tag{2}$$

V_{cmax} describes the maximal rate of carboxylation for Rubisco, and K_c and K_o are the Michaelis-Menten constants for Rubisco reacting with CO_2 and oxygen, respectively. These parameters describe the enzymatic properties of Rubisco, which are temperature dependent. C_c and O denote the CO_2 and oxygen partial

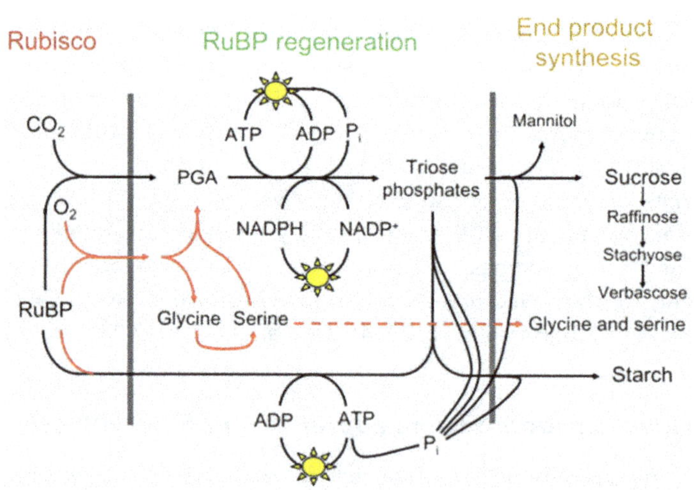

Figure 1 Scheme showing some of the processes that affect photosynthetic rate (Sharkey et al., 2007, Figure 1).

pressure at the chloroplast, and Γ^* is the CO_2 compensation point, defined as the CO_2 concentration level at which carbon uptake through photosynthesis is balanced by CO_2 released through photorespiration. Finally, the calculation accounts for leaf-level daytime respiration, R_d, which is the respiratory release of CO_2 excluding that from photorespiration.

A_j describes the CO_2 assimilation rate limited by RuBP regeneration and is calculated based on the electron transport rate within the light reaction of photosynthesis that generates ATP and NADPH to fuel RuBP regeneration (Eqn. 3):

$$A_j = J\frac{C_c - \Gamma^*}{4C_c + 8\Gamma^*} - R_d \qquad (3)$$

$$J = \frac{\alpha_{LL} I_{abs} + J_{max} - \sqrt{(\alpha_{LL} I_{abs} + J_{max})^2 - 4\theta J_{max} \alpha_{LL} I_{abs}}}{2\theta} \qquad (4)$$

J describes the electron transport rate that is a function of light absorbed in photosynthetically active radiation, usually modeled as shown in Eqn. (4). J depends on the maximum capacity of electron transport (J_{max}), the amount of light absorbed (I_{abs}), the quantum yield of electron transport under light-limiting conditions (α_{LL}), and an empirical parameter θ that represents the convexity of the J to I_{abs} response curve.

Finally, A_p represents photosynthetic rates limited by TPU and is calculated based on the rate at which photosynthetic end products are being utilized and transported out of the chloroplast (Eqn. 5):

$$A_p = 3TPU - R_d \qquad (5)$$

in which the photosynthetic rate is determined by TPU and R_d.

2.1.3 Scaling photosynthesis from leaf to canopy

The main question to consider when attempting to scale photosynthesis from a single-leaf level up to a canopy level is the distribution of light within the canopy. Monsi and Saeki (1953) published the very first mathematical model for canopy-level photosynthesis, applying the Beer-Lambert optic law to describe light attenuation within a plant canopy affected by leaf area index and the optical properties of the leaves. This concept of radiation penetration through the canopy set the foundation for several canopy photosynthesis models that divided the plant canopy into multiple layers with different light-level and leaf-angle classes and calculated the canopy-level photosynthetic rate by numerically integrating photosynthetic rates throughout the canopy layers (de Wit, 1965; Duncan et al., 1967; Monsi and Saeki, 1953; Monteith, 1965).

Further improvements of the original Monsi and Saeki model included acknowledging the importance of direct versus diffuse light within the canopy (de Wit, 1965; Duncan et al., 1967; Spitters et al., 1986). Subsequent observations also revealed the difference in photosynthetic capacities between sunlit and shaded leaves. Specifically, photosynthetic rates typically increased linearly with light levels for shaded leaves, while these responses were often saturated in sunlit leaves (Boardman, 2003). This allowed the multilayer model to be simplified by classifying the leaves within a canopy into two types, sunlit and shaded leaves, while still being able to accurately capture the canopy-level photosynthetic rate (de Pury and Farquhar, 1997). Methods exist to further simplify the canopy structure into just a single photosynthetic layer (known as the big leaf model). While computationally efficient, the big leaf model was prone to overestimating canopy-level photosynthesis, particularly for dense canopies. Introducing empirical coefficients or fitting with observations helped correct for this issue, but applications tend to be limited to the calibrated conditions (de Pury and Farquhar, 1997).

Alternatively, Functional-Structural Plant Models (FSPM) took a different approach in the level of complexity when modeling the plant architecture. Detailed descriptions of the plant structure, including the branching system, leaf angle, and leaf curvature allow for a more realistic representation of the plant canopy and the light distribution within it (Allen et al., 2005; Godin and Sinoquet, 2005; Vos et al., 2010). Such approaches require significantly more work to compute parameters, but can be very useful, especially when asking research questions that link plant function and structure. Further information on FSPMs can be found in Chapter 2 by Evers and Marcelis.

2.1.4 Recent advances

First published in 1980, the FvCB photosynthesis model has gained widespread impact through time (Farquhar et al., 2001). Its elegance in capturing fundamental mechanistic processes within photosynthesis has led to broad applications in terrestrial, ecosystem, and Earth system modeling communities (Farquhar et al., 2001; Field, 2002; Rogers et al., 2017; Sellers et al., 1997). Nevertheless, the FvCB model has not been widely adopted within the crop modeling community (Yin and Struik, 2009) likely due to perceived complexity in the model structure and the level of required parameter information when compared to the commonly used RUE approach (Cannell and Thornley, 1998). RUE remains a useful approach in crop simulation models, especially due to its simple form. Factors that are known to affect RUE, such as CO_2 concentrations, can also be accounted for if explicitly calibrated (White et al., 2011). However, RUE remains a largely empirical approach, and its ability to capture dynamic interactions between different environmental effects on photosynthesis and biomass production can be limited, especially when projecting conditions under a changing climate in which current relationships between crops and environmental factors may not remain the same.

Since the development of the FvCB model, sophisticated portable instruments for leaf gas-exchange measurements have become readily available, providing much more information to aid the parameterization process for the FvCB model, and composites of estimated parameter values now exist within the literature for a wide range of C_3 and C_4 plant species (Bellasio et al., 2017; Wullschleger, 1993; Yin and Struik, 2009). In addition, it is commonly assumed that enzyme kinetic parameters of Rubisco are relatively conserved between plant species and crop cultivars that share the same photosynthetic pathway (Sharkey et al., 2007; von Caemmerer, 2000), which can reduce the number of unknown parameters within the model. However, the FvCB model was not intended to capture all the steps within photosynthesis; rather, it synthesizes key components in order to simulate the steady-state photosynthetic rate. Recent advances in process-based photosynthesis models have incorporated more dynamic and explicit mechanisms, with an increasing interest in representing the whole photosynthesis process.

A mechanistically detailed and comprehensive photosynthetic model requires a series of interconnected biochemical and biophysical reactions that dynamically shift under different environmental conditions; these models are regulated by specific conditions at the photosynthetic site, such as pH levels, redox state, concentrations of key ions, etc. (Zhu et al., 2013). Developing such model requires extensive knowledge of the photosynthetic system as well as computational power and numerical algorithms to solve for it, and such methods have recently started to become more available. Laisk et al. (2006) developed a photosynthesis model for C_3 plants that described in detail both the light-dependent and independent reactions. The model compartmentalized a leaf into the thylakoid membranes, thylakoid lumen, chloroplast stroma, and the cytosol, while tracking the intermediate compounds that occur throughout the photosynthesis process through a system of differential equations. This approach showed the growing interest in mechanistically modeling the photosynthetic process as a whole. In 2013, Zhu and colleagues published a mechanistic and dynamic photosynthesis modeling platform, *e-photosynthesis*, that built upon several previous works on this topic (Laisk et al., 2006; Zhu et al., 2005, 2007, 2013) and is likely one of the most comprehensive models of its kind to date.

The detailed mechanisms captured within such models make them great tools for exploring breeding and genetic modification targets to improve photosynthesis, with implications for improving yield for major food crops. However, the level of information in such models requires expansive parameter sets and considerable computation power, which can pose challenges when incorporated into a whole-plant crop model structure. Nevertheless, studies have proposed ideas of scaling such models up to canopy levels (Zhu et al., 2012). While this process remains informationally and computationally challenging, new tools and approaches have been proposed to tackle the complex interactions between different processes at varying time scales as well as the difficulty in parameterizing and validating such models (Zhu et al., 2012). If successful, a complete model linking molecular-level mechanisms to canopy level can be used as a powerful tool both for hypothesis testing and for crop engineering.

2.2 Models of stomatal conductance

Stomata have long been a subject of interest for scientists. Early documentation can be dated back to 1898, when Francis Darwin, Charles Darwin's son, published detailed observations on stomatal responses to various environmental variables (Darwin, 1898). Up to this day, stomatal behavior and the processes behind them still remain an active field of research; while the primary signals controlling stomatal opening and closing are understood, a full mechanistic understanding of these responses still remains a work in progress, especially in stomatal hydraulic responses (Buckley, 2016, 2017). Because of the tightly coupled relationship between CO_2 uptake for photosynthesis and water loss from transpiration through stomata, modeling stomatal conductance is critical for mechanistic representation of plant carbon-water relations and also to fully utilize the FvCB photosynthesis model (Eqn. 1) for which $[CO_2]$ inside the leaf (C_i) or at the chloroplast (C_c) is the driving variable. We will summarize here several modeling approaches with varying degrees of complexity that attempt to capture these stomatal responses.

2.2.1 Empirical approaches

One of the earliest attempts to model stomatal conductance (g_s) was an empirical approach published by Jarvis (1976). The model described stomatal conductance of illuminated leaves as a function of leaf temperature (T_l), vapor pressure deficit (VPD), leaf water potential (ψ_l), and ambient CO_2 concentrations (C_a), and modeled their responses as independently multiplicative (Eqn. 6). The simplicity of this model made it easy to incorporate into whole-plant models, but the model provided little mechanistic information on stomatal function, thus requiring parameterization under new environmental conditions. Observed interactions between the model inputs also have led to criticism of its assumption of independent model input (Tardieu et al., 1996).

$$g_s = f_1(Q)\cdot f_1(T_l)\cdot f_1(VPD)\cdot f_1(C_a)\cdot f_1(\psi_l) \quad (6)$$

Around the same time, Cowan and Farquhar (1977) approached the problem through a different school of thought. They described stomatal functioning from a top-down method with concepts that largely stem from economic theories of optimality and the general idea that stomatal responses are selected over evolutionary time to achieve an optimal behavior of maximizing carbon gained through photosynthesis (A) while minimizing water lost through transpiration (E). This concept can be described by minimizing the integrated sum of the following equation between time-steps t1 and t2:

$$\int_{t_1}^{t_2} \left(E(t) - \lambda A(t)\right) dt \tag{7}$$

Where the empirical λ parameter describes the marginal cost (in this case, water lost) for carbon gain. A few studies have attempted to implement this theory to model stomatal behavior (Lloyd et al., 2002; Mencuccini, 2003). However, overall usage of such a modeling approach was limited, mostly due to the ambiguity in parameterizing λ, the difficulty in determining the time interval in which to capture the optimal stomatal behavior (t1-t2 in Eqn. 7), and the model's inaccuracy in capturing stomatal responses to CO_2 concentrations.

2.2.2 Semi-empirical approaches

The initial proposal of an optimality-based stomatal conductance model was not widely applied, but various leaf-level observations supported its prediction of a conservative ratio of intercellular to ambient CO_2 concentrations (Wong et al., 1979). These observations helped shape the development of a semi-empirical (or semi-mechanistic) stomatal model developed by Ball et al. (1987), which has since become one of the most commonly used stomatal conductance models up to this day, and is often referred to as the Ball-Berry model.

The Ball-Berry model was developed based on leaf-level observations, typically showing stomatal conductance correlated with photosynthetic rate (Wong et al., 1979) and modified by relative humidity. This led to a simple model formulation that describes g_s as a function of net photosynthetic rate (A), the CO_2 concentration at the leaf surface (c_s), and the relative humidity at the leaf surface (h_s; Eqn. 8):

$$g_s = g_o + m\frac{Ah_s}{c_s} \tag{8}$$

The two free parameters within the model, g_o and m, represent minimal stomatal conductance when A equals zero and an empirical parameter that describes the sensitivity of stomatal conductance to environmental variables, respectively. The model has been further modified by Leuning in 1995 by incorporating the concept of a CO_2 compensation point including respiration (Γ), modifying the original relative humidity response into VPD (D_s), which has been shown to better describe the stomatal response to humidity (Aphalo and Jarvis, 1991), and including an additional free parameter, D_o (Eqn. 9):

$$g_s = g_o + \frac{mA}{(c_s - \Gamma)\left(1 + \frac{D_s}{D_o}\right)} \tag{9}$$

Note that g_s is not defined when c_s equals Γ (Eqn. 9). Similarly, g_s is not defined when c_s equals zero in Eqn. (8).

While still largely empirical, the Ball-Berry model and its variations were elegant and simple in form and easy to test and parameterize with leaf gas-exchange measurements. When using it to dynamically predict stomatal conductance from environmental variables, the Ball-Berry model required an input of photosynthetic rate calculated through a separate model (e.g. often the FvCB photosynthesis model). Thus, the model structure represents the strong dependence and coupling between photosynthesis and stomatal conductance. We describe the coupling process in further detail in Section 2.3.

The simplicity and mechanistic behavior of this coupled g_s-A approach has led to a wide range of applications (Egea et al., 2011). However, a common criticism of the approach relates to the empirical parameter m, also referred to as the Ball-Berry slope. The m parameter within the model represents a composite sensitivity of g_s to photosynthetic rate, humidity, CO_2 concentrations, and temperature (indirectly through photosynthesis). The lack of a specific biological meaning within this term can require re-parameterization whenever applying the model for a new species or under new environmental conditions. Miner et al. (2017) surveyed m values published through the past three decades, which showed a wide range of variability within and between plant functional groups. These results pointed out some outstanding debates on whether m changes under adaptation to a change of environmental conditions such as elevated CO_2 levels and water stress and whether the wide range of values we see in the literature represents the true variability within this value, or whether it stems from the range of different measurement methods.

Recently, Medlyn et al. (2011) demonstrated that a model directly analogous to the Ball-Berry model and its variations can be derived and approximated by combining the optimal stomatal behavior theory (Cowan and Farquhar, 1977) with parts of the FvCB photosynthesis model (Farquhar et al., 1980). They termed this model as the unified stomatal model (Eqns. 10 and 11):

$$g_s \approx g_0 + 1.6\left(1 + \frac{g_1}{\sqrt{D}}\right)\frac{A}{C_a} \tag{10}$$

$$g_1 \propto \sqrt{\Gamma^* \lambda} \quad (11)$$

The notations are modified slightly from its original publication here for consistency with those previously presented in the Ball-Berry model (8–9). Parameter g_1 is directly proportional to both the CO_2 compensation point without dark respiration (Γ^*) and the marginal cost of carbon gain (λ), giving parameter g_1 a biological meaning that was lacking in the similar *m* parameter within the Ball-Berry type models (Eqns. 8 and 9). The CO_2 compensation point is generally assumed to be constant across C_3 species and is shown to increase with leaf temperature (Bernacchi et al., 2001). Thus, under a given leaf temperature, λ determines parameter g_1. When originally proposed by Cowan and Farquhar (1977), the difficulty in estimating λ experimentally hindered the application of the optimal stomatal behavior theory. This new framework (Eqn. 10) allowed for an easier estimation of λ by fitting the model with stomatal behavior observations.

The unified stomatal conductance model (also known as the Medlyn model (MED)) combined existing empirical data with theory-based stomata behavior and has since been applied to large-scale Earth System Models (Franks et al., 2018). Franks and colleagues (2017) also compared the unified stomatal conductance model with the Ball-Berry model and a water use efficiency (WUE) stomatal conductance model published by Bonan et al. (2014). These comparisons show that when properly calibrated for, the g_s models responded similarly to different environmental perturbations (i.e. radiation, CO_2 concentration, and temperature), with the exception of VPD responses under low and high VPD conditions in the Ball-Berry model (Fig. 2). The sensitivity of g_s to these environmental perturbations can be interpreted by visualizing the slope at specific points of the response curve; for example, g_s responses are most sensitive to changes in CO_2 at lower CO_2 concentrations.

2.2.3 More mechanistic approaches

Stomatal conductance increases when an influx of water causes guard cells to become turgid, facilitating the exchange of carbon dioxide and water vapor. This water influx into guard cells can result either from passive water flow following a water potential gradient or through osmotic water movement actively regulated by guard cell ion transport and carbon metabolism (Franks et al., 1998, 2001). Rather than capturing observed relationships between g_s and environmental factors, more process-oriented models focus on several known mechanisms relating to stomatal behavior, such as plant hydraulics, guard cell turgor pressure, water stress, and ABA signaling. Current understanding of the mechanisms behind stomatal responses is still an active field of study. Damour et al. (2010) provides a good summary of a selection of modeling approaches developed to capture stomatal behavior.

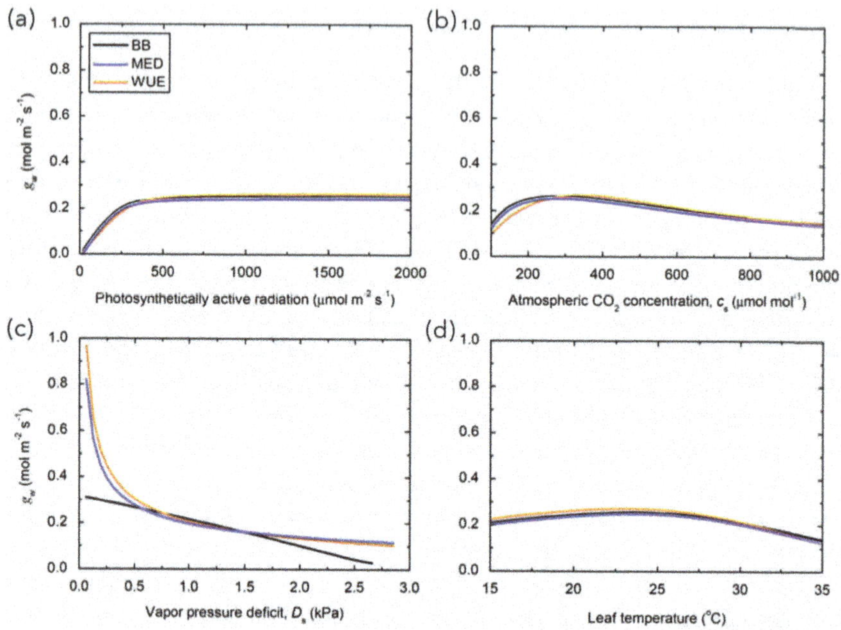

Figure 2 Simulation of stomatal conductance (g_{sw}) in response to photosynthetically active radiation (a), ambient CO_2 concentration at the leaf surface, c_s (b), vapor pressure deficit at the leaf surface, D_s (c), and leaf temperature (d). Shown results are results for the Ball-Berry model (BB), the unified stomatal conductance model (MED, Medlyn et al., 2011), and the water use efficiency model (WUE, Bonan et al., 2014). (Franks et al., 2017, Figure 4).

Detailed mechanistic knowledge of stomatal functioning has expanded in recent years, leading to various attempts to incorporate such knowledge into process-based stomatal models (Bellasio et al., 2017). A hydromechanical and biochemical stomatal model published by Buckley et al. (2003) provides a good example of such a process-based modeling approach. The model predicted g_s based on the balance between the opposing hydromechanical and biochemical forces in and around the guard cells: the hydromechanical force resulted from the turgor pressure in epidermal cells adjacent to guard cells, varying due to changes in water supply and demand; the biochemical force resulted from the guard cell osmotic pressure, which the authors hypothesized to be proportional with ATP concentration within the cell. The role of guard cell osmotic pressure in stomatal responses has long been established in the stomatal physiology literature. The guard cell osmotic pressure is set by the osmotic gradient created through solute influx, part of which is generated through the H^+-ATPases proton pump (Tominaga et al., 2001). ATP drives the proton pumps to create and maintain the osmotic pressure, thus the model by Buckley et al. (2003) described guard cell osmotic pressure as a function of guard cell ATP concentration. To capture guard cell ATP concentration, Buckley et al. (2003) adopted the model proposed by Farquhar and Wong (1984), in which ATP concentration is described by ATP production through the light reaction electron transport chain, and ATP consumption through the Calvin cycle and the photorespiration pathway. The authors based this statement on the assumption that ATP concentrations are controlled by similar processes in guard cells and mesophyll cells, but acknowledge the conflicting evidence that exists on Calvin cycle activity within guard cells (Cardon and Berry, 1992; Lawson et al., 2002; Outlaw Jr., 1989). ATP concentration within the guard cell was thus linked with photosynthetic activity and responded to environmental perturbations such as changes in temperature, light level, and CO_2 concentrations.

In addition to capturing the stomatal responses to key environmental perturbations, the mechanism-oriented stomatal conductance model proposed by Buckley et al. (2003) also described the transient stomatal opening response often observed when plants experience a drop in humidity in the surrounding air (often termed the 'wrong way' stomatal response). This response is caused by the mechanical advantage of the adjacent epidermal cells over guard cells, such that when water is lost through both cell types, reduced backpressure imposed on the guard cells from the epidermal cells causes a net stomatal opening response. Most empirical and semi-empirical stomatal conductance models fail to capture this response, but the mechanistic nature of the hydromechanical and biochemical approaches accounted for this transient response by explicitly describing the mechanical advantage of epidermal cells. This transient state is later overcome through osmotic adjustments within the guard cell resulting in a steady-state stomatal closure response to water loss or drop in humidity.

A process-based approach (e.g. Buckley et al., 2003) allows for a greater mechanistic understanding of the controls on leaf gas-exchange and can be structured to accommodate new knowledge on stomatal functions to move forward from the currently dominant empirical modeling paradigms (Buckley, 2017). However, there have been limited applications of such modeling approaches in whole-plant crop models due to the complexity and species-specific biophysical parameters that require involved measurements to obtain. Attempts to streamline the model to include fewer and more-easily measurable parameters while maintaining the mechanistic information would be key to facilitate a broader usage of such models (Buckley et al., 2012; Diaz-Espejo et al., 2012; Rodriguez-Dominguez et al., 2016).

2.3 Coupling photosynthesis, stomatal conductance, and transpiration

Various approaches exist to model photosynthesis and g_s independently, but the two processes are interdependent at the leaf level. Describing photosynthesis, transpiration, and stomatal conductance simultaneously captures the tight coupling between the gas-exchange of CO_2 and water vapor. Throughout this process, water lost through transpiration also regulates leaf temperature, further affecting photosynthetic rate through the temperature dependence of the enzyme activities. Coupling the biochemical process of photosynthesis with stomata-regulated gas-exchange is therefore critical to incorporate these processes into a whole-plant modeling framework. Important to note here is that the actual rate of CO_2 assimilation (A) that the FvCB model (the CO_2 demand function) predicts only what can be attained when the supply of CO_2 at the site of carboxylation is known after accounting for the resistances (e.g. stomatal and mesophyll) along the diffusional pathway. The supply side of CO_2 assimilation is described by the Fick's law of diffusion:

$$A = g_t\left(C_a - C_c\right) \tag{12}$$

Where g_t is the total conductance along the CO_2 diffusion pathway that accounts for boundary layer (g_b), stomatal (g_s), and mesophyll (g_m) conductances. C_a and C_c are the CO_2 concentrations in the air and at the chloroplast, respectively. The solution where the demand function (Eqn. 1) meets the supply function (Eqn. 12) can be obtained by coupling a photosynthesis model (e.g. FvCB model) with a stomatal conductance model (e.g. Ball-Berry model) and other conductance terms (i.e. g_b and g_m) using a numerical method (e.g. Kim and Lieth, 2003; Tuzet et al., 2003) or by solving for them analytically (Baldocchi, 1994).

Collatz et al. (1991) proposed to model the coupling of gas and energy exchange at the leaf surface by simultaneously solving for three key components: net photosynthetic rate, stomatal conductance, and leaf surface temperature (Fig. 3). The authors adopted the FvCB model to describe A (Eqn. 1), the Ball-Berry model to describe g_s (Eqn. 8), and an energy balance equation that describes the leaf surface energy exchange by considering surrounding environmental factors (e.g. solar radiation, air temperature, wind speed) and the aerodynamic characteristics of the leaf. The three equations can be considered interdependent at the leaf level; the FvCB model requires knowledge on stomatal conductance to determine the CO_2 concentration at the carboxylation site (C_c, Eqns. 2 and 3), as well as knowledge on leaf temperature since the enzymatic performance rates described within the model are temperature dependent; the Ball-Berry model requires knowledge on the net photosynthetic rate, which is one of the inputs to calculate stomatal conductance (Eqn. 8); and the leaf energy balance equation requires knowledge on stomatal conductance, which determines how easily heat can be lost through transpiration. Solving the equations iteratively eventually can lead to simultaneous answers of the photosynthetic rate, stomatal conductance, and leaf temperature when the solutions converge. Kim and Lieth (2003) illustrated a computational framework

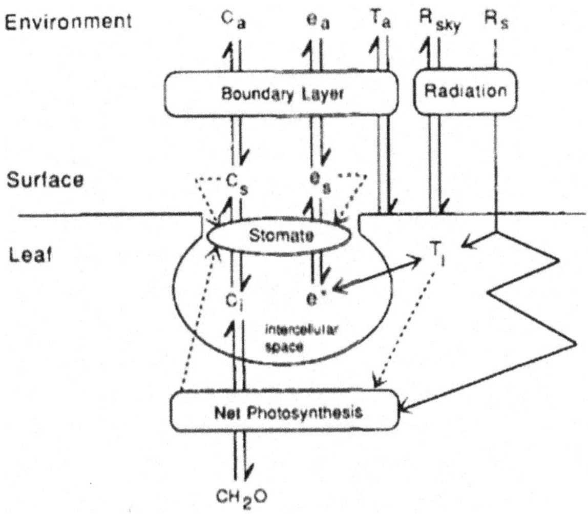

Figure 3 Conceptual diagram of the coupling of gas and energy. Fluxes are shown as solid lines and regulatory interactions are shown as dashed lines. *c*, *e*, and *T* stand for CO_2, H_2O vapor concentrations and temperature, respectively. The subscripts *a*, *s* and *l*, refer to properties in ambient air, at the leaf surface and within the leaf, respectively. R_{sky} specifies the long-wave length radiation input from and lost to the sky and R_{solar} represents solar radiation (Collatz et al., 1991, Figure 1).

of the iterative coupling process for the three model components, applying the coupled model to greenhouse rose crop. An analytical method also exists to simplify and approximate the solution (Baldocchi, 1994).

A coupled model provides mechanistic information often not represented by a more empirical approach and has been widely used among various communities such as ecosystem models, flux tower data analysis, and Earth System Models (Farquhar et al., 2001). However, a majority of crop models adopt the RUE method to describe biomass gain and yield, which is structurally a very different modeling approach compared to a coupled photosynthesis and stomatal model; simply incorporating such an approach is often non-feasible. In addition, concerns regarding computation complexity and parameter availability have led to resistance in adopting a coupled approach in the modeling community for newly developed crop models. Regardless, successful examples of models that adopt a coupling approach exist for major crops such as maize, rice, and wheat (Durand et al., 2018; Kim et al., 2012; Li et al., 2015; Palosuo et al., 2011). Detailed measurements of species-specific enzymatic information have also become more available, making it easier to gather parameter inputs for the FvCB model. A recent study published by Hsiao et al. (2019) provides an example of using a comprehensive process-based maize crop model that implemented a coupled photosynthesis-stomatal conductance modeling approach to quantify the independent yield impacts that come from warming and increased VPD in maize in the US Corn Belt. The coupling of photosynthesis, stomatal conductance, and leaf energy balance within the model led to a clear representation of tradeoffs between carbon gained and water lost throughout the growing season, which allowed for a mechanistic understanding of how maize plants responded to increasing temperature and VPD perturbations and provided insights on how these responses could further interact with rising CO_2 levels (Hsiao et al., 2019).

A coupled approach allows models to capture the potential feedback that boundary conditions can impose on gas-exchange, as well as the regulation of leaf temperature through transpiration and the temperature impacts that is then imposed on photosynthesis and stomatal conductance. This provides a more comprehensive view of the soil-plant-atmosphere continuum system in which plants are linked with their surrounding environment.

3 Modeling respiration and assimilate partitioning

3.1 Respiration

In addition to photosynthesis for assimilation of carbon, respiration is another important physiological process that determines plant growth. Respiration breaks down the photosynthates (e.g., sucrose) to provide energy (e.g., ATP and NAD(P)H) for biochemical processes needed for growth and maintenance of the living cells and provides C skeletons for biosynthesis. The functional view of respiration gave rise to the Growth and Maintenance Respiration Paradigm (GMRP; Amthor, 2000; McCree, 1970), which models respiration in two functional components: (1) *growth respiration* to provide energy (ATP), reducing power (NAD(P)H), and C skeletons for building new biomass and (2) *maintenance respiration* to support energy needed for the biochemical processes for maintaining protein turnovers, cellular components, and ion-gradients of existing biomass. Several good reviews are available on GMRP and how it is used in crop models (Amthor, 2000; McCree, 1970; Thornley, 2011; Thornley and Cannell, 2000). In its basic form, GMRP is represented by the following equation (Amthor, 2000):

$$R = R_G + R_m = g_R G + m_R W \tag{13}$$

Where R is respiration rate (mol CO_2 s^{-1}), R_G is growth respiration rate, R_M is maintenance respiration rate, G is growth rate (g new biomass s^{-1}), W is living biomass (g dry-matter), g_R is a growth respiration coefficient (mol CO_2 g new biomass^{-1}), and m_R is maintenance respiration coefficient (mol CO_2 g living biomass^{-1}). Mechanistic representation of the maintenance respiration related to protein and cellular components turnovers and the maintenance of intercellular ion-gradients has not been straightforward (Thornley, 2011). On the other hand, growth respiration can be represented somewhat mechanistically through a 'pathway analysis method' by accounting for substrates used, pathways of biosynthesis, and composition of biomass produced in growth (Amthor, 2000). The growth respiration parameter: g_R can be further decomposed with respect to the type of subprocesses, substrates, pathways, and products involved (Amthor, 2000; Penning de Vries et al., 1974). Approaches that account for the components of g_R have been implemented in the PBCMs that are currently in use (Boote et al., 1998, 2002). A comprehensive theoretical representation and interpretation of the GMRP based model behavior can be found in Thornley and Cannell (2000).

A simpler approach to model respiration is based on empirical observations that $R:P_g$, the ratio between daily respiration (R) and daily gross photosynthesis (P_g), remains fairly constant in the long term in many species (Gifford, 2003). This simplified approach to use $R:P_g$ to model respiration is implemented in a number of plant production models (Gifford, 2003).

By separating substrate and structural carbon pools as state variables responding to photosynthesis, growth respiration, senescence, and recycling, Thornley (2011) demonstrated theoretically that maintenance respiration as used in GMRP may be an emergent property that need not be modeled as a specific process. This analysis by Thornley (2011) also illustrated that, depending on two key parameters, namely, growth yield and recycled fraction, the $R:P_g$ ratio would remain conservative as found in empirical observations and reviewed by Gifford (2003).

When modeling respiration, there are several potential sources of error, some of which are routinely considered in today's models, others of which cannot be easily incorporated as our understanding of the mechanisms involved is still limited. While photosynthesis takes place only in tissues with chloroplasts, respiration occurs in all living cells with mitochondria. This distinction is important in modeling because all photosynthetic tissues also respire but not all respiring tissues photosynthesize. Recognizing this distinction, process-based models often simulate gross photosynthesis with respect to total plant leaf area while respiration is calculated for organ or plant biomass at the whole-plant level. Another complicating factor in modeling respiration is the difference between day respiration and night respiration rates in leaves. The leaf respiratory CO_2 efflux in light is lower than that in the darkness but the mechanisms contributing to this effect are not fully understood (Tcherkez et al., 2017). An accurate accounting of day respiration may not be feasible at this point because of our limited understanding of the mechanisms involved and limited means to measure it accurately (Tcherkez et al., 2017). The difference in day and night respiration rates is currently not commonly acknowledged or adequately addressed in most PBCMs.

3.2 Partitioning of photosynthates into organs and other sinks

Another important process that determines plant growth, biomass accumulation, and eventually crop yield is the partitioning of photosynthates to different plant organs. Using a simple model of plant growth based on relative growth rate (*r*) used in classical growth analysis, Monsi (1960) intuitively illustrated the significance of the intricate relationships among net whole-plant CO_2 assimilation (*A*), whole-plant respiration (*R*), and the fraction of assimilates partitioned to photosynthetic (*m*) and non-photosynthetic tissues (*n*) in the whole plant by defining the relative growth rate (*r*) as:

$$r = k(mA - nR) \tag{14}$$

Where *k* is CO_2 conversion factor to biomass that can be extended to different plant functional types. This analysis demonstrated that multiple pathways contribute to maintaining the plant growth rate through the coordination of assimilate production, assimilate allocation to photosynthetic tissues, and assimilate loss through respiration.

Despite considerable advances made in understanding the individual processes of photosynthesis, respiration, sugar metabolism, long-distance transport, and cell growth (Marcelis et al., 1998), the integration of these processes into a biomass-partitioning model is limited in crop simulation models that operate at the crop level. Marcelis et al. (1998) provides an outstanding review of modeling approaches for dry-matter partitioning that is still relevant to this date. Briefly, most commonly used method is *descriptive allometry*. It is a primarily empirical approach that partitions photosynthates for biomass production across different plant parts or organs according to predetermined partitioning coefficients (Marcelis et al., 1998). These partitioning coefficients are empirically determined and often vary with crop developmental stage as a function of thermal unit accumulation (Goudriaan and van Laar, 1994).

More mechanistic approaches address biomass partitioning through *sink regulations* (Marcelis et al., 1998). One such approach is to prioritize dry-matter partitioning among plant organs by their sink strength, which is defined as the potential capacity of an organ for assimilate accumulation (Marcelis, 1996). This *sink priority* approach relies on the simplifying assumption that the source (or transport) path is non-dominant in regulating dry-matter partitioning (Marcelis et al., 1998). Another approach that accounts for the *transport-resistance* network from source to sink is to simulate the phloem transport of assimilates based on Ernst Munch's hypothesis of bulk flow, in which assimilates are driven by an osmotically generated pressure gradient between source and sink organs and the kinetics of assimilate unloading at the sink (Minchin et al., 1993). A notable expansion of this transport-resistance modeling approach was introduced by Daudet et al. (2002). Daudet and colleagues coupled the movement of assimilates with water fluxes between source and sink organs through a network of resistive pathways connecting the phloem and xylem. However, despite its theoretical foundation, the adoption of transport-resistance approaches in crop models has been slow. This is partly because of the difficulty in estimating the model parameters, but also because of inherent limitations in models that describe organs as bulk compartments (Minchin and Lacointe, 2005). Recent advances in functional-structural modeling that explicitly address organ-level functions and whole-plant interactions across organs made it possible to incorporate the transport-resistance concept into a crop model (e.g. Barillot et al., 2016). Further mechanistic layers, not yet incorporated into crop models, are currently being considered. For example, Chang and Zhu (2017) highlighted the need to develop a molecular-systems model of source-sink interaction. This would facilitate computational exploration of the source-sink relationship and hypothesis testing, which, when translated into real plants, could be used to increase crop yield. Assimilate partitioning is also influenced by temperature and temperature gradients within a whole plant (Sperling et al., 2017; Zwieniecki et al., 2015).

In addition to respiring photosynthates and allocating them to sink organs within the plant (e.g. grains, fruits, roots), most plants also allocate considerable amounts of carbohydrates to microbes that are symbiotic or otherwise (e.g. mycorrhizae, rhizobia and endophytes) as well as to the rhizosphere in the form of root exudates that attract and feed plant-growth-promoting rhizobacteria. These external sinks are likely to compete with internal sinks, like growing plant organs, for carbon. It is also likely that the cost-benefit ratio of symbiosis will vary depending on growth conditions and resource availability. For example, the estimated carbon cost allocated to arbuscular mycorrhizae (AM) can be as much as 20% of the daily carbon fixed in photosynthesis (Lambers et al., 2008, p. 418). Similarly, the carbon cost for legume-rhizobia symbiosis may be similar to that of AM in N-limited environments (Lambers et al., 2008, p. 435). The estimate of carbon cost for legume-rhizobia symbiosis is likely to be lower in crop systems where N is supplemented as a fertilizer. Some PBCMs explicitly account for the carbon cost of symbiotic N fixation in legumes (Boote et al., 2008). To gain an accurate representation of assimilates used for growth and yield, and to better understand the effects of changing CO_2 concentrations and nutrient (e.g. N) dynamics on crops, this practice needs to be extended to all crop models, especially those where symbiotic relationships are critical for the crop or cropping system, as is true of legumes or organic agriculture.

The carbon cost estimates of endophytes, which are either symbiotic bacteria or fungi living inside plants, are currently unknown. However, plant respiration increases significantly after endophytic inoculation in rice, illustrating the influence of endophytes on the carbon metabolism of host plants (Rho et al., 2018). While symbionts and root exudates to the rhizosphere can be significant external carbon sinks, they remain as *missing carbons* that are not explicitly accounted for in current crop models. It will be critical to address this hole in the future for an accurate accounting of carbon and nutrient dynamics, and their reciprocal costs and benefits for maintaining the symbiotic relationship.

4 Modeling phenology: the timing of crop developmental events

The empirical relationship between growth temperature and the rate of plant processes such as organ emergence, leaf expansion, and timing to germination or anthesis has been modeled for over a century with new models emerging regularly. Beginning in the early twentieth century, researchers also noted the role that other environmental factors such as photoperiod (day length) and vernalization (extended period of cold temperatures) had on the timing of phenological events as well as the interactive influences these factors had with ambient temperature. Accurate models are crucial today, as uncertainty in estimating such things as the rate of leaf development or final leaf number as a function of temperature, can affect day of flowering estimates by several days (Jamieson et al., 1998; Xue et al., 2004), impacting the ability of models to predict and for growers to effectively plan. Here, we overview several common approaches used to model temperature influence alone and in conjunction with photoperiod and vernalization on phenology. We, then, discuss recent efforts to improve the versatility and predictive capacity of these models.

4.1 History of phenology models

4.1.1 Temperature models

The timing of phenological events like germination, onset of the reproductive phase, and seed set are dependent upon the rate of development which is strongly influenced by growing-season temperature. The influence of temperature on the rate of development was observed and modeled early. One of the earliest models was calculation of accumulated thermal units - or the number of degrees above a baseline temperature accumulated each day - to a predetermined threshold value to predict timing of phenological events. In this case, thermal units are accumulated more quickly during warm growing seasons than in cool ones. This is termed Growing Degree Days (GDD), and the equation appears as below:

$$GDD = \left[\frac{(T_{max} - T_{min})}{2}\right] - T_{base} \tag{15}$$

T_{max} and T_{min} are daily maximum and minimum temperatures, respectively (as represented by McMaster and Wilhelm, 1997). In addition to the base temperature (T_{base}) at or below which the crop development ceases, these models may also include an optimal temperature (T_{opt}) above which the accrual of thermal units does not increase. For an interesting discussion on the ways in which the basic GDD equation can be interpreted see McMaster and Wilhelm (1997).

This form of thermal unit accumulation, while simplistic and easy to use, did not entirely capture the observed plant response, and several variants of the basic GDD model have been proposed (see review by Yin et al., 1995). For example, Gilmore and Rogers proposed a bilinear model in which the rate of development increased linearly between T_{base} and T_{opt}, and then declined linearly beyond the optimum temperature, rather than ceasing (Gilmore and Rogers, 1958). Others have proposed an additional maximum or critical temperature (T_{crit}), beyond which growth does cease. These three temperatures (T_{base}, T_{opt}, and T_{crit}) were coined the cardinal temperatures, and the temperatures between T_{base} and T_{opt} called sub-optimal and between T_{opt} and T_{crit} supra-optimal (Garcia-Huidobro et al., 1982; Roberts and Summerfield, 1987; Yin et al., 1995). Recently, a similar concept called Killing Degree Days (KDD) was coined to describe the damaging effects of extreme heat on crops (Butler and Huybers, 2015). KDD accumulated when the daily maximum temperature (T_{max}) exceeded T_{opt}.

$$KDD = T_{max} - T_{opt} \tag{16}$$

While some studies have demonstrated little variation across cultivars in the cardinal temperatures for many processes (Porter and Gawith, 1999), what values the cardinal temperatures should take is a debate that began early and continues today (Boote et al., 2013; Lehenbauer, 1914). The values that cardinal temperatures take are dependent on multiple factors such as the cultivars and temperature ranges used for training, the shape of the temperature-response function used, or the phase of development being assessed. For example, White and colleagues noted that estimates for T_{base} ranged nearly 6°C depending on whether a one-segment-linear, two-segment-linear, or quadratic equation was used (White et al., 2012). Further, in a study of eight wheat cultivars in three different locations in Georgia, USA, the base temperature varied up to 8°C across cultivars within a single developmental phase (Salazar-Gutierrez et al., 2013). The base temperatures differed across developmental phases as well. From planting to heading they ranged from 3.1 to 8.1°C across the eight cultivars, whereas they ranged between 10.6 and 18.4°C during heading to harvest maturity. The cardinal temperatures, especially those characterizing growth and survival at low temperatures, can also vary depending on the 'hardening' or exposure to cold, non-freezing temperatures that plants have experienced prior to exposure to extreme cold temperatures (Porter and Gawith, 1999). Even multi-part linear models could not capture some aspects of the behavior of temperature-dependent development, at least for some genotypes. Yin and colleagues noted that as temperatures increased from the base temperature, the rate of development or expansion for cassava and maize appeared curvilinear - increasing with temperature increase (as shown in Fig. 4), and that this was typical of multiple plant processes (Yin et al., 1995).

To account for this behavior, they proposed using the Beta function (Abramowitz and Stegun, 1965) to describe development rate as a function of temperature, as is shown in the simplified variant used in some PBCMs (Kim et al., 2012; Yan and Hunt, 1999) in Fig. 4 and Eqn. (17).

$$r(T) = R_{max}\left(\frac{T_{crit} - T}{T_{crit} - T_{opt}}\right)\left(\frac{T}{T_{opt}}\right)^{\frac{T_{opt}}{(T_{crit} - T_{opt})}} \quad (17)$$

The temperature-dependent rate of development is $r(T)$, R_{max} is the maximum rate of development, and T is an input of the mean temperature at a given timestep. T_{opt} and T_{crit} take on the same meanings as they do in the GDD variants above: T_{opt} is the temperature at which $r(T) = R_{max}$ and T_{crit} is the temperature

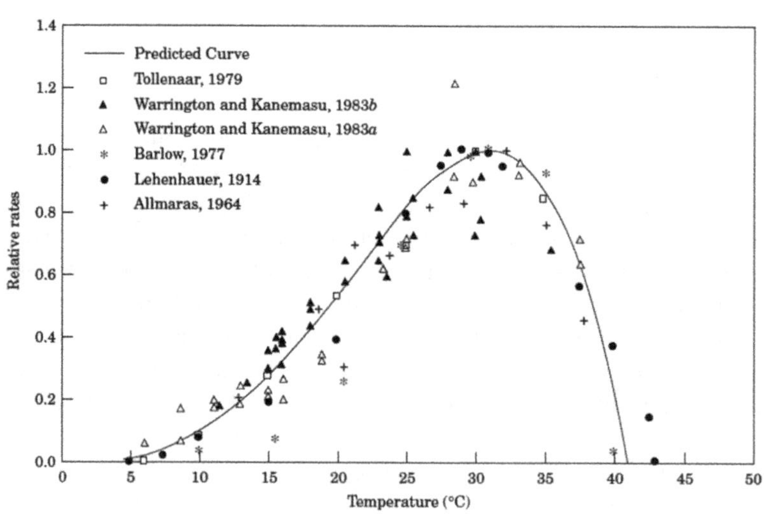

Figure 4 Measured relative rates of all development or growth of maize, together with predicted relative rates based on a single curve with T_{crit}=41 and T_{opt}=31°C (Yan and Hunt, 1999; Figure 3).

above which development stops. The Beta function was advantageous in that it could describe the temperature-dependent behavior of developmental rate across several datasets, but had fewer parameters than other non-linear models and, therefore, was practical for use in the field (for review of these other non-linear variants see Yin et al., 1995). However, some have noted that choice of T_{opt} heavily influences estimates of T_{base} and T_{crit}, and suggest that the Beta function is not a logical choice for simulating biological processes (Andreucci et al., 2016). (The following model comparisons, discussed later in this section, provide an excellent review of the different temperature models in use: Kumudini et al., 2014; Wang et al., 2017).

How temperature is incorporated into these models varies. Early variants sometimes used the mean temperature to a specific stage of development, but this poses problems if the daily temperatures exceed the optimum temperature (Lawn et al., 1995; Yan and Wallace, 1998). To improve on this, modelers used the daily average temperature to calculate rate of development per day (Hunt and Pararajasingham, 1995). Now, in an effort to better capture the effects of diurnal temperature fluctuations, some models calculate GDDs based on 3-hour or 1-hour intervals over a 24-hour period, which uses a sine function to span the daily maximum and minimum temperatures to mimic a diurnal pattern (e.g. Lizaso et al., 2005). Others use the average hourly temperature (e.g. Kim et al., 2012). Other variants may give different weights to daytime and nighttime temperatures or describe the temperature-dependent rate of development differently for day and night. For example, daytime temperatures better described the temperature-dependent rate of flowering in *Arabidopsis* lines across several planting dates and locations. The researchers thus gave either a minimal or zero weighting to nighttime temperatures (Chew et al., 2012; Wilczek et al., 2010). In contrast, the Crop Heat Unit (CHU) model (Brown, 1960, 1969) uses a linear and quadratic function to describe nighttime and daytime temperature effects, respectively, in a combined function. The base temperatures also differ, being 4.4 and 10°C. Interestingly, the CHU model best predicted timing of maize anthesis when compared to several linear and nonlinear models describing either the empirical relationship between temperature and anthesis or an underlying process. All of these other models treated daytime and nighttime temperatures the same (Kumudini et al., 2014).

Additionally, there has been some debate as to whether to use air, canopy, soil, or meristem temperature. Traditionally, air temperature is the easiest to acquire as there are weather data stations spread across the globe. However, meristem and canopy temperature can differ from the air temperature, and this could affect interpretations of the temperature-dependent developmental rate. For example, Jamieson and colleagues showed that the wheat leaf appearance rate was linearly related to soil temperature 2 cm below the surface near the level of the meristem (Jamieson et al., 1998). They proposed that differences in model predictions based on sowing date and air temperature were likely due to how much the air and soil temperature differed - with the least difference occurring in winter.

Models can be trained to describe the relationship between temperature and any one of multiple layers of plant organismal organization such as the timing of a target process like germination or onset of anthesis (Garcia-Huidobro et al., 1982; Yan and Wallace, 1998), the rate of an underlying process like phyllochron appearance or leaf emergence (Jamieson et al., 1998; Kumudini et al., 2014), or gene transcription or enzymatic functioning (Parent et al., 2010). The latter two categories could be related to higher-level target processes by, for example, simulating onset of anthesis once a target leaf number or level of gene product is reached (Brown et al., 2013; Kumudini et al., 2014).

4.1.2 Models incorporating photoperiod and vernalization

In the early 1900s, agronomists began to recognize the influence that daylength (or photoperiod; Garner and Allard, 1920, 1923; Tournois, 1914) and winter chilling (or vernalization; Coville, 1920; Schafer et al., 1921) had on some strains and cultivars, as well as the interactions among these two factors and growing-season temperature (Adams, 1924; Garner and Allard, 1930). In short, some species or cultivars flower only once the photoperiod has either exceeded or fallen below a given number of hours. Some species or cultivars require a period of cold winter temperatures, which if not experienced causes them to remain in a vegetative state or to produce only a few flowers during the growing season. Both the photoperiod and vernalization requirements can be reduced, and sometimes offset, as plant age (Dingkuhn et al., 2008; Wang et al., 1995). (For a concise review on the ways used to incorporate photoperiod and vernalization into models see Yan and Wallace, 1998).

The influence of these factors is commonly modeled very simply by either increasing or decreasing the rate of thermal time accumulation using a modifier that falls between zero and one, as in the following two forms used by Weir and colleagues (Eqn. 18) and McMaster and colleagues (Eqn. 19), respectively (McMaster et al., 2008; Weir et al., 1984).

$$R = f(T)f(P)f(V) \tag{18}$$

$$DTDU = GDD \cdot \min(VF, PF) \tag{19}$$

R is the rate of development and $DTDU$ is the daily thermal development units accumulated. P and V are photoperiod and vernalization and VF and PF are vernalization and photoperiod factors. Vernalization is, of course, included or excluded as relevant to a given species. For example, for a species that transitions to reproduction most quickly in long photoperiods, the photoperiod modifier would equal one when day lengths range above the maximum critical day length and the rate of thermal time accumulation would be at its maximum. Below the critical day length, the modifier would fall below one, slowing the rate of thermal time accumulation until a base daylength is reached. As with the temperature response, the influences of photoperiod and vernalization can be modeled in linear or non-linear forms.

There are a few intriguing model forms, not widely in use, which could add insight to plant phenological responses. A model describing tree phenology, which requires a period of winter chilling like vernalization before ending endodormancy, accumulates 'chilling units' below the threshold temperatures (e.g. Chung et al., 2011). Another early work assumes that, in addition to its effect on the rate of development, temperature will affect the underlying molecular processes associated with the photoperiod response, as shown for Eqn. (20) (Yan and Wallace, 1998).

$$R = f(T) + f(TP) \tag{20}$$

R is rate of development and TP indicates temperature modification of the photoperiod effect. This model accurately simulated the responses of several species of both long- and short-photoperiod-requiring varieties.

4.2 Recent advances in improving temperature-response functions

Recently through ensemble modeling exercises, crop modelers generated a new non-linear function based on the recent observation that plant processes from rates of cell division to phenological development followed the same Arrhenius-like response curve (Parent et al., 2010). Incorporating this function, along with another to describe Q10, into all 29 models vastly improved their predictive capacity across the multiple temperature regimes (Wang et al., 2017). These models were compilations of several different plant processes in addition to phenological development, and the authors noted that the interactions of these separate temperature-response functions can compound model error (Wang et al., 2017).

This group and others also recognized continued uncertainty in predicting plant responses in two areas: at higher temperatures, especially those ranging above optimum, and in the developmental rates post anthesis. For example, the datasets used in assessing phenology functions in maize models (Kumudini et al., 2014) did not contain sufficient high temperature data above optimum. Additionally, their work also highlights the potential need to derive separate models for different developmental phases. For example, the model showing the greatest precision across those tested for maize (e.g. CHU model) incorporated separate temperature functions for pre- and post-silking development (Kumudini et al., 2014). Similarly, the previously mentioned ensemble of 29 models containing the newly derived temperature functions accurately simulated most observed plant responses between 0 and 30 or 35°C. However, simulated development after anthesis was only accurate up to 25°C (Wang et al., 2017).

4.2.1 Incorporating genetic and molecular mechanisms

We now have over a century of phenology model development, and over much of that time, the emphasis has been on using and improving thermal-time approaches. However, in the last several years, models have begun to incorporate gene- or QTL-specific parameters. For example, Bogard and colleagues replaced two empirically derived terms, which described both the vernalization and photoperiod effects, with parameters derived from QTL analysis of 210 field-grown wheat lines (Bogard et al., 2014). In this case, they first parameterized those terms through empirical analysis of heading date in several different conditions in the field, and then related the variation across parameter space to genetic markers. They were able to derive unambiguous parameter vectors for each genotype, in part, they propose, because they used just two parameters. They were not able to do so in a three-parameter trial (Bogard et al., 2014). A similar effort using 5266 maize lines to parameterize four different parameters in CERES-Maize model, found that multiple parameter sets were equally effective at predicting anthesis dates, additionally many of the parameters values were heavily influenced by planting location, calling into question whether the parameters indeed described underlying genetic characteristics (Lamsal et al., 2018).

There has been parallel movement toward incorporating the molecular regulatory pathways influencing phenology, as informed by studies using a model species, *Arabidopsis thaliana*. These models parallel thermal time models in that they incorporate accumulation of gene product to a threshold. For example, Brown and colleagues used an approach for wheat based on recognition that photoperiod and vernalization influence the final leaf number at which transition to reproduction occurs; temperature influences the rate of leaf growth and gene-product accumulation (Brown et al., 2013). They simulated *VRN3* accumulation to a threshold in wheat, at which the reproductive transition is induced, with the rate of accumulation being influenced by both photoperiod and vernalization (*VRN3* is functionally like the key floral promoting gene, *FT*, in *Arabidopsis*; Brown et al., 2013). Additionally, in their model, the transcription rate of all gene products was greater under higher temperatures based on the idea that temperature should impact kinetic rates of gene expression (Yan and Wallace, 1996). In another example, Wenden and colleagues demonstrated that the threshold requirement for accumulation of a mobile floral signal could be genetically determined based on known genetic variants of the *LATE FLOWERING* locus in Pea (Wenden et al., 2009), which is a homologue of a flowering repressor gene, *TFL1*, in *Arabidopsis* which acts at the shoot apical meristem. Finally, Kinmonth-Schultz and colleagues recognized the influence temperature may have on whole-plant *FT* accumulation in *Arabidopsis* through its influence on leaf emergence and expansion as *FT* is expressed in the leaves (Kinmonth-Schultz et al., 2019). They modeled both direct temperature influence on *FT* gene transcription (coupled with photoperiodic control) and temperature influence of *FT* gene transcription through the rate of leaf tissue production. This concept needs further testing and validation in a broader range of conditions; however, this approach may enable future linking of model modules that simulate leaf emergence and growth to modules simulating the phenology of anthesis.

4.2.2 Molecular mechanisms as a means to address model uncertainty

Models incorporating molecular mechanisms hold promise as a means to address the uncertainty in phenology prediction that still exists. The molecular mechanisms driving phenological shifts in response to both ambient and extreme temperatures alone and through interactions with photoperiod are well known in *Arabidopsis*. Additionally, more recent work has shown striking parallels between - and, therefore, the applicability of - mechanisms in *Arabidopsis* compared to those in economically relevant crop species (for review see Distelfeld et al., 2009; Kobayashi and Weigel, 2007; Song et al., 2015). Within many of these crop species, work has also advanced rapidly, such that we are gaining not only a grasp of key gene players, but the intricacies of their regulation (for review see Hill and Li, 2016).

If gene-product accumulation, indeed, underlies the plant behaviors currently described through thermal-time approaches, this opens several doors into understanding and better modeling plant phenological behaviors in numerous conditions. For example, just as the threshold of an accumulated floral signal was altered by genetic alleles (Wenden et al., 2009), it is highly possible that environmental factors could also influence the threshold by altering the activity of genes regulating the sensitivity to floral inducing signals. Further, it is logical that the 'threshold' exists within cells of the shoot apical meristem, which, after traveling from the leaves, is the site of action of *FT*. Consistent with this proposal, both *SVP* and *TFL* genes act at the shoot apical meristem to inhibit the ability of *FT* to induce flowering in *Arabidopsis*, and both are highly sensitive to ambient temperature (Kaufmann et al., 2010; Kim et al., 2013; Lee et al., 2007; Wickland and Hanzawa, 2015). A proof of concept of this proposal, still using thermal time to alter the threshold, was demonstrated by Kinmonth-Schultz and colleagues (2019).

5 Conclusion and future trends

As our understanding of the complexity of plant processes and responses to the environment has increased over time, so have our models in an effort to improve their range of explanatory and predictive capacity. This effort has yielded a tremendous amount of materials that can be evaluated through comparisons to determine the best approaches for given situations and to assess where work still needs to be done in model improvement.

5.1 Modeling photosynthesis and stomatal conductance

With expanding knowledge on mechanisms that govern photosynthesis and stomatal conductance, how to strike the right balance of complexity becomes the key question in whole-plant crop models. Several of the commonly used approaches in modeling photosynthesis and stomatal conductance have been published decades ago. There is a great potential to incorporate updated mechanistic knowledge into these models, especially in terms of stress and acclimation responses. We also emphasize the importance of treating the process of gas-exchange as a whole, in which carbon gained at the leaf level is coupled with water lost from the soil through the canopy into the atmosphere, linked by a realistic representation of stomatal response to drought stress. In addition, rigorous testing and validation of models under a broad range of environmental conditions along with systematic comparisons between modeling methods would greatly benefit the crop modeling community as a whole.

5.2 Modeling respiration and assimilate partitioning

Compared to photosynthesis models, respiration models used in crop modeling are empirical rather than mechanistic. The GMRP is a primary modeling approach but many models also use the constant $R{:}P_g$ approach for its simplicity. The components determining maintenance respiration in GMRP remains semi-empirical and difficult to interpret mechanistically. On the other hand, modeling approaches for growth respiration using the substrate and biosynthetic pathways remain more robust. New insights continue to emerge from derivatives and applications of GMRP. Similar to respiration, the advances in modeling assimilate partitioning have been somewhat slower than other processes such as photosynthesis and flowering phenology. Most crop models use *descriptive allometry*, in which empirically determined partitioning coefficients are used across the crop phenological stages. Sink regulation-based models have more mechanistic foundations, but their incorporation into full crop models has been slow. Rapid adoption of specialized models representing the current state of disciplinary knowledge into submodules of full crop model is much needed for modeling these processes. Finally, the carbon balance in crop models should address the missing carbons that are allocated to symbionts and other external sinks that are not part of plant biomass.

5.3 Modeling phenology

Early on, growers and agronomists noted the predictable changes in developmental rates with growing-season temperatures and very quickly also the relationship between temperature and other environmental variables like photoperiod or the length of exposure to vernalizing temperatures. How to model the influence of these phenomena is an ongoing debate that is not yet resolved. Recent comparisons of large suites of models should provide a framework to begin to resolve this debate (Kumudini et al., 2014; Wang et al., 2017), so too should our increasing understanding of the molecular processes underlying plant environmental responses. Complex models parameterized for each protein and molecular interaction are not yet feasible for widespread use, as fitting each parameter for a new cultivar is labor and computationally intensive. However, our current understanding of molecular processes underlying plant developmental responses could still assist in generating and testing new theories that can be incorporated into phenology models to reflect the mechanisms involved and improve their predictive capability. Crop modelers and agronomists need to work with molecular biologists to clarify questions that remain, such as how the rates of gene-product responses are coordinated across the whole-plant and across different processes within a single cultivar. If so, cultivar-specific responses may become easier to predict.

In addition to the plant processes covered in this chapter, there are other key processes fundamental for improving crop models such as root growth, reproductive physiology (e.g. grain filling), nutrient dynamics, and crop nutritional quality. Previous studies provide comprehensive and critical reviews of modeling some plant processes not covered in this chapter, for example, root growth (Dunbabin et al., 2013; Dupuy et al., 2010) and grain filling (Boote et al., 2018). More attention and rapid advances are critically needed for modeling crop nutrient dynamics and nutritional quality for improving food security while preserving health benefits in a changing climate (Asseng et al., 2019; Soares et al., 2019).

6 Where to look for further information

- An overview of PBCMs:
  - Boote, K. J., Jones, J. W., White, J. W., Asseng, S., and Lizaso, J. I., 2013. Putting mechanisms into crop production models. *Plant Cell Environ.* 36, 1658–72.
- For a brief history and perspective on photosynthesis models:
  - Farquhar, G. D., von Caemmerer, S., and Berry, J. A., 2001. Models of photosynthesis. *Plant Physiol.* 125, 42–5.
- R package for gas-exchange modeling (https://bitbucket.org/remkoduursma/plantecophys/):
  - Duursma, R. A., 2015. Plantecophys – An R package for analysing and modelling leaf gas exchange data. *PLoS ONE* 10(11), e0143346.
- Global leaf-gas exchange database (https://bitbucket.org/gsglobal/leafgasexchange/):
  - Lin et al., 2015. Optimal stomatal behaviour around the world. *Nature Clim. Change* 5(5), 459–64.
- For a comprehensive, regularly updated database on known molecular flowering control pathways: Flowering Interactive Database (http://www.phytosystems.ulg.ac.be/florid/).
- Reviews of parallel flowering phenology mechanisms in crops:
  - Song, Y.-H., Shim, J.-S., Kinmonth-Schultz, H. A., and Imaizumi, T., 2015. Photoperiodic flowering: Time measurement mechanisms in leaves. *Annu. Rev. Plant Biol.* 66, 441–64.
  - Hill, C. B. and Li, C., 2016. Genetic architecture of flowering phenology in cereals and opportunities for crop improvement. *Front. Plant Sci.* 7, 1906. https://doi.org/10.3389/fpls.2016.01906.
- Comprehensive crop model comparisons of temperature responses:
  - Wang, E. et al., 2017. The uncertainty of crop yield projections is reduced by improved temperature response functions. *Nature Plants* 3, 17102. https://doi.org/10.1038/nplants.2017.102.
  - Kumudini, S. et al., 2014. Predicting maize phenology: Intercomparison of functions for developmental response to temperature. *Agron. J.* 106, 2087–97. https://doi.org/10.2134/agronj14.0200.

7 References

Abramowitz, M., and Stegun, I. A. 1965. *Handbook of Mathematical Function with Formulas, Graphs, and Mathematical Tables*. Dover Publications, New York.

Acock, B., and Reddy, V. R. 1997. Designing an object-oriented structure for crop models. *Ecol. Model.* 94, 33–44.

Adams, J. 1924. Does light determine the date of heading out in winter wheat and winter rye? *Am. J. Bot.* 11, 535–9. https://doi.org/10.2307/2435629.

Allen, M. T., Prusinkiewicz, P., and DeJong, T. M. 2005. Using L-systems for modeling source-sink interactions, architecture and physiology of growing trees: the L-PEACH model. *New Phytol.* 166, 869–80. https://doi.org/10.1111/j.1469-8137.2005.01348.x.

Amthor, J. S. 2000. The McCree-de Wit-Penning de Vries-Thornley respiration paradigms: 30 years later. *Ann. Bot.* 86, 1–20.

Andreucci, M. P., Moot, D. J., Black, A. D., and Sedcole, R. 2016. A comparison of cardinal temperatures estimated by linear and nonlinear models for germination and bulb growth of forage brassicas. *Eur. J. Agron.* 81, 52–63. https://doi.org/10.1016/j.eja.2016.08.010.

Aphalo, P. J., and Jarvis, P. G. 1991. Do stomata respond to relative humidity? *Plant Cell Environ.* 14, 127–32.

Asseng, S., Ewert, F., Rosenzweig, C., Jones, J. W., Hatfield, J. L., Ruane, A. C., Boote, K. J., Thorburn, P. J., Rötter, R. P., Cammarano, D., Brisson, N., Basso, B., Martre, P., Aggarwal, P. K., Angulo, C., Bertuzzi, P., Biernath, C., Challinor, A. J., Doltra, J., Gayler, S., Goldberg, R., Grant, R., Heng, L., Hooker, J., Hunt, L. A., Ingwersen, J., Izaurralde, R. C., Kersebaum, K. C., Müller, C., Naresh Kumar, S., Nendel, C., O'Leary, G., Olesen, J. E., Osborne, T. M., Palosuo, T., Priesack, E., Ripoche, D., Semenov, M. A., Shcherbak, I., Steduto, P., Stöckle, C., Stratonovitch, P., Streck, T., Supit, I., Tao, F., Travasso, M., Waha, K., Wallach, D., White, J. W., Williams, J. R., and Wolf, J. 2013. Uncertainty in simulating wheat yields under climate change. *Nat. Clim. Change* 3, 827–32. https://doi.org/10.1038/nclimate1916.

Asseng, S., Martre, P., Maiorano, A., Rötter, R. P., O'Leary, G. J., Fitzgerald, G. J., Girousse, C., Motzo, R., Giunta, F., Babar, M. A., Reynolds, M. P., Kheir, A. M. S., Thorburn, P. J., Waha, K., Ruane, A. C., Aggarwal, P. K., Ahmed, M., Balkovič, J., Basso, B., Biernath, C., Bindi, M., Cammarano, D., Challinor, A. J., Sanctis, G. D., Dumont, B., Rezaei, E. E., Fereres, E., Ferrise, R., Garcia-Vila, M., Gayler, S., Gao, Y., Horan, H., Hoogenboom, G., Izaurralde, R. C., Jabloun, M., Jones, C. D., Kassie, B. T., Kersebaum, K.-C., Klein, C., Koehler, A.-K., Liu, B., Minoli, S., Martin, M. M. S., Müller, C., Kumar, S. N., Nendel, C., Olesen, J. E., Palosuo, T., Porter, J. R., Priesack, E., Ripoche, D., Semenov, M. A., Stöckle, C., Stratonovitch, P., Streck, T., Supit, I., Tao, F., Velde, M. V. der, Wallach, D., Wang, E., Webber, H., Wolf, J., Xiao, L., Zhang, Z., Zhao, Z., Zhu, Y., and Ewert, F. 2019. Climate change impact and adaptation for wheat protein. *Glob. Change Biol.* 25, 155–73. https://doi.org/10.1111/gcb.14481.

Baldocchi, D. 1994. An analytical solution for coupled leaf photosynthesis and stomatal conductance models. *Tree Physiol.* 14, 1069–79.

Ball, J. T., Woodrow, I. E., and Berry, J. A. 1987. A model predicting stomatal conductance and its contribution to the control of photosynthesis under different environmental conditions, in: Biggins, J. (Ed.), *Progress in Photosynthesis Research*. Springer, Dordrecht, pp. 221–4.

Barillot, R., Chambon, C., and Andrieu, B. 2016. CN-Wheat, a functional-structural model of carbon and nitrogen metabolism in wheat culms after anthesis. I. Model description. *Ann. Bot.* 118, 997–1013. https://doi.org/10.1093/aob/mcw143.

Bassu, S., Brisson, N., Durand, J.-L., Boote, K. J., Lizaso, J., Jones, J. W., Rosenzweig, C., Ruane, A. C., Adam, M., Baron, C., Basso, B., Biernath, C., Boogaard, H., Conijn, S., Corbeels, M., Deryng, D., Sanctis, G., Gayler, S., Grassini, P., Hatfield, J., Hoek, S., Izaurralde, C., Jongschaap, R., Kemanian, A. R., Kersebaum, C. K., Kim, S.-H., Kumar, N. S., Makowski, D., Müller, C., Nendel, C., Priesack, E., Pravia, M., Sau, F., Shcherbak, I., Tao, F., Teixeira, E., Timlin, D., and Waha, K. 2014. How do various maize crop models vary in their responses to climate change factors? *Glob. Change Biol.* 20, 2301–20.

Bellasio, C., Quirk, J., Buckley, T. N., and Beerling, D. 2017. A dynamic hydro-mechanical and biochemical model of stomatal conductance for C4 photosynthesis. *Plant Physiol.* 175, 104–19. https://doi.org/10.1104/pp.17.00666.

Bernacchi, C. J., Singsaas, E. L., Pimentel, C., Portis Jr, A. R., and Long, S. P. 2001. Improved temperature response functions for models of Rubisco-limited photosynthesis. *Plant Cell Environ.* 24, 253-9. https://doi.org/10.1111/j.1365-3040.2001.00668.x.

Boardman, N. K. 2003. Comparative Photosynthesis of Sun and Shade Plants. *Ann. Rev. Plant Physiol.* 28, 355–77. https://doi.org/10.1146/annurev.pp.28.060177.002035.

Bogard, M., Ravel, C., Paux, E., Bordes, J., Balfourier, F., Chapman, S. C., Le Gouis, J., and Allard, V. 2014. Predictions of heading date in bread wheat (*Triticum aestivum* L.) using QTL-based parameters of an ecophysiological model. *J. Exp. Bot.* 65, 5849–65. https://doi.org/10.1093/jxb/eru328.

Bonan G. B., Williams M., Fisher R. A., and Oleson K. W. 2014 Modeling stomatal conductance in the earth system: linking leaf water-use efficiency and water transport along the soil-plant-atmosphere continuum. *Geosci. Model Dev.*, 7, 2193-2222.

Boote, K. J., Jones, J. W., and Hoogenboom, G. 1998. Simulation of crop growth: CROPGRO model, in: R. M. Peart and R. B. Curry (Eds), *Agricultural Systems Modeling and Simulation*. Marcel Dekker, New York, pp. 651-92.

Boote, K. J., Minguez, M. I., and Sau, F. 2002. Adapting the CROPGRO legume model to simulate growth of faba bean. *Agron. J.* 94, 743-56.

Boote, K. J., Hoogenboom, G., Jones, J. W., and Ingram, K. T. 2008. Modeling nitrogen fixation and its relationship to nitrogen uptake in the CROPGRO model, in: Ma, L., Ahuja, L. R., and Bruulsema, T. (Eds), *Quantifying and Understanding Plant Nitrogen Uptake for Systems Modeling*. CRC Press, Boca Raton, FL, pp. 219–43.

Boote, K. J., Jones, J. W., White, J. W., Asseng, S., and Lizaso, J. I. 2013. Putting mechanisms into crop production models. *Plant Cell Environ.* 36, 1658–72. https://doi.org/10.1111/pce.12119.

Boote, K. J., Prasad, V., Allen, L. H., Singh, P., and Jones, J. W. 2018. Modeling sensitivity of grain yield to elevated temperature in the DSSAT crop models for peanut, soybean, dry bean, chickpea, sorghum, and millet. *Eur. J. Agron.* 100, 99–109. https://doi.org/10.1016/j.eja.2017.09.002.

Brown, D. M. 1960. Soybean ecology: I. Development-temperature relationships from controlled-environment studies. *Agron. J.* 52, 493–6. https://doi.org/10.2134/agronj1960.00021962005200090001x.

Brown, D. M. 1969. Heat units for corn in Southern Ontario. *Ontario Dept. Agric. Food*, AGDEX 111/31, 4.

Brown, H. E., Jamieson, P. D., Brooking, I. R., Moot, D. J., and Huth, N. I. 2013. Integration of molecular and physiological models to explain time of anthesis in wheat. *Ann. Bot.* 112, 1683–703. https://doi.org/10.1093/aob/mct224.

Buckley, T. N. 2016. Stomatal responses to humidity: Has the "black box" finally been opened? *Plant Cell Environ.* 39, 482–4. https://doi.org/10.1111/pce.12651.

Buckley, T. N. 2017. Modeling stomatal conductance. *Plant Physiol.* 174, 572-82. https://doi.org/10.1104/pp.16.01772.

Buckley, T. N., Mott, K. A., and Farquhar, G. D. 2003. A hydromechanical and biochemical model of stomatal conductance. *Plant Cell Environ.* 26, 1767-85. https://doi.org/10.1046/j.1365-3040.2003.01094.x.

Buckley, T. N., Turnbull, T. L., and Adams, M. A. 2012. Simple models for stomatal conductance derived from a process model: Cross-validation against sap flux data. *Plant Cell Environ.* 35, 1647-62. https://doi.org/10.1111/j.1365-3040.2012.02515.x.

Butler, E. E., and Huybers, P. 2015. Variations in the sensitivity of US maize yield to extreme temperatures by region and growth phase. *Environ. Res. Lett.* 10. https://doi.org/10.1088/1748-9326/10/3/034009.

Cannell, M. G. R., and Thornley, J. H. M. 1998. Temperature and CO_2 responses of leaf and canopy photosynthesis : A clarification using the non-rectangular hyperbola model of photosynthesis. *Ann. Bot.* 82, 883-92.

Cardon, Z. G., and Berry, J. 1992. Effects of O_2 and CO_2 concentration on the steady-state fluorescence yield of single guard cell pairs in intact leaf discs of *Tradescantia albifloral*. *Plant Physiol.* 99, 1238-44.

Chang, T. G., and Zhu, X. G. 2017. Source-sink interaction: A century old concept under the light of modern molecular systems biology. *J. Exp. Bot.* 68, 4417-31. https://doi.org/10.1093/jxb/erx002.

Chew, Y. H., Wilczek, A. M., Williams, M., Welch, S. M., Schmitt, J., and Halliday, K. J. 2012. An augmented *Arabidopsis* phenology model reveals seasonal temperature control of flowering time. *New Phytol.* 194, 654-65. https://doi.org/10.1111/j.1469-8137.2012.04069.x

Chung, U., Mack, L., Yun, J. I., and Kim, S.-H. 2011. Predicting the Timing of Cherry Blossoms in Washington, DC and Mid-Atlantic States in Response to Climate Change. *PLoS ONE* 6, e27439. https://doi.org/10.1371/journal.pone.0027439.

Collatz, G. J., Ball, J. T., Grivet, C., and Berry, J. A. 1991. Physiological and environmental regulation of stomatal conductance, photosynthesis and transpiration : A model that includes a laminar boundary layer. *Agric. For. Meteorol.* 54, 107-36.

Coville, F. V. 1920. The influence of cold in stimulating the growth of plants. *J. Agric. Res.* 20, 151-60.

Cowan, I. R., and Farquhar, G. D. 1977. Stomatal function in relation to leaf metabolism and environment. *Symp. Soc. Exp. Biol.* 31, 471-505.

Damour, G., Simonneau, T., Cochard, H., and Urban, L. 2010. An overview of models of stomatal conductance at the leaf level. *Plant Cell Environ.* 33, 1419-38. https://doi.org/10.1111/j.1365-3040.2010.02181.x.

Darwin, F. 1898. Observations on stomata. *Philos. Trans. Royal Soc. B* 190, 531-621. https://doi.org/10.1098/rstb.1898.0009.

Daudet, F. A., Lacointe, A., Gaudillère, J. P., and Cruiziat, P. 2002. Generalized Münch coupling between sugar and water fluxes for modelling carbon allocation as affected by water status. *J. Theor. Biol.* 214, 481-98. https://doi.org/10.1006/jtbi.2001.2473.

de Pury, D. G. G., and Farquhar, G. D. 1997. Simple scaling of photosynthesis from leaves to canopies without the errors of big-leaf models. *Plant Cell Environ.* 20, 537-57.

de Wit, C. T. 1965. *Photosynthesis of Leaf Canopies, Agricultural Research Reports*. PUDOC, Wageningen.

Diaz-Espejo, A., Buckley, T. N., Sperry, J. S., Cuevas, M. V., de Cires, A., Elsayed-Farag, S., Martin-Palomo, M. J., Muriel, J. L., Perez-Martin, A., Rodriguez-Dominguez, C. M., Rubio-Casal, A. E., Torres-Ruiz, J. M., and Fernández, J. E. 2012. Steps toward an improvement in process-based models of water use by fruit trees: A case study in olive. *Agric. Water Manag.* 114, 37-49. https://doi.org/10.1016/j.agwat.2012.06.027.

Dingkuhn, M., Kouressy, M., Vaksmann, M., Clerget, B., and Chantereau, J. 2008. A model of sorghum photoperiodism using the concept of threshold-lowering during prolonged appetence. *Eur. J. Agron.* 28, 74–89. https://doi.org/10.1016/j.eja.2007.05.005

Distelfeld, A., Li, C., and Dubcovsky, J. 2009. Regulation of flowering in temperate cereals. *Opin. Plant Biol., Genome Stud. Mol. Genet.* 12, 178–84. https://doi.org/10.1016/j.pbi.2008.12.010

Dunbabin, V. M., Postma, J. A., Schnepf, A., Pages, L., Javaux, M., Wu, L., Leitner, D., Chen, Y. L., Rengel, Z., and Diggle, A. J. 2013. Modelling root-soil interactions using three-dimensional models of root growth, architecture and function. *Plant Soil* 372, 93–124. https://doi.org/10.1007/s11104-013-1769-y.

Duncan, W. G., Loomis, R. S., Williams, W. A., and Hanau, R. 1967. A model for simulating photosynthesis in plant communities. *Hilgardia* 38, 181–205. https://doi.org/10.3733/hilg.v38n04p181.

Dupuy, L., Gregory, P. J., and Bengough, A. G. 2010. Root growth models: Towards a new generation of continuous approaches. *J. Exp. Bot.* 61, 2131–43. https://doi.org/10.1093/jxb/erp389.

Durand, J.-L., Delusca, K., Boote, K. J., Lizaso, J., Manderscheid, R., Weigel, H., Ruane, A. C., Rosenzweig, C., Jones, J., Ahuja, L., Anapalli, S., Basso, B., Baron, C., Bertuzzi, P., Biernath, C., Deryng, D., Ewert, F., Gaiser, T., Gayler, S., Heinlein, F., Kersebaum, K., Kim, S.-H., Müller, C., Nendel, C., Olioso, A., Priesack, E., Villegas, J., Ripoche, D., Rötter, R. P., Seidel, S. I., Srivastava, A., Tao, F., Timlin, D., Twine, T., Wang, E., Webber, H., and Zhao, Z. 2018. How accurately do maize crop models simulate the interactions of atmospheric CO2 concentration levels with limited water supply on water use and yield? *Eur. J. Agron.* 100, 67–75.

Egea, G., Verhoef, A., and Vidale, P. L. 2011. Towards an improved and more flexible representation of water stress in coupled photosynthesis-stomatal conductance models. *Agric. For. Meteorol.* 151, 1370–84. https://doi.org/10.1016/j.agrformet.2011.05.019.

Farquhar, G. D., and Wong, S. C. 1984. An empirical model of stomatal conductance. *Aust. J. Plant Physiol.* 11: 191–210.

Farquhar, G. D., von Caemmerer, S., and Berry, J. A. 1980. A biochemical model of photosynthetic CO2 assimilation in leaves of C3 species. *Planta* 149, 78–90.

Farquhar, G. D., von Caemmerer, S., and Berry, J. A. 2001. Models of photosynthesis. *Plant Physiol.* 125, 42–5.

Field, C. B. 2002. Plant physiology of the "Missing" carbon sink. *Plant Physiol.* 125, 25–8. https://doi.org/10.1104/pp.125.1.25.

Fleisher, D. H., Condori, B., Quiroz, R., Alva, A., Asseng, S., Barreda, C., Bindi, M., Boote, K. J., Ferrise, R., Franke, A. C., Govindakrishnan, P. M., Harahagazwe, D., Hoogenboom, G., Kumar, S. N., Merante, P., Nendel, C., Olesen, J. E., Parker, P. S., Raes, D., Raymundo, R., Ruane, A. C., Stockle, C., Supit, I., Vanuytrecht, E., Wolf, J., and Woli, P. 2017. A potato model intercomparison across varying climates and productivity levels. *Glob. Change Biol.* 23, 1258–81. https://doi.org/10.1111/gcb.13411.

Franks, P. J., Cowan, I. R., and Farquhar, G. D. 1998. A study of stomatal mechanics using the cell pressure probe. *Plant Cell Environ.* 21, 94–100. https://doi.org/10.1046/j.1365-3040.1998.00248.x.

Franks, P. J., Buckley, T. N., Shope, J. C., and Mott, K. A. 2001. Guard cell volume and pressure measured concurrently by confocal microscopy and the cell pressure probe. *Plant Physiol.* 125, 1577–84. https://doi.org/10.1104/pp.125.4.1577.

Franks, P. J., Berry, J. A., Lombardozzi, D. L., and Bonan, G. B. 2017. Stomatal function across temporal and spatial scales: Deep-time trends, land-atmosphere coupling and global models. *Plant Physiol.* 174, 583–602. https://doi.org/10.1104/pp.17.00287.

Franks, P. J., Bonan, G. B., Berry, J. A., Lombardozzi, D. L., Holbrook, N. M., Herold, N., and Oleson, K. W. 2018. Comparing optimal and empirical stomatal conductance models for application in Earth system models. *Glob. Change Biol.* 5708–23. https://doi.org/10.1111/gcb.14445.

Garcia-Huidobro, J., Monteith, J. L., and Squire, G. R. 1982. Time, temperature and germination of Pearl Millet (*Pennisetum typhoides* S. & H.): I. constant temperature. *J. Exp. Bot.* 33, 288–96. https://doi.org/10.1093/jxb/33.2.288.

Garner, W. W., and Allard, H. A. 1920. Effect of the relative length of day and night and other factors of the environment on growth and reproduction in plants. *J. Agric. Res.* 18, 553–606.

Garner, W. W., and Allard, H. A. 1923. Further studies in photoperiodism, the response of the plant to relative length of day and night. *J. Agric. Res.* 23, 871–969.

Garner, W. W., and Allard, H. A. 1930. Photoperiodic response of soybeans in relation to temperature and other environmental factors. *J. Agic. Res.* 41, 719–35.

Gifford, R. M. 2003. Plant respiration in productivity models: Conceptualisation, representation and issues for global terrestrial carbon-cycle research. *Funct. Plant Biol.* 30, 171–86. https://doi.org/10.1071/fp02083

Gilmore, E. C., and Rogers, J. S. 1958. Heat units as a method of measuring maturity in corn. *Agron. J.* 50, 611–15. https://doi.org/10.2134/agronj1958.00021962005000100014x

Godin, C., and Sinoquet, H. 2005. Functional-structural plant modelling. *New Phytol.* 166, 705–8.

Goudriaan, J., and van Laar, H. H. 1994. *Modelling Potential Crop Growth Processes: Textbook with Exercises, Current Issues in Production Ecology: 2.* Kluwer Academic Publishers, Dordrecht.

Harley, P. C., and Sharkey, T. D. 1991. An improved model of C3 photosynthesis at high CO_2: Reversed O_2 sensitivity explained by lack of glycerate reentry into the chloroplast. *Photosynth. Res.* 27, 169–78. https://doi.org/10.1007/BF00035838.

Hill, C. B., and Li, C. 2016. Genetic architecture of flowering phenology in cereals and opportunities for crop improvement. *Front. Plant Sci.* 7. https://doi.org/10.3389/fpls.2016.01906.

Hsiao, J., Swann, A. L. S., and Kim, S.-H. 2019. Maize yield under a changing climate: The hidden role of vapor pressure deficit. *Agric. For. Meteorol.* 279. https://doi.org/10.1016/j.agrformet.2019.107692.

Hunt, L. A., and Pararajasingham, S. 1995. CROPSIM-WHEAT: A model describing the growth and development of wheat. *Can. J. Plant Sci.* 75, 619–32.

Jamieson, P. D., Brooking, I. R., Semenov, M. A., and Porter, J. R. 1998. Making sense of wheat development: a critique of methodology. *Field Cops Res.* 55, 117–27. https://doi.org/10.1016/S0378-4290(97)00072-5.

Jarvis, P. G. 1976. The interpretation of the variations in leaf water potential and stomatal conductance found in canopies in the field. *Philos. Trans. R. Soc. Lond. Ser. B-Biol. Sci.* 273, 593–610. https://doi.org/10.1098/rstb.1976.0035.

Kaufmann, K., Wellmer, F., Muiño, J. M., Ferrier, T., Wuest, S. E., Kumar, V., Serrano-Mislata, A., Madueño, F., Krajewski, P., Meyerowitz, E. M., Angenent, G. C., and Riechmann,

J. L. 2010. Orchestration of floral initiation by APETALA1. *Science* 328, 85-9. https://doi.org/10.1126/science.1185244.

Kim, S.-H., and Lieth, J. H. 2003. A coupled model of photosynthesis, stomatal conductance and transpiration for a rose leaf (*Rosa hybrida* L.). *Ann. Bot.* 91, 771-81. https://doi.org/10.1093/aob/mcg080.

Kim, S.-H., Yang, Y., Timlin, D. J., Fleisher, D. H., Dathe, A., Reddy, V. R., and Staver, K. 2012. Modeling temperature responses of leaf growth, development, and biomass in maize with MAIZSIM. *Agron. J.* 104, 1523-37. https://doi.org/10.2134/agronj2011.0321.

Kim, W., Park, T. I., Yoo, S. J., Jun, A. R., and Ahn, J. H. 2013. Generation and analysis of a complete mutant set for the *Arabidopsis FT/TFL1* family shows specific effects on thermo-sensitive flowering regulation. *J. Exp. Bot.* 64, 1715-29. https://doi.org/10.1093/jxb/ert036.

Kinmonth-Schultz, H. A., MacEwen, M. J. S., Seaton, D. D., Millar, A. J., Imaizumi, T., and Kim, S.-H. 2019. An explanatory model of temperature influence on flowering through whole-plant accumulation of FLOWERING LOCUS T in *Arabidopsis thaliana*. *Silico Plants* 1. https://doi.org/10.1093/insilicoplants/diz006.

Kobayashi, Y., and Weigel, D. 2007. Move on up, it's time for change - mobile signals controlling photoperiod-dependent flowering. *Genes Dev.* 21, 2371-84. https://doi.org/10.1101/gad.1589007.

Kumudini, S., Andrade, F. H., Boote, K. J., Brown, G. A., Dzotsi, K. A., Edmeades, G. O., Gocken, T., Goodwin, M., Halter, A. L., Hammer, G. L., Hatfield, J. L., Jones, J. W., Kemanian, A. R., Kim, S.-H., Kiniry, J., Lizaso, J. I., Nendel, C., Nielsen, R. L., Parent, B., Stöckle, C. O., Tardieu, F., Thomison, P. R., Timlin, D. J., Vyn, T. J., Wallach, D., Yang, H. S., and Tollenaar, M. 2014. Predicting maize phenology: Intercomparison of functions for developmental response to temperature. *Agron. J.* 106, 2087-97. https://doi.org/10.2134/agronj14.0200.

Laisk, A., Eichelmann, H., and Oja, V. 2006. C3 photosynthesis in silico. *Photosynth. Res.* 90, 45-66. https://doi.org/10.1007/s11120-006-9109-1.

Lambers, H., Chapin, F. S., and Pons, T. L. 2008. *Plant Physiological Ecology*, 2nd ed. Springer, New York.

Lamsal, A., Welch, S. M., White, J. W., Thorp, K. R., and Bello, N. M. 2018. Estimating parametric phenotypes that determine anthesis date in *Zea mays*: Challenges in combining ecophysiological models with genetics. *PLoS ONE* 13, e0195841.

Lawn, R. J., Summerfield, R. J., Ellis, R. H., Qi, A., Roberts, E. H., Chay, P. M., Brouwer, J. B., Rose, J. L., and Yeates, S. J. 1995. Towards the reliable prediction of time to flowering in six annual crops. VI. Applications in crop improvement. *Exp. Agric.* 31, 89-108.

Lawson, T., Oxborough, K., Morison, J. I. L., and Baker, N. R. 2002. Responses of photosynthetic electron transport in stomatal guard cells and mesophyll cells in intact leaves to light, CO_2, and humidity. *Plant Physiol.* 128, 52-62. https://doi.org/10.1104/pp.010317.

Lee, J. H., Yoo, S. J., Park, S. H., Hwang, I., Lee, J. S., and Ahn, J. H. 2007. Role of *SVP* in the control of flowering time by ambient temperature in *Arabidopsis*. *Genes Dev.* 21, 397-402. https://doi.org/10.1101/gad.1518407.

Lehenbauer, P. A. 1914. Growth of maize seedlings in relation to temperature. *Physiol. Res.* 1, 247-88.

Leuning, R. 1995. A critical appraisal of a combined stomatal-photosynthesis model for C3 plants. *Plant, Cell & Environment,* 18, 339-335.

Li, T., Hasegawa, T., Yin, X., Zhu, Y., Boote, K., Adam, M., Bregaglio, S., Buis, S., Confalonieri, R., Fumoto, T., Gaydon, D., Marcaida, M., Nakagawa, H., Oriol, P., Ruane, A. C., Ruget, F., Singh, B., Singh, U., Tang, L., Tao, F., Wilkens, P., Yoshida, H., Zhang, Z., and Bouman, B. 2015. Uncertainties in predicting rice yield by current crop models under a wide range of climatic conditions. *Glob. Change Biol.* 21, 1328–41. https://doi.org/10.1111/gcb.12758.

Lin, Y. S., Medlyn B. E., Duursma R. A., Prentice I. C., Wang H., Baig S., Eamus D., de Dios V. R., Mitchell P., Ellsworth D. S., Op de Beeck M., Wallin G., Uddling J., Tarvainen L., Linderson M. L., Cernusak L. A., Nippert J. B., Ocheltree T., Tissue D. T., Martin-St Paul N. K., Rogers A., Warren J. M., De Angelis P., Hikosaka K., Han Q. M., Onoda Y., Gimeno T. E., Barton C. V. M., Bennie J., Bonal D., Bosc A., Low M., Macinins-Ng C., Rey A., Rowland L., Setterfield S. A., Tausz-Posch S., Zaragoza-Castells J., Broadmeadow M. S. J., Drake J. E., Freeman M., Ghannoum O., Hutley L. B., Kelly J. W., Kikuzawa K., Kolari P., Koyama K., Limousin J. M., Meir P., da Costa A. C. L., Mikkelsen T. N., Salinas N., Sun W. and Wingate L. 2015. Optimal stomatal behaviour around the world. *Nat. Clim. Change* 5, 459–464.

Lizaso, J. I., Batchelor, W. D., and Westgate, M. E. 2003. A leaf area model to simulate cultivar-specific expansion and senescence of maize leaves. *Field Crops Res.* 80, 1–17.

Lizaso, J. I., Batchelor, W. D., Boote, K. J., and Westgate, M. E. 2005. Development of a leaf-level canopy assimilation model for CERES-Maize. *Agron. J.* 97, 722–33.

Lloyd, J., Shibistova, O., Zolotoukhine, D., Kolle, O., Arneth, A., Wirth, C., Styles, J. M., Tchebakova, N. M., and Schulze, E. D. 2002. Seasonal and annual variations in the photosynthetic productivity and carbon balance of a central Siberian pine forest. *Tellus B Chem. Phys. Meteorol.* 54, 590–610. https://doi.org/10.1034/j.1600-0889.2002.01487.x.

Marcelis, L. F. M. 1996. Sink strength as a determinant of dry matter partitioning in the whole plant. *J. Exp. Bot.* 47, 1281–91. https://doi.org/10.1093/jxb/47.Special_Issue.1281.

Marcelis, L. F. M., Heuvelink, E., and Goudriaan, J. 1998. Modelling biomass production and yield of horticultural crops: A review. *Sci. Hortic.* 74, 83–111. https://doi.org/10.1016/S0304-4238(98)00083-1.

McCree, K. J. 1970. *An Equation for the Rate of Respiration of White Clover Grown Under Controlled Conditions.* PUDOC, Wageningen, pp. 221–9.

McMaster G. S. and Wilhelm W. W. 1997. Growing degree-days: one equation, two interpretations. *Agric.For. Meteorol.* 87, 291–300.

McMaster, G. S., White, J. W., Hunt, L. A., Jamieson, P. D., Dhillon, S. S., and Ortiz-Monasterio, J. I. 2008. Simulating the influence of vernalization, photoperiod and optimum temperature on wheat developmental rates. *Ann. Bot.* 102, 561–9. https://doi.org/10.1093/aob/mcn115.

Medlyn, B. E., Duursma, R. A., Eamus, D., Ellsworth, D. S., Prentice, I. C., Barton, C. V. M., Crous, K. Y., De Angelis, P., Freeman, M., and Wingate, L. 2011. Reconciling the optimal and empirical approaches to modelling stomatal conductance. *Glob. Change Biol.* 17, 2134–44. https://doi.org/10.1111/j.1365-2486.2010.02375.x

Mencuccini, M. 2003. The ecological significance of long-distance water transport: Short-term regulation, long-term acclimation and the hydraulic costs of stature across plant life forms. *Plant Cell Environ.* 26, 163–82. https://doi.org/10.1046/j.1365-3040.2003.00991.x.

Minchin, P. E. H., and Lacointe, A. 2005. New understanding on phloem physiology and possible consequences for modelling long-distance carbon transport. *New Phytol.* 166, 771-9. https://doi.org/10.1111/j.1469-8137.2005.01323.x.

Minchin, P. E. H., Thorpe, M. R., and Farrar, J. F. 1993. A simple mechanistic model of phloem transport which explains sink priority. *J. Exp. Bot.* 44, 947-55. https://doi.org/10.1093/jxb/44.5.947.

Miner, G. L., Bauerle, W. L., and Baldocchi, D. D. 2017. Estimating the sensitivity of stomatal conductance to photosynthesis: A review. *Plant Cell Environ.* 40, 1214-38. https://doi.org/10.1111/pce.12871.

Monsi, M. 1960. Dry-matter reproduction in plants 1. Schemata of dry-matter reproduction. *Bot. Mag.* 73, 81-90.

Monsi, M., and Saeki, T. 1953. Uber den Lichtfaktor in den Pflanzengesellschaften und Seine Bedeutung für die Stoffproduktion. *Jpn. J. Bot.* 14, 22-52.

Monteith, J. L. 1965. Light distribution and photosynthesis in field crops. *Ann. Bot.* 29, 17-37.

Monteith, J. L. 1972. Solar radiation and productivity in tropical ecosystems. *J. Appl. Ecol.* 9, 747-66.

Monteith, J. L. 1977. Climate and efficiency of crop production in Britain. *Philos. Trans. R. Soc. Lond. Ser. B-Biol. Sci.* 281, 277-94.

Outlaw Jr., W. H. 1989. Critical examination of the quantitative evidence for and against photosynthetic CO2 fixation by guard cells. *Physiol. Plant.* 77, 275-81. https://doi.org/10.1111/j.1399-3054.1989.tb04981.x.

Palosuo, T., Kersebaum, K. C., Angulo, C., Hlavinka, P., Moriondo, M., Olesen, J. E., Patil, R. H., Ruget, F., Rumbaur, C., Takáč, J., Trnka, M., Bindi, M., Çaldag, B., Ewert, F., Ferrise, R., Mirschel, W., Saylan, L., Šiška, B., and Rötter, R. 2011. Simulation of winter wheat yield and its variability in different climates of Europe: A comparison of eight crop growth models. *Eur. J. Agron.* 35, 103-14. https://doi.org/10.1016/j.eja.2011.05.001.

Parent, B., Turc, O., Gibon, Y., Stitt, M., and Tardieu, F. 2010. Modelling temperature-compensated physiological rates, based on the co-ordination of responses to temperature of developmental processes. *J. Exp. Bot.* 61, 2057-69. https://doi.org/10.1093/jxb/erq003.

Penning de Vries, F. W. T., Brunsting, A. H. M., and Van Laar, H. H. 1974. Products, requirements and efficiency of biosynthesis a quantitative approach. *J. Theor. Biol.* 45, 339-77. https://doi.org/10.1016/0022-5193(74)90119-2.

Porter, J. R., and Gawith, M. 1999. Temperatures and the growth and development of wheat: A review. *Eur. J. Agron.* 10, 23-36. https://doi.org/10.1016/S1161-0301(98)00047-1.

Rho, H., Doty, S. L., and Kim, S.-H. 2018. Estimating microbial respiratory CO2 from endophytic bacteria in rice. *Plant Signal. Behav.* 13, e1500067. https://doi.org/10.1080/15592324.2018.1500067.

Roberts, E. H., and Summerfield, R. J. 1987. Measurement and prediction of flowering in annual crops, in J. G. Atherton (Ed.), *Manipulation of Flowering*. Butterworths, London, UK, pp. 17-50.

Rodriguez-Dominguez, C. M., Buckley, T. N., Egea, G., de Cires, A., Hernandez-Santana, V., Martorell, S., and Diaz-Espejo, A. 2016. Most stomatal closure in woody species under moderate drought can be explained by stomatal responses to leaf turgor. *Plant Cell Environ.* 39, 2014-26. https://doi.org/10.1111/pce.12774.

Rogers, A., Medlyn, B. E., Dukes, J. S., Bonan, G., von Caemmerer, S., Dietze, M. C., Kattge, J., Leakey, A. D. B., Mercado, L. M., Niinemets, Ü., Prentice, C. I., Serbin, S. P., Sitch, S., Way, D. A., and Zaehle, S. 2017. A roadmap for improving the representation of photosynthesis in Earth system models. *New Phytol.* 213, 22-42.

Salazar-Gutierrez, M. R., Johnson, J., Chaves-Cordoba, B., and Hoogenboom, G. 2013. Relationship of base temperature to development of winter wheat. *Int. J. Plant Prod.* 7, 741-62.

Schafer, E. G., Edwin G., Gaines, E. F., Edward F., Barbee, O. E. 1921. Wheat production as influenced by variety, time of seeding, and source of seed. Issue 159 of Bulletin. State College of Washington, Agricultural Experiment Station, p. 34.

Sellers, P. J., Dickinson, R. E., Randall, D. A., Betts, A. K., Hall, F. G., Berry, J. A., Collatz, G. J., Denning, A. S., Mooney, H. A., Nobre, C. A., Sato, N., Field, C. B., and Henderson-Sellers, A. 1997. Modeling the exchanges of energy, water, and carbon between continents and the atmosphere. *Science* 275, 502-9. https://doi.org/10.1126/science.275.5299.502.

Sharkey, T. D., Bernacchi, C. J., Farquhar, G. D., and Singsaas, E. L. 2007. Fitting photosynthetic carbon dioxide response curves for C3 leaves. *Plant Cell Environ.* 30, 1035-40. https://doi.org/10.1111/j.1365-3040.2007.01710.x.

Soares, J. C., Santos, C. S., Carvalho, S. M. P., Pintado, M. M., and Vasconcelos, M. W. 2019. Preserving the nutritional quality of crop plants under a changing climate: Importance and strategies. *Plant Soil*. https://doi.org/10.1007/s11104-019-04229-0.

Song, Y.-H., Shim, J.-S., Kinmonth-Schultz, H. A., and Imaizumi, T. 2015. Photoperiodic flowering: Time measurement mechanisms in leaves. *Annu. Rev. Plant Biol.* 66, 441-64.

Sperling, O., Silva, L. C. R., Tixier, A., Théroux-Rancourt, G., and Zwieniecki, M. A. 2017. Temperature gradients assist carbohydrate allocation within trees. *Sci. Rep.* 7, 3265. https://doi.org/10.1038/s41598-017-03608-w.

Spitters, C. J. T., Toussaint, H., and Goudriaan, J. 1986. Separating the diffuse and direct component of global radiation and its implications for modeling canopy photosynthesis.1. Components of incoming radiation. *Agric. For. Meteorol.* 38, 217-29.

Tardieu, F., Lafarge, T., and Simonneau, Th. 1996. Stomatal control by fed or endogenous xylem ABA in sunflower: Interpretation of correlations between leaf water potential and stomatal conductance in anisohydric species. *Plant Cell Environ.* 19, 75-84.

Tcherkez, G., Gauthier, P., Buckley, T. N., Busch, F. A., Barbour, M. M., Bruhn, D., Heskel, M. A., Gong, X. Y., Crous, K. Y., Griffin, K., Way, D., Turnbull, M., Adams, M. A., Atkin, O. K., Farquhar, G. D., and Cornic, G. 2017. Leaf day respiration: Low CO2 flux but high significance for metabolism and carbon balance. *New Phytol.* 216, 986-1001. https://doi.org/10.1111/nph.14816.

Thornley, J. H. M. 2011. Plant growth and respiration re-visited: maintenance respiration defined - it is an emergent property of, not a separate process within, the system - and why the respiration: Photosynthesis ratio is conservative. *Ann. Bot.* 108, 1365-80. https://doi.org/10.1093/aob/mcr238.

Thornley, J. H. M., and Cannell, M. G. R. 2000. Modelling the components of plant respiration: Representation and realism. *Ann. Bot.* 85, 55-67.

Thornley, J. H. M., and Johnson, I. A. 1990. *Plant and Crop Modelling: A Mathematical Approach to Plant and Crop Physiology*. Clarendon Press and Oxford University Press, Oxford and New York.

Tominaga, M., Kinoshita, T., and Shimazaki, K. 2001. Guard-cell chloroplasts provide ATP required for H+ pumping in the plasma membrane and stomatal opening. *Plant Cell Physiol.* 42, 795–802. https://doi.org/10.1093/pcp/pce101.

Tournois, J. 1914. Études sur de la sexualité du houblon. *Ann. Sci. Nat.* 19, 49–191.

Tuzet, A., Perrier, A., and Leuning, R. 2003. A coupled model of stomatal conductance, photosynthesis and transpiration. *Plant Cell Environ.* 26, 1097–116.

von Caemmerer, S. 2000. *Biochemical Models of Leaf Photosynthesis*. CSIRO Publishing, Collingwood, Victoria, Australia.

von Caemmerer, S., and Furbank, R. T. 1999. Modeling C4 photosynthesis, in: Sage, R. F., and Monson, R. K. (Eds), *C4 Plant Biology*. Toronto, pp. 173–211.

Vos, J., Evers, J. B., Buck-Sorlin, G. H., Andrieu, B., Chelle, M., and de Visser, P. H. B. 2010. Functional-structural plant modelling: A new versatile tool in crop science. *J. Exp. Bot.* 61, 2101–15. https://doi.org/10.1093/jxb/erp345.

Wang, S.-Y., Ward, R. W., Ritchie, J. T., Fischer, R. A., and Schulthess, U. 1995. Vernalization in wheat I. A model based on the interchangeability of plant age and vernalization duration. *Field Crops Res.* 41, 91–100. https://doi.org/10.1016/0378-4290(95)000 06-C.

Wang, E., Martre, P., Zhao, Z., Ewert, F., Maiorano, A., Rötter, R. P., Kimball, B. A., Ottman, M. J., Wall, G. W., White, J. W., Reynolds, M. P., Alderman, P. D., Aggarwal, P. K., Anothai, J., Basso, B., Biernath, C., Cammarano, D., Challinor, A. J., De Sanctis, G., Doltra, J., Dumont, B., Fereres, E., Garcia-Vila, M., Gayler, S., Hoogenboom, G., Hunt, L. A., Izaurralde, R. C., Jabloun, M., Jones, C. D., Kersebaum, K. C., Koehler, A.-K., Liu, L., Müller, C., Naresh Kumar, S., Nendel, C., O'Leary, G., Olesen, J. E., Palosuo, T., Priesack, E., Eyshi Rezaei, E., Ripoche, D., Ruane, A. C., Semenov, M. A., Shcherbak, I., Stöckle, C., Stratonovitch, P., Streck, T., Supit, I., Tao, F., Thorburn, P., Waha, K., Wallach, D., Wang, Z., Wolf, J., Zhu, Y., and Asseng, S. 2017. The uncertainty of crop yield projections is reduced by improved temperature response functions. *Nature Plants* 3, 17102. https://doi.org/10.1038/nplants.2017.102.

Warren Wilson, J. 1967. Ecological data on dry-matter production by plants and plant communities, in: Bradley, E. F., Denmead, O. T. (Eds.), *Collection and Processing of Field Data*. John Wiley, New York, pp. 77–123.

Weir, S. Y., Ward, R. W., Ritchie, J. T., Fischer, R. A., and Schulthess, U. 1984. A winter wheat crop simulation model without water or nutrient limitations. *J. Agric. Sci. Camb.* 102, 371–82.

Wenden, B., Dun, E. A., Hanan, J., Andrieu, B., Weller, J. L., Beveridge, C. A., and Rameau, C. 2009. Computational analysis of flowering in pea (*Pisum sativum*). *New Phytol.* 184, 153–67. https://doi.org/10.1111/j.1469-8137.2009.02952.x.

White, J. W., Hoogenboom, G., Kimball, B. A., and Wall, G. W. 2011. Methodologies for simulating impacts of climate change on crop production. *Field Crops Res.* 124, 357–68.

White, J. W., Kimball, B. A., Wall, G. W., and Ottman, M. J. 2012. Cardinal temperatures for wheat leaf appearance as assessed from varied sowing dates and infrared warming. *Field Crops Res.* 137, 213–20. https://doi.org/10.1016/j.fcr.2012.08.013.

Wickland, D. P., and Hanzawa, Y. 2015. The *FLOWERING LOCUS T/TERMINAL FLOWER 1* gene family: functional evolution and molecular mechanisms. *Mol Plant* 8, 983–97. https://doi.org/10.1016/j.molp.2015.01.007.

Wilczek, A. M., Burghardt, L. T., Cobb, A. R., Cooper, M. D., Welch, S. M., and Schmitt, J. 2010. Genetic and physiological bases for phenological responses to current and predicted climates. *Philos. Trans. Roy. Soc. B Biol. Sci.* 365, 3129–47. https://doi.org/10.1098/rstb.2010.0128.

Wong, S. C., Cowan, R., and Farquhar, G. D. 1979. Stomatal conductance correlates with photosynthetic capacity. *Nature* 282, 424–6.

Wullschleger, S. D. 1993. Biochemical limitations to carbon assimilation in C3 plants – A retrospective analysis of the A/Ci curves from 109 species. *J. Exp. Bot.* 44, 907–20.

Xue, Q., Weiss, A., and Baenziger, P. S. 2004. Predicting leaf appearance in field-grown winter wheat: Evaluating linear and non-linear models. *Ecol. Model.* 175, 261–70. https://doi.org/10.1016/j.ecolmodel.2003.10.018.

Yan, W., and Hunt, L. A. 1999. An equation for modelling the temperature response of plants using only the cardinal temperatures. *Ann. Bot.* 84, 607–14.

Yan, W., and Wallace, D. H. 1996. A model of photoperiod × temperature interaction effects on plant development. *Crit. Rev. Plant Sci.* 15, 63–96.

Yan, W., and Wallace, D. H. 1998. Simulation and prediction of plant phenology for five crops based on photoperiod × temperature interaction. *Ann. Bot.* 81, 705–16. https://doi.org/10.1006/anbo.1998.0625.

Yin, X., and Struik, P. C. 2009. C3 and C4 photosynthesis models: An overview from the perspective of crop modelling. *NJAS-Wagen J. Life Sci.* 57, 27–38. https://doi.org/10.1016/j.njas.2009.07.001.

Yin, X., Kropff, M. J., McLaren, G., and Visperas, R. M. 1995. A nonlinear model for crop development as a function of temperature. *Agric For. Meteorol.* 77, 1–16. https://doi.org/10.1016/0168-1923(95)02236-Q.

Zhu, X. G., Govindjee, Baker, N. R., DeSturler, E., Ort, D. R., and Long, S. P. 2005. Chlorophyll a fluorescence induction kinetics in leaves predicted from a model describing each discrete step of excitation energy and electron transfer associated with Photosystem II. *Planta* 223, 114–33. https://doi.org/10.1007/s00425-005-0064-4.

Zhu, X.-G., de Sturler, E., and Long, S. P. 2007. Optimizing the Distribution of Resources between Enzymes of Carbon Metabolism Can Dramatically Increase Photosynthetic Rate: A Numerical Simulation Using an Evolutionary Algorithm. Plant Physiol. 145, 513–26. https://doi.org/10.1104/pp.107.103713.

Zhu, X. G., Song, Q., and Ort, D. R. 2012. Elements of a dynamic systems model of canopy photosynthesis. *Curr. Opin. Plant Biol.* 15, 237–44. https://doi.org/10.1016/j.pbi.2012.01.010.

Zhu, X. G., Wang, Y., Ort, D. R., and Long, S. P. 2013. e-photosynthesis: A comprehensive dynamic mechanistic model of C3 photosynthesis: From light capture to sucrose synthesis. Plant Cell Environ. 36, 1711–27. https://doi.org/10.1111/pce.12025.

Zwieniecki, M. A., Tixier, A., and Sperling, O. 2015. Temperature-assisted redistribution of carbohydrates in trees. *Am. J. Bot.* 102, 1216–18. https://doi.org/10.3732/ajb.1500218.

Chapter 2

Functional–structural plant modeling of plants and crops

Jochem B. Evers and Leo F. M. Marcelis, Wageningen University, The Netherlands

1 Introduction

Crop models have been created and applied in the agricultural and horticultural sciences already for decades. Their range of application extends from prediction of crop yields for given weather, soil and greenhouse conditions (Marcelis et al., 1998; Keating et al., 2003; Jones et al., 2003), to detailed scientific analysis of ecophysiological processes occurring in crop canopies (Poorter et al., 2013; Yin and Struik, 2017). Over the years, crop models have become robust tools that are capable of simulating production and yields of a wide range of different crop types, (future) climate conditions, soil properties and management factors (reviewed in Keating and Thorburn, 2018). Despite this versatility, there are a number of applications for which crop models are not directly suitable, simply because they were never designed with such applications in mind.

In typical crop models, crop performance is calculated based on resource acquisition (light capture, water and nutrient uptake) by the crop as a whole. This means that questions related to the interactions between plants that make up the canopy cannot be readily addressed using a crop model. For typical purposes of crop models this is not necessarily an issue, but it does prevent the use of the model to address scientific questions on growth and development of individual plants to consequences for whole canopy performance. This also means that in crop models, plant traits are typically integrated and defined as canopy traits. The most prominent example of this is the light-extinction coefficient often used in crop models to calculate light capture by the crop. The

http://dx.doi.org/10.19103/AS.2019.0061.02

value of this coefficient integrates plant traits such as leaf angle distribution, leaf curvature, leaf optical properties and internode length, which means that most crop models cannot be readily used to analyze any of these component plant traits and their effects on crop performance individually.

Additionally, crop models have typically not been designed to represent spatial heterogeneity, such as non-uniform crop canopies. Oftentimes they simulate a crop canopy as a one-dimensional layer or collection of layers, above- and below-ground, without spatial heterogeneity in horizontal directions. This means that non-uniform canopies are difficult to simulate. Such non-uniformity is seen in species or cultivar mixtures and intercrop systems (Brooker et al., 2015), but also in canopies with spatial variation in growth conditions, wide paths in between plant rows or large variation in moment of emergence. On top of that, plants tend to adapt their growth and development to their local conditions, which means that in heterogeneous canopies plant phenotypical development may differ from location to location in the plot (e.g. Zhu et al., 2016), adding to overall crop performance. Next to heterogeneous canopies, heterogeneous environments like greenhouse setups with additional lighting above and even within the crop are hard to capture in traditional crop models. Additionally, pruning of plants or trees to improve light distribution, growth or quality are difficult to include in traditional crop models (Allen et al., 2005; Buck-Sorlin et al., 2011).

For research questions that deal with crop performance in relation to plant-plant interactions, plant architecture and phenotypic plasticity, as well as heterogeneity in canopy and environment, functional-structural plant (FSP) models have been developed (Godin and Sinoquet, 2005; Vos et al., 2010; Evers et al., 2018). The essence of FSP models is the simulation of the relationship between plant functioning and plant structure in 3D, with explicit feedback between plant growth and development on the one hand and environmental drivers on the other (Vos et al., 2010). This feedback is crucial to the functionality of FSP models: for instance, light capture by leaves determines the production of growth substrates and eventual growth of leaves, stems and other plant organs resulting in a change in plant structure, but this change itself determines to what extent light can be captured. Here, the 3D aspect of FSP models plays an important role, since growth of organs like leaves and stems may result in different (self) shading patterns depending on their positioning in 3D space, affecting light interception and ultimately growth. The same is true for other resources such as water or nitrogen distribution in the soil in relation to root system architecture. This feedback makes FSP models appropriate tools to simulate crop vegetation performance emergent from the underlying processes related to interaction between plants.

Typically, FSP models that produce output at the level of the whole plant and the plant canopy, contain three main components: growth (the increase in biomass and size of organs), development (the creation of new organs over time, vegetative and reproductive) and architecture (the 3D shape of organs). Other computational models for plants exist that produce output at different integration levels, such as production of leaves at the shoot apical meristem (Smith et al., 2006) or development of the shape of an individual leaf (Runions et al., 2017). For the purpose of this chapter, these types of models will not be considered. In this chapter, FSP models that simulate individual plants interacting with each other in plant stands will be introduced, including the processes they typically include, and their applications. The focus of this chapter will be on annual non-woody crop species, but concepts extend to simulation of perennial as well as woody species. Importantly, this chapter does not describe a particular FSP model or a particular modeling platform, but rather a generic approach toward modeling plant interactions and canopy performance using FSP modeling. Whether or not specific processes need to be included in such a model, and how they should be implemented, depends on the research question at hand and on the modeling platform chosen (e.g. Karwowski and Prusinkiewicz, 2003; Cournède et al., 2008; Hemmerling et al., 2008; Pradal et al., 2008), and thus differs from case to case.

2 Simulating growth

Growth, the increase in biomass and size of organs in FSP models is the result of a loop that includes light capture determined by incoming light and the distribution of the leaf area in the canopy; assimilation of CO_2 driven by the amount of light captured, and dependent on leaf nitrogen and water content, temperature and CO_2 concentration; and organ biomass increase as a result of the balance between the amount of assimilates available and the demand for assimilates. Subsequently, the 3D distribution of leaf area is updated based on the growth of individual leaves, internodes and other organs that might further expose or shade leaf area. The updated leaf area distribution is subsequently input to calculate light capture in the next time step of the model, closing the loop.

2.1 Light capture

In FSP models, the capture of photosynthetically active radiation (PAR) by a leaf to drive leaf photosynthesis is the result of the orientation and position of the leaf in the canopy, the amount and direction of incoming radiation, and the optical properties of the leaf itself as well as the surrounding leaves and other objects (their absorptance, reflectance and transmittance for PAR). The incoming radiation is defined by the light sources and their position in the scene. Depending on the question to be addressed with the FSP model, choices need to be made regarding the level of realism of the light sources. This may range from one light source in zenith emitting a fixed amount of PAR, to more realistic conditions with a combination of light sources that represent direct sunlight and diffuse skylight at an intensity and location depending on day of the year and location on the globe. For greenhouse studies, additional light sources may be added representing greenhouse lamps above or in the canopy (Buck-Sorlin et al., 2011; de Visser et al., 2014). For a typical field crop situation, an arc of light sources can be used that represent the daily course of the sun (Fig. 1). Diffuse light can be represented by a dome of light sources, arranged in rings at different heights (Chelle and Andrieu, 1999; Evers et al., 2010). The amount of PAR emitted by the light sources can be taken from actual weather data, or can be approximated mathematically (Goudriaan and Van Laar, 1994).

The radiation actually reaching a leaf or other (photosynthesizing) object in the scene, is calculated using techniques such as ray tracing (Cieslak et al., 2008; Hemmerling et al., 2008) or radiosity (Borel et al., 1991; Chelle and Andrieu, 1998). While the techniques differ substantially in the way they work,

Figure 1 Bird's-eye view of a field of simulated maize plants (30 rows of 120 plants each) showing the dome and arc of light sources that emit the radiation, which the plants capture and use for growth.

their aim is the same: estimating the distribution of radiation absorbed in the canopy, considering the optical properties of all objects in the scene (Chelle et al., 2007; Evers et al., 2010; Sarlikioti et al., 2011; Evers and Bastiaans, 2016). In many FSP models, this means that all leaves, stems and other objects in the scene are to be provided with values for reflectance, transmittance and absorptance for PAR. These optical properties determine the interaction with the light as it travels from the sources: it may be absorbed ready to be used in photosynthesis, or it may be scattered and captured by a different object, or not captured at all. In most FSP models, these optical properties are included as parameters derived from spectrophotometric measurements. Alternatively, organ optical properties may be calculated as a function of age and the associated leaf physiological characteristics (Coussement et al., 2018b).

2.2 Carbon assimilation

The result of the calculation of leaf light capture is the 3D distribution of absorbed PAR within the canopy. Since for each leaf the amount of PAR absorbed is known, the amount of CO_2 it can assimilate can be readily calculated. The most widely used method to do so in FSP models is the Farquhar-Von Caemmerer-Berry (FvCB) leaf photosynthesis model (Farquhar et al., 1980) and its derivatives (e.g. Yin and Struik, 2009). The FvCB model calculates the photosynthesis rate of a leaf or any photosynthetic organ based on the amount of PAR absorbed and a range of environmental variables and organ traits such as photosynthetic nitrogen content. The latter is known to scale with the fraction of incoming radiation that is absorbed by the organ and which is exponentially decreasing with depth in the canopy (Niinemets and Anten, 2009). This would eliminate the direct need to simulate the mechanisms underlying nitrogen distribution in the plant. Depending on the research question to be addressed using the FSP model, nitrogen uptake and distribution can be included at various levels of detail (Bertheloot et al., 2008; Barillot et al., 2016; Louarn and Faverjon, 2018). Due to its robust performance, the FvCB routines to calculate organ photosynthesis have been implemented in many FSP modeling studies (e.g. Evers et al., 2010; Buck-Sorlin et al., 2011; Sarlikioti et al., 2011; Wiechers et al., 2011; Xu et al., 2011). However, depending on the research question to be addressed using the model, a simpler implementation may be sufficient using a leaf-level photosynthesis-light-response curve, in which the amount of absorbed radiation is directly converted into an amount of assimilated CO_2, without calculating the underlying biochemistry. Usually, no regulation of photosynthesis by carbon demand (sink strength) is taken into consideration, given the large amount of uncertainty on the subject.

Either of these methods will provide the amount of carbon assimilated per photosynthetic organ. Depending on the plant species simulated, a choice needs to be made regarding the level of pooling of the assimilated carbon available for growth. In the case of the simulation of a small annual species at a model time step of one day, one could assume all carbon assimilated within the time step of the model could potentially reach any growing organ in the plant. In such a case, assimilated carbon can be pooled at the plant level. For larger species, or for smaller model time steps, the assumption on the transport reach of assimilated carbon may not hold anymore, which would require carbon pooling to occur at smaller units such as at the branch or even phytomer level. The level of assimilate pooling is heavily determined by additional modeling choices regarding assimilate allocation to growing organs, for instance related to the presence of any sink priority and/or the consideration of the simulation of the actual assimilate transport itself (Minchin and Lacointe, 2005; Gu et al., 2018). For this chapter, the assumption is that assimilate pooling at the plant level is sufficient, no sink prioritization occurs, and translocation is not simulated. Typically, the amount of assimilates in the pool is converted into an amount of growth substrates (new biomass) by using a fixed conversion rate. From this pool of substrates, a certain amount needs to be deducted to represent maintenance respiration. Typically, the maintenance respiration is defined as a fraction of the current plant biomass.

2.3 Organ sink strength for carbon

After having calculated the distribution of light capture, assimilate production and ultimately the amount of assimilates available for growth, the next step is to determine how those assimilates are allocated to different organs (reviewed in Marcelis and Heuvelink, 2007). In many crop models, allocation of assimilates is calculated based on empirically derived partitioning parameters that are input to the model (Marcelis, 1993). In a number of FSP models the allocation of assimilates to individual growing organs is simulated based on organ sink strength and the balance between assimilate supply and demand (Allen et al., 2005; Evers et al., 2010; Evers and Bastiaans, 2016; De Vries et al., 2018; Bongers et al., 2018). The idea behind this concept is that the amount of assimilates allocated to a growing organ is calculated every time step as a function of its sink demand, the amount of available assimilates, and, importantly, the presence and strength of other sinks in the same plant. Since organ sink strength, assimilate availability and the amount of within-plant competition for assimilates changes continuously, assimilate allocation is calculated every time step. Thus, the fractions of assimilates partitioned to organs eventually become an output rather than an input of the model, and the amount of assimilates received by an organ is the result of competition with other growing organs for the substrates available.

Organ sink strength can be represented by the organ potential growth rate, that is the growth rate an organ would show in case its demand for assimilates would be satisfied all the time (Marcelis, 1996). Since organ biomass typically follows a sigmoidal growth pattern, a bell-shaped curve resulting from taking the first derivative of such a sigmoidal shape can represent organ potential growth rate. A widely used example is the beta growth function (Yin et al., 2003), which is defined by parameters for growth duration (for how long is an organ a sink for assimilates), the organ age at which growth rate is highest (usually halfway its growth duration) and the maximum biomass an organ could potentially accumulate. These parameters need to be estimated from experimental data. Normally potential growth rate is expressed as a function of thermal time, making organ sink strength depend on temperature. Finally, the sink strength calculation should consider growth respiration, that is the amount of assimilates needed for the process of growth itself. Typically, growth respiration is defined as a fixed fraction with which the potential growth rate is increased, to take into account the assimilate cost of growth.

2.4 Organ growth

Based on the organ sink strengths and the amount of assimilates available, the actual allocation of assimilates to the growing organs can be calculated making use of the concept of relative sink strength (Heuvelink, 1996), which is the fraction of total plant sink strength an organ represents. For example, if the relative sink strength of a leaf is 0.1, the sink strength of this leaf equals 10% of the total plant sink strength. Using this fraction, the absolute amount of assimilates available to the organ can be calculated by taking this fraction of the total amount of assimilates available in the plant. After this, there are two possibilities that are as follows: (1) the amount of assimilates available to the organ is larger than the absolute sink strength (the demand) of the organ. This means that the organ can grow at its potential rate, as overall assimilate acquisition was plentiful and/or sink strength of the rest of the plant was apparently relatively low. (2) the amount of assimilates available to the organ is smaller than the demand of the organ. This would indicate that the plant is experiencing shading, and/or the plant contains many or strong sinks for carbon. This option results in organ growth lower than the potential rate; depending on the species under consideration, priority may be given to particular organ types. As this evaluation is done at every time step, actual organ growth may be at its potential rate at one time, but a lower rate several time steps later, depending on appearance of new sinks (covered in the next section) and changes in the radiation conditions the plant experiences, notably shading by neighboring plants. In case the amount of available assimilates exceeds total demand of the plant, the assimilates could be stored in an assimilate pool and made available for later time steps of the model. Quantitative knowledge on this storage and whether and for how long it will remain available is limited.

The amount of substrates an organ receives needs to be converted in an increase in organ size. For a leaf, in most FSP models the newly acquired biomass is multiplied with a value representing the specific leaf area (SLA, in m^2 area g^{-1} biomass) to determine the size increment of the leaf. This SLA value may be a fixed parameter or may be a function of leaf age, leaf position in the plant and/or environmental factors like light, temperature and the degree of water or nitrogen availability. For an internode a similar calculation can be made using a specific internode length (SIL) parameter (in m length g^{-1} biomass, but can also be expressed in terms of internode volume instead of length). As organ sizes are updated every time step, the distribution of leaf area changes continuously, feeding back on the interception of light that drives growth. This cycle of light-driven source-sink based plant growth enables the simulation of plants competing for light, and the consequences for growth and performance of the individual plants and the 3D canopy as a whole. This is one of the main selling points of FSP modeling that set it apart from other modeling approaches. Of added value here is that responses to competition in the form of phenotypic plasticity, next to the consequences in terms of organ biomass and size, can be included as well. For example, well-known shade-avoidance responses such as changes in SLA and SIL which lead to more elongated stems and petioles and thinner leaves in conditions of competition (Pierik and de Wit, 2014; Ballare and Pierik, 2017) have been rewarding subjects of study using FSP modeling in explaining how competition between plants determines their performance (Kahlen and Stuetzel, 2011; Bongers et al., 2018). Accurately simulating organ and plant growth in canopies may require inclusion of shade-avoidance signals in an FSP model.

2.5 Roots and soil resources

FSP models in which plant growth is driven by light capture and temperature are well developed and established. In such models, the root system is usually not represented, or represented without architectural detail as one sink organ demanding assimilates. This means the root system only affects the source/sink balance of the plant, but does not function as a source of nutrients or water. However, rather independently of the above-ground FSP models, models of root system architecture have been developed with the purpose of capturing root architectural development in relation to genotype-specific traits, soil properties and uptake of nutrients and water (reviewed in Dunbabin et al., 2013). In such models, architectural development of the root system is considered at high levels of detail, but oftentimes the shoot part is reduced to a one-dimensional source of assimilates and sink for nutrients and water (e.g. Postma et al., 2017). FSP models that include both root and shoot part in a balanced way at the same level of architectural and functional detail have been reported

Figure 2 Simulated mixture of generic cereal and dicot plants in a strip intercrop design.

(Drouet and Pagès, 2007) but do not allow to scale up to the vegetation or crop level, although progress has been described (Louarn and Faverjon, 2018). To be able to better understand the functioning of plants growing in canopies as a result of competition for resources like light, water and nutrients, and to explore hypotheses on the mechanisms underlying competitive outcomes, an integrated modeling approach capturing both above and below-ground processes and whole-plant growth will be essential, especially for complex mixed-species canopies (Fig. 2) (Evers et al., 2019). As it stands, whole-plant FSP modeling of above and below-ground processes that allows scaling up from the organ to the field level is still in its infancy, and thus the remainder of this chapter will focus on shoot canopy modeling only.

3 Simulating development

3.1 Organ production

In FSP models, development refers to the creation of new plant organs over time. Once created, these plant organs become sinks for assimilates allowing them to grow, and some also become source organs, providing assimilates for growth. Thus, the dynamics of organ creation heavily influence the source/sink balance of the plant over time and should therefore be represented accurately. Typically, above-ground organ creation in FSP models is captured by relatively simple rules that describe the production of new phytomers at the shoot apical meristem, of which one or more are present within the architecture of the plant shoot. These rules are formalized in grammars such as L-systems (Lindenmayer, 1968; Prusinkiewicz and Lindenmayer, 1990) or other similar, derived or generalized grammars (de Reffye et al., 1988; Kurth and Sloboda, 1997; Hemmerling et al., 2008).

In the simplest L-system implementation, a vegetative plant shoot is described as a series of phytomers, which consist of an internode and one or more leaves (depending on the plant species) with a lateral meristem in the leaf axil, and an apical meristem on top. Production of new phytomers over time by the apex is governed by a rule that describes the addition of a new phytomer in between the top phytomer and the apex, extending the chain of phytomers by one: *Apex → Internode [Bud] [Leaf] Apex*, where the square brackets indicate the axillary bud and leaf fork off the main shoot. The timing of the application of this rule, that is the rate of production of new phytomers, is determined by a time interval called the plastochron, which is an input parameter of the model and thus needs to be measured or derived experimentally. Essentially, such a production rule describes the output of mechanistic models that simulate phytomer production based on the underlying physiological mechanisms involving hormonal activity (Smith et al., 2006; van Mourik et al., 2012; O'Connor et al., 2014). Given that the purpose of FSP models that simulate plant competition and canopy growth normally does not include prediction of organ production as a function of underlying processes, descriptive L-system-like rules suffice. Rate of phytomer production is usually strongly temperature dependent and plastochron is thus often well described by a temperature sum. If some flexibility in organ production in relation to eco physiological factors is required, an intermediate approach that captures coordination between production, appearance and growth of organs (e.g. organ n appears when organ n-2 has become mature) without including detailed physiology can also be considered (Zhu et al., 2014).

For shoot development normally a rule is included that describes the apical switch to flowering. The apex switching to flowering can be described by a similar rule as the one producing vegetative phytomers, describing the production of one or more flowers, along with a new leaf depending on whether the species is determinate or indeterminate, for example: *Apex → Internode Flower* for determinate and *Apex → Internode [Bud] [Leaf] [Flower] Apex* for indeterminate species. The exact composition of these rules depends on the species. The moment of the generative switch can be made a function of, for example the number of vegetative phytomers produced, plant age, and/or day length.

3.2 Shoot branching

Branch production can be represented by the simplest of rules: *Bud → Apex*, because once the switch from a dormant bud into an active apex has been done, the further development of the branch is determined by its apex and thus described by the rule on vegetative phytomer production also used for the main stem. Therefore, the moment this rule is applied greatly determines the architecture of the plant: new branches are built from phytomers that contain axillary buds themselves, and non-limited bud break would result in an exponentially developing architecture. Therefore, in FSP models it is vital that bud break is restricted by default and only allowed in case certain criteria apply. In many species bud break is a particularly plastic trait governed by a host of environmental and physiological factors (Domagalska and Leyser, 2011; Barbier et al., 2019), which for the same genotype may result in branch production widely differing between different conditions such as plant population densities. This argues against the use of one single parameter determining the rate of branch production, in contrast to the use of a single plastochron value driving phytomer production rate. To simulate bud break in FSP shoot models, a number of different approaches can be adopted that capture the variation and plasticity in branch production (Evers and Vos, 2013):

1 A purely empirical approach entails the simulation of bud break based on measured probability distributions. The probabilities of appearance of certain branches or cohorts of branches can be gathered from experimental data and used in an FSP model as input. The fate of each bud (stay dormant or break) will then be determined during simulation based on the probability distributions put in. This approach is fine in case the plant canopies simulated do not differ from the canopies from which the bud break data was derived in terms of, for example population density or environmental conditions. In this way no form of plasticity in bud break is considered, limiting its application to very specific situations such as the virtual recreation of real canopies.
2 A more flexible approach toward bud break can be implemented in FSP models using response curves, relating an environmental or physiological variable like light intensity, spectral composition, or plant resource status, to the probability of bud break (Luquet et al., 2006; Evers et al., 2007). Such a response curve will need to be calibrated using data from experiments in which the driving variable is manipulated and the resulting bud break is observed. An FSP model that simulates bud break using such a response curve, will predict branch appearance and number based on the local conditions experienced by the plant. For example, plants simulated at a high population density will produce fewer branches than at a lower density, because at a high density canopy light levels drop more quickly, the spectral composition of the light changes more quickly, and the plant resource status will go down more quickly. Importantly, in contrast to the approach using probabilities, the

response-curve approach contains a feedback loop: the newly produced branches alter the light availability and composition (through shading) and plant resource status (due to the production of new sink and source organs), which has consequences for further branch production.

3 Response curves relating bud break to an internal (such as ratio supply/demand of assimilates) or external variable (such as spectral composition of intercepted light), essentially capture all underlying ecophysiological processes in one relationship. If the research question requires a deeper understanding of how the bud break response is regulated, these ecophysiological processes need to be incorporated in the model. The response curve will then become an output rather than an input of the model. For instance, bud break can be simulated based on the auxin activity dynamics in a developing plant (Prusinkiewicz et al., 2009). If such physiological regulation is coupled to an environmental factor such as the spectral composition of light sensed by plant photoreceptors, the branching pattern of a plant will emerge from the simulation as a function of vegetation density, the associated changes in light spectral composition, photoreceptor activity and the coupled auxin production and transport parameters (Evers et al., 2011). This approach to simulate the variation and plasticity in branch production captures most of the realism in the regulation of branching, yet is only required when the aim of the modeling study is to increase understanding of how branching is regulated in relation to the consequences for plant functioning. Given the extra computational weight of incorporating such physiological mechanisms, response curves are in many cases a good compromise between realism and efficiency.

Whichever method is chosen to simulate bud break, the resulting growth of organs in the branch (leaves, internodes, flowers, fruits) obey the source/sink rules as outlined in Section 2, 'Simulating growth'. No distinction is made between main stem or branch organs regarding their growth rules. This way the carbon-balance in the plant is honored.

3.3 Organ death

For accurate simulation of leaf area development over time, the entire life span of organs needs to be considered, which includes leaf senescence and death, flower and fruit abortion, as well as branch abortion for certain species. Leaf death can be simulated at various degrees of complexity and realism, depending on the research question. The simplest approach is the removal of a leaf once a certain criterion or set of criteria are met, such as leaf age, light level or photosynthesis rate. On the other end of the spectrum, the process of leaf ageing and the associated changes in optical properties (Coussement et al., 2018b) and in leaf architecture can be included in the model. This has implications for the distribution of light in the canopy due to the changes in light absorption and scattering, as well as in the opening up of the canopy at the location of the dying canopy layer, feeding back on plant growth and potentially on the mechanisms that govern developmental processes such as the production of new branches. Additionally, the relocation of nitrogen and thus photosynthetic capacity from a dying leaf to a newly created leaf can be considered.

Abortion of branches occurs in important annual cereal species, and is a key process in the determination of final plant and crop performance. Senescence of such cereal branches has been linked to light signals and plant nutrient status (Sparkes et al., 2006). Analogous to bud break, branch abortion can therefore be implemented in an FSP model as a function of physiological or environmental variables, using a response curve or using a description of deeper ecophysiological regulatory mechanisms. In contrast to leaf death, the abortion of branches has larger implications for the remaining canopy as multiple organs are removed simultaneously, feeding back to plant and canopy growth.

In fruit or flower crops, abortion of flowers or fruits is an important determinant of the final yield and product quality. Simulation of fruit or flower abortion can be done analogous to that of buds or branches with a key role for temperature and source/sink balance (e.g. Marcelis et al., 2004). Abortion of fruit and flowers shows strong plasticity and is highly variable. In a conventional crop model, Wubs et al. (2011) simulated this variation by using survival analysis, an approach that could also be used for FSP models.

4 Simulating architecture

In FSP models that simulate plant growth and development in canopy settings, organs appear according to the rules of development, they accumulate biomass and grow in size according to the rules of growth, but a final set of rules needs to be in place that determines the shape and orientation of each organ. Together with the plant 'skeleton' provided by the branching rules, organ shape and orientation determine the overall architecture of the plant and the structure of the canopy. Similar to bud break, organ shape and orientation can be included as empirical parameters or parameter distributions, as well as model variables that respond to environmental factors.

4.1 Organ shape

Similar to organ creation, the physiological mechanisms underlying the shape of organs such as leaves can be captured in FSP models (Bilsborough et al., 2011; Runions et al., 2017). For the simulation of canopy performance as a function of plant-plant interactions however, organ shape is normally simulated using simple geometrical descriptions. The most prominent organ in a canopy, the leaf, comes in a wide variation of shapes and levels of complexity, ranging from simple leaves with smooth edges and a flat surface to heavily curved and bent shapes with serrated, lobed or otherwise curvy edges. They come either as simple leaves consisting of a petiole and a blade, to complex compound leaves with multiple leaflets of different shapes and relative sizes on a rachis of variable length. Therefore, how leaf shape can be captured in an FSP model is determined by the species to be simulated. In all cases, the choice can be made between mathematically describing the shape of the leaves or leaflets (e.g. Evers et al., 2005; Watanabe et al., 2005), or using lists of coordinates that together represent a mesh of polygons representing the leaf shape (Coussement et al., 2018c; De Vries et al., 2018; Bongers et al., 2018). In the former case, any mathematical description that captures the shape and curvature of the leaf will be appropriate. In the latter case, any spatial resolution (number of polygons) that captures the shape and curvature of the leaf will be appropriate. In both cases, however, care needs to be taken regarding the realism aimed for when representing the leaves: a higher level of realism may not necessarily result in more accurate simulation output but may increase computational requirements. A sensitivity analysis may provide insight into the extent to which a more accurate leaf representation leads to more accurate simulation results. Finally, the shape of other organs such as internodes, petioles and peduncles are oftentimes represented by simple cylinders, and representation of flower and fruit geometry is very much species dependent. Similar to organ size, organ shape may also change over time and/or depend on the position on the stem, which can be taken into account by making organ shape a function of plant age and/or phytomer rank.

4.2 Organ orientation

The orientation of leaves, that is the direction of the leaf primary axis, is determined by (1) the insertion angle of the leaf, which is the angle between the leaf petiole and the internode it is connected to and (2) the phyllotactic angle, which is the angle between two consecutive leaves on a stem. Both angles are key determinants of plant architecture, and together with organ 3D shape, they have a substantial effect on the distribution of light capture in a crop canopy. Leaf insertion angle and shape may differ greatly between genotypes of the same species, with important repercussions for light capture and photosynthetic characteristics, for example in rice (Chen et al., 2014b), and thus plant performance.

These angles can be included in the FSP model as leaf properties that may be fixed, or change over time with leaf age or leaf rank. Additionally, both leaf insertion and phyllotactic angles can be plastic traits, responding to signals related to the density of the vegetation and thus competition. An example of plasticity of leaf insertion angle is hyponasty, which is another well-known shade-avoidance response (de Wit et al., 2012) aiming at improving light capture when in competition. Also the angle between consecutive leaves on a stem can be sensitive to competition signals and adopt a more favorable orientation. This does not change the phyllotactic angle itself, which is determined at the level of the shoot apex, but rather it reorients the leaf through petiole (Kahlen et al., 2008) or sheath (Maddonni et al., 2002) bending during leaf growth. Depending on the research question, such responses can be included in FSP models by relating the leaf angle to light or other signals (Kahlen et al., 2008; Pantazopoulou et al., 2017).

4.3 Overall plant architecture

In the end, the shape of the plant over time is an integrated result of organ size determined by growth, organ number, and organ geometry represented by their shape and orientation. The number of organs does not only pertain to the rate at which new leaves and internodes are produced by the apical meristem of a shoot, but also by the branching pattern of the plant driven by bud break, which enables new shoots to be formed and the number of organs to increase further. In the case an FSP model is equipped with both provisions to simulate light-driven growth, as well as provisions to simulate phenotypic plasticity in internode and petiole length, leaf angle and branch production (e.g. De Vries et al., 2018), all resulting changes in plant architecture feed back to the light that drives growth, and to the light signals that elicit plastic responses.

5 Conclusion and future trends

As with any simulation modeling approach, FSP modeling is about finding the balance between what to include and what to leave out, determining the level of realism of the model. FSP models have been created for a wide variety of applications, and even the category of FSP models treated in this chapter contains quite some variation in the choice of ecophysiological processes and environmental factors. This shows that the choice of what to include and what to leave out very much depends on the type of question to be answered. For example, in Zhu et al. (2015) light capture of the plant system needed to be analyzed in order to answer the research question. This means that the model did not need to incorporate mechanistic growth of organs. Instead the increase in size of organs was represented using relatively simple empirical rules. In contrast, Faverjon et al. (2019) required their model to simulate carbon and nitrogen metabolism to address the research question on interspecific competition in grasslands, leading to a more mechanistic modeling approach. This 'tailored' modeling principle, in which the complexity of the model used is matched with the requirements of the research question, is particularly important in FSP modeling: incorporating unnecessary mechanisms (1) comes with additional parameter requirement and (2) may lead to decreased computational efficiency, both of which are already substantial in FSP models, and (3) may obscure model outcomes due to inherent stochasticity of the included mechanism or other sources of variation. This emphasizes the relevance of building a conceptual model, in which the relevant processes and factors are included to address a research question, before starting the actual implementation of the model in a modeling platform.

Regarding simulation of above-ground plant growth and development, most FSP models use light as the only driving variable that is calculated at the leaf level. Other environmental drivers, such as temperature, humidity and CO_2 concentration are normally considered to be the same throughout the plant canopy and often even the same as above the canopy. In reality, canopies show gradients in those environmental variables (Qian et al., 2012; Savvides et al., 2013), formed by the dynamic feedback relationships with canopy architecture, CO_2 assimilation, and transpiration of water. Being able to simulate all key microclimatic drivers at the same level of detail as light, would allow for more accurate calculation of leaf photosynthesis. This would include taking into account the climatic conditions directly around the leaf (Chelle, 2005) as well as leaf temperature instead of air temperature. The latter would also allow for better prediction of plant growth and development, since organ sink strength as well as the production of new organs by the apex is driven by the temperature of the organ (e.g. Savvides et al., 2016), not air temperature. Additionally, being able to simulate microclimatic variables other than light at the level of the organ would open up possibilities to apply FSP models in the domain of plant disease ecology. Infection probabilities of crop diseases highly depend on local leaf conditions such as humidity, temperature and associated conditions such as leaf wetness. Simulating such conditions would allow for the exploration of the relation between plant architecture, physiology, and environmental conditions leading to (i) local plant infection, (ii) disease spread, and (iii) consequences for crop performance.

Further novel applications of FSP modeling are to be expected in the domain of greenhouse horticulture, within which it has been applied for, for example leaf and stem trait optimization in tomato (Fig. 3) (Sarlikioti et al., 2011), light signal responses and photosynthesis limitations in cucumber (Kahlen and Stuetzel, 2011; Chen et al., 2014a) and optimization of rose management

Figure 3 An example of tomato plant simulated using FSP modeling. In the 'open structure' phenotype, internode and petioles were longer than in the 'dense structure' variety, leading to an estimated 10% higher crop photosynthesis in a canopy situation. Source: adapted from Sarlikioti et al. (2011).

and production (Buck-Sorlin et al., 2011). Advancements in the domain of greenhouse LED technology bring up new question regarding the optimization of both the spectral composition of the radiation in relation to plant responses, as well as the placement of the light sources also within the crop canopies. These kind of questions can benefit from significant progress made in the simulation of LEDs and full light spectra for FSP models (Hitz et al., 2019). Especially in combination with the simulation of other microclimatic variables as mentioned above, this would also open up the possibilities to optimize greenhouse energy use and crop performance simultaneously.

FSP models have great potential, especially when combined with experimental work providing data for model calibration and validation, but also as a means to test model predictions. Although FSP models until now have mostly served as scientific tools, progress is to be expected in improving crop management strategies (plant pattern, application of light sources, pruning strategies, biocide application, etc.) and in the identification of favorable plant traits informing breeding, especially when the whole-plant approach will be adopted simulating above and below-ground growth and interactions. Further focus on as yet underrepresented plant-water relationships (Coussement et al., 2018a) and nutrient distribution within the plant will contribute to that. The visual output of FSP models, which essentially is a side-effect of simulating plants in 3D, can be highly instrumental in visually demonstrating the effects of treatments or management options (e.g. plant spacing, pruning) on plant growth and architecture to stakeholders, as well as in teaching plant biology. This way, FSP modeling fills a niche next to more traditional crop models, in focusing on those questions that really require the consideration of individual plants and their interactions.

There are opportunities to improve conventional crop models that do not consider the 3D structure of plants using information from FSP models, and the other way around. For instance, relationships that emerge from FSP modeling can potentially be used to improve empirical relationships normally used in crop models (Marcelis et al., 1998; Keating et al., 2003; Jones et al., 2003). For example, FSP models could provide relationships between light extinction and crop stage over time for canopies with a heterogeneous structure due to non-uniform plant emergence, species mixtures or weed infestation. Subsequently, such relationships could be incorporated into crop models allowing yield predictions to be done more accurately. Also, output from crop models can be used to improve FSP models by generating validation datasets to compare to FSP model output. This way, both types of modeling can strengthen each other, improving the quality of the predictions made.

6 References

Allen, M. T., Prusinkiewicz, P. and DeJong, T. M. 2005. Using L-systems for modeling source-sink interactions, architecture and physiology of growing trees: the L-PEACH model. *The New Phytologist* 166(3), 869–80. doi:10.1111/j.1469-8137.2005.01348.x.

Ballaré, C. L. and Pierik, R. 2017. The shade-avoidance syndrome: multiple signals and ecological consequences. *Plant, Cell and Environment* 40(11), 2530–43. doi:10.1111/pce.12914.

Barbier, F. F., Dun, E. A., Kerr, S. C., Chabikwa, T. G. and Beveridge, C. A. 2019. An update on the signals controlling shoot branching. *Trends in Plant Science* 24(3), 220–36. doi:10.1016/j.tplants.2018.12.001.

Barillot, R., Chambon, C. and Andrieu, B. 2016. CN-Wheat, a functional-structural model of carbon and nitrogen metabolism in wheat culms after anthesis. II. Model evaluation. *Annals of Botany* 118, 1015–31. doi:10.1093/aob/mcw144.

Bertheloot, J., Andrieu, B., Fournier, C. and Martre, P. 2008. A process-based model to simulate nitrogen distribution within wheat (*Triticum aestivum*) plant during grain filling. *Functional Plant Biology* 35(10), 781–96. doi:10.1071/FP08064.

Bilsborough, G. D., Runions, A., Barkoulas, M., Jenkins, H. W., Hasson, A., Galinha, C., Laufs, P., Hay, A., Prusinkiewicz, P. and Tsiantis, M. 2011. Model for the regulation of *Arabidopsis thaliana* leaf margin development. *Proceedings of the National Academy of Sciences of the United States of America* 108(8), 3424–9. doi:10.1073/pnas.1015162108.

Bongers, F. J., Pierik, R., Anten, N. P. R. and Evers, J. B. 2018. Subtle variation in shade avoidance responses may have profound consequences for plant competitiveness. *Annals of Botany* 121(5), 863–73. doi:10.1093/aob/mcx151.

Borel, C. C., Gerstl, S. A. W. and Powers, B. J. 1991. The radiosity method in optical remote sensing of structural 3-D surfaces. *Remote Sensing of the Environment* 36, 13–44.

Brooker, R. W., Bennett, A. E., Cong, W. F., Daniell, T. J., George, T. S., Hallett, P. D., Hawes, C., Iannetta, P. P., Jones, H. G., Karley, A. J., Li, L., McKenzie, B. M., Pakeman, R. J., Paterson, E., Schöb, C., Shen, J., Squire, G., Watson, C. A., Zhang, C., Zhang, F., Zhang, J. and White, P. J. 2015. Improving intercropping: a synthesis of research in agronomy, plant physiology and ecology. *The New Phytologist* 206(1), 107–17. doi:10.1111/nph.13132.

Buck-Sorlin, G. H., de Visser, P. H. B., Henke, M., Sarlikioti, V., van der Heijden, G. W., Marcelis, L.F.M. and Vos, J. 2011. Towards a functional-structural plant model of cut-rose - simulation of light environment, light absorption, photosynthesis and interferences with the plant structure. *Annals of Botany* 108(6), 1121–34. doi:10.1093/aob/mcr190.

Chelle, M. 2005. Phylloclimate or the climate perceived by individual plant organs: what is it? How to model it? What for? *The New Phytologist* 166(3), 781–90. doi:10.1111/j.1469-8137.2005.01350.x.

Chelle, M. and Andrieu, B. 1998. The nested radiosity model for the distribution of light within plant canopies. *Ecological Modelling* 111(1), 75–91. doi:10.1016/S0304-3800(98)00100-8.

Chelle, M. and Andrieu, B. 1999. Radiative models for architectural modelling. *Agronomie* 19(3–4), 225–40. doi:10.1051/agro:19990304.

Chelle, M., Evers, J. B., Combes, D., Varlet-Grancher, C., Vos, J. and Andrieu, B. 2007. Simulation of the three-dimensional distribution of the red:far-red ratio within crop canopies. *The New Phytologist* 176(1), 223–34. doi:10.1111/j.1469-8137.2007.02161.x.

Chen, T. W., Henke, M., de Visser, P. H. B., Buck-Sorlin, G., Wiechers, D., Kahlen, K. and Stützel, H. 2014a. What is the most prominent factor limiting photosynthesis in different layers of a greenhouse cucumber canopy? *Annals of Botany* 114(4), 677–88. doi:10.1093/aob/mcu100.

Chen, C. P., Sakai, H., Tokida, T., Usui, Y., Nakamura, H. and Hasegawa, T. 2014b. Do the rich always become richer? Characterizing the leaf physiological response of the high-yielding rice cultivar Takanari to free-air CO_2 enrichment. *Plant and Cell Physiology* 55(2), 381–91. doi:10.1093/pcp/pcu009.

Cieslak, M., Lemieux, C., Hanan, J. and Prusinkiewicz, P. 2008. Quasi-Monte Carlo simulation of the light environment of plants. *Functional Plant Biology* 35(10), 837–49. doi:10.1071/FP08082.

Cournède, P. H., Mathieu, A., Houllier, F., Barthélémy, D. and de Reffye, P. 2008. Computing competition for light in the GREENLAB model of plant growth: a contribution to the study of the effects of density on resource acquisition and architectural development. *Annals of Botany* 101(8), 1207–19. doi:10.1093/aob/mcm272.

Coussement, J., De Swaef, T., Lootens, P., Roldán-Ruiz, I. and Steppe, K. 2018a. Introducing turgor-driven growth dynamics into functional-structural plant models. *Annals of Botany* 121(5), 849–61. doi:10.1093/aob/mcx144.

Coussement, J., Henke, M., Lootens, P., Roldán-Ruiz, I., Steppe, K. and De Swaef, T. 2018b. Modelling leaf spectral properties in a soybean functional-structural plant model by integrating the prospect radiative transfer model. *Annals of Botany* 122(4), 669–76. doi:10.1093/aob/mcy105.

Coussement, J., Steppe, K., Lootens, P., Roldan-Ruiz, I. and De Swaef, T. 2018c. A flexible geometric model for leaf shape descriptions with high accuracy. *Silva Fennica* 52(2). doi:10.14214/sf.7740.

de Reffye, P., Edelin, C., Françon, J., Jaeger, M. and Puech, C. 1988. Plant models faithful to botanical structure and developmentr. *ACM SIGGRAPH Computer Graphics* 22(4), 151–8. doi:10.1145/378456.378505.

de Visser, P. H., Buck-Sorlin, G. H. and van der Heijden, G. W. 2014. Optimizing illumination in the greenhouse using a 3D model of tomato and a ray tracer. *Frontiers in Plant Science* 5, 48. doi:10.3389/fpls.2014.00048.

De Vries, J., Poelman, E. H., Anten, N. P. R. and Evers, J. B. 2018. Elucidating the interaction between light competition and herbivore feeding patterns using functional-structural plant modelling. *Annals of Botany* 121(5), 1019–31. doi:10.1093/aob/mcx212.

De Wit, M., Kegge, W., Evers, J. B., Vergeer-van Eijk, M. H., Gankema, P., Voesenek, L. A. and Pierik, R. 2012. Plant neighbor detection through touching leaf tips precedes phytochrome signals. *Proceedings of the National Academy of Sciences of the United States of America* 109(36), 14705–10. doi:10.1073/pnas.1205437109.

Domagalska, M. A. and Leyser, O. 2011. Signal integration in the control of shoot branching. *Nature Reviews. Molecular Cell Biology* 12(4), 211–21. doi:10.1038/nrm3088.

Drouet, J. L. and Pagès, L. 2007. GRAAL-CN: a model of GRowth, Architecture and ALlocation for Carbon and Nitrogen dynamics within whole plants formalised at the organ level. *Ecological Modelling* 206(3–4), 231–49. doi:10.1016/j.ecolmodel.2007.03.036.

Dunbabin, V. M., Postma, J. A., Schnepf, A., Pagès, L., Javaux, M., Wu, L., Leitner, D., Chen, Y. L., Rengel, Z. and Diggle, A. J. 2013. Modelling root-soil interactions using three-dimensional models of root growth, architecture and function. *Plant and Soil* 372(1–2), 93–124. doi:10.1007/s11104-013-1769-y.

Evers, J. B. and Bastiaans, L. 2016. Quantifying the effect of crop spatial arrangement on weed suppression using functional-structural plant modelling. *Journal of Plant Research* 129(3), 339–51. doi:10.1007/s10265-016-0807-2.

Evers, J. B. and Vos, J. 2013. Modelling branching in cereals. *Frontiers in Plant Science* 4, 399. doi:10.3389/fpls.2013.00399.

Evers, J. B., Vos, J., Fournier, C., Andrieu, B., Chelle, M. and Struik, P. C. 2005. Towards a generic architectural model of tillering in Gramineae, as exemplified by spring wheat (*Triticum aestivum*). *The New Phytologist* 166(3), 801–12. doi:10.1111/j.1469-8137.2005.01337.x.

Evers, J. B., Vos, J., Chelle, M., Andrieu, B., Fournier, C. and Struik, P. C. 2007. Simulating the effects of localized red:far-red ratio on tillering in spring wheat (*Triticum aestivum*) using a three-dimensional virtual plant model. *The New Phytologist* 176(2), 325–36. doi:10.1111/j.1469-8137.2007.02168.x.

Evers, J. B., Vos, J., Yin, X., Romero, P., van der Putten, P. E. L. and Struik, P. C. 2010. Simulation of wheat growth and development based on organ-level photosynthesis and assimilate allocation. *Journal of Experimental Botany* 61(8), 2203–16. doi:10.1093/jxb/erq025.

Evers, J. B., van der Krol, A. R., Vos, J. and Struik, P. C. 2011. Understanding shoot branching by modelling form and function. *Trends in Plant Science* 16(9), 464–7. doi:10.1016/j.tplants.2011.05.004.

Evers, J. B., Letort, V., Renton, M. and Kang, M. Z. 2018. Computational botany: advancing plant science through functional-structural plant modelling. *Annals of Botany* 121(5), 767–72. doi:10.1093/aob/mcy050.

Evers, J. B., van der Werf, W., Stomph, T. J., Bastiaans, L. and Anten, N. P. R. 2019. Understanding and optimizing species mixtures using functional-structural plant modelling. *Journal of Experimental Botany* 70(9), 2381–8. doi:10.1093/jxb/ery288.

Farquhar, G. D., von Caemmerer, S. and Berry, J. A. 1980. A biochemical model of photosynthetic CO_2 assimilation in leaves of C3 species. *Planta* 149(1), 78–90. doi:10.1007/BF00386231.

Faverjon, L., Escobar-Gutiérrez, A., Litrico, I., Julier, B. and Louarn, G. 2019. A generic individual-based model can predict yield, nitrogen content, and species abundance in experimental grassland communities. *Journal of Experimental Botany* 70(9), 2491-504. doi:10.1093/jxb/ery323.

Godin, C. and Sinoquet, H. 2005. Functional-structural plant modelling. *The New Phytologist* 166(3), 705–8. doi:10.1111/j.1469-8137.2005.01445.x.

Goudriaan, J. and Van Laar, H. H. 1994. *Modelling Potential Crop Growth Processes*. Kluwer Academic Publishers, Dordrecht.

Gu, S., Zhang, L., Yan, Z., Werf, W. van der and Evers, J. B. 2018. Quantifying within-plant spatial heterogeneity in carbohydrate availability in cotton using a local-pool model. *Annals of Botany* 121(5), 1005–17. doi:10.1093/aob/mcx210.

Hemmerling, R., Kniemeyer, O., Lanwert, D., Kurth, W. and Buck-Sorlin, G. H. 2008. The rule-based language XL and the modelling environment GroIMP illustrated with simulated tree competition. *Functional Plant Biology* 35(10), 739–50. doi:10.1071/FP08052.

Heuvelink, E. 1996. Re-interpretation of an experiment on the role of assimilate transport resistance in partitioning in tomato. *Annals of Botany* 78(4), 467–70. doi:10.1006/anbo.1996.0143.

Hitz, T., Henke, M., Graeff-Hönninger, S. and Munz, S. 2019. Three-dimensional simulation of light spectrum and intensity within an LED growth chamber. *Computers and Electronics in Agriculture* 156, 540–8. doi:10.1016/j.compag.2018.11.043.

Jones, J. W., Hoogenboom, G., Porter, C. H., Boote, K. J., Batchelor, W. D., Hunt, L. A., Wilkens, P. W., Singh, U., Gijsman, A. J. and Ritchie, J. T. 2003. The DSSAT cropping system model. *European Journal of Agronomy* 18(3–4), 235–65. doi:10.1016/S1161-0301(02)00107-7.

Kahlen, K. and Stützel, H. 2011. Modelling photo-modulated internode elongation in growing glasshouse cucumber canopies. *The New Phytologist* 190(3), 697–708. doi:10.1111/j.1469-8137.2010.03617.x.

Kahlen, K., Wiechers, D. and Stuetzel, H. 2008. Modelling leaf phototropism in a cucumber canopy. *Functional Plant Biology* 35(10), 876–84. doi:10.1071/FP08034.

Karwowski, R. and Prusinkiewicz, P. 2003. Design and implementation of the L+C modeling language. *Electronic Notes in Theoretical Computer Science* 86(2), 134–52. doi:10.1016/S1571-0661(04)80680-7.

Keating, B. A. and Thorburn, P. J. 2018. Modelling crops and cropping systems—evolving purpose, practice and prospects. *European Journal of Agronomy* 100, 163–76. doi:10.1016/j.eja.2018.04.007.

Keating, B. A., Carberry, P. S., Hammer, G. L., Probert, M. E., Robertson, M. J., Holzworth, D., Huth, N. I., Hargreaves, J. N. G., Meinke, H., Hochman, Z., McLean, G., Verburg, K., Snow, V., Dimes, J. P., Silburn, M., Wang, E., Brown, S., Bristow, K. L., Asseng, S., Chapman, S., McCown, R. L., Freebairn, D. M. and Smith, C. J. 2003. An overview of APSIM, a model designed for farming systems simulation. *European Journal of Agronomy* 18(3–4), 267–88. doi:10.1016/S1161-0301(02)00108-9.

Kurth, W. and Sloboda, B. 1997. Growth grammars simulating trees - an extension of L-systems incorporating local variables and sensitivity. *Silva Fennica* 31(3), 285–95. doi:10.14214/sf.a8527.

Lindenmayer, A. 1968. Mathematical models for cellular interactions in development. I. Filaments with one-sided inputs. *Journal of Theoretical Biology* 18(3), 280–99. doi:10.1016/0022-5193(68)90079-9.

Louarn, G. and Faverjon, L. 2018. A generic individual-based model to simulate morphogenesis, C-N acquisition and population dynamics in contrasting forage legumes. *Annals of Botany* 121(5), 875–96. doi:10.1093/aob/mcx154.

Luquet, D., Dingkuhn, M., Kim, H., Tambour, L. and Clement-Vidal, A. 2006. EcoMeristem, a model of morphogenesis and competition among sinks in rice. 1. Concept, validation and sensitivity analysis. *Functional Plant Biology* 33(4), 309–23. doi:10.1071/FP05266.

Maddonni, G. A., Otegui, M. E., Andrieu, B., Chelle, M. and Casal, J. J. 2002. Maize leaves turn away from neighbors. *Plant Physiology* 130(3), 1181–9. doi:10.1104/pp.009738.

Marcelis, L. F. M. 1993. Simulation of biomass allocation in greenhouse crops - a review. *Acta Horticulturae*, 49–68. doi:10.17660/ActaHortic.1993.328.3.

Marcelis, L. F. M. 1996. Sink strength as a determinant of dry matter partitioning in the whole plant. *Journal of Experimental Botany* 47, 1281–91. doi:10.1093/jxb/47.Special_Issue.1281.

Marcelis, L. F. M. and Heuvelink, E. 2007. Concepts of modelling carbon allocation among plant organs. In: Vos, J., Marcelis, L. F. M., de Visser, P. H. B., Struik, P. C. and Evers, J. B. (Eds), *Functional-Structural Plant Modelling in Crop Production*. Springer, Dordrecht, The Netherlands, pp. 103–11.

Marcelis, L. F. M., Heuvelink, E. and Goudriaan, J. 1998. Modelling biomass production and yield of horticultural crops: a review. *Scientia Horticulturae* 74(1–2), 83–111. doi:10.1016/S0304-4238(98)00083-1.

Marcelis, L. F. M., Heuvelink, E., Baan Hofman-Eijer, L. R., Den Bakker, J. and Xue, L. B. 2004. Flower and fruit abortion in sweet pepper in relation to source and sink strength. *Journal of Experimental Botany* 55(406), 2261–8. doi:10.1093/jxb/erh245.

Minchin, P. E. H. and Lacointe, A. 2005. New understanding on phloem physiology and possible consequences for modelling long-distance carbon transport. *The New Phytologist* 166(3), 771–9. doi:10.1111/j.1469-8137.2005.01323.x.

Niinemets, Ü. and Anten, N. P. R. 2009. Packing the photosynthetic machinery: from leaf to canopy. In: Laisk, A., Nedbal, L. and Govindjee (Eds), *Photosynthesis in Silico*. Springer, Dordrecht, The Netherlands, 363–99. doi:10.1007/978-1-4020-9237-4_16.

O'Connor, D. L., Runions, A., Sluis, A., Bragg, J., Vogel, J. P., Prusinkiewicz, P. and Hake, S. 2014. A division in PIN-mediated auxin patterning during organ initiation in grasses. *PLoS Computational Biology* 10(1), e1003447. doi:10.1371/journal.pcbi.1003447.

Pantazopoulou, C. K., Bongers, F. J., Küpers, J. J., Reinen, E., Das, D., Evers, J. B., Anten, N. P. R. and Pierik, R. 2017. Neighbor detection at the leaf tip adaptively regulates upward leaf movement through spatial auxin dynamics. *Proceedings of the National Academy of Sciences of the United States of America* 114(28), 7450–5. doi:10.1073/pnas.1702275114.

Pierik, R. and de Wit, M. 2014. Shade avoidance: phytochrome signalling and other aboveground neighbour detection cues. *Journal of Experimental Botany* 65(11), 2815–24. doi:10.1093/jxb/ert389.

Poorter, H., Anten, N. P. and Marcelis, L. F. M. 2013. Physiological mechanisms in plant growth models: do we need a supra-cellular systems biology approach? *Plant, Cell and Environment* 36(9), 1673–90. doi:10.1111/pce.12123.

Postma, J. A., Kuppe, C., Owen, M. R., Mellor, N., Griffiths, M., Bennett, M. J., Lynch, J. P. and Watt, M. 2017. OpenSimRoot: widening the scope and application of root architectural models. *The New Phytologist* 215(3), 1274–86. doi:10.1111/nph.14641.

Pradal, C., Dufour-Kowalski, S., Boudon, F., Fournier, C. and Godin, C. 2008. OpenAlea: a visual programming and component-based software platform for plant modelling. *Functional Plant Biology* 35(10), 751–60. doi:10.1071/FP08084.

Prusinkiewicz, P. and Lindenmayer, A. 1990. *The Algorithmic Beauty of Plants*. Springer-Verlag, New York.

Prusinkiewicz, P., Crawford, S., Smith, R. S., Ljung, K., Bennett, T., Ongaro, V. and Leyser, O. 2009. Control of bud activation by an auxin transport switch. *Proceedings of the National Academy of Sciences of the United States of America* 106(41), 17431–6. doi:10.1073/pnas.0906696106.

Qian, T., Dieleman, J. A., Elings, A., Gelder, A. de, Kooten, O. van and Marcelis, L. F. M. 2012. Vertical temperature gradients in the semi-closed greenhouses: occurrence and effects. *Acta Horticulturae* 927(927), 59–66. doi:10.17660/ActaHortic.2012.927.5.

Runions, A., Tsiantis, M. and Prusinkiewicz, P. 2017. A common developmental program can produce diverse leaf shapes. *The New Phytologist* 216(2), 401–18. doi:10.1111/nph.14449.

Sarlikioti, V., de Visser, P. H. B. and Marcelis, L. F. M. 2011. Exploring the spatial distribution of light interception and photosynthesis of canopies by means of a functional-structural plant model. *Annals of Botany* 107(5), 875–83. doi:10.1093/aob/mcr006.

Savvides, A., van Ieperen, W., Dieleman, J. A. and Marcelis, L. F. M. 2013. Meristem temperature substantially deviates from air temperature even in moderate environments: is the magnitude of this deviation species-specific? *Plant, Cell and Environment* 36(11), 1950–60. doi:10.1111/pce.12101.

Savvides, A., Dieleman, J. A., Van Ieperen, W. and Marcelis, L. F. M. 2016. A unique approach to demonstrating that apical bud temperature specifically determines leaf initiation rate in the dicot *Cucumis sativus*. *Planta* 243(4), 1071–9. doi:10.1007/s00425-015-2464-4.

Smith, R. S., Guyomarc'h, S., Mandel, T., Reinhardt, D., Kuhlemeier, C. and Prusinkiewicz, P. 2006. A plausible model of phyllotaxis. *Proceedings of the National Academy of Sciences of the United States of America* 103(5), 1301–6. doi:10.1073/pnas.0510457103.

Sparkes, D. L., Holme, S. J. and Gaju, O. 2006. Does light quality initiate tiller death in wheat? *European Journal of Agronomy* 24(3), 212–7. doi:10.1016/j.eja.2005.08.003.

van Mourik, S., Kaufmann, K., van Dijk, A. D. J., Angenent, G. C., Merks, R. M. H. and Molenaar, J. 2012. Simulation of organ patterning on the floral meristem using a polar auxin transport model. *PLoS ONE* 7(1), e28762. doi:10.1371/journal.pone.0028762.

Vos, J., Evers, J. B., Buck-Sorlin, G. H., Andrieu, B., Chelle, M. and de Visser, P. H. B. 2010. Functional-structural plant modelling: a new versatile tool in crop science. *Journal of Experimental Botany* 61(8), 2101–15. doi:10.1093/jxb/erp345.

Watanabe, T., Hanan, J. S., Room, P. M., Hasegawa, T., Nakagawa, H. and Takahashi, W. 2005. Rice morphogenesis and plant architecture: measurement, specification and the reconstruction of structural development by 3D architectural modelling. *Annals of Botany* 95(7), 1131–43. doi:10.1093/aob/mci136.

Wiechers, D., Kahlen, K. and Stützel, H. 2011. Dry matter partitioning models for the simulation of individual fruit growth in greenhouse cucumber canopies. *Annals of Botany* 108(6), 1075–84. doi:10.1093/aob/mcr150.

Wubs, A. M., Heuvelink, E., Marcelis, L. F. M. and Hemerik, L. 2011. Quantifying abortion rates of reproductive organs and effects of contributing factors using time-to-event analysis. *Functional Plant Biology* 38(5), 431–40. doi:10.1071/FP10249.

Xu, L., Henke, M., Zhu, J., Kurth, W. and Buck-Sorlin, G. H. 2011. A functional-structural model of rice linking quantitative genetic information with morphological development and physiological processes. *Annals of Botany* 107(5), 817-28. doi:10.1093/aob/mcq264.

Yin, X. and Struik, P. C. 2009. C3 and C4 photosynthesis models: an overview from the perspective of crop modelling. *NJAS - Wageningen Journal of Life Sciences* 57(1), 27–38. doi:10.1016/j.njas.2009.07.001.

Yin, X. and Struik, P. C. 2017. Can increased leaf photosynthesis be converted into higher crop mass production? A simulation study for rice using the crop model GECROS. *Journal of Experimental Botany* 68(9), 2345–60. doi:10.1093/jxb/erx085.

Yin, X., Goudriaan, J., Lantinga, E. A., Vos, J. and Spiertz, H. J. 2003. A flexible sigmoid function of determinate growth. *Annals of Botany* 91(3), 361–71. doi:10.1093/aob/mcg029.

Zhu, J., Andrieu, B., Vos, J., Van der Werf, W., Fournier, C. and Evers, J. B. 2014. Towards modelling the flexible timing of shoot development: simulation of maize organogenesis based on coordination within and between phytomers. *Annals of Botany* 114(4), 753–62. doi:10.1093/aob/mcu051.

Zhu, J., Van der Werf, W., Anten, N. P. R., Vos, J. and Evers, J. B. 2015. The contribution of phenotypic plasticity to complementary light capture in plant mixtures. *The New Phytologist* 207(4), 1213–22. doi:10.1111/nph.13416.

Zhu, J., van der Werf, W., Vos, J., Anten, N. P. R., van der Putten, P. E. L. and Evers, J. B. 2016. High productivity of wheat intercropped with maize is associated with plant architectural responses. *Annals of Applied Biology* 168(3), 357–72. doi:10.1111/aab.12268.

Chapter 3

Improving modeling of nutrient cycles in crop cultivation

Upendra Singh and Cheryl Porter, International Fertilizer Development Center (IFDC) and University of Florida, USA

1 Introduction

Agricultural productivity depends on crops receiving adequate amounts of essential nutrients from the atmosphere, soils, and/or supplied fertilizers and manures. Through the biogeochemical cycling of nutrients, fertile soils supply the following essential nutrients to plants: nitrogen (N), phosphorus (P), potassium (K), sulfur (S), calcium (Ca), magnesium (Mg), iron (Fe), zinc (Zn), copper (Cu), manganese (Mn), boron (B), molybdenum (Mo), cobalt (Co), and nickel (Ni). Deficiency of any of these nutrients results in lower productivity. In Europe, North America, and many parts of Asia, the agricultural practice of depleting soil nutrient reserves (nutrient mining) for farming ceased several decades ago. Unfortunately, nutrient mining continues in many developing countries, particularly in Sub-Saharan Africa. The impact of such practices can be seen in Sub-Saharan Africa's low rate of productivity (FAO, 2019). Mineral fertilizers are applied to soils to supplement plants' nutrient needs for adequate production. In addition to replenishing soil nutrients and increasing crop production, fertilizers facilitate the adoption of improved seed varieties, which have significantly increased yield potential; however, improved seed varieties also have higher nutrient requirements. Varieties are currently being bred to improve the nutritional quality of harvested crops. The potential of these

http://dx.doi.org/10.19103/AS.2019.0061.05

improved varieties cannot be realized unless they are grown under optimum conditions, particularly under balanced plant nutrition. For the inherently nutrient-poor or degraded soils of the tropics, fertilizer use increases the production of both biomass and yields. This additional biomass will supplement the supply and maintenance of soil organic matter (SOM) and its positive effects on soil properties and nutrient use efficiency. An important component of modeling the nutrient cycle is the role and management of fertilizers in improving crop productivity and their impact on the environment via different loss processes. Fertilizers will remain a critical component of food production, as they are responsible for at least half of the global food production (Erisman et al., 2008).

The organic matter in soils can also provide an important source of nutrients for plants. As the organic matter decomposes, nutrients are released in inorganic forms, which may be available for plant uptake, depending on the pH and other chemical properties of the soil. SOM also improves the efficiency of applied fertilizers by improving the soil's water-holding capacity and nutrient retention. Agricultural systems that sustain SOM decomposition over many years must provide annual additions of organic matter to the soil, for example, from crop residues or the addition of manure, in order to offset the decomposition rates. In the tropics, organic matter decomposition often occurs faster than organic inputs are added to the soil, and organic matter depletion and degraded soils are the result.

In addition to the more prevalent nutrient deficiencies and low nutrient-retention capacities in the highly weathered soils of the tropics, high rainfall and temperatures further aggravate the soil nutrient status, leading to higher nutrient losses. Hence, crop simulation models that simulate the nutrient cycle must integrate various processes into the soil-crop-atmosphere continuum that determines crop growth and production and nutrient dynamics. Modeling the nutrient cycle also requires more input data and data that are less readily available. The tropical regions that can gain the most from nutrient modeling are, unfortunately, data poor. This also implies that soil nutrient models do not get tested and evaluated as much as crop growth models.

This chapter describes nutrient cycle modeling for N, P, and K and is not meant to be a review of existing nutrient models. A number of the nitrogen models have been reviewed by Nieder and Benbi (2008) and the P models by Barber (1995), Claassen and Steingrobe (1999), and Tinker and Nye (2000). Most of these models were designed to predict the effects of N and P on crop growth response. However, potassium models, most adapted from Barber and Cushman (1981), were concerned with calculating crop K uptake. Given the importance of N in global food production and its impact on the environment, the major emphasis has been on N modeling.

2 Modeling nitrogen dynamics

An estimated 59.0 million tons of N fertilizers were applied to cereals in 2014–2015. This represents 57.6% of the world's N fertilizer consumption, with 19.3%, 18.5%, and 15.2% of global N consumption going to maize, wheat, and rice, respectively (Heffer et al., 2017). Future projections of the human population, agricultural intensification, biofuel crops, and meat consumption demand an ever-increasing global production of nitrogen fertilizers. Together with human-driven increases in crop biological nitrogen fixation, the Haber-Bosch process of converting nitrogen gas into ammonia and N fertilizers has more than doubled the global production of reactive nitrogen compared to pre-industrial levels (Galloway et al., 2008). However, Springmann et al. (2018) argue that the use of N fertilizers for food production surpassed the safe planetary boundary in 2010, so an additional 50–52% increase in N fertilizer use, as expected by 2050, will have an even greater negative impact on the environment. N fertilizers primarily impact greenhouse gas (GHG) emissions, specifically nitrous oxide (N_2O) emissions, and contribute 1.2% toward global agricultural GHG emissions (Angle et al., 2017). Thus, to operate within a safe planetary boundary and continue to meet the increasing food demands, nitrogen use efficiency (NUE) needs to increase from the current 30–50% (Ladha et al., 2005) to 75–80%.

N modeling can play a crucial role in achieving the above-mentioned needs through efficient N fertilizer management and effective recycling of reactive N. It is not surprising that more than 50 soil N balance models exist. These models focus on various facets of the N transformation process, ranging from process-based to site-specific calibrated to empirical models (Shaffer et al., 2001). The more widely used models that are capable of process-based simulations of soil carbon (C) and N dynamics are DSSAT (Godwin and Singh, 1998), APSIM (Keating et al., 2003), DAYCENT (Kelly et al., 2000), DNDC (Li et al., 1994), DRAINMOD-Forest (Tian et al., 2012), EPIC (Williams et al., 1989), GECROS (Biernath et al., 2011), SUCROS (Stenger et al., 1999), FASSET (Berntsen et al., 2003), InfoCrop (Aggarwal et al., 2006), MAIZSIM (Kim et al., 2012), MONICA (Nendel et al., 2011), NFLOOD (Reddy et al., 1990), APSIM-Oryza (Gaydon et al., 2012), PlantSys (Jongschaap, 2007), RIWER (Jing et al., 2010), RICEWNB (Antonopoulos, 2010), RZWQM (Ahuja et al., 2000), and SALUS (Basso et al., 2010). Other models that simulate crop N response but do not simulate soil C dynamics are CropSys (Stockle et al. 2003), HERMES (Kersebaum, 2007), Sirius (Jamieson et al., 1998), and STICS (Brisson et al., 2003). Additional information on the current status of modeling C and N dynamics in soils is available in the review chapter by Nieder and Benbi (2008).

2.1 Soil carbon modeling

Most soil C cycle models have a 'system boundary' drawn around the soil column; the state variables are the soil C stored in each soil layer. These are compartmentalized into SOM and fresh organic matter (FOM). The application of organic matter amendments, senescence of roots and other plant matter, and any crop residues that are not removed from the field at harvest all help increase FOM in the soil system. FOM is reduced by decomposition, which moves C into SOM pools and releases CO_2 into the atmosphere. SOM is also reduced by decomposition accompanied by the equivalent release of CO_2 into the atmosphere. However, a tillage event can redistribute and mix the organic matter within the soil profile (Fig. 1). Major differences in soil C modeling are associated with the number of SOM pools, their relative size, and the decomposition rates. The data requirement and the lack of available data for soil C modeling increase with the increasing number of soil C pools. Various methods for measuring and estimating soil C pools are available (Basso et al., 2011). Basic soil C model inputs, in terms of initializing soil C pools, are FOM from previous crop residues (aboveground and roots); additional external organic amendments that are transported into the field (animal manures, green manure, etc.); and soil organic C content, total N content, and C:N ratio for each soil layer.

For any given FOM or SOM pool, mineralization or decomposition of the organic matter results in a net release of nutrients (N, P, and S) and a release of CO_2 into the atmosphere. Immobilization is also associated with the decay of residues and occurs when nutrients such as N, P, and S are incorporated into microorganisms and temporarily rendered unavailable to crops. FOM decay is driven by the C:N ratio of the residue; the proportion of carbohydrates, cellulose, and lignin; and, in some cases, the polyphenolic content. In general,

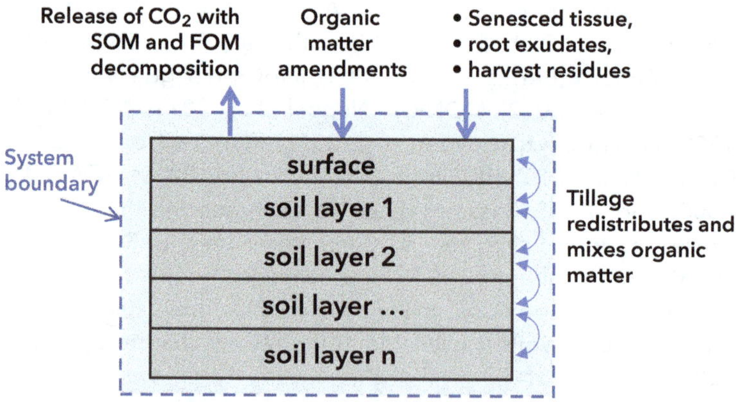

Figure 1 Components of soil C balance routine.

FOM with a low C:N ratio has a higher decomposition rate. However, pigeon pea residue, in spite of its lower C:N ratio compared to maize residue, has a lower mineralization rate due to its high polyphenol content, which acts as an anti-metabolite to the microorganisms (Singh et al., 2001). The soil's mineral N content through the 'temporary' immobilization of N lowers the C:N ratio of FOM in a given soil layer, which in turn allows decomposition to continue. Other factors that influence organic matter breakdown are soil temperature and soil moisture. The decomposition rate of FOM increases by up to 3% per °C increase under ideal soil moisture and C:N ratios (Godwin and Singh, 1998). Generally, the rate of decomposition of organic matter pools increases linearly from air-dry soil to field capacity moisture content soil; however, above the saturated soil moisture content, the decay rate declines to a relative rate of 35–50%. For any given environmental condition, as illustrated by the simulated response in Fig. 2, SOM pools play a critical role in dictating the decomposition rate and the role soils play as a CO_2 sink or source. More stable forms of SOM decompose more slowly. SOM breakdown has a positive impact on crop growth by releasing nutrients such as N, P, and S, but it also has a negative impact due to the emission of CO_2 into the atmosphere.

As expected, under anaerobic (oxygen-depleted) conditions with lower SOM breakdown, organic C storage is an important component of the global C cycle in flooded/wetland soils, and their management impacts global warming and ozone depletion (Whiting, 1994). Microbial biomass makes up a significant

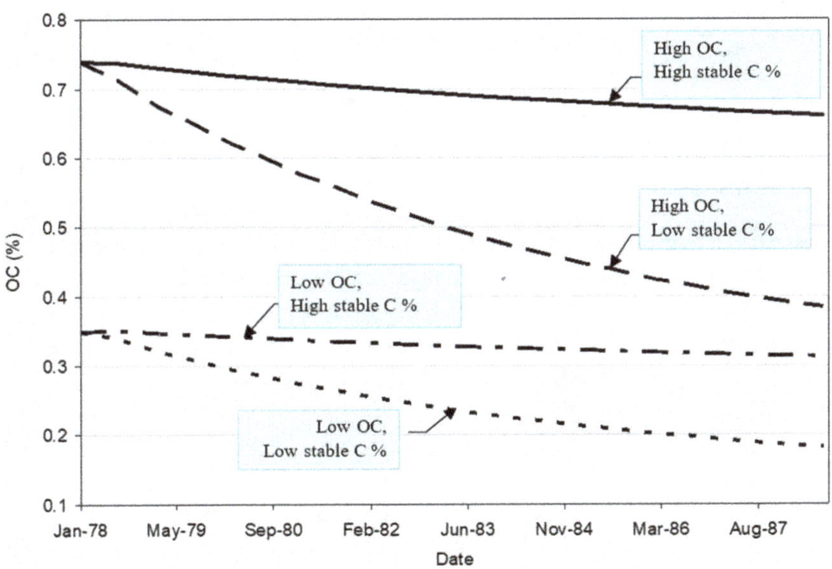

Figure 2 Effects of total organic C and stable C on the decomposition of organic matter (expressed as OC%).

Published by Burleigh Dodds Science Publishing Limited, 2026.

amount of C in wetlands. Under anaerobic conditions, SOM mineralization results in the release of nutrients and the emission of CO_2 and methane (CH_4) (Matthews et al., 2000). Long-term microbial activity that is sustainable occurs under anaerobic conditions in which the microbial respiration is supported by electron acceptors with a lower reduction potential, such as sulfates (SO_4^{2-}) and bicarbonates (HCO_3^-). Methanogenesis and sulfate reduction are dominant processes for anaerobic decomposition in wetlands and coastal ecosystems. In addition to the reduction of CO_2/HCO_3^-, methane formation occurs due to the reduction of acetates. Soluble C substrates also facilitate heterotrophic denitrification of nitrates (NO_3^-) to nitrous oxide (N_2O) and nitrogen gas (N_2).

2.2 Soil and floodwater nitrogen (N) transformations

2.2.1 Ammonium-N

The mineralization and immobilization processes are described using the above C pools and partitioning, with some models explicitly taking into account the role of microbes. Most soil N dynamic models express the mineralization rate (RM_i) as Michaelis-Menten kinetics:

$$RM_i = (\mu_{max} C_i)/(k_s + C_i) \tag{1}$$

where C_i is the carbon content of pool i, μ_{max} is the maximum mineralization rate, and k_s is half the saturation content. For models using first-order kinetics, $(\mu_{max})/(k_s + C_i)$ is simplified to k_1, the first-order rate coefficients:

$$RM_i = k_1 C_i \tag{2}$$

For zero-order kinetics, the equation can further be simplified where the mineralization rate is dependent only on k_0, zero-order rate coefficient:

$$RM_i = k_0 \tag{3}$$

As discussed in the previous section, the amount of N mineralized is dependent on the C:N ratios of the pools in each layer, soil temperature, and soil moisture/water-filled pore space (oxygen supply). Simpler models generally have a pre-defined C:N ratio. Thus, the mineralization rate under anaerobic conditions is slower. Ammonium-N (NH_4-N) is the mineralized product:

$$\text{SOM pools} \underset{\text{immobilization}}{\overset{\text{mineralization}}{\rightleftarrows}} NH_4\text{-N} \tag{4}$$

The transport of NH_4-N across soil layers and floodwater (when present) is via diffusion. The diffusion of NH_4-N across soil layers driven by concentration gradients is described by Fick's law and can be in either the upward or downward direction for a given soil layer. Diffusion of NH_4-N is also influenced by the cation exchange capacity of a soil and the soil buffering power of NH_4-N. NH_4-N may reside in the soil solution (exchangeable or plant-available NH_4-N) or be adsorbed into clay particles. Some models also simulate adsorbed NH_4-N in clay particles (Li et al., 1994). NH_4-N adsorption and desorption are complex and site-specific processes, and the measurement method to discriminate between native (non-available) and recently adsorbed (plant-available) NH_4-N needs further research.

Under anaerobic conditions, N remains in the form of NH_4-N for uptake by the rice crop. N partitioning between NH_4-N and NH_3-N and its susceptibility to volatilization loss are regulated by the soil and floodwater properties, such as temperature, pH (Fig. 3), and wind speed. For soils with a pH < 8, ammonia volatilization loss should be negligible; however, when urea is applied as N fertilizer, its hydrolysis can result in a soil pH > 9 surrounding the urea granules. Under flooded conditions, floodwater pH is one of the key regulators of NH_3 volatilization (Godwin and Singh, 1998). The trend of floodwater pH peaking during the day due to the consumption of CO_2 by algal photosynthetic activity and declining at night with the release of CO_2 is captured by CERES-Rice and APSIM-ORYZA (Gaydon et al., 2012). Thus, NH_4-N concentration and volatilization loss are also regulated by floodwater depth. While most N model components operate on a daily time-step, some processes such as changes in floodwater pH and ammonia volatilization loss are modeled at sub-daily to hourly time-steps.

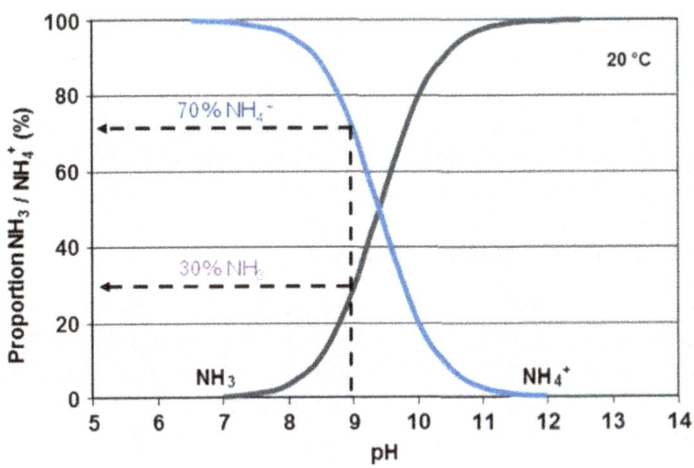

Figure 3 Changes in the proportion of NH_3-N/NH_4-N due to pH. Source: based on Denmead et al. (1982).

Published by Burleigh Dodds Science Publishing Limited, 2026.

2.2.2 Urea-N

Urea is the most widely used N fertilizer, with a worldwide production rate of 170 million tons in 2017. It is popular due to its high N content (46.4%), easy handling (non-flammable and non-explosive properties), and low price per unit of N. Urea-N hydrolysis to NH_4-N catalyzed by urease enzymes leads to increased soil pH and favorable conditions for volatilization (Eq. (5)). Hence, it is not surprising that volatilization is a major N loss mechanism in agriculture. In rice fields, 20–50% of applied N could be lost through NH_3-N volatilization caused by the combined effects of urea hydrolysis and algal activity in floodwater with a pH > 9.

(5)

Incorporating urea into the soil column can reduce volatilization loss compared to surface broadcast application. The practice of urea application followed by irrigation for dryland crops is also an effective way to minimize ammonia volatilization loss. Under lowland (flooded) rice field conditions, 50–60% of applied urea still enters the floodwater, despite incorporation into the soil (Vlek and Craswell, 1981), with the potential for volatilization loss. Thus, model simulations may substantially deviate from reality if the model assumes that none of the urea incorporated into the soil diffuses into the floodwater. On the other hand, urea deep placement (UDP) is highly effective in preventing diffusion of urea-N and ammonium-N into the floodwater, resulting in negligible ammonia volatilization loss, as evident from Fig. 4 (IFDC, 2017).

Urease inhibitors such as N-(n-butyl) thiophosphoric triamide (NBTPT), N-(n-butyl) phosphoric triamide (NBPT), phenyl phosphoryldiamidate (PPDA), and boric acid temporarily inhibit or slow urea hydrolysis and can significantly lower NH_3-N and NH_4-N concentration and the increase in soil and floodwater pH, thereby reducing ammonia volatilization loss (Fig. 4).

2.2.3 Nitrification and denitrification

Nitrification is the oxidation of ammonium to nitrite and then to nitrate under aerobic conditions facilitated by various nitrifying organisms. Very few models consider the growth and death of nitrifiers and denitrifiers and, thus, simulate nitrification and denitrification processes using Michaelis-Menten kinetics or first-order kinetics (Godwin and Singh, 1998; Li et al., 2004).

$$NH_3 + O_2 + 2H^+ + 2e^- \xrightarrow{AMO} NH_2OH + H_2O \xrightarrow{HAO\ (\rightarrow NO,\ N_2O)} NO_2^- + 0.5O_2 \xrightarrow{OXRD} NO_3^- \quad (6)$$

Published by Burleigh Dodds Science Publishing Limited, 2026.

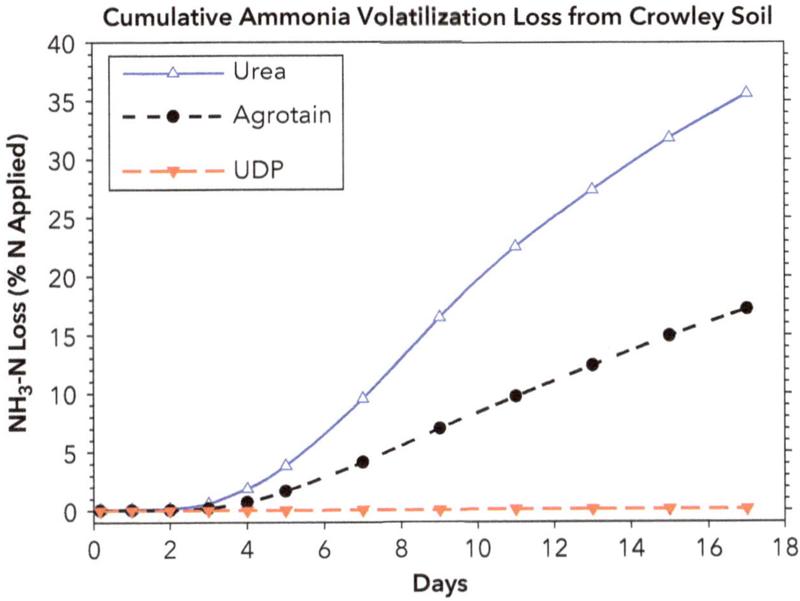

Figure 4 Effect of urea deep placement (UDP) and urease inhibitor – NBTPT (Agrotain®) compared to the broadcast application of urea on volatilization loss.

The conversion of NH_3-N/NH_4-N to hydroxyl amine (NH_2OH) as an oxidation-reduction process by *Nitrosomonas* sp. is catalyzed by monooxygenase enzyme (AMO), followed by conversion to nitrite (NO_2^-) in the presence of hydroxylamine oxidoreductase enzyme (HAO). The final step is carried out by *Nitrobacter* in the presence of nitrite oxidoreductase enzyme, converting nitrite to nitrate-N (NO_3-N). The rate coefficients of the kinetics can be constant or regulated processes limited by additional factors, such as pH, water-filled pore space, soil temperature, and a lag effect to accommodate unfavorable conditions and wetting and drying episodes. These are generally described by index-factors, ranging from zero (no activity) to unity (most active). In most models, nitrification ceases within 24 h under flooded conditions. Nitrification in the 'oxidized' soil layer under flooded conditions is insignificant and ignored by most N models. The high greenhouse warming potential of N_2O at 265 CO_2 equivalent (Myhre et al., 2013) makes nitrification process in the oxidized layer and/or its subsequent denitrification an important contributor of N_2O emission.

The denitrification process involves the reduction of NO_3-N to N_2 and N_2O gases in soils and flooded conditions where anaerobic conditions persist. Modeling of the denitrification process is empirical at best, given the spatial and temporal variability of anaerobic conditions, soil temperature, complexity of C substrates, and the presence of electron acceptors such as NO_3^-, Fe^{3+}, SO_4^{2-}, and HCO_3^- in the soil. In process-based models, the SOM modules simulate soil C in FOM, in humic matter (SOM), and as CO_2 released as a byproduct of the decomposition. These variables are then used for further calculations of denitrification and N-gas losses. The mathematical simulation can be expressed as zero-order, first-order, or Michaelis-Menten kinetics. A generalized first-order denitrification rate coefficient (k_{denit}), as used in RZWQM (Ma and Shaffer, 2001), can be expressed as:

$$k_{denit} = k_d T_{denit} [N_{wet} + MF_{denit} \left(t - N_{wet}\right) \quad (7)$$

where k_d is rate constant, T_{denit} is the effect of soil temperature on denitrification, MF_{denit} is the soil moisture factor's effect on denitrification based on water-filled pore space, N_{wet} is the number of wet days with rainfall or irrigation, and t is time in days. The first-order denitrification rate coefficient can be modified by soil-soluble carbon fraction (Rolston et al., 1980), which is an energy source for some denitrifiers. Denitrification loss from any given soil layer is then determined as a function of k_{denit} and the NO_3-N content.

In soil N models that simulate N_2O emission, denitrified N is further partitioned between N_2 and N_2O where the ratio of N_2:N_2O ($R_{N_2:N_2O}$) is dictated by functions involving soil nitrate ($F_r(NO_3)$), soil CO_2 or respiration rate ($F_r(CO_2)$), and water-filled pore space and the number of previous wet days ($F_r(WFP)$):

$$R_{N_2:N_2O} = \text{minimum}\left[F_r\left(NO_3\right), F_r\left(CO_2\right)\right] F_r\left(WFP\right) \quad (8)$$

In addition to the denitrification of NO_3-N, N_2O production and emission can occur during the nitrification process (Eq. (6)). N_2O emission from agricultural fields is one of the major GHG emissions. As with ammonia volatilization loss, losses due to N_2O emissions and NO_3-N leaching losses can be managed and reduced with the use of nitrification inhibitors. Inhibitors such as dicyandiamide (DCD), 3,4-dimethylpyrazole phosphate (DMPP), nitrapyrin, and acetylene have been very effective in reducing N losses, particularly N_2O emissions and nitrate leaching (Linquist et al., 2013; Abalos et al., 2014; Feng et al., 2016).

2.2.4 Movement of N

The transport of dissolved urea-N and NO_3-N across soil layers and floodwater occurs via mass flow (percolation). N transport simulation is dependent on the associated soil water balance modules. The movement of urea-N and NO_3-N and the diffusion of NH_4-N are represented at a complexity that varies between models. As discussed, the diffusion of NH_4-N across soil layers is driven by concentration gradients of NH_4-N, cation exchange capacity of the soil, and the soil buffering power of NH_4-N. Mass flow in most upland N models is,

Published by Burleigh Dodds Science Publishing Limited, 2026.

generally, the movement of dissolved urea and nitrate (with water) from one soil layer to the layer below. The implicit assumption in the model is that all urea-N or nitrate-N present in a particular layer is uniformly and instantaneously in solution in all of the water in that layer. This assumption will not hold for macropore water movement, for example, soils with large cracks. If needed, the DSSAT models use a nitrate retention function to model NO_3-N leaching in soils with variably charged surfaces and anion-retention capacities. The retention capacity for nitrate-N is estimated for subsoils as a function of delta pH (pH in KCl minus pH in water) and SOM content (Singh et al., 2002).

A major application of soil N models is their ability to predict N leaching losses, since maximizing crop responses to N management should not occur at the expense of high N leaching losses (McGechan and Wu, 2001; Hansen et al., 2001; Basso et al., 2016; Pathak, 2004; Gaydon et al., 2017). The long-term sustainability of both food security and the environment has set new goals for N management.

2.2.5 Enhanced efficiency N fertilizers

Enhanced efficiency N fertilizers with urease inhibitors, nitrification inhibitors, and controlled-release coatings improve NUE by reducing N losses and improving the synchrony of N release with crop N demand (Fig. 5) (Trenkel,

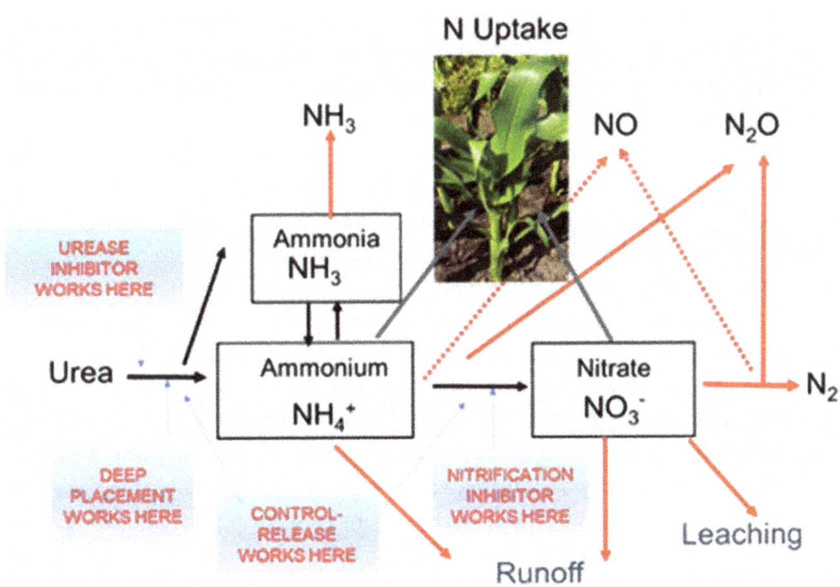

Figure 5 Role of inhibitors, control-release coatings, and urea deep placement on the N transformation process.

Published by Burleigh Dodds Science Publishing Limited, 2026.

2010). UDP reduces N losses and improves plant N availability by preventing the diffusion of N into floodwater. With an increasing emphasis in improving NUE, the European Union has stipulated that, after 2020, urea cannot be sold without the incorporation of urease- and nitrification-inhibitor compounds. The endorsement of the international code of conduct for the sustainable use and management of fertilizers, which aims to guarantee an effective and efficient use of fertilizers, also highlights the global importance of fertilizers in food production and food security and their impact on the environment (FAO, 2019, http://www.fao.org/3/mz476en/mz476en.pdf).

The above-mentioned changes in N management and N fertilizers demand that the existing N models incorporate the effects of inhibitors, controlled-release coatings, and slow-release N fertilizers on soil and plant N dynamics. This will help improve the synchrony between soil and fertilizer N supply and crop N demand and reduce N losses.

2.3 Crop nitrogen (N) modeling

Optimal N management involves synchronizing the timing of soil N supply via mineralization and fertilization (as discussed earlier) with crop N demand to promote optimum N uptake. Crop N uptake is influenced by environmental and crop factors. As discussed in previous sections, N supply is influenced by environmental factors such as the soil temperature, soil moisture, soil pH, and management of N fertilizers and organic amendments. Root morphology, root architecture, and root length density regulate the N supply to crops. All of these factors interact with each other, to influence not just N supply, but also N demand by crops. Modeling N uptake provides a process-based approach to deal with these complex interactions. In process-based models developed to simulate crop yield, the soil and floodwater N model is coupled with the crop growth and phenology model to simulate crop N uptake. Thus, the prediction of crop N uptake depends directly on a model's ability to correctly simulate crop duration, growth stages, crop growth, root distribution, soil water balance, and soil and floodwater dynamics. Hence, input errors, simulation errors, and assumptions from other sub-models would significantly influence the performance of crop N models. Models that were developed to estimate the overall N balances, in general, use a simpler approximation of N uptake (Chowdary et al., 2004; Liang et al., 2007). N uptake has also been described by Michaelis-Menten kinetics where the maximum rate of N uptake is limited by an index-factor for leaf area index (LAI), root distribution, and temperature (Antonopoulos, 2010). Crop N demand can be estimated from a logistic growth curve in simpler models. Ma et al. (2008) provide a comprehensive review of a wide range of plant N uptake models for cereals, legumes, oilseeds, and tuber crops.

2.3.1 Nitrogen (N) supply

In most process-based models, the supply of soil N to plant roots is determined by plant available N (NH_4-N and NO_3-N), rate of water transpiration, and the diffusion of soil N to root surfaces (Shaffer et al., 2001). In the DSSAT model, the potential N supply from the soil profile (N_supply), which is the sum of nitrate and ammonium supplied from all soil layers, integrates the effect of rooting density, NH_4-N and NO_3-N concentrations, and soil water (Godwin and Singh, 1998). When N_supply for any given day is higher than crop N demand (N_demand), the ratio of N_demand to N_supply (less than 1) is used to reduce the N supply proportionally from each soil layer for NH_4-N and NO_3-N. The above normally occurs when plants are young (lower N demand) or after the application of N fertilizers (higher N supply). In legumes, when N_supply is less than N_demand, carbohydrates are metabolized to meet the crop N demand via N-fixation as dictated by nodule mass and species-defined nitrogenase activity (Boote et al., 2008). The N-fixation rate is influenced by temperature, soil water deficit, soil aeration (water logging), and plant reproductive age. For all crops, when N_supply (soil and/or N-fixation) is less than N_demand, vegetative tissues are grown at lower N concentrations. If such conditions persist, N deficiency symptoms arise. In most models, N deficiency results in LAI reduction, reduced photosynthesis (growth and yield reduction), and accelerated senescence. Prolonged N deficiency can also affect crop development (delayed vegetative development or accelerated maturity).

2.3.2 Nitrogen (N) demand

Crop N demand differs with the growth stage of a crop, as dictated by the growth of given organs (leaf blades, stalks, stems, ears/panicles/pods, and grains) and their critical tissue N concentration. As shown in Fig. 6, N uptake for hybrid maize changes over the duration of growth, with extremely high N demand from the tassel initiation (V6-8) to anthesis (R1) growth stages. For maximum use efficiency, the N supply from soils and fertilizers must match the crop N demand. N demand in most models is driven by the plant growth rate, growth stage, and tissue N status as a function of the growth stage (Godwin and Singh, 1998; Probert et al., 1998; Ahuja et al., 2000; Williams, 1995). Models differ in terms of N partitioning, ranging from a single plant N pool to shoot, root, and grain N pools to more complex models with leaf, roots, stem, ears (pods), and grain N pools.

The N status is generally expressed as the N concentration (weight basis) and content per unit area, for example, leaf N. A daily attempt is made to keep leaf, stem, root, and so on tissue N concentrations at a pre-defined critical

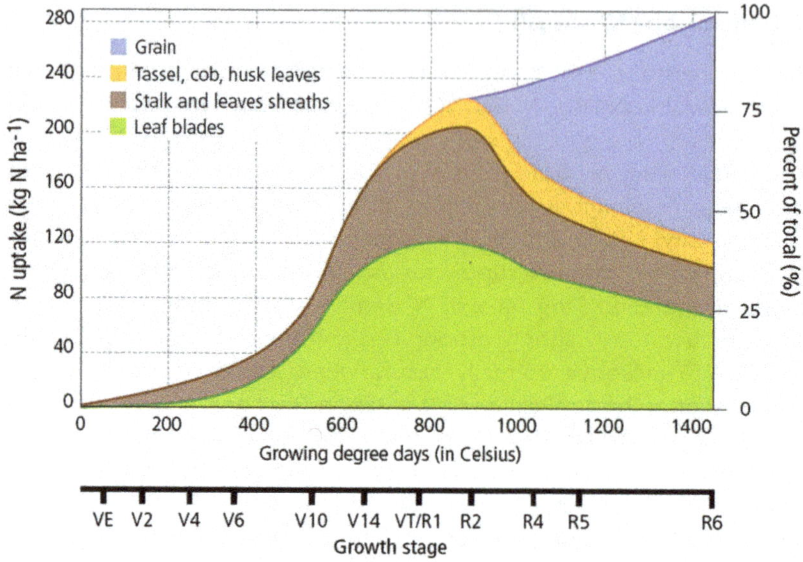

Figure 6 Nutrient uptake pattern of modern hybrid maize. Source: adapted from Bender et al. (2013).

concentration ($N_{critical}$). $N_{critical}$ is the lowest concentration at which maximum (non-stressed) growth occurs. As tissue N concentrations (N_{actual}) fall below $N_{critical}$ and growth processes are affected, while $N_{actual} > N_{critical}$ leads to luxury N consumption. The DSSAT N model also uses a minimum tissue N concentration (N_{min}), below which all growth ceases. $N_{critical}$ and N_{min} are crop specific and differ with growth stages, as illustrated for rice in Fig. 7. An index for N deficiency for a given organ on any given day (NFAC) can be determined based on the actual tissue N concentration compared to the pre-defined concentrations:

$$NFAC = 1.0 - \left[\left(N_{critical} - N_{actual}\right)/\left(N_{critical} - N_{min}\right)\right] \quad (9)$$

Since all plant growth processes are affected differently by N stress, several N stress indices have been estimated based on NFAC. These indices model the effect of N stress on leaf expansion, plant height, photosynthesis, senescence, crop development, and so on. For example, N stress can shorten the duration of grain-filling, which can further reduce grain yield (Singh et al., 1999).

Total crop N demand is driven by the N demand for new growth and the demand to overcome any deficiency in the existing biomass (organs). New growth N demand (New_N_demand) is determined from potential new growth (Potential_growth) and $N_{critical}$:

$$New\_N\_demand = Potential\_growth * N_{critical} \quad (10)$$

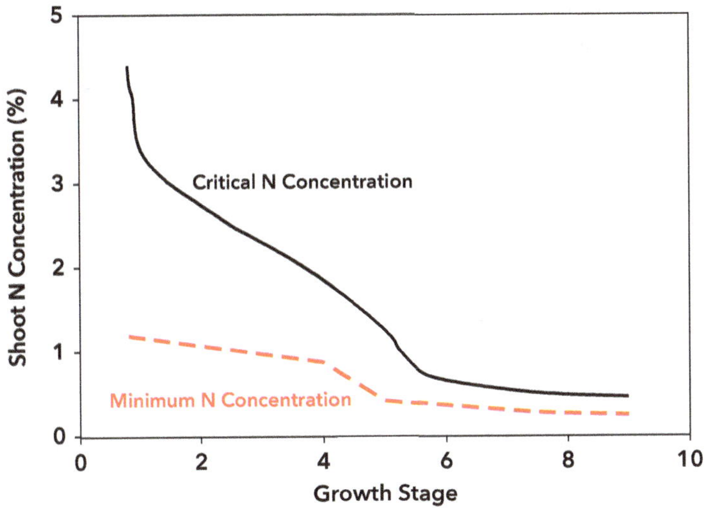

Figure 7 Changes in critical and minimum shoot N concentration in rice at given growth stages.

When New_N_demand is not met by N supply (N_supply) from soil and/or N-fixation, then mobilization of N can occur from older organs. Priority for use of mobilized N is grains > ears (shells) > vegetative organs. If New_N_demand is not met, potential growth will be reduced for that day while maintaining N concentration at $N_{critical}$. The N demand to overcome deficiency in any existing organ (DNDEM) is calculated for each organ as:

$$DNDEM = Mass_{organ} * \left(N_{critical} - N_{actual}\right) \tag{11}$$

Total crop N demand (N_demand) is a summation of all deficiency demands from various plant organs and new growth demand. Based on N_demand and N_supply, N uptake from each layer can be adjusted when N_supply > N_demand. However, when N_supply < N_demand, N deficiency exists, and a nitrogen stress effect is simulated.

3 Modeling phosphorus dynamics

Phosphorus is one of the most limiting nutrients for crop growth in tropical soils. Its availability to plants is strongly dictated by soil properties, environmental factors such as temperature and rainfall, and management. Management of P inputs and rationalizing P fertilizer recommendations are crucial for improving P use efficiency - which is currently at less than 20–30% for initial P application - and reducing the risk of P losses to water bodies. Total P fertilizer consumption in 2014–2015 was 20.2 million tons; 44% of all global P application occurred

Published by Burleigh Dodds Science Publishing Limited, 2026.

on cereals and 16% on oilseed crops (soybean, oil palm, etc.) (Heffer et al., 2017). Unlike N fertilizers, P is a nonrenewable resource that is dependent on a finite supply of phosphate rock (van Kauwenbergh, 2010; Cordell et al., 2009). P fertilizer recommendations based on extractable soil P are highly dependent on the method of extraction. These methods have been developed for specific soil types based on soil pH levels and the presence of calcium carbonate, iron (Fe), or aluminum (Al) oxides. Since P has a strong residual effect that is soil dependent, long-term experiments are required to understand and predict the depletion and build-up of soil P pools through continuous cultivation and fertilization (Johnston, 1997). Since increased P loads in water bodies can have detrimental effects on aquatic life, various P dynamics models are in use for simulating P transport and exchange in watersheds, aquatic sediments, wetlands, stormwater treatments, aquaponics, and so on (Basu et al., 2018; Dahl and Pers, 2004; Sinclair et al., 2016; Walker and Kadlec, 2011; Wang et al., 2003). Since soils are often the main source of agricultural P loss through runoff, the modeling of soil P processes, the dynamics within and among soil P pools, and the effect of abiotic and biotic factors on soil P pool size and their transformation will contribute toward both watershed applications and plant P nutrition. Dynamics of plant-available soil P can be modeled using the mass conservation approach (Morel et al., 2014). Process-based simulation models, such as APEX (Gassman et al., 2010), AnnAGNPS (Yuan et al., 2011), DSSAT (Dzotsi et al., 2010), and APSIM (Keating et al., 2003) can assess agricultural P loss, soil P dynamics, and plant P response.

3.1 Modeling soil phosphorus (P) processes

Quantifying soil P content and distribution among chemical forms requires soil P models to capture the complex effects and interactions of soil, environment, time, and management factors. In intensified agriculture, soil inorganic P (P_i) dominates the soil P dynamics; however, in low-input tropical agroecosystems, organic P (P_o) can constitute more than 35% of the total P in tropical soils (Abekoe and Tiessen, 1998). Hence, simulating the soil P dynamics that accounts for both P_i and P_o pools is important. Although plant roots absorb water-soluble P as orthophosphate ions P_i, $H_2PO_4^-$, and HPO_4^{2-}, almost 99% of absorbed P_i is derived from the soil solid phase (Morel et al., 2014). Several processes, including the diffusion of P_i at the solid-to-solution interface, the dissolution of phosphate minerals by acidification, soil buffering capacity, the mineralization of organic P compounds, and soil temperature and moisture, govern the P_i release from soil constituents to the soil solution.

Mineralization of P_o from various SOM pools follows a similar process as N mineralization, with the C:P ratio dictating whether mineralization or immobilization of P occurs for a given SOM pool and soil layer. The net P

Published by Burleigh Dodds Science Publishing Limited, 2026.

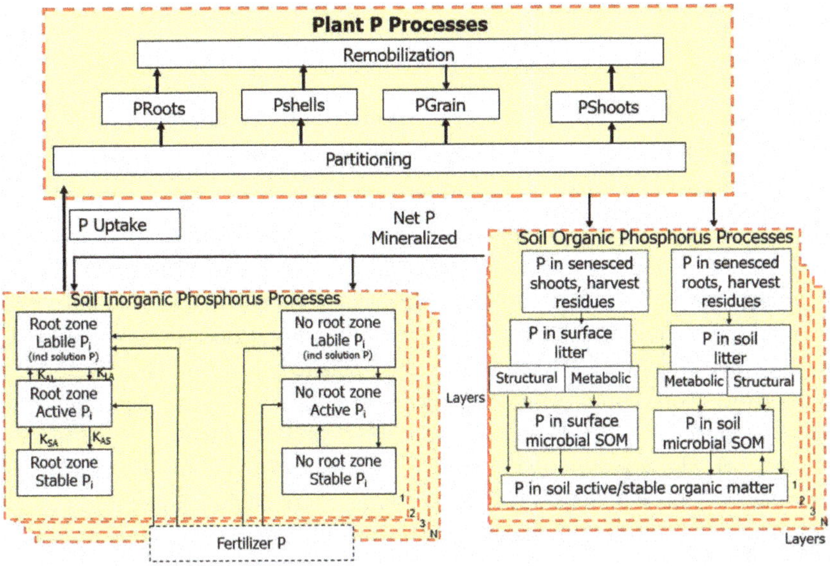

Figure 8 Generic soil and plant P model. The 'no-root-zone' pools are distant, for example, away from the effective P-absorbing radius of roots, but this pool shrinks as root length density increases.

mineralization for a given soil layer is determined from the difference of P mineralized and immobilized from different SOM pools; it is then added to the respective P_i labile pool layer (Fig. 8). The size and the dynamics of P in the inorganic and organic pools directly influence the concentration of P in the soil solution. The initialization of soil P model, both P_i and P_o pools, is perhaps the most critical input requirement because such data are rarely readily available or, if available, different extraction procedures are used. The choice of chemical extraction methods for soil P determination is dictated by soil properties. Empirical functions based on Jones et al. (1984), Sharpley et al. (1984, 1989), Singh (1985), and Vadas and White (2010) have been used to initialize P_i and P_o so that different extraction methods can be correlated to resin-extractable P. The fractionation of P_i in different pools (solution, labile, active, and stable) can be determined from the extraction of soil P with a concoction of different chemical extracts (Hedley et al., 1982). However, the fractionation procedure is complex and such data are scarce. Empirical functions based on the type of soil (calcareous, highly weathered, slightly weathered, volcanic) and its physical and chemical properties, including clay and organic matter content, have been used to derive solution, labile, active, and stable P_i pools, their respective transformation rate constants, and the proportion of sorbed and dissolved P (Fig. 8). The soil solution P pool thus captures the differences due to the above-mentioned four soil categories and the effect of soil water factor (SWF), texture (F_Texture), and organic carbon (F_OC) within each soil category:

$$\text{Soil Solution P}(\text{mg/kg}) = \text{P}_i\text{Labile} \times \text{PFraction} \times \text{F\_OC} \times \text{Min}(\text{F\_Texture, SWF}) \quad (12)$$

where PFraction is 0.05 for calcareous soils, 0.02 for slightly weathered soils, 0.015 for highly weathered soils, and 0.008 for volcanic soils (Dzotsi et al., 2010).

The solution P pool is subject to (i) uptake by plants, (ii) fixation by soil colloids, (iii) additions of P fertilizer, and (iv) mineralization from P_o. Since P mobility due to diffusion is restricted, P available for plant uptake is dependent on the simulation of the root 'zone'. Leaching and runoff P losses are generally modeled based on solution and labile P_i pools. A fraction (20–40%) of P leached from the top layer can be lost as runoff, while the remainder is added to the labile P_i pool in the second layer (Vadas et al., 2012). Depending on the complexity of P models, P leaching loss is simulated as a fraction from each layer or all from the bottom-most layer. P leaching losses are of concern in sandy soils.

In agricultural soils, the labile P_i pool size, particularly for the top 0–20 cm layers (depending on the tillage practice), is highly dependent on P fertilizer application. Most P models assume P fertilizers are water soluble and are immediately added to the labile and solution P_i pools. However, given the low P fertilizer use efficiency, potential leaching losses on sandy soils and conversion of applied P fertilizers to less-available active and stable P_i pools emphasize the need to explore opportunities for the direct application of phosphate rocks and other less-soluble P fertilizers (Smalberger et al., 2006).

3.2 Modeling crop phosphorus (P) processes

The supply of soil P for crop P uptake is highly dependent on the presence of roots; therefore, labile P_i, including solution P, should be within the vicinity (1–5 mm) of the roots. Unlike N, which is highly mobile, P supply is highly restricted to nearby roots. The magnitude of 'root zone' and 'non-root zone' P is modeled for each layer to determine the available P for root uptake (Fig. 8). As the root length volume per soil layer increases, the proportion of 'root zone' P pool is expected to increase and 100% of the soil volume may come in contact with the roots (Fig. 9). The potential P supply from a layer is determined as a function of labile P_i concentration and the effective root length volume (root length volume × root-fraction).

Modeling crop P demand is similar to crop N (as discussed previously). Daily demand (P_demand) is calculated for each plant part as the amount of P required to bring P concentration from the actual (P_{actual}) in each of the

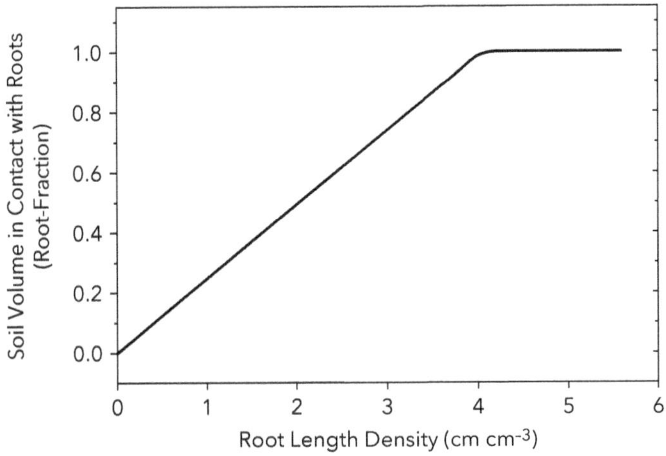

Figure 9 Fraction of soil volume in contact with roots as the function of root length volume (also dependent on time).

plant parts ($Mass_{organ}$) up to the critical ($P_{critical}$), plus P required for new growth (New_P_demand):

$$P\_demand = Mass_{organ} * \left(P_{critical} - P_{actual}\right) \quad (13)$$

$$New\_P\_demand = Potential\_growth * P_{critical} \quad (14)$$

The P demand is met through P soil supply and/or mobilization of plant P pools (Fig. 8). If the P soil supply after remobilization of P is less than P demand, then P concentrations in plant parts are reduced inducing P stress. P stress factors are determined based on similar principles as N stress. Models simulating both N and P uptake also compute the maximum and minimum plant N:P ratio on a daily basis; this information is then used to limit P uptake if on any day the actual N:P ratio is below the minimum.

Modeling soil and crop P dynamics is challenging given that there are different soil P pools; there is no standardized procedure for 'identifying" and "quantifying' these pools, and there are soil P chemical forms that are highly specific to the soil type. P uptake is challenged by ensuring that the root model is reliable because P uptake occurs only within a very close proximity of the roots. Simulating the effects of P stress also has its challenges, since P stress has a strong impact on crop duration; P stress generally prolongs the duration. Depending on the severity of P deficiency, leaf appearance rate and crop development rate are reduced, resulting in anthesis delay by up to a month (Singh, 2008). If the impact on delay is not considered, P-stressed plants may compensate, and economic yields may not be different from unstressed plants with a normal (shorter) duration.

4 Modeling potassium dynamics

Potassium (K), just like N and P, is a macronutrient with important functions for crop growth. It activates over 60 enzymes, promotes photosynthesis, regulates stomatal opening, improves N utilization, promotes the transport of assimilates, and increases growth and crop yield. Total K fertilizer consumption in 2014–2015 was 27.8 million tons, and, unlike N and P, almost 23% of K was applied to oilseed crops, and 12.2% was applied to soybean plants (Heffer et al., 2017). Global K application on cereals was at 36% and 14%, 11%, and 7% on maize, rice, and wheat, respectively. K ions are mobile and, therefore, can be leached along the soil profile. Simulation models have been important research tools in studies involving complex and interactive processes of water flow and K transport through the soil profile as well as the effects of management practices on agricultural production and the environment. The high solubility of K has contributed to many studies simulating the movement of water and solutes (Beckett, 1964a,b; Adiscott and Talibudeen, 1969; Barber, 1985, 1995; Frissel, 1978; Dessougi and Claassen, 2001; Šimůnek et al., 2008; Zeng and Brown, 2000). In addition to K solute movement, modeling the root effects on K uptake (Brüggemann et al., 1999; Ahn et al., 2004; Reintanz et al., 2002; Samal et al., 2010; Zhang et al., 2007; Trehan and Claassen, 1998; Dessougi et al., 2002; Steingrobe and Claassen, 2000; Rengel and Damon, 2008) and release characteristics has seen major efforts (Hosseinpur and Motaghian, 2013; Springob and Richter, 1998; Moritsuka et al., 2004).

Most soil K and plant K models are generally focused on specific research or management tasks, for example, solute transport, water quality/waste management, and rhizosphere interactions. Such models, as are cited earlier, simulate K uptake at roots and root hairs level; the nutrient uptake into the root is dependent on the nutrient concentration in the soil solution at the root surface and can be described by a modified Michaelis-Menten equation. The transport of nutrients to roots in the solution phase is modeled by mass flow and diffusion (Barber, 1985). The release of nutrients from the soil solid phase into the solution phase is governed by sorption and desorption processes. The quantity/intensity (Q/I) ratio of potassium has been used to model soil K status. In this approach, immediately available K is related to the intensity factor, the reserve of the non-exchangeable K is referred to as the quantity factor, and the renewal capacity is applied to the buffering capacity (Evangelou et al., 1994; Roy et al., 1991; Hamdan et al., 1999; Wang et al., 2004). Soil properties such as CEC, SOM content, pH, and K mineralogy affect K release, K buffering capacity, and K-fixation characteristics.

Comprehensive soil-plant K models have been developed to capture the effect of K management on crop growth, plant K, and soil K (Greenwood and Karpinets, 1997; Zhang et al., 2007; Koós and Kovács, 2008; Scanlan

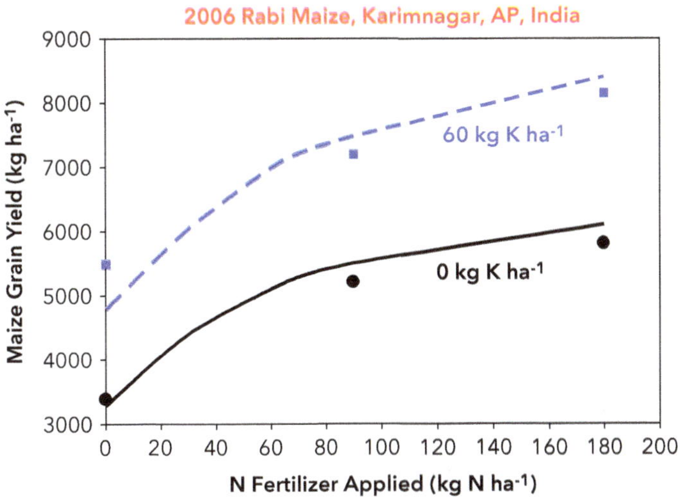

Figure 10 Comparison of simulated and observed maize grain yield response to N and K.

et al., 2015a,b; Singh et al., 2007). Both DSSAT (Singh et al., 2007) and APSIM (Scanlan et al., 2015a) models utilize existing soil-plant-environment modules of crop growth and development, soil and plant water balance, soil and carbon dynamics, and couplings of these processes to simulate the effect of K. The DSSAT-K model has been evaluated for maize, rice, and cotton, and the APSIM-K model has been evaluated for wheat. The plant K uptake in these models is driven by the soil K supply and crop K demand as previously described for plant N and P models. The uptake of K is also dependent on the equilibrium between root and soil solution K concentration. In contrast to slower P mobility and greater rooting dependence in simulating P uptake, K mobility occurs up to twice the distance from roots compared to P. The photosynthetic assimilation rate and water-use efficiency are modified according to shoot K concentration. Crop responses have been evaluated on a few experiments with well-characterized sites (Fig. 10). Additional research to capture the effect of K fertilization on plant abiotic (low- and high-temperature tolerance and drought) and biotic stresses will help highlight some of the key roles of K nutrition.

Generic soil K models, as shown in Fig. 11, account for the flow of K from fertilizers, crop residues, K-fixing soils, and weathering of minerals to exchangeable and solution K pools. The outflows from these 'available' pools include plant K uptake, K leaching loss, and K-fixation. Model input for the initialization of soil K pools and transport coefficients is not as challenging as for the soil P model. However, more experiments conducted on a wide range of soils that are also well characterized will help expand pedotransfer functions

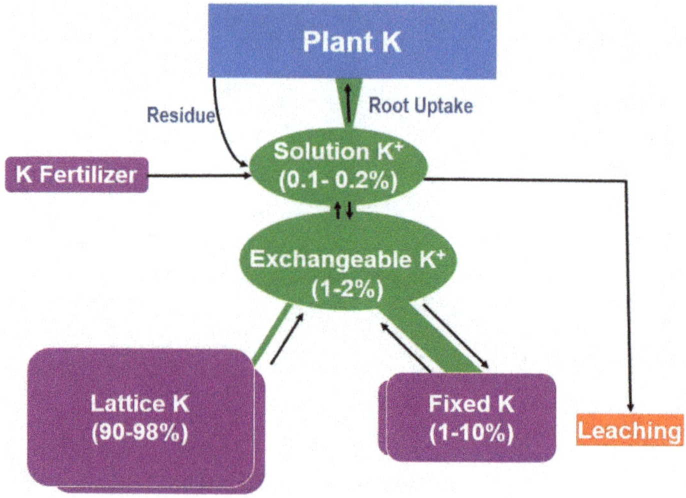

Figure 11 Generic soil K model.

and their utility for quantifying K pools and transfer coefficients without the need for complex data to run K models.

5 Conclusion

The ability to model soil and plant nutrient dynamics is an important component of research involving agricultural systems that are sustainable while addressing the current challenges of scarce resources, climate change and variability, increasing population, and environmental degradation.

The current capability of many agricultural and soil models to predict nitrogen dynamics is well established and generally robust. Inorganic N dynamics is routinely quantified with predictability when compared to measured data. These models are ideally suited for the study of fertilizer use efficiency and resource optimization. Modern fertilizers that allow for the controlled and slow release of N, fertilizer deep placement, and inhibitors for urease and nitrification processes are a relatively new addition to modeling capabilities, but they are increasingly being addressed by the agricultural modeling community.

Agricultural production models that include nutrient dynamics and chemical transport are used to study environmental quality, including N, P, and K leaching into shallow groundwaters. The ability of models to predict the release of GHGs from crop production systems is an important contribution to climate change research and the development of climate change mitigation options.

Published by Burleigh Dodds Science Publishing Limited, 2026.

Organic C and N dynamics are also well characterized in many current models. Unfortunately, challenges related to the initialization of soil organic C pools are often a problem due to a lack of data. However, the main deficiency in current models is in handling other nutrients, including sulfur (S), calcium (Ca), magnesium (Mg), iron (Fe), zinc (Zn), copper (Cu), manganese (Mn), boron (B), molybdenum (Mo), cobalt (Co), and nickel (Ni). The main reason for this deficiency is a lack of data to both develop models for nutrient transformation, solubility, and transport in the soil and to quantify the plant response to nutrient deficiency. Perhaps the most limiting factor is the information on soil properties that drive the nutrient transformation processes and their interactions. For those models that do exist, the lack of data from agronomic studies to parameterize and initialize the models limits the usefulness of the models.

6 Where to look for further information

The use and development of nutrient dynamics models in research has been primarily promoted through training programs, particularly in the developing world. Hence in the developing world, and specifically in Sub-Saharan Africa emphasis should be placed on the inclusion of agricultural systems modeling in the regular curricula of the universities, understanding the limitations of models, and using models in generating ideas and insights, instead of uncritically using model results. The existing nutrient modeling effort can be substantially improved by conducting more research and model evaluation under adverse environmental and soil conditions. The various focus crops and themes within the Agricultural Model Intercomparison and Improvement Project (AgMIP) have challenged modelers to evaluate crop and nutrient dynamics model under a wide range of conditions. AgMIP (https://agmip.org/) also provides useful information and contacts for the existing models.

7 References

Abalos, D., Jeffery, S., Sanz-Cobena, A., Guardia, G. and Vallejo, A. 2014. Meta-analysis of the effect of urease and nitrification inhibitors on crop productivity and nitrogen use efficiency. *Agric. Ecosyst. Environ.* 189, 136–44. doi:10.1016/j.agee.2014.03.036.

Abekoe, M. K. and Tiessen, H. 1998. Phosphorus forms, lateritic nodules and soil properties along a hillslope in Northern Ghana. *Catena* 33(1), 1–15. doi:10.1016/S0341-8162(98)00063-0.

Adiscott, T. M. and Talibudeen, O. 1969. The buffering capacity of potassium reserves in soils. *Potash Rev.* 4, 45.

Aggarwal, P. K., Kalra, N., Chander, S. and Pathak, H. 2006. InfoCrop: a dynamic simulation model for the assessment of crop yields losses due to pests and environmental impact of agro-ecosystems in tropical environments. I. Model description. *Agric. Syst.* 89(1), 1–25. doi:10.1016/j.agsy.2005.08.001.

Ahn, S. J., Ryoung, S. and Daniel, P. S. 2004. Expression of KT/KUP genes in *Arabidopsis* and the role of root hairs in $K^{[+]}$ uptake. *Plant Physiol.* 134, 1135–45. doi:10.1104/pp.103.034660.

Ahuja, L. R., Rojas, K. W., Hanson, J. D., Shaffer, M. J. and Ma, L. 2000. *Root Zone Water Quality Model: Modeling Management Effects on Water Quality and Crop Production*. Water Resources Publications, Littleton, CO.

Angle, J. S., Singh, U., Dimkpa, C. O., Bindraban, P. S. and Hellums, D. T. 2017. Role of fertilizers for climate-resilient agriculture. Proceedings no. 802. International Fertiliser Society, London, UK. Available at: http://fertiliser-society.org/societyproceedings/proceedings-802/c-23/p-1022.

Antonopoulos, V. Z. 2010. Modelling of water and nitrogen balances in the ponded water and soil profile of rice fields in Northern Greece. *Agric. Water Manag.* 98(2), 321–30. doi:10.1016/j.agwat.2010.08.026.

Barber, S. A. 1985. Potassium availability at the soil-root interface and factors influencing potassium uptake. In: Munson, R. D. (Ed.), *Potassium in Agriculture*. ASA-CSSA-SSSA, Madison, WI, pp. 309–24.

Barber, S. A. 1995. *Soil Nutrient Bioavailability: Mechanistic Approach* (2nd edn.). Wiley, New York, p. 398.

Barber, S. A. and Cushman, J. 1981. Nitrogen uptake model for agronomic crops. In: Iskandar, I. R. (Ed.), *Modeling Wastewater Renovation-Land Treatment*. Wiley-Interscience, New York, pp. 382–409.

Basso, B., Cammarano, D., Troccoli, A., Chen, D. and Ritchie, J. T. 2010. Long-term wheat response to nitrogen in a rainfed Mediterranean environment: field data and simulation analysis. *Eur. J. Agron.* 33(2), 132–8. doi:10.1016/j.eja.2010.04.004.

Basso, B., Gargiulo, O., Paustian, K., Robertson, G. P., Porter, C., Grace, P. R. and Jones, J. W. 2011. Procedures for initializing soil organic carbon pools in the DSSAT-Century model for agricultural systems. *Soil Sci. Soc. Am. J.* 75(1), 69–78. doi:10.2136/sssaj2010.0115.

Basso, B., Liu, L. and Ritchie, J. T. 2016. A comprehensive review of the CERES-Wheat, - Maize and -Rice models' performances. *Adv. Agron.* 136, 27–132. doi:10.1016/bs.agron.2015.11.004.

Basu, N. B., Van Meter, K. J. and Van Cappellen, P. 2018. Beyond the mass balance: modeling legacy phosphorus dynamics in a Great Lakes Watershed. Abstract presented at *American Geophysical Union, Fall Meeting*.

Beckett, P. H. T. 1964a. Studies on soil K: (a) confirmation of the ratio law: measurement of K potential. *J. Soil Sci.* 5, 1–8.

Beckett, P. H. T. 1964b. Studies on soil K: (b) the immediate Q/I relations of labile K in the soil. *J. Soil Sci.* 15(1), 9–23. doi:10.1111/j.1365-2389.1964.tb00240.x.

Bender, R., Haegele, J., Ruffo, M. and Below, F. 2013. Modern corn hybrids' nutrient uptake patterns. *Better Crop* 97(1), 7–10.

Berntsen, J., Petersen, B. M., Jacobsen, B. H., Olesen, J. E. and Hutchings, N. J. 2003. Evaluating nitrogen taxation scenarios using the dynamic whole farm simulation model FASSET. *Agric. Syst.* 76(3), 817–39. doi:10.1016/S0308-521X(02)00111-7.

Biernath, C., Gayler, S., Bittner, S., Klein, C., Högy, P., Fangmeier, A. and Priesack, E. 2011. Evaluating the ability of four crop models to predict different environmental impacts on spring wheat grown in open-top chambers. *Eur. J. Agron.* 35(2), 71-82. doi:10.1016/j.eja.2011.04.001.

Boote, K. J., Hoogenboom, G., Jones, J. W. and Ingram, K. T. 2008. Modeling N-fixation and its relationship to N uptake in the CROPGRO model. In: Ma, L., Ahuja, L. and Bruulsema, T. (Eds), *Quantifying and Understanding Plant Nitrogen Uptake for Systems Modeling*. Taylor & Francis Group LLC, Boca Raton, FL.

Brisson, N., Gary, C., Justes, E., Roche, R., Mary, B., Ripoche, D., Zimmer, D., Sierra, J., Bertuzzi, P., Burger, P., Bussière, F., Cabidoche, Y. M., Cellier, P., Debaeke, P., Gaudillère, J. P., Hènault, C., Maraux, F., Seguin, B. and Sinoquet, H. 2003. An overview of the crop model STICS. *Eur. J. Agron.* 18(3-4), 309-32. doi:10.1016/S1161-0301(02)00110-7.

Brüggemann, L., Dietrick, P., Becker, D., Dreyer, I., Palme, K. and Hedrich, R. 1999. Channel-mediated high-affinity K^+ uptake into guard cells from *Arabidopsis*. *Proc. Natl Acad. Sci. U. S. A.* 96(6), 3298-302. doi:10.1073/pnas.96.6.3298.

Chowdary, V. M., Rao, N. H. and Sarma, P. B. S. 2004. A coupled soil water and nitrogen balance model for flooded rice fields in India. *Agric. Ecosyst. Environ.* 103(3), 425-41. doi:10.1016/j.agee.2003.12.001.

Claassen, N. and Steingrobe, B. 1999. Mechanistic simulation models for a better understanding of nutrient uptake from soil. In: Rengel, Z. (Ed.), *Mineral Nutrition of Crops – Fundamental Mechanisms and Implications*. Haworth, New York, pp. 327-67.

Cordell, D., Drangert, J. O. and White, S. 2009. The story of phosphorus: global food security and food for thought. *Glob. Environ. Change* 19(2), 292-305. doi:10.1016/j.gloenvcha.2008.10.009.

Dahl, M. and Pers, B. C. 2004. Comparison of four models simulating phosphorus dynamics in Lake Vänern, Sweden. *Hydrol. Earth Syst. Sci.* 8(6), 1153-63. doi:10.5194/hess-8-1153-2004.

Denmead, O., Freney, R., Simpson, R. and Simpson, J. 1982. Dynamics of ammonia volatilization during furrow irrigation of maize. *Soil Sci. Soc. Am. J.* 46(1). doi:10.2136/sssaj1982.03615995004600010029x.

Dessougi, H. I. and Claassen, N. 2001. Use of a mechanistic model to predict potassium uptake in the field by sugar beet and wheat and its depletion in the rhizosphere. In: Horst, W. J. Schenk, M. K., Bürkert, A., Claassen, N., Flessa, H., Frommer, W. B., Goldbach, Olfs, H.-W., Römheld, V., Sattelmacher, B., Schmidhalter, U., Schubert, S., Wirén, N. v. and Wittenmayer, L. (Eds), *Plant Nutrition, Developments in Plant and Soil Sciences* (vol. 92). Springer, Dordrecht, pp. 582-3.

Dessougi, H. I., Claassen, N. and Steingrobe, B. 2002. Potassium efficiency mechanisms of wheat, barley, and sugar beet grown on a K fixing soil under controlled conditions. *J. Plant Nutr. Soil. Sci.* 165(6), 732-7. doi:10.1002/jpln.200290011.

Dzotsi, K. A., Jones, J. W., Adiku, S. G. K., Naab, J. B., Singh, U., Porter, C. H. and Gijsman, A. J. 2010. Modeling soil and plant phosphorus within DSSAT. *Ecol. Modell.* 221(23), 2839-49. doi:10.1016/j.ecolmodel.2010.08.023.

Erisman, J. W., Sutton, M. A., Galloway, J., Klimont, Z. and Winiwarter, W. 2008. How a century of ammonia synthesis changed the world. *Nat. Geosci.* 1(10), 636-9. doi:10.1038/ngeo325.

Evangelou, V. P., Wang, J. and Phillips, R. E. 1994. New developments and perspectives on soil potassium quantity/intensity relationships. *Adv. Agron.* 52, 173-227. doi:10.1016/S0065-2113(08)60624-0.

Published by Burleigh Dodds Science Publishing Limited, 2026.

Feng, J., Li, F., Deng, A., Feng, X., Fang, F. and Zhang, W. 2016. Integrated assessment of the impact of enhanced-efficiency nitrogen fertilizer on N_2O emission and crop yield. *Agric. Ecosyst. Environ.* 231, 218–28. doi:10.1016/j.agee.2016.06.038.

Food and Agriculture Organization (FAO) of the United Nations. 2019. *The International Code of Conduct for the Sustainable Use and Management of Fertilizers*. FAO, Rome, Italy.

Frissel, M. J. 1978. *Cycling of Mineral Nutrients in Agricultural Ecosystems*. Elsevier, Amsterdam, Netherlands.

Galloway, J. N., Townsend, A. R., Erisman, J. W., Bekunda, M., Cai, Z., Freney, J. R., Marinelli, L. A., Seitzinger, S. P. and Sutton, M. A. 2008. Transformation of the nitrogen cycle: recent trends, questions, and potential solutions. *Science* 320(5878), 889–92.

Gassman, P. W., Williams, J. R., Wang, X., Saleh, A., Osei, E., Hauck, L. M., Izaurralde, R. C. and Flowers, J. D. 2010. The Agricultural Policy/Environmental Extender (Apex) Model: an emerging tool for landscape and watershed environmental analyses. *Trans. ASABE* 53, 711–40.

Gaydon, D. S., Probert, M. E., Buresh, R. J., Meinke, H. and Timsina, J. 2012. Modelling the role of algae in rice crop nutrition and soil organic carbon maintenance. *Eur. J. Agron.* 39, 35–43. doi:10.1016/j.eja.2012.01.004.

Gaydon, D. S., Balwinder-Singh, Wang, E., Poulton, P. L., Ahmad, B., Ahmed, F., Akhter, S., Ali, I., Amarasingha, R., Chaki, A. K., Chen, C., Choudhury, B. U., Darai, R., Das, A., Hochman, Z., Horan, H., Hosang, E. Y., Kumar, P. V., Khan, A. S. M. M. R., Laing, A. M., Liu, L., Malaviachichi, M. A. P. W. K., Mohapatra, K. P., Muttaleb, M. A., Power, B., Radanielson, A. M., Rai, G. S., Rashid, M. H., Rathanayake, W. M. U. K., Sarker, M. M. R., Sena, D. R., Shamim, M., Subash, N., Suriadi, A., Suriyagoda, L. D. B., Wang, G., Wang, J., Yadav, R. K. and Roth, C. H. 2017. Evaluation of the APSIM model in cropping systems of Asia. *Field Crop. Res.* 204, 52–75. doi:10.1016/j.fcr.2016.12.015.

Godwin, D. C. and Singh, U. 1998. Nitrogen balance and crop response to nitrogen in upland and lowland cropping systems. In: Tsuji, G. Y., Hoogenboom, G. and Thornton, P. K. (Eds), *Understanding Options for Agricultural Production, Systems Approaches for Sustainable Agricultural Development* (vol. 7). Springer, Dordrecht, pp. 55–77. doi:10.1007/978-94-017-3624-4_4.

Greenwood, D. J. and Karpinets, T. V. 1997. Dynamic model for the effects of K-fertilizer on crop growth, K-uptake and soil-K in arable cropping. 1. Description of the model. *Soil Use Manag.* 13(4), 178–83. doi:10.1111/j.1475-2743.1997.tb00582.x.

Hamdan, J., Burnham, C. P. and Ruhana, B. 1999. Evaluation of quantity-intensity of potassium in deeply weathered soil profile develops over granite from relationships Peninsular Malaysia. *Commun. Soil. Sci. Plant Anal.* 30(17–18), 2311–21. doi:10.1080/00103629909370375.

Hansen, S., Thirup, C., Refsgaard, J. C. and Jensen, L. S. 2001. Modeling nitrate leaching at different scales – application of the DAISY model. In: Shaffer, M. J., Ma, L. and Hansen, S. (Eds), *Modeling Carbon and Nitrogen Dynamics for Soil Management*. Lewis Publishers, Boca Raton, FL, pp. 511–48.

Hedley, M. J., Stewart, J. W. B. and Chauhan, B. S. 1982. Changes in inorganic soil phosphorus fractions induced by cultivation practices and by laboratory incubations. *Soil Sci. Soc. Am. J.* 46(5), 970–6. doi:10.2136/sssaj1982.03615995004600050017x.

Heffer, P., Gruère, A. and Roberts, T. 2017. *Assessment of Fertilizer Use by Crop at the Global Level*. International Fertilizer Association and International Plant Nutrition Institute, Paris, France.

Hosseinpur, A. R. and Motaghian, H. R. 2013. Application of kinetic models in describing soil potassium release characteristics and their correlations with potassium extracted by chemical methods. *Pedosphere* 23(4), 482-92. doi:10.1016/S1002-0160(13)60041-7.

International Fertilizer Development Center. 2017. *Rapid Introduction and Market Development for Urea Deep Placement Technology for Lowland Transplanted Rice: A Reference Guide*. International Fertilizer Development Center, Muscle Shoals, AL.

Jamieson, P. D., Semenov, M. A., Brooking, I. R. and Francis, G. S. 1998. Sirius: a mechanistic model of wheat response to environmental variation. *Eur. J. Agron.* 8(3-4), 161-79. doi:10.1016/S1161-0301(98)00020-3.

Jing, Q., Van Keulen, Hv and Hengsdijk, H. 2010. Modeling biomass, nitrogen and water dynamics in rice-wheat rotations. *Agric. Syst.* 103(7), 433-43. doi:10.1016/j.agsy.2010.04.001.

Johnston, A. E. 1997. The value of long-term field experiments in agricultural, ecological, and environmental research. *Adv. Agron.* 59, 291-333. doi:10.1016/S0065-2113(08)60057-7.

Jones, C. A., Cole, C. V., Sharpley, A. N. and Williams, J. R. 1984. A simplified soil and plant phosphorus model: 1. Documentation. *Soil Sci. Soc. Am. J.* 48(4), 800-5. doi:10.2136/sssaj1984.03615995004800040020x.

Jongschaap, R. E. E. 2007. Sensitivity of a crop growth simulation model to variation in LAI and canopy nitrogen used for run-time calibration. *Ecol. Modell.* 200(1-2), 89-98. doi:10.1016/j.ecolmodel.2006.07.015.

Keating, B. A., Carberry, P. S., Hammer, G. L., Probert, M. E., Robertson, M. J., Holzworth, D., Huth, N. I., Hargreaves, J. N. G., Meinke, H., Hochman, Z., McLean, G., Verburg, K., Snow, V., Dimes, J. P., Silburn, M., Wang, E., Brown, S., Bristow, K. L., Asseng, S., Chapman, S., McCown, R. L., Freebairn, D. M. and Smith, C. J. 2003. An overview of APSIM, a model designed for farming systems simulation. *Eur. J. Agron.* 18(3-4), 267-88. doi:10.1016/S1161-0301(02)00108-9.

Kelly, R. H., Parton, W. J., Hartman, M. D., Stretch, L. K., Ojima, D. S. and Schimel, D. S. 2000. Intra- and interannual variability of ecosystem processes in shortgrass steppe: new model, verification, simulations. *J. Geophys. Res. Atmos.* 105, 93-100.

Kersebaum, K. C. 2007. Modelling nitrogen dynamics in soil-crop systems with HERMES. *Nutr. Cycl. Agroecosys* 77(1), 39-52. doi:10.1007/s10705-006-9044-8.

Kim, S.-H., Yang, Y., Timlin, D. J., Fleisher, D. H., Dathe, A., Reddy, V. R. and Staver, K. 2012. Modeling temperature responses of leaf growth, development, and biomass in maize with MAIZSIM. *Agron. J.* 104(6), 1523-37. doi:10.2134/agronj2011.0321.

Koós, S. and Kovács, G. J. 2008. Modelling of seasonal dynamics of nitrogen, phosphorus and potassium in the rooted zone. *Cereal Res. Commun.* 36, 635-8. Supplement: Proceedings of the VII. Alps-Adria Scientific Workshop, 28 April - 2 May 2008, Stara Lesna, Slovakia (June 2008).

Ladha, J. K., Pathak, H., Krupnik, T. J., Six, J. and van Kessel, C. 2005. Efficiency of fertilizer nitrogen in cereal production: retrospects and prospects. *Adv. Agron.* 87, 85-156. doi:10.1016/S0065-2113(05)87003-8.

Li, C., Frolking, S. E., Harris, R. C. and Terry, R. E. 1994. Modelling nitrous oxide emissions from agriculture: a Florida case study. *Chemosphere* 28(7), 1401–15. doi:10.1016/0045-6535(94)90081-7.

Li, C., Mosier, A., Wassman, R., Cai, Z., Zheng, X., Huang, Y., Tsuruta, H., Boonjawat, J. and Lantin, R. 2004. Modeling greenhouse gas emissions from rice-based production systems: sensitivity and scaling. *Glob. Biogeochem. Cycl.* 18(GB1043), 1-19.

Liang, X. Q., Chen, Y. X., Li, H., Tian, G. M., Ni, W. Z., He, M. M. and Zhang, Z. J. 2007. Modeling transport and fate of nitrogen from urea applied to a near-trench paddy field. *Environ. Pollut.* 150(3), 313–20. doi:10.1016/j.envpol.2007.02.003.

Linquist, B. A., Liu, L., van Kessel, C. and van Groenigen, K. J. 2013 Enhanced efficiency nitrogen fertilizers for rice systems: meta-analysis of yield and nitrogen uptake. *Field Crop. Res.* 154, 246–54. doi:10.1016/j.fcr.2013.08.014.

Ma, L. and Shaffer, M. J. 2001. A review of carbon and nitrogen processes in nine U.S. soil nitrogen dynamics models. In: Shaffer, M. J., Ma, L. and Hansen, S. (Eds), *Modeling Carbon and Nitrogen Dynamics for Soil Management*. Lewis Publishers, Boca Raton, FL, pp. 55–102.

Ma, L., Ahuja, L. R. and Bruulsema, T. W. (Eds). 2008. *Quantifying and Understanding Plant Nitrogen Uptake for Systems Modeling*. Taylor and Francis Group LLC, Boca Raton, FL, 320pp.

Matthews, R. B., Wassmann, R. and Arah, J. 2000. Using a crop/soil simulation model and GIS techniques to assess methane emissions from rice fields in Asia. I. Model development. *Nutr. Cycl. Agroecosyst.* 58(1/3), 141–59. doi:10.1023/A:1009894619446.

McGechan, M. B. and Wu, L. 2001. A review of carbon and nitrogen processes in European soil nitrogen dynamics models. In: Shaffer, M. J., Ma, L. and Hansen, S. (Eds), *Modeling Carbon and Nitrogen Dynamics for Soil Management*, Lewis Publishers, Boca Raton, FL, pp. 103–72.

Morel, C., Ziadi, N., Messiga, A., Belanger, G., Denoroy, P., Jeangros, B., Jouany, C., Fardeau, J., Mollier, A., Parent, L., Proix, N., Rabeharisoa, L. and Sinaj, S. 2014. Modeling of phosphorus dynamics in contrasting agroecosystems using long-term field experiments. *Can. J. Soil Sci.* 94(3), 377–87. doi:10.4141/cjss2013-024.

Moritsuka, N., Yanai, J. and Kosaki, T. 2004. Possible processes releasing nonexchangeable K from the rhizosphere of maize. *Plant Soil* 258(1), 261–8. doi:10.1023/B:PLSO.0000016556.79278.7f.

Myhre, G., Shindell, D., Bréon, F.-M., Collins, W., Fuglestvedt, J., Huang, J., Koch, D., Lamarque, J.-F., Lee, D., Mendoza, B., Nakajima, T., Robock, A., Stephens, G., Takemura, T. and Zhang, H. 2013. Anthropogenic and natural radiative forcing. In: Stocker, T. F., Qin, D., Plattner, G. K., Tignor, M., Allen, S. K., Boschung, J., Nauels, A., Xia, Y., Bex, V. and Midgley, P. M. (Eds), *Climate Change 2013: the Physical Science Basis. Contribution of Working Group I to the Fifth Assessment Report of the Intergovernmental Panel on Climate Change*. Cambridge University Press, Cambridge, UK and New York, NY.

Nieder, R. and Benbi, D. K. 2008. Modeling carbon and nitrogen dynamics in the soil-plant-atmosphere system. In: *Carbon and Nitrogen in the Terrestrial Environment*. Springer, Dordrecht, pp. 307–42.

Nendel, C., Berg, M., Kersebaum, K. C., Mirschel, W., Specka, X., Wegehenkel, M., Wenkel, K. O. and Wieland, R. 2011. The MONICA model: testing predictability for crop growth, soil moisture and nitrogen dynamics. *Ecol. Modell.* 222(9), 1614–25. doi:10.1016/j.ecolmodel.2011.02.018.

Published by Burleigh Dodds Science Publishing Limited, 2026.

Pathak, J. 2004. Methods to reduce soil erosion and nutrient losses in Kavre district, central Nepal. *Himal. J. Sci.* 2(3), 42–5.

Probert, M. E., Dimes, J. P., Keathing, B. A., Dalal, R. C. and Strong, W. M. 1998. APSIM's water and nitrogen modules and simulation of the dynamics of water and nitrogen in fallow systems. *Agric. Sys.* 56(1), 1-28. doi:10.1016/S0308-521X(97)00028-0.

Reddy, K. R., Rao, P. S. C. and Jessup, R. E. 1990. Transformation and transport of ammonium nitrogen in a flooded organic soil. *Ecol. Modell.* 51(3-4), 205–16. doi:10.1016/0304-3800(90)90065-O.

Reintanz, B., Szyoki, A., Ivashikina, N., Ache, P., Godde, M., Becker, D., Palme, K. and Hedrich, R. 2002. AtKC1, a silent *Arabidopsis* potassium channel (x-subunit modulates root hair K^+ influx). *Proc. Natl Acad. Sci. U. S. A.* 99(6), 4079–84.

Rengel, Z. and Damon, P. N. 2008. Crops and genotypes differ in efficiency of potassium uptake and use. *Physiol. Plant.* 133(4), 624–36. doi:10.1111/j.1399-3054.2008.01079.

Rolston, D. E., Sharpley, A. N., Toy, D. W., Hoffman, D. L. and Broadbent, F. E. 1980. Denitrification as affected by irrigation frequency of a field soil. EPA -600/2-80-06. United States Environmental Protection Agency, ADA, Ada, OK.

Roy, H. K., Kumar, A. and Sarkar, A. K. 1991. Quantity/intensity relations of potassium in a representative acid sedentary. *J. Indian Soc. Soil Sci.* 39, 175–7.

Samal, D., Kovar, J. L., Steingrobe, B., Sadana, U. S., Bhadoria, P. S. and Claassen, N. 2010. Potassium uptake efficiency and dynamics in the rhizosphere of maize (*Zea mays* L.), wheat (*Triticum aestivum* L.), and sugar beet (*Beta vulgaris* L.) evaluated with a mechanistic model. *Plant Soil* 332(1–2), 105–21. doi:10.1007/s11104-009-0277-6.

Scanlan, C. A., Bell, R. W. and Brennan, R. F. 2015a. Simulating wheat growth response to potassium availability under field conditions in sandy soils. I. Model development. *Field Crop. Res.* 178, 125–34. doi:10.1016/j.fcr.2015.03.019.

Scanlan, C. A., Bell, R. W. and Brennan, R. F. 2015b. Simulating wheat growth response to potassium availability under field conditions in sandy soils. II. Effect of subsurface potassium on grain yield response to potassium fertilizer. *Field Crop. Res.* 178, 125–34. doi:10.1016/j.fcr.2015.03.019.

Shaffer, M. J., Ma, L. and Hansen, S. (Eds). 2001. *Modeling Carbon and Nitrogen Dynamics for Soil Management*. Lewis Publishers, Boca Raton, FL.

Sharpley, A. N., Jones, C. A., Gray, C. and Cole, C. V. 1984. A simplified soil and plant phosphorus model: 2. Prediction of labile, organic, and sorbed phosphorus. *Soil Sci. Soc. Am. J.* 48(4), 805–9. doi:10.2136/sssaj1984.03615995004800040021x.

Sharpley, A. N., Singh, U., Uehara, G. and Kimble, J. 1989. Modeling soil and plant phosphorus dynamics in calcareous and highly weathered soils. *Soil Sci. Soc. Ami J.* 53(1), 153–8. doi:10.2136/sssaj1989.03615995005300010029x.

Šimůnek, J., van Genuchten, M. T. H. and Sejna, M. 2008. Development and applications of the Hydrus and STANMOD software packages and related codes. *Vadose Zone J.* 7(2), 587-600. doi:10.2136/vzj2007.0077.

Sinclair, A. C., Rob, C., Jamieson, R. J., Gordon, A. M., Hart, W. and Hebb, D. 2016. Modeling impacts of residential and agricultural development and beneficial management practice scenarios on phosphorus dynamics in a small watershed. *Trans. ASABE* 59(1), 63-79. doi:10.13031/trans.59.10842.

Singh, U. 1985. A crop growth model for predicting corn (*Zea mays* l.) performance in the tropics. Ph.D. thesis. University of Hawaii, Honolulu, HI.

Singh, U. 2008. Expanding the use of dynamic soil -crop models with field-generated inputs. Final consultancy report submitted to the University of Florida. IFDC, Muscle Shoals, AL.

Singh, U., Wilkins, P. W., Chude, V. and Oikeh, S. 1999. Predicting the effect of nitrogen deficiency on growth duration and yield. In: Robert, P. C., Rust, R. H. and Larson, W. E. (Eds), *Precision Agriculture*. ASA-CSSA-SSSA, Madison, WI, pp. 1379–93. doi:10.2134/1999.precisionagproc4.c40b.

Singh, U., Giller, K. E., Palm, C. A., Ladha, J. K. and Breman, H. 2001. Synchronizing N release from organic residues: opportunities for integrated management of N. *Sci. World J.* 1, 880–6. doi:10.1100/tsw.2001.361.

Singh, U., Wilkens, P. W., Baethegen, W. E. and Bontkes, T. S. 2002. Decision support tools for improved resource management and agricultural sustainability. In: Ahuja, L. R., Ma, L. and Howell, T. A. (Eds), *Agricultural System Models in Field Research and Technology Transfer*. Lewis Publishers, Boca Raton, FL, pp. 91–117.

Singh, U., Wilkens, P. W., Gopi, R. and Potter, C. 2007. Development and application of a potash decision support system. *2007 Annual Meetings Abstracts*. ASA/CSSA/SSSA.

Smalberger, S., Singh, U., Chien, S. S., Wilkens, P. W. and Henao, J. 2006. A phosphate rock decision support system for evaluating agronomic effectiveness and economic feasibility of phosphate rick for direct application. *Agron. J.* 98, 471–83.

Springmann, M., Clark, M., Mason-D'Croz, D., Wiebe, K., Bodirsky, B. L., Lassaletta, L., de Vries, W., Vermeulen, S. J., Herrero, M., Carlson, K. M., Jonell, M., Troell, M., DeClerck, F., Gordon, L. J., Zurayk, R., Scarborough, P., Rayner, M., Loken, B., Fanzo, J., Godfray, H. C. J., Tilman, D., Rockström, J. and Willett, W. 2018. Options for keeping the food system within environmental limits. *Nature* 562(7728), 519–25. doi:10.1038/s41586-018-0594-0.

Springob, G. and Richter, J. 1998. Measuring interlayer potassium release rates from soil minerals. II. A percolation procedure to study the influence of the variable "solute K" in the <1–10 μM range. *Z. Pflanzenernaehr. Bodenk.* 161(3), 315–22. doi:10.1002/jpln.1998.3581610320.

Steingrobe, B. and Claassen, N. 2000. Potassium dynamics in the rhizosphere and K efficiency of crops. *J. Plant Nutr. Soil Sci.* 163(1), 101–6. doi:10.1002/(SICI)1522-2624(200002)163:1<101::AID-JPLN101>3.0.CO;2-J.

Stenger, R., Priesack, E., Barkle, G. and Sperr, C. 1999. Expert-N, a tool for simulating nitrogen and carbon dynamics in the soil-plant-atmosphere system. In: Tomer, M., Robinson, M. and Gielen, G. (Eds), *NZ Land Treatment Collective, Proceedings of Technical Session 20: Modelling of Land Treatment Systems*. New Plymouth, New Zealand, pp. 19–28.

Stockle, C. O., Donatelli, M. and Nelson, R. 2003. CropSyst, a cropping systems simulation model. *Eur. J. Agron.* 18(3–4), 289–307. doi:10.1016/S1161-0301(02)00109-0.

Tian, S., Youssef, M. A., Skaggs, R. W., Amatya, D. M. and Chescheir, G. M. 2012. DRAINMOD-FOREST: integrated modeling of hydrology, soil carbon and nitrogen dynamics, and plant growth for drained forests. *J. Environ. Qual.* 41(3), 764–82. doi:10.2134/jeq2011.0388.

Tinker, P. B. and Nye, P. H. 2000. *Solute Movement in the Rhizosphere*. Oxford University Press, London, p. 444.

Trehan, S. P. and Claassen, N. 1998. External K requirement of young plants of potato, sugar beet, and wheat in flowing culture resulting from different internal requirements and uptake efficiency. *Potato Res.* 41(3), 229-37. doi:10.1007/BF02358193.

Trenkel, M. E. 2010. *Slow- and Controlled-Release and Stabilized Fertilizers: an Option for Enhancing Nutrient Use Efficiency in Agriculture*. International Fertilizer Industry Association (IFA), Paris, France.

Vadas, P. A. and White, M. J. 2010. Validating soil phosphorus routines in the SWAT model. *Trans. ASABE* 53(5), 1469-76. doi:10.13031/2013.34897.

Vadas, P. A., Joern, B. C. and Moore Jr., P. A. 2012. Simulating soil phosphorus dynamics for a phosphorus loss quantification tool. *J. Environ. Qual.* 41(6), 1750-7. doi:10.2134/jeq2012.0003.

Van Kauwenbergh, S. 2010. World phosphate reserves and resources. Technical Bulletin. International Fertilizer Development Center, Muscle Shoals, AL.

Vlek, P. L. and Craswell, E. T. 1981. Ammonia volatilization from flooded soils. *Fert. Res.* 2(4), 227-45. doi:10.1007/BF01050196.

Walker, W. W. and Kadlec, R. H. 2011. Modeling phosphorus dynamics in Everglades wetlands and stormwater treatment areas. *Critic. Rev. Environ. Sci. Technol.* 41 (Suppl.), 430-46. doi:10.1080/10643389.2010.531225.

Wang, H., Appan, A. and Gulliver, J. S. 2003. Modeling of phosphorus dynamics in aquatic sediments: I–model development. *Water Res.* 37(16), 3928-38. doi:10.1016/S0043-1354(03)00304-X.

Wang, J. J., Harrell, D. L. and Bell, P. F. 2004. Potassium buffering characteristics of three soils low in exchangeable potassium. *Soil. Sci. Soc. Am. J.* 68(2), 654-61. doi:10.2136/sssaj2004.0654.

Whiting, G. J. 1994. CO_2 exchange in the Hudson Bay lowlands: community characteristics and multispectral reflectance properties. *J. Geophys. Res.* 99(D1), 1519-28. doi:10.1029/93JD01833.

Williams, J. R. 1995. The EPIC model. In: Singh, V. P. (Ed.), *Computer Models of Watershed Hydrology*. Water Resources Publications, Highlands Ranch, CO, pp. 909-1000.

Williams, J. R., Jones, C. A., Kiniry, J. R. and Spanel, D. A. 1989. The EPIC crop growth model. *Trans. ASABE* 32(2), 0497-511. doi:10.13031/2013.31032.

Yuan, Y. P., Bingner, R. L., Locke, M. A., Stafford, J. and Theurer, F. D. 2011. Assessing the long-term impact of phosphorus fertilization on phosphorus loadings using AnnAGNPS. *Int. J. Environ. Res. Public Health* 8(6), 2181-99. doi:10.3390/ijerph8062181.

Zhang, K., Greenwood, D. J., White, P. J. and Burns, I. G. 2007. A dynamic model for the combined effects of N, P and K fertilizers on yield and mineral composition; description and experimental test. *Plant Soil* 298(1-2), 81-98. doi:10.1007/s11104-007-9342-1.

Zeng, Q. and Brown, P. H. 2000. Soil potassium mobility and uptake by corn under differential soil moisture regimes. *Plant Soil* 221(2), 121-34. doi:10.1023/A:1004738414847.

Published by Burleigh Dodds Science Publishing Limited, 2026.

Chapter 4

Improving modelling of water cycles in crop cultivation

Claudio O. Stöckle, Washington State University, USA; and Francisco Meza, Pontificia Universidad Católica de Chile, Chile

1 Introduction

Water use by plants and net primary productivity are two sides of the same coin. Under unstressed water conditions the stomata remain open, maximizing canopy conductance, and facilitating latent heat exchange as well as carbon dioxide uptake. From the energy budget standpoint, as more energy is utilized in evapotranspiration, canopy temperature remains close to ambient temperature, and thus cell respiration and crop development can occur at normal rates. On the contrary, if water in the soil is limited, canopy conductance is diminished as stomata tend to close. This in turn decreases the rate of carbon assimilation via photosynthesis. In addition, the surplus in available net radiation has to be balanced. Because restricted vapor exchange affects latent heat flux, an increase in canopy temperature (normally well above air temperature) is observed, accelerating the rate of crop development and increasing cell respiration.

The relationship between soil available water, crop evapotranspiration and yield represents a cornerstone that usually allow us to explain the

http://dx.doi.org/10.19103/AS.2019.0061.06

difference between 'good' and 'bad' weather years, as well as the difference between agroecological zones in terms of yield potential and the effect of irrigation on agricultural productivity. A wide range of equations are available to estimate biomass productivity. A simple quantitative approximation can be found in the original work of Doorenbos and Kassam (1979) revisited by Steduto et al. (2012), where the effect of the ratio between attainable yield to potential yield was modeled as a linear function between actual and potential evapotranspiration.

Since then, the actual amount of water transpired by a crop has been considered a key variable to predict its total biomass and yield. It is for this reason that, regardless of the level of complexity when representing biophysical processes in the soil-plant-atmospheric continuum, modelling the water cycle, particularly the evapotranspiration process, is a critical component of every crop simulation model.

2 Crop water availability and its link to weather, soil, and management at point, field, and watershed scales

Actual evapotranspiration depends on observed weather conditions, soil water availability, and crop specific attributes related to solar radiation interception and the ability to extract water from the soil. The relevance of each factor depends on the scale and level of human intervention. While at point scale micrometeorological conditions and agricultural management are the most important variables to explain crop water availability, at farm scale the spatial heterogeneity of soils (i.e. texture, slope, depth, and bulk density) must be accounted to correctly represent the relationship between crop productivity and water. At the level of watersheds or districts, however, large scale hydrological processes become important to predict regional crop yields.

The simulation of the soil water balance is at the center of the relationship between evapotranspiration and productivity. Processes such as infiltration, surface and subsurface runoff, water distribution, drainage, water uptake by roots are represented by specific equations and algorithms that vary in their level of detail depending on the scale of analysis and available information. The scope of this chapter is to present and discuss the main approximations used by crop simulation models and challenges to adequately estimate soil water availability, crop evapotranspiration, and crop water uptake.

3 Determination of the potential and actual crop evapotranspiration

The atmospheric evaporative demand accounts for the energy available to evaporate water per unit of land area, and it is central for defining crop water demand. This energy drives crop transpiration (mostly through plant stomata) and soil water evaporation, as well as evaporation of water intercepted by crop canopy and residues. If the water available for evaporation is less than the demand, the extra energy heats the surface and surrounding air. In the context of crop modelling, evapotranspiration (ET) refers to the water vapor flux from a unit area of cropped land, including all the sources mentioned.

To estimate the ET demand from a well-watered crop/soil land unit area, the combination of the energy balance and vapor and heat transfer equations is the most bio-physically complete approach.

The energy balance can be written as:

$$R_n - \lambda E - H - G = 0 \tag{1}$$

where

R_n = Net radiation (MJ m^{-2} s^{-1})
G = Soil heat flux ≈ 0 (MJ m^{-2} s^{-1})
H = Sensible heat flux (MJ m^{-2} s^{-1})
λE = Latent heat flux (MJ m^{-2} s^{-1})

The transfer equations represent the steady state exchange of sensible heat (H) and vapor (λE) between a cropped surface and the atmosphere. We will follow here the standard approach of treating the cropped surface as a 'big leaf', with implications for actual cropland surfaces discussed later.

Sensible heat fluxes can be expressed as:

$$H = \frac{\rho C_p (T_s - T_a)}{r_a} \tag{2}$$

T_s = Canopy temperature (°C)
T_a = Air temperature (°C)
r_a = aerodynamic resistance (s m^{-1})
ρ = Atmospheric density (kg m^{-3})
C_p = Specific heat capacity at constant pressure (MJ kg^{-1} C^{-1})

On the other hand, the steady state exchange of vapor (latent heat) may be expressed as:

$$\lambda E = \frac{\rho C_p (e^0(T_s) - e_a)}{\gamma [r_a + r_s]} \tag{3}$$

e_a = Atmospheric vapor pressure (kPa)
$e^0(T_s)$ = Saturation vapor pressure (Kpa) at temperature T_s (°C)
γ = Psychrometric constant (kPa C^{-1})
r_s = canopy resistance (s m^{-1})

The net radiation (R_n) term can be measured or estimated from solar radiation and temperature. The soil heat flux term (G) can be estimated from temperature or as a fraction of R_n, but for applications using daily or larger time steps G can be assumed as zero.

Because R_n, H, and λE depend on surface temperature, the combination of Eqs. (1)-(3) requires iteration (Lascano and van Bavel, 2007), which is often not practical for implementation in crop models.

Penman (1948) introduced a transformation that allows the elimination of the surface temperature by using an approximation of the slope of the saturation vapor pressure as a function of temperature.

$$e^0(T_s) - e_a = \left[e^0(T_a) - e_a\right] + \Delta\left(T_s - T_a\right) \tag{4}$$

where Δ is calculated as:

$$\Delta = \frac{e^0(T_s) - e^0(T_a)}{T_s - T_a} \tag{5}$$

The vapor pressure deficit (VPD) of the atmosphere is:

$$\text{VPD} = e^0(T_a) - e_a \tag{6}$$

Combining Eqs. (3), (4), and (6) yields the following expression for λE

$$\lambda E = \frac{\rho C_p\left[\text{VPD} + \Delta\left(T_s - T_a\right)\right]}{\gamma\left[r_a + r_s\right]} \tag{7}$$

Finally, combining (2), (7), and (1), and after some algebraic manipulation, the Penman-Monteith (P-M) form of the combination method is obtained.

$$\lambda E = \frac{\Delta\left(R_n - G\right) + \rho C_p \text{VPD}/r_a}{\Delta + \gamma\left[1 + r_s/r_a\right]} \tag{8}$$

Dividing λE in MJ m^{-2} s^{-1} by the latent heat of vaporization λ (MJ kg^{-1}) gives evapotranspiration in kg m^{-2} s^{-1}, which is equivalent to mm s^{-1}.

For λE in Eq. (8) to represent the maximum atmospheric evaporative demand for specific crops (to be referred to as maximum crop evapotranspiration, *ETmax*), an estimation of r_s and r_a is required, resistances whose values change dynamically. For empirical simplifications (see discussion below), suitable coefficients are needed to account for this dynamic. It must be recognized that Eq. (8) works best for dense canopies that approximate a homogeneous big leaf. However, canopies of annual crops are essentially negligible at emergence, they can intercept 95% of incident radiation at full development, and during rapid growth the evaporating surface changes continuously with varying proportions of crop canopy and soil surfaces, each with different temperatures and resistances. Indeed, the big leaf assumption implicit in the P-M equation is not valid in sparse canopies. Accounting for these dual surfaces has been attempted (Shuttleworth and Wallace, 1985), but with difficulties in the parameterization and proper representation of crop, soil, and aerodynamic resistances.

Practical engineering approximations developed for irrigation and water management can be useful for inclusion in crop models. The Food and Agriculture Organization (FAO) introduced a version of the P-M equation (FAO56-PM) for daily ET calculations for a hypothetical short-grass reference crop (ET_o) with specified parameters (Allen et al., 1998). The American Society of Civil Engineering (ASCE, 2005) implemented a similar approach parameterized for an alfalfa reference crop (ET_r), which more closely resembles many agricultural crops. Both approaches have been shown to compare well with daily ET determined for these crop surfaces using lysimeters, and they can be used to estimate ET_{max} by multiplying these standardized values by single or dual crop coefficients, the former integrating crop transpiration and soil water evaporation into one coefficient and the latter splitting the coefficient to represent the two different sources of water vapor (Allen et al., 1998). Another approach, amenable for use in crop models, is based on the definition of a hypothetical maximum crop coefficient ($K_{cx} = ET_{cx} / ET_o$) representing a situation where the crop has reached maximum height and the canopy foliage is assumed to intercept all available solar radiation so that ET_{cx} corresponds solely to water vapor fluxes originated from foliage. The value of K_{cx} represents mainly differences in aerodynamic resistance, and to a lesser extent in stomatal resistance between the crop under the assumed condition and the short-grass reference crop. In this approach, daily K_c values (K_{cd}) are determined as a function of the current fraction of solar radiation intercepted by the green canopy (F_{sri_gc}), so that $K_{cd} = 1 + (K_{cx} - 1) \times F_{sri\_cg}$. In this formulation, when $F_{sri\_gc} = 1$ (hypothetical) then $K_{cd} = K_{cx}$, and $K_{cd} = 1$ when $F_{sri\_gc} = 0$ (bare soil). The daily value of ET_{max} is given by $ET_o \times K_{cd}$, where both terms are calculated for the given day of interest. The value of daily ET_{max} can be partitioned into T_{max} (maximum transpiration) and E_{max} (maximum soil water evaporation), where $T_{max} = ET_{max} \times F_{sri\_gc}$, and $E_{max} = ET_{max} \times (1 - F_{sri\_c}) \times (1 - F_{sri\_r})$ where F_{sri_c} is the total solar radiation interception by the canopy including green and senesced portions, and F_{sri_r} is interception by residues or other elements shading the soil. During crop senescence (or significant residue mulching), $ET_{max} > T_{max} + E_{max}$ because solar radiation intercepted by senescing portions of the canopy and by residues constitute excess energy warming the canopy, residues, and surrounding area. The actual transpiration and actual soil water evaporation are a function of T_{max} and E_{max}, respectively, and soil water availability, as will be discussed later.

The use of the P-M equation (Eq. (8)) requires daily (or shorter period) weather data including maximum temperature, minimum temperature, solar radiation, maximum and minimum relative humidity or dew-point temperature, and wind speed. For conditions where not all required weather variables are available, or to reduce parameterization, simplifications are necessary. One such simplification is the Penman equation (Penman, 1948, 1963), originally derived for open water surfaces, and later adapted to crops. In this equation, $r_s = 0$, and a calibrated aerodynamic wind function ($f(U) = \rho C_p/r_a$) is introduced so that the P-M equation (8) becomes:

$$\lambda E = \frac{\Delta(R_n - G) + f(U)VPD}{\Delta + \gamma} \tag{9}$$

If we assume that λE depends mainly on radiation supply, the advection term (the right-hand side term in the numerator of Eq. (9)) can be neglected and a correlation constant (α = 1.26 - 1.70) is used as multiplier of the radiation term. These simplifications yield the Priestley-Taylor equation (Priestley and Taylor, 1972).

$$\lambda E = \alpha \frac{\Delta(R_n - G)}{\Delta + \gamma} \tag{10}$$

If we further assume that $G \approx 0$ when averaged over a few days, and that R_n is proportional to solar radiation (R_s) and air temperature (T_a), a general λE expression used by so-called radiation methods is obtained.

$$\lambda E = \alpha (T_a + b) R_s \tag{11}$$

One step further includes an approximation of R_s based on R_a, the extraterrestrial solar radiation, resulting in the Hargreaves equation (Hargreaves and Samani, 1985), which only requires air temperature.

$$\lambda E = a(T_{mean} + b)(T_{max} - T_{min})^{0.5} R_a \tag{12}$$

where coefficients a and b are 0.0023 and 17.8, respectively (Hargreaves and Samani, 1985).

Although simplifications relax the input requirements increasing applicability, they introduce uncertainty, decrease accuracy, and require local calibration. However, many of these simplifications are used in crop models.

Several studies have compared different ET equations with field observations. For example, López-Urrea et al. (2006) evaluated seven methods for calculating average daily ET_o in a semiarid climate by comparison with measurements obtained with a continuous weighing lysimeter from 2000 to 2002 in Albacete, Spain. The FAO56-PM equation was shown to have the best performance of all methods evaluated including the FAO24 versions of the Penman equation I and II, Blaney-Criddle, and Radiation (Dorenboos and

Pruitt, 1977), Penman-1963 (1963), and Hargreaves-1985 (Hargreaves and Samani, 1985), with the latter being the second best despite its simplicity. FAO24 Penman I and II and Blaney-Criddle overpredicted ET_o (30 to 60%), while Penman (1963) underpredicted (34%). Allen (1986) evaluated ten forms of the combination evapotranspiration equation including Penman (Penman, 1948) with six variations of the aerodynamic wind function, two equations including resistance terms - Penman-Monteith and Thom-Oliver, the latter using the Penman (1948) wind function but including surface roughness (Thom and Oliver, 1977), and the Priestley-Taylor equation. These ten equations were compared with precision-weighing lysimeter measurements at three US locations including irrigated arid (alfalfa), humid rainfed (tall grass), and Mediterranean (short grass) sites. The author concluded that the Penman-Monteith model provided the most reliable and consistent daily estimates of alfalfa and grass reference ET at the three locations, while the original Penman and Priestley versions underestimated ET in the arid environment (Allen, 1986). In another study using weighing lysimeter data with short grass as reference (ET_o), Benli et al. (2010) compared FAO56-PM, Priestley-Taylor, FAO24 Radiation, Hargreaves-1985, and Blaney-Criddle equations at Ankara, Turkey, a semiarid highland environment, during three seasons (2000 to 2002). The best performance was obtained by the FAO56-PM followed by Hargreaves-1985. FAO-24 Radiation and Blaney-Criddle overpredicted, and Priestley-Taylor underpredicted the lysimeter seasonal ET_o.

In summary, the adequate performance of Penman-Monteith has been demonstrated in several studies, and the implementations by FAO-56 (Allen et al., 1998) and ASCE (ASCE, 2005) are now considered standard. The use of the simplifications of the combination method discussed above to determine ET in crop models is likely to introduce errors that will propagate to other portions of the model. When weather data is limited, resorting to simplifications may be the only option available, but calibration of the coefficients used in these simplifications should be performed using complete weather data sets from neighboring or similar climate locations. Another option is to estimate the missing weather variables. For example, using two years of complete weather, Stöckle et al. (2004) used ClimGen (Castellvi and Stöckle, 2001) to estimate daily solar radiation and VPD from maximum and minimum temperature, and generated wind speed based on a Weibull distribution. These authors compared FAO56-PM ET_o based on temperature data (P-M ET_o*-Est*) with calculations where all required weather variables were available (P-M ET_o) at five locations representing tropical, temperate, semiarid, and arid climates. They found that P-M ET_o*-Est* compared poorly to acceptable with P-M ET_o for daily periods, and acceptably to well for weekly periods in the arid and semiarid locations. The performance was poorer in humid locations. Comparisons were also conducted using Hargreaves-1985 with calibrated a and b coefficients (Eq.

(12)), showing a similar performance as P-M ET_o-*Est*, but without the need to estimate solar radiation and VPD.

To illustrate the need for calibration in the performance of simpler models, Priestley-Taylor (PT) and Hargreaves (HG) approximations of ET_o, with and without calibration, were compared to FAO56-PM (PM) ET_o in seven world locations with a wide range of VPD conditions (Table 1). In the case of HG, only the coefficient a was calibrated. Sensitivity was low to changes of the coefficient b (17.8), and calibration changing a, b and the power coefficient 0.5 simultaneously resulted in many solutions with not much relationship to the different environments, a fact discussed by Hargreaves and Allen (2003). Although PT did not require calibration in Prosser WA and HG in Pullman WA and Temple TX, the performance of both equations improved with calibrated coefficients in all other locations.

There is continued interest in evaluating the impact on ET of elevated atmospheric carbon dioxide concentration (CO_2) and associated global warming. Because changes in CO_2 affect stomatal resistance, with resistance increasing as CO_2 increases, transpiration rates are also affected. To model this response, the Penman-Monteith ET equation, which explicitly includes stomatal resistance, is suitable. Experimental observation of CO_2 effects on stomatal resistance are available in the literature (e.g. Morison, 1985; Allen, 1990).

4 Determination of soil water availability in crop models

At the center of the vast majority of dynamic simulation models is the correct representation of soil water availability for crops under variable environmental conditions. The soil water balance is a particular case of the law of mass conservation or continuity equation, stating that changes in soil volumetric water content over a period of time are equal to the difference between the amount of water added and the amount of water withdrawn. For the one-dimensional analysis commonly used in crop models, the amount of water supplied to the soil comes from precipitation and irrigation, while the amount of water loss from the soil corresponds to the evaporation from the soil surface layers, plant transpiration, and drainage beyond the rooting zone. Thus, the amount of water available for the subsequent time step is the algebraic sum of the components of this budget equation. Aside from plant water uptake, that will be addressed in a following subsection, the most common (and challenging) processes to be represented in crop models are infiltration, soil evaporation, and redistribution of water within the soil. Some crop models include additional processes to better represent the dynamics of soil water availability. One example is the simulation of canopy interception of precipitation and irrigation (normally sprinkler or center pivot types). In this case, a fraction of the amount of water from rainfall

Table 1 Performance of the Priestley-Taylor (PT) and Hargreaves (HG) evapotranspiration (ET) equations, with (Cal.) and without (Orig.) calibration, compared to the Penman-Monteith equation (PM ET_o) as reference at seven locations with contrasting weather

Location	VPD kPa	WS m/s	PM ETo (mm)		PT alpha	PT RMSE (mm/d)	PT PT-ET (mm)	PT PT/PM	HG a	HG RMSE (mm/d)	HG HG-ET (mm)	HG HG/PM
Landriano,	0.85	1.13	387	Orig	1.26	0.75	449	1.16	0.00230	0.83	447	1.16
Italy				Cal.	1.09	0.19	387	1.00	0.00202	0.54	392	1.01
Pullman, WA,	1.14	2.09	541	Orig	1.26	0.74	508	0.94	0.00230	0.70	549	1.01
USA				Cal.	1.33	0.7	536	0.99	0.00230	0.70	549	1.01
Prosser, WA,	1.33	1.54	588	Orig	1.26	0.39	584	0.99	0.00230	0.92	667	1.13
USA				Cal.	1.27	0.39	587	1.00	0.00201	0.58	584	0.99
Temple, TX,	1.34	2.69	639	Orig	1.26	0.61	607	0.95	0.00230	0.92	637	1.00
USA				Cal.	1.32	0.56	636	1.00	0.00230	0.92	637	1.00
Ankara, Turkey	1.75	1.76	610	Orig	1.26	1.05	523	0.86	0.00230	0.78	643	1.05
				Cal.	1.48	0.72	614	1.01	0.00218	0.73	610	1.00
Aleppo, Syria	2.48	4.79	1096	Orig	1.26	3.70	696	0.64	0.00230	2.91	817	0.75
				Cal.	1.99	1.46	1101	1.00	0.00306	1.80	1088	0.99
Maricopa, AZ,	2.91	2.18	919	Orig	1.26	1.57	761	0.83	0.00230	0.94	848	0.92
USA				Cal.	1.52	0.89	915	1.00	0.00248	0.75	917	1.00

Parameters calibrated were alpha (PT equation) and a (HG equation).

or irrigation falls in the canopy and is evaporated back to the atmosphere, another fraction is intercepted by crop residues, and the remainder reaches the soil and can be infiltrated (Stockle et al., 1994). This process is particularly relevant in climates where precipitation is relatively small and occasional, and/or in locations where wind speed is particularly high, as there is a substantial difference between rainfall recorded and the amount of water that either infiltrates or generates runoff. Another hydrological process that is particularly relevant in cold climates corresponds to snow/ice melting and subsequent infiltration. This process has received comparatively less attention by the crop modelling community, being more frequent in hydrological models, particularly when representing seasonal streamflow changes in snow driven basins (Yates et al., 2005) or in permafrost soils (see an example in Zhang et al., 2010).

Soils present a large spatial heterogeneity that results in variable soil physical properties among sites and in depth within a specific site. To better describe processes that require specific soil and crop parameters (i.e. water retention, water redistribution, plant uptake, etc.), crop models divide the soil profile into small fractions or layers. In this way, water movement within the soil, that is governed by differential equations (Richards-Darcy), is either solved numerically using finite difference approaches or replaced by simple semi-empirical equations. The total length of the soil unit that will be considered in the water budget equation is determined by crop rooting depth, which changes during the growing season as plants develop. However, maximum root depth is probably a better descriptor because water and nutrients are truly not lost from the soil until they are beyond the reach of fully developed roots. In fact, when crops in rotation are considered, probably the descriptor is the maximum root depth of the crop with deepest roots.

4.1 Infiltration

Infiltrability or infiltration capacity refers to the maximum rate at which water can be absorbed by a soil under a given condition. If the supply rate surpasses infiltrability, excess water will either accumulate on the soil's surface or become runoff. Some empirical approximations to address this feature are the Horton's equation (1940) and the Kostiakov's equation (1932) further expanded by Lewis (1937). Another simple formulation corresponds to the Green and Ampt model (1911). These methods require knowledge of specific soil parameters that vary with soil moisture and/or are restricted to certain type of conditions such as ponded flow.

As an alternative, most crop models rely on the USDA-SCS Curve Number Method (USDA, 2004) to determine runoff and infiltration. This method attempts to account for site specific conditions that affect the rate of infiltration such as texture, slope, and tillage.

$$RO = \begin{cases} 0 & P \le Ia \\ \dfrac{25.4(P - Ia)^2}{(P - Ia + S)} & P > Ia \end{cases} \qquad (13)$$
$$S = \frac{1000}{CN} - 10$$

where RO is the runoff (mm day^{-1}), P is rainfall (in day^{-1}), Ia is the initial abstraction or losses before runoff begins (in day^{-1}) and is set to $0.2S$, S is the potential maximum soil moisture retention after runoff begins (in day^{-1}) and CN is the empirical curve number value (in^{-1}) with values ranging from 0 to 100 depending on the hydrological condition.

Infiltration is assumed to be the difference between precipitation reaching the soil and runoff. If irrigation is applied, the amount of water supplied is added to the rainfall in that particular day to account for infiltration. Literature shows one case in which a Green and Ampt model is used once runoff is generated based on the USDA-SCS method (Liang et al., 2016).

4.2 Soil evaporation

Soil evaporation, the water flux from the top soil surface to the atmosphere, is affected by atmospheric evaporative demand, soil water availability, soil properties, and other factors such as tillage and crop and residue cover. Mechanistic models that estimate one-dimensional fluxes of heat and water in the soil using a network of nodes and numerical iterative solutions of the resulting set of finite difference equations are available (e.g. Flerchinger and Saxton, 1989; Stockle and Campbell, 1989), but these approaches are seldom used in crop models.

Most crop models follow simple approaches that divide soil water evaporation in two stages: energy limited (Stage I) and water limited (Stage II), with evaporation involving a top soil of specified depth, often 0.1 m, which is decreased (sandy soils) or increased (clay soils) depending on soil texture. Soil drying during Stage I depends on the atmospheric evaporative demand, and thus is affected by the choice of ET equation and proper accounting for soil shading by crop canopy and residues. Soil drying during Stage II is simulated based on a threshold soil water content at the onset of this stage, and by different methods describing the soil evaporation rate reduction as a function of time or soil water content (Ritchie, 1972; Campbell and Diaz, 1988).

4.3 Soil water distribution

Water infiltrating into the soil must be redistributed in the soil profile. The most mechanistic models describe water transport after dividing the soil into layers, with properties concentrated at nodes distributed throughout the soil profile of interest, and solving the set of finite difference equations representing fluxes throughout the network using numerical methods (Flerchinger and Saxton, 1989; Stockle and Campbell, 1989). Cascade or tipping bucket is the most common empirical method used to represent water movement between soil layers. With some variations, the general idea is that water exceeding field capacity (also known as drainage upper limit) infiltrates to the following layer. This can be limited by saturated hydraulic conductivity (K_{sat}) (Jones et al., 2003), or quantified according to an empirical drainage coefficient that depends on Ksat and a non-linear function of the soil water content above field capacity (Barrios-Gonzales, 1999). Another approach to represent the process is by cascading the infiltrating amount of water instantaneously within the soil profile leaving soil layers at field capacity until the infiltrating water is exhausted (Boogard et al., 2011).

Water contribution from a shallow water table is sometimes an important source of water for the soil water budget relevant to crops. Most models do not consider this contribution and the ones that do differ substantially when treating upward flow or capillarity rise. While some use the Richards' equation (e.g. Campbell, 1985), others use empirical approaches that depend on the presence of groundwater table or rely on water diffusivity parameters. It is important to consider that the evaluation of fluctuations of water table depth requires hydrologic modelling that is usually outside of the scope of crop models.

A comparative table of the soil evaporation, cascade methods and upward flow as well as the most important equations involved is presented by Castañeda-Vera et al. (2015).

5 Modelling crop water uptake

Crop water uptake and transpiration are normally taken as equivalent since changes in plant water storage are negligible compared with the transpiration flux. Modelling crop water uptake requires evaluation of the equilibrium between the atmospheric evaporative demand for transpiration (T_{max}) and the soil water supply. Because Tmax can be met by crop water uptake plus evaporation, if any, of water intercepted by the canopy (C_{wi}), the amount of water that must be supplied from roots to the sites of evaporation in the sub-stomatal cavities should equal $T_{max} - C_{wi}$ if water stress is to be avoided. If T_{max} is larger than the rate of water supply by roots, stomatal closure will occur reducing transpiration even if the soil water content is equivalent to its maximum water holding capacity.

There are many models available to simulate crop water uptake, including a wide range of complexity (e.g. Dardanelli et al., 2004; Jara and Stockle,

1999; Annandale et al., 2000; Feddes et al., 2001; Li et al., 2001; Jarvis, 2011). A good review of crop water uptake models is presented by Wang and Smith (2004). These authors identified the simplification of the modelling of plant root systems as an unresolved issue, with emphasis placed on uptake functions by roots normally assumed to be uniformly distributed in soil layers, with the same capacity to uptake water, and with root density always enough for water uptake. Many models resort to the use of empirical root activity coefficients and/or water uptake reduction factors as a function of soil moisture. In addition, actual rooting depth is more dynamic than the usual assumption of a fixed root depth for a crop species and cultivar. These limitations may not be critical under conditions of good water supply but become more significant if the available water for crop uptake is low.

A few studies have compared crop water uptake algorithms. For example, van den Berg (2002) compared three simple approaches, concluding that different methods of calculating water uptake may lead to considerably different results. An additional problem with these differences is that they are propagated to other modules in crop models such as crop biomass production, leaf area index, or irrigation demand, which in turn may feedback to the calculation of water uptake/transpiration. Jara and Stockle (1999) compared three models with different levels of process detail with maize crop water uptake data. The three models adequately described maize water use under fully irrigated conditions, but the more mechanistic approach incorporating water potential and resistances in the soil-plant-atmosphere pathway outperformed the other two models under severe water stress.

Camargo and Kemanian (2016) compared the water uptake methods implemented in six crop models, which ranged from simple empiric approaches to more mechanistic approaches based on the water potential gradient and root distribution in the soil-plant system. They compared the six water uptake algorithms in scenarios with different evaporative demand, soil texture, and water distribution with depth, and found that each method responded differently to these scenarios which affected the onset of water stress, the cumulative water uptake, the shape of the drying front under a constant transpiration demand, and the response to high transpiration demand. Some of the models were even sensitive to changes in the thickness of the soil layers with depth (Camargo and Kemanian, 2016).

The water flow from soil layers to the sites of evaporation in the sub-stomatal cavities can be readily approximated using an electric analog based on plant-soil water potential gradient and plant conductance (Campbell, 1985). During the second half of last century and earlier, plant biophysicists and physiologists experimentally observed and developed theory and models of crop water use and plant responses to water based on a soil-plant-atmosphere continuum (SPAC) framework. Campbell et al. (1976), for example, presented and tested an early SPAC-based model describing soil-plant-water relationships for potatoes. Limited data to parameterize models (e.g. short-term complete weather data, soil hydraulic properties, plant-water relationship descriptions for a large list of cultivated plants, etc.), computer speed limitations, and insufficient interdisciplinary participation have been a traditional barrier to the embracing of SPAC approaches by the crop modelling community. Thus, simplifications have been used to enable progress including the use of soil water content instead of soil water potential to drive water use, representation of root water uptake based on plant available soil water (PAW) limits, onset of water stress based on set points of PAW with limited or no consideration of water vapor atmospheric demand, and other simplifications requiring empirical coefficients. Approaches have also been proposed avoiding the problem all together by relating crop water use to biomass production and transpiration-use efficiency. The latter, which almost completely decouples water use from the biophysics of evapotranspiration, have been shown to have the potential for large errors in the estimation of crop water use (Basso and Ritchie, 2018).

The application of SPAC concepts is amenable to short-term hourly or sub-hourly time steps to accommodate the diurnal variation of weather and stomatal responses. However, evidence is available showing that daily time steps can be used without sacrificing accuracy; in fact, it can be argued that short-term time step implementations of SPAC do not guarantee better model performance compared to field observations (Stockle and Jara, 1998). It is important to note that, compared to the commonly used water content-based approaches, daily SPAC-based models do not require taxing additional input data (Camargo and Kemanian, 2016). More importantly, they largely reduce the use of coefficients needed to mimic crop uptake responses to variable weather, water supply, and stomatal adjustments to water stress.

Based on the SPAC implementation in Stockle and Jara (1998), the following example illustrates how weather, soil water, and stomatal responses influence crop water uptake. To simplify the presentation, the soil profile is divided in 10-cm layers of uniform properties and soil water content; the weather is represented by atmospheric transpiration demands (ET_{max}) fluctuating between 3 and 8 mm/day in 1 mm/day increments; and the crop is assumed to be fully developed and with a rooting depth of 1.5 m. The soil profile distribution of root fraction was approximated by setting a maximum fraction in the top soil layer, with the fraction decreasing linearly with depth to reach a value of zero at the tip of the root depth. The profile sum of all the root fractions in each soil layer is 1.0. Although the root fraction distribution used in this example is often observed in soil water extraction profiles, it represents one of several possible distributions. However, this choice of root fraction distribution does not change the points illustrated in this example.

The maximum crop capacity for water uptake before stomatal closure was set at 7 mm/day. The term plant available water (PAW) in Fig. 1 is defined as PAW = (WC - PWP)/(FC - PWP), where WC is water content, FC is field capacity (WC available for crop uptake at 24 to 72 hours after irrigation to allow gravity-driven

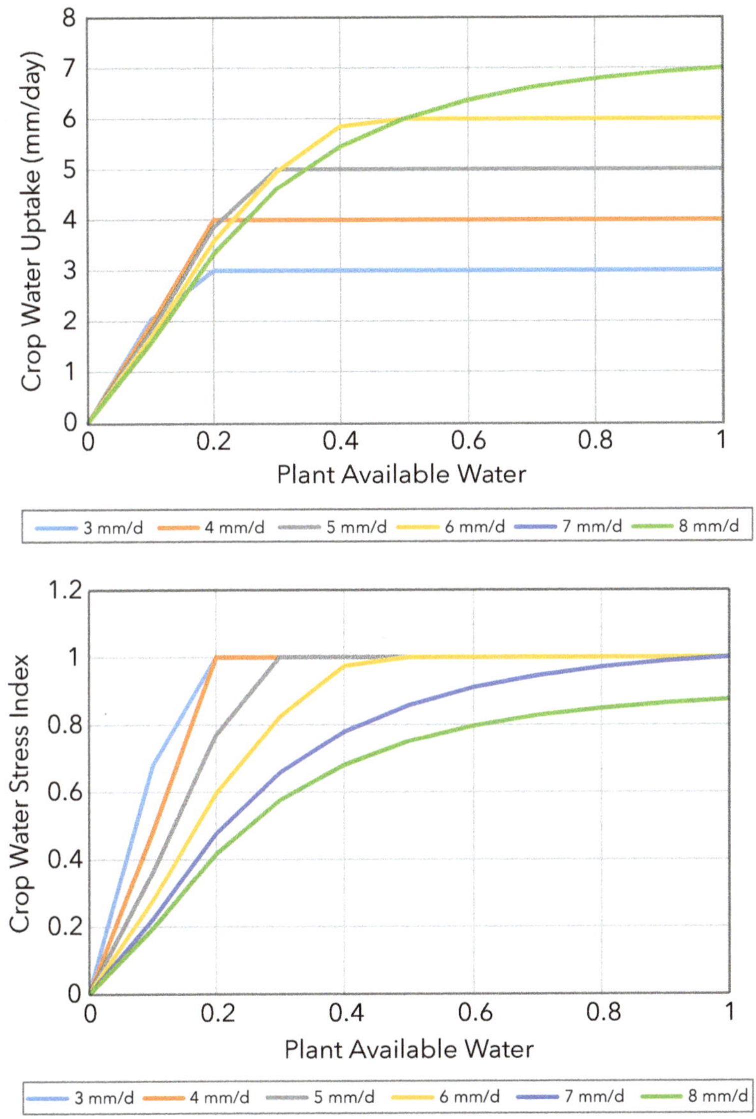

Figure 1 Crop water uptake (top) and crop water stress index (bottom) as a function of plant available water and six ET_{max} levels for a crop that will begin experiencing stomatal closure at an ET_{max} of 7 mm/day.

soil water redistribution), and PWP is the permanent wilting point (lower limit of WC extractable by crops). As seen in the figure, crop water uptake fluctuates with PAW and transpiration demand. At high atmospheric evaporative demand (ET_{max} of 7 and 8 mm/day), which in these cases are equal or greater than the assumed maximum crop capacity for water uptake, actual water uptake decreases immediately after soil PAW begins to decrease from its maximum value of 1.0 (Note: the lines for 7 and 8 mm/day overlap so that only one line is shown), implying onset of water stress. The other feature is that, at lower ET_{max}, crop uptake can meet the evaporative demand even while PAW decreases, but eventually reaching a point where uptake begin to decrease (onset of water stress). Only six levels of ET_{max} and 10 points of PAW were used for the calculations shown in the figure, but indeed values fluctuate continuously with weather and soil water availability so that the point of stress onset can fluctuate with the same soil water content. This behavior is hard to reproduce with crop uptake models based solely on water content, with no consideration of evaporative demand or resistance to flow through the soil-plant-atmosphere system.

The crop water stress index (CWSI) in Fig. 1 is a normalized index where one is no stress and zero is full stress. Values of CWSI are used in crop models to reduce unstressed biomass production due to water deficit, and therefore they are important in the simulation of growth and yield. Given the difficulty in defining coefficients to approximate the type of responses illustrated in the figure, crop water uptake models that do not use SPAC concepts are likely to incorporate uncertainty that will affect canopy growth simulation and feedback to water use. For additional discussion comparing water content versus water potential-based crop water uptake models, readers are referred to Camargo and Kemanian (2016).

6 Integration of soil water balance components in crop models

Equation (14) shows all components of the soil water balance of a control volume representing a small surface area of a field (say 1 m2) and a depth given by the maximum water extraction depth of the crop(s) of interest:

$$\frac{dW}{dt} = P + I + R_{in} + L_{in} + D_{in} - E_s - E_c - E_r - R_o - L_o - D_o \tag{14}$$

where dW/dt is the change of soil water mass per unit time, and the right-hand side terms are the rates of fluxes affecting this change, including precipitation (P), irrigation (I), runoff water in (R_{in}), soil lateral flow in (L_{in}), water table flux in (D_{in}), soil water evaporation (E_s), evaporation of water intercepted by the canopy (E_c), evaporation of water intercepted by residues (E_r), runoff water out (R_o), soil lateral flow out (L_o), and deep drainage (D_o).

Several fluxes in Eq. (14) are not accounted for in crop models, which typically are only concerned with one-dimensional fluxes with a few exceptions (e.g. Ward et al., 2018). Runoff is often included using empirical relations (USDA, 2004), and mainly applied to rainfall and not irrigation, and water movement across the landscape is neglected. Accounting for water table fluxes would require coupling crop models with multi-dimensional hydrologic models and appropriate parameterization of the spatially variable soil properties. Overall, the implication is that robust treatment of the soil water balance is complex making the application of crop models to the field scale challenging.

The application of Eq. (14) also has other issues seldom accounted for in crop models. Evaporation of water intercepted by residues and crop canopies contribute errors to the soil water balance. Water intercepted by residues does not infiltrate into the soil and shading reduces soil water evaporation. The evaporation of water intercepted by crop canopies utilizes energy and reduces the atmospheric demand for crop water uptake. Other issues are the influence of different irrigation methods, which affect canopy and residue interception of water (above canopy sprinkler systems), the partitioning of irrigation water losses such as evaporation losses during irrigation, soil water losses, runoff, and even the volume of soil participating in crop water uptake (e.g. drip irrigation).

7 Crop water use and atmospheric carbon dioxide concentration

Representation of the influence of atmospheric carbon dioxide concentration [CO_2] on crop growth and water use is often a source of variation in crop models. Although models coupling stomatal conductance, transpiration, and photosynthesis can readily accommodate responses to changes in [CO_2] (e.g. Wang and Leuning, 1998; Dai et al., 2004; Kremer et al., 2008), their parameters are difficult to determine, physiological and biophysical responses must be evaluated at short time intervals (hourly or sub-hourly), and there are uncertainties in translating the amount of CO_2 fixed via photosynthesis into biomass production. These factors have limited their inclusion in crop models, which often resort to simplifications based on observations that relate biomass production to radiation capture by crop canopies via a radiation-use efficiency (RUE) parameter (Monteith, 1977), or to crop transpiration via a transpiration-use efficiency (TUE) parameter (Tanner and Sinclair, 1983). The challenge for these simpler models is determining changes in RUE and TUE representing biomass production response to [CO_2], and linking biomass production to concurrent changes in crop water use resulting from the [CO_2] effect on stomatal conductance.

The magnitude of changes in RUE and TUE with [CO_2] can be observed experimentally, but data is scarce and do not cover the large number of crop species of interest in agricultural production. In addition, experiments in enclosures representing the bulk of the data available cannot be easily translated to field conditions without assumptions and uncertainties (Long et al., 2006). Data from field experiments (Free-air CO_2 enrichment experiments, FACE) are useful but limited in number and the levels of [CO_2] tested, typically ~350 and ~550 to 660 ppm. Projections indicate that [CO_2] will likely reach between 600 and 900 ppm by the end of this century, with the response of biomass production changing from linear to non-linear, and then approaching a saturation level of minimal additional response at ~900 ppm. Crop models use different mathematical approximations to represent the biomass response to [CO_2], fluctuating from linear to non-linear formulations, and with variation in the data sources used to fit these formulations. This often leads to different results in projections of crop production with elevated [CO_2], a matter of concern deserving careful evaluation by the community of modelers.

Even more challenging is the determination of crop transpiration response to [CO_2]. It is well known that elevated [CO_2] will normally result in stomatal closure and reduced transpiration, partially counteracted by concurrent increase in canopy temperature (Allen, 1990). Ainsworth and Rogers (2007) presented a meta-analysis of the mean responses of stomatal conductance to elevated [CO_2] in FACE studies, with the averaged ambient and elevated [CO_2] at ~366 ppm and 567 ppm, respectively, resulting in an average reduction of 22% across all species. The average reduction in C3 crops was ~25% (42 data points), and ~31% in C4 crops (11 data points). Regarding elevated [CO_2] effect on evapotranspiration, a summary of FACE results for six crops showed a reduction of ET fluctuating from 5 to 15% with elevated [CO_2] of ~550 ppm compared to ambient (~350 ppm), with one data point for cotton at 1% (Leakey et al., 2009). ET includes both transpiration and soil water evaporation, so that transpiration reduction must be a bit higher (soil water evaporation is not affected by CO_2).

An approximation to the change in crop transpiration with [CO_2] can be obtained by calculating the Penman-Monteith water vapor flux in [Eq.8] with (a) decreased canopy conductance (increased canopy resistance) due to elevated [CO_2] and (b) with the canopy conductance at a reference [CO_2], and dividing the two quantities (a/b) to obtain a transpiration adjustment factor (F_T). This can be represented as follows:

$$F_T = \frac{\Delta + \gamma\left(r_{co} + r_a\right)/r_a}{\Delta + \gamma\left(r_{ce} + r_a\right)/r_a} \tag{15}$$

where r_{co} and r_{ce} are the reference and elevated CO_2 canopy resistance, respectively. This simplification cannot readily accommodate the increase of foliage temperature due to reduced transpiration, requiring the use of the

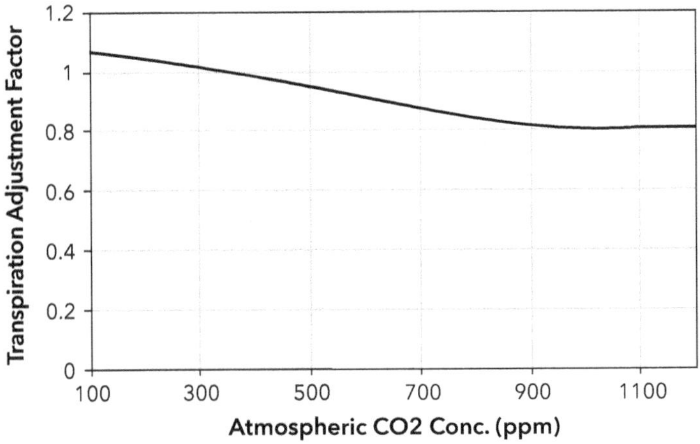

Figure 2 Transpiration adjustment factor as a function of atmospheric CO_2 concentration calculated for soybeans using data from Allen (1990) and Eq. (15).

iterative solution of the energy balance Eq. (1) that explicitly accounts for canopy temperature. Kimball et al. (2002) reported leaf temperature in FACE experiments to increase by an average of 0.8°C and 0.6°C for cotton and wheat, respectively grown at 550 ppm (ambient at 360 ppm).

Figure 2 shows an example of F_T as a function of $[CO_2]$ in the range from 100 to 1100 ppm, which was prepared using a function of stomatal conductance versus $[CO_2]$ fitted to soybeans data (Allen, 1990), and setting r_{co} at a reference $[CO_2]$ of 360 ppm at which $F_T = 1.0$. The transpiration adjustment can be substantial, leveling at ~900 ppm. However, temperature increase due to global warming will increase ET_{max}, counteracting to a varying degree the $[CO_2]$ effect depending on conditions.

Integration of $[CO_2]$ effects on crop biomass production and water use into a comprehensive and standardized approach for use in crop models has not been attempted. The problem is compounded by the lack of FACE information allowing sufficient data points to fit RUE and TUE responses to $[CO_2]$ and to fit stomatal conductance responses in the entire $[CO_2]$ range between ambient and 1000 ppm. This is a significant limitation in the use of crop models in the projection of yield and water use under elevated $[CO_2]$ scenarios.

8 Conclusion

Adequate representation of the water cycle in cropping systems remains as one of the major concerns for crop model developers as it is involved in many critical processes such as the determination of the energy balance and crop temperature, the determination of crop productivity, and the movement of nutrients in the soil and their uptake.

Even though the water budget equation is rather simple, the scarcity of measurements and data to specify the parameters and evaluate the component processes, as well as the myriad empirical and mechanistic approaches available, makes it difficult for crop modelers to decide on suitable approaches to incorporate in their models. This chapter presents a summary of the most common approaches that are included in crop simulation models, the parameterization of the equations involved, and the main limitations currently observed.

One aspect needing attention regards the wide variety of methods that exist to represent atmospheric demand for water (usually synthesized by or called reference evapotranspiration). If data is not available to calculate the effect of environmental variables on water demand via the Penman-Monteith equation, it is of utmost importance that the alternative equations undergo a process of calibration at local to regional scales. The same applies when using soil and crop parameterization schemes that try to provide with the best possible data replacing observations and in situ measurements. Reducing uncertainty in the water cycle of crop simulation models must follow a systematic approach that starts with the best available input information to avoid cascade effects that could make results questionable and limit the use of models.

9 Future trends

Modelling water fluxes in the soil-crop-atmosphere system is still a challenging task. As in any other subcomponent of a crop model, increasing realism and obtaining high degrees of accuracy are paramount. However, as the crop modelling user community grows, and new systems have to be represented, especially to evaluate potential impacts of climate change and design suitable adaptation strategies, new topics emerge. Here we briefly present some of the most salient aspects that will shape the research agenda in the coming years.

- Effective parameterization schemes, internal calibration, and adaptation to data scarce conditions.

  One important barrier for the adoption of models, especially when representing the components of the agrohydrological cycle, is the tremendous disparity in data availability among locations and agricultural systems. While some regions have dense networks of automatic weather stations that record all the relevant variables to compute the Penman-Monteith equations, others have little access to reliable data, either because of the lack of stations and/or specific measurements (some of the only measure temperature and/or precipitation), or because of the lack of policies and resources to facilitate distribution and dissemination. Weather generators (i.e. computing algorithms able to produce synthetic series of weather variables with the same statistical properties of the original data series) can be an effective tool to generate data and allow the use of models. However, there are still many issues to be solved in weather generators such as internal calibration routines to obtain parameters of the probability distribution functions used, and their change according to large scale features that modulate climate such as the El Niño Southern Oscillation and/or climate change trends. At the same level, the methods used to determine evapotranspiration need to be locally calibrated and/or incorporate regionalization schemes that improve their ability to reproduce crop water demands.

The same applies to soil parameters. Since crop models require detailed characterization of spatially heterogeneous soil physical properties, and those variables are obtained in specialized laboratories, users usually face the problem of parameter selection. New data mining techniques can be used to examine data bases and infer soil physical properties from routinely available data (following the spirit of the traditional pedo-transfer functions).

- Modelling water cycle in complex topography with significant water runoff and subsurface lateral flow.

  Precision agriculture principles rely on the ability to effectively represent spatial heterogeneity and design specific management practices to optimize crop productivity on space. One important source of variability in soils and yield potential relates to undulated landscape that influences subsurface water flow. Traditional approximations of water cycle simulations focus on one-dimensional movement, failing to represent the dynamics of crop productivity and water use in complex terrain. Parallel computing techniques have been applied to deal with multiple grid simulations that need to be connected and exchange information from one another, to better represent cropping systems grown under these conditions (Ward et al., 2018).

  Another interesting application of such array of interconnected models could be the ex-ante evaluation of the benefits of rainwater harvesting techniques applied to support farming systems in drylands. This could be achieved by designing a grid of fallow land in the upper section of a small watershed and crops at the foothills.

- Coupling crop and hydrological models for water resources allocation.

  Historical arrangements, inflexible water allocation structures and regulations have created complex water distribution systems with strong degree of dependence and vulnerability. These systems normally perform well when water resources are sufficient but cause tremendous distress to water user communities when water resources become scarce. How to effectively distribute water resources is always a big problem that requires reliable modelling tools to explore solutions. Because irrigation decisions, both related to water management as well as to the type of irrigation system used, influence water availability for communities located downstream as they determine the magnitude of the return flows, it is necessary to accurately estimate crop water demands and the yield impact that water shortages can have on the crops grown in specific irrigation locations. This type of information can be generated integrating hydrological models with crop simulation models in a way that the information on water availability and water use becomes exchangeable in every simulation time step.

- Data assimilation from remote sensing. Forecasting real time water demand and crop productivity.

  Nowadays, the use of satellite information has positively influenced our ability to obtain relevant biophysical data from remote areas and study their behavior. Particularly, remote sensing (RS) has been used to estimate actual evapotranspiration for different purposes, such as consumptive water use, determination of crop coefficients and characterization of water use patterns. A promising area of research and development for the coming years corresponds to the use of data assimilation techniques (see Ines et al., 2013 as an example) to correct the estimation of the evolution of critical crop simulation parameters as well as to obtain the most recently available data of the meteorological conditions to reduce the inherent uncertainty in crop outcomes. Literature describes different approaches to combine the information, ranging from the direct use of RS in estimating crop performance (Prasad et al., 2006) and seasonal ET (Sánchez et al., 2015), to the incorporation of RS data as forcing mechanisms using data assimilation procedures to determine total seasonal growth and yield (Olioso et al., 2005). These types of technological advances can be further developed to produce operational forecasts of crop yields and water use by crops updated on a regular basis (i.e. every time new remote sensing or weather data is obtained).

The modelling of the different aspects of the hydrological cycle affecting cultivated lands has been the subject matter of enormous amounts of books and journal papers back to at least the beginning of the last century. Readers should be readily able to find further information on this topic. However, future research should place more emphasis on improving the modelling of crop water use, critical to agricultural lands, and an area where large uncertainties exist. Another area of fruitful research is the application of remote sensing and data analytics to complement models for near real-time assessment of water use and management in agriculture.

10 References

Ainsworth, E. A. and Rogers, A. 2007. The response of photosynthesis and stomatal conductance to rising [CO_2]: mechanisms and environmental interactions. *Plant Cell Environ.* 30(3), 258–70. doi:10.1111/j.1365-3040.2007.01641.x.

Allen, R. G. 1986. A Penman for all seasons. *J. Irrig. Drain. Eng.* 112(4), 348–68. doi:10.1061/(ASCE)0733-9437(1986)112:4(348).

Allen, L. H. 1990. Plant responses to rising carbon dioxide and potential interactions with air pollutants. *J. Environ. Qual.* 19(1), 15–34. doi:10.2134/jeq1990.0047242500190 0010002x.

Allen, R. G., Pereira, L. S., Raes, D. and Smith, M. 1998. *Crop Evapotranspiration: Guidelines for Computing Crop Water Requirements.* FAO Irrigation and Drainage Paper No. 56. FAO, Rome.

American Society of Civil Engineering. 2005. *The ASCE Standardized Reference Evapotranspiration Equation.* ASCE-EWRI Task Committee Report, 59p.

Annandale, J. G., Campbell, G. S., Olivier, F. C. and Jovanovic, N. Z. 2000. Predicting crop water uptake under full and deficit irrigation: an example using pea (*Pisum sativum* L. cv. Puget). *Irrig. Sci.* 19(2), 65–72. doi:10.1007/s002710050002.

Barrios Gonzales, J. M. 1999. *Comparative Study of Three Approaches in the Computation of a Soil Water Balance.* Master dissertation IUPWARE. KU Leuven University, Leuven, Belgium.

Basso, B. and Ritchie, J. T. 2018. Evapotranspiration in high-yielding maize and under increased vapor pressure deficit in the US Midwest. *Agric. Environ. Lett.* 3(1). doi:10.2134/ael2017.11.0039.

Benli, B., Bruggeman, A., Oweis, T. and Ustun, H. 2010. Performance of Penman-Monteith FAO56 in a semiarid highland environment. *J. Irrig. Drain. Eng.* 136(11), 757–65. doi:10.1061/(ASCE)IR.1943-4774.0000249.

Boogaard, C. A., Van Diepen, R. P., Rötter, J. M. C. A., Cabrera, H. H., VanLaar, H. L., DeWit, H. L. and TeRoller, J. A. 2011. *WOFOST Control Centre 1.8 and WOFOST 7.1.3.* Wageningen.

Camargo, G. G. T. and Kemanian, A. R. 2016. Six crop models differ in their simulation of water uptake. *Agric. For. Metereol.* 220, 116–29. doi:10.1016/j.agrformet.2016.01.013.

Campbell, G. S. 1985. *Soil Physics with BASIC: Transport Models for Soil-Plant Systems.* Elsevier, Amsterdam.

Campbell, G. S. and Diaz, R. 1988. Simplified soil-water balance models to predict crop transpiration. In: Bidinger, F. R. and Johanses, C. (Eds), *Drought Research Priorities for the Dryland Tropics.* International Crops Research Institute for the Semi-Arid Tropics, Patancheru, India, 220p.

Campbell, M. D., Campbell, G. S., Kunkel, R. and Papendick, R. I. 1976. A model describing soil-plant-water relations for potatoes. *Amer. Potato J.* 53(12), 431–42. doi:10.1007/BF02852657.

Castañeda-Vera, A., Leffelaar, P. A., Álvaro-Fuentes, J., Cantero-Martínez, C. and Mínguez, M. I. 2015. Selecting crop models for decision making in wheat insurance. *Eur. J. Agron.* 68, 97–116. doi:10.1016/j.eja.2015.04.008.

Castellvi, F. and Stockle, C. O. 2001. Comparing the performance of WGEN and ClimGen in the generation of temperature and solar radiation. *Trans. ASAE* 44(6), 1683–7. doi:10.13031/2013.7038.

Dai, Y., Dickinson, R. E. and Wang, Y. P. 2004. A two-big-leaf model for canopy temperature, photosynthesis, and stomatal conductance. *J. Climate* 17(12), 2281–99. doi:10.1175/1520-0442(2004)017<2281:ATMFCT>2.0.CO;2.

Dardanelli, J. L., Ritchie, J. T., Calmon, M., Andriani, J. M. and Collino, D. J. 2004. An empirical model for root water uptake. *Field Crops Res.* 87(1), 59–71. doi:10.1016/j.fcr.2003.09.008.

Doorenbos, J. and Kassam, A. H. 1979. *Yield Response to Water*. FAO Irrigation and Drainage Paper No. 33. FAO, Rome, 193p.

Doorenbos, J. and Pruitt, W. O. 1977. *Crop Water Requirements*. Irrigation and Drainage Paper No. 24. FAO, Rome.

Feddes, R. A., Hoff, H., Bruen, M., Dawson, T., de Rosnay, P., Dirmeyer, P., Jackson, R. B., Kabat, P., Kleidon, A., Lilly, A. and Pitman, A. J. 2001. Modeling root water uptake in hydrological and climate models. *Bull. Amer. Meteor. Soc.* 82(12), 2797–809. doi:10.1175/1520-0477(2001)082<2797:MRWUIH>2.3.CO;2.

Flerchinger, G. N. and Saxton, K. E. 1989. Simultaneous heat and water model of a freezing snow-residue-soil system I. Theory and development. *Trans. ASAE* 32, 565–71.

Green, W. H. and Ampt, G. A. 1911. Studies on soil physics: 1. Flow of air and water through soils. *J. Agric. Sci.* 4, 1–24.

Hargreaves, G. H. and Allen, R. G. 2003. History and evaluation of Hargreaves evapotranspiration equation. *J. Irrig. Drain. Eng.* 129(1), 53–63. doi:10.1061/(ASCE)0733-9437(2003)129:1(53).

Hargreaves, G. H. and Samani, Z. A. 1985. Reference crop evapotranspiration from temperature. *Appl. Eng. Agric.* 1(2), 96–9. doi:10.13031/2013.26773.

Horton, R. E. 1940. An approach toward a physical interpretation of infiltration capacity. *Soil Science Society of America Proceedings* 5, 399–417.

Ines, A. V. M., Das,N. N., Hansen, J. W. and Njoku, E. G. 2013. Assimilation of remotely sensed soil moisture and vegetation with a crop simulation model for maize yield prediction. *Remote Sensing of Environment* 138, 149–64.

Jara, J. and Stöckle, C. O. 1999. Simulation of water uptake in maize using different levels of process detail. *Agron. J.* 91(2), 256–65. doi:10.2134/agronj1999.00021962009100020013x.

Jarvis, N. J. 2011. Simple physics-based models of compensatory plant water uptake: concepts and eco-hydrological consequences. *Hydrol. Earth Syst. Sci.* 15(11), 3431–46. doi:10.5194/hess-15-3431-2011.

Jones, J. W., Hoogenboom, G., Porter, C. H., Boote, K. J., Batchelor, W. D., Hunt, L. A., Wilkens, P. W., Singh, U., Gijsman, A. J. and Ritchie, J. T. 2003. The DSSAT cropping system model. *Eur. J. Agron.* 18(3–4), 235–65. doi:10.1016/S1161-0301(02)00107-7.

Kimball, B. A., Kobayashi, K. and Bindi, M. 2002. Responses of agricultural crops to free-air CO_2 enrichment. *Adv. Agron.* 77, 293–368.

Kostiakov, A. N. 1932. On the dynamics of the coefficient of water-percolation in soils and on the necessity of studying it from a dynamic point of view for purposes of amelioration. *Transactions of 6th Committee International Society of Soil Science*, Russia, Part A, pp. 17–21.

Kremer, C., Stöckle, C. O., Kemanian, A. R., Howel, T., Ahuja, L. R., Reddy, V. R., Saseendran, S. A. and Yu, Q. 2008. A canopy transpiration and photosynthesis model for evaluating simple crop productivity models. In: Ahuja, L. (Ed.), *Response of Crops to Limited Water: Understanding and Modeling Water Stress Effects on Plant Growth*

Processes. Advances in Agricultural Systems Modeling Series 1. ASA, CSSA, SSSA, Madison, WI, 435p. doi:10.2134/advagricsystemmodel1.c6.

Lascano, R. J. and van Bavel, C. H. M. 2007. Explicit and recursive calculation of potential and actual evapotranspiration. *Agron. J.* 99(2), 585–90. doi:10.2134/agronj2006.0159.

Leakey, A. D. B., Ainsworth, E. A., Bernacchi, C. J., Rogers, A., Long, S. P. and Ort, D. R. 2009. Elevated CO_2 effects on plant carbon, nitrogen, and water relations: six important lessons for FACE. *J. Exp. Bot.* 60(10), 2859–76. doi:10.1093/jxb/erp096.

Lewis, M. R. 1937. The rate of infiltration of water in irrigation practice. *EOS Trans. Am. Geophys. Union* 18(2), 361–8. doi:10.1029/TR018i002p00361.

Li, K. Y., De Jong, R. and Boisvert, J. B. 2001. An exponential root-water-uptake model with water stress compensation. *J. Hydrol.* 252(1–4), 189–204. doi:10.1016/S0022-1694(01)00456-5.

Liang, H., Hu, K., Batchelor, W. D., Qi, Z. and Li, B. 2016. An integrated soil-crop system model for water and nitrogen management in North China. *Sci. Rep.* 6, Article 25755. doi:10.1038/srep25755.

Long, S. P., Ainsworth, E. A., Leakey, A. D. B., Nösberger, J. and Ort, D. R. 2006. Food for thought: lower than expected crop yield stimulation with rising CO_2 concentrations. *Science* 312(5782), 1918–21. doi:10.1126/science.1114722.

López-Urrea, R., Martín de Santa Olalla, F., Fabeiro, C. and Moratalla, A. 2006. Testing evapotranspiration equations using lysimeter observations in a semiarid climate. *Agric. Water Manage.* 85(1–2), 15–26. doi:10.1016/j.agwat.2006.03.014.

Monteith, J. L. 1977. Climate and crop efficiency of crop production in Britain. *Phil. Trans. R. Soc. London. B, Biol. Sci.* 281, 277–329.

Morison, J. I. L. 1985. Sensitivity of stomata and water use efficiency to high CO_2. *Plant Cell Environ.* 8(6), 467–74. doi:10.1111/j.1365-3040.1985.tb01682.x.

Olioso, A., Inoue, Y., Ortega-Farias, S., Demarty, J., Wigneron, J.-P., Braud, I., Jacob, F., Lecharpentier, P., Hotel, C., Calvet, J.-C. and Brisson, N. 2005. Future directions for advanced evapotranspiration modeling: assimilation of remote sensing data into crop simulation models and SVAT models. *Irrig. Drain. Syst.* 19(3–4), 377–412. doi:10.1007/s10795-005-8143-z.

Penman, H. L. 1948. Natural evaporation from open water, bare soil and grass. *Proc. R. Soc. London A* 193(1032), 120–45. doi:10.1098/rspa.1948.0037.

Penman, H. L. 1963. *Vegetation and hydrology. Soil Science* 96(5). doi:10.1097/00010694-196311000-00014.

Prasad, A. K., Chai, L., Singh, R. P. and Kafatos, M. 2006. Crop yield estimation model for Iowa using remote sensing and surface parameters. *International Journal of Applied Earth Observation and Geoinformation* 8, 26–33.

Priestley, C. H. B. and Taylor, R. J. 1972. On the assessment of surface heat flux and evaporation using large-scale parameters. *Mon. Wea. Rev.* 100(2), 81–92. doi:10.1175/1520-0493(1972)100<0081:OTAOSH>2.3.CO;2.

Ritchie, J. T. 1972. Model for predicting evaporation from a row crop with incomplete cover. *Water Resour. Res.* 8(5), 1204–13. doi:10.1029/WR008i005p01204.

Sánchez, J. M., López-Urrea, R., Doña, C., Caselles, V., González-Piqueras, J. and Niclòs, R. 2015. Modeling evapotranspiration in a spring wheat from thermal radiometry: crop coefficients and *E/T* partitioning. *Irrig. Sci.* 33(6), 399–410. doi:10.1007/s00271-015-0476-2.

Shuttleworth, W. J. and Wallace, J. S. 1985. Evaporation from sparse crops - an energy combination theory. *Q. J. R. Meteorol. Soc.* 111(469), 839–55. doi:10.1002/qj.49711146910.

Steduto, P., Hsiao, T. C., Fereres, E. and Raes, D. 2012. *Crop Yield Response to Water*. FAO Irrigation and Drainage Paper No. 66. FAO, Rome. Available at: http://www.fao.org/docrep/016/i2800e/i2800e00.htm.

Stöckle, C. O. and Campbell, G. S. 1989. Simulation of crop response to water and nitrogen: an example using spring wheat. *Trans. ASAE* 32, 66–74.

Stöckle, C. O. and Jara, J. 1998. Modeling transpiration and soil water content from a corn (*Zea Maize* L.) field: 20 min vs. daytime integration step. *Agric. For. Meteorol.* 92(2), 119-30. doi:10.1016/S0168-1923(98)00086-0.

Stöckle, C. O., Martin, S. A. and Campbell, G. S. 1994. CropSyst, a cropping systems model: water/nitrogen budgets and crop yield. *Agric. Syst.* 46(3), 335–59. doi:10.1016/0308-521X(94)90006-2.

Stöckle, C. O., Kjelgaard, J. and Bellocchi, G. 2004. Evaluation of estimated weather data for calculating Penman-Monteith reference crop evapotranspiration. *Irrig. Sci.* 23(1), 39–46. doi:10.1007/s00271-004-0091-0.

Tanner, C. B. and Sinclair, T. R. 1983. Efficient water use crop production: research or re-search? In: Taylor, H. M., Jordan, W. R. and Sinclair, T. R. (Eds), *Limitations to Efficient Water Use in Crop Production*. ASA-CSSA-SSSA, Madison, WI, pp. 1-27.

Thom, A. S. and Oliver, H. R. 1977. On Penman's equation for estimating regional evaporation. *Q. J. Royal Met. Soc.* 103(436), 345–57. doi:10.1002/qj.49710343610.

USDA. 2004. Estimation of direct runoff from storm rainfall. In: *National Engineering Handbook*. United States Department of Agriculture (USDA), pp. 1–51.

van den Berg, M., Driessen, P. M. and Rabbinge, R. 2002. Water uptake in crop growth models for land use systems analysis: II. Comparison of three simple approaches. *Ecol. Modell.* 148(3), 233–50. doi:10.1016/S0304-3800(01)00436-7.

Wang, Y.-P. and Leuning, R. 1998. A two-leaf model for canopy conductance, photosynthesis and partitioning of available energy. I: model description and comparison with a multi-layered model. *Agric. For. Meteorol.* 91(1–2), 89–111. doi:10.1016/S0168-1923(98)00061-6.

Wang, E. and Smith, C. J. 2004. Modeling the growth and water uptake function of plant root systems: a review. *Aust. J. Agric. Res.* 55(5), 501–23. doi:10.1071/AR03201.

Ward, N. K., Maureira, F., Stöckle, C. O., Brooks, E. S., Painter, K. M., Yourek, M. A. and Gasch, C. K. 2018. Simulating field-scale variability and precision management with a 3D hydrologic cropping systems model. *Precis. Agric.* 19(2), 293–313. doi:10.1007/s11119-017-9517-6.

Yates, D., Sieber, J., Purkey, D. and Huber Lee, A. 2005. WEAP21: a demand, priority, and preference driven water planning model: part 1, model characteristics. *Water Int.* 30(4), 487–500. doi:10.1080/02508060508691893.

Zhang, Y., Carey, S. K., Quinton, W. L., Janowicz, J. R., Pomeroy, J. W. and Flerchinger, G. N. 2010. Comparison of algorithms and parameterisations for infiltration into organic-covered permafrost soils. *Hydrol. Earth Syst. Sci.* 14(5), 729–50. doi:10.5194/hess-14-729-2010.

Chapter 5

Improving crop pest/disease modeling

J. M. Fernandes, Embrapa/Universidade de Passo Fundo, Brazil; W. Pavan, Universidade de Passo Fundo, Brazil; D. Pequeno, International Maize and Wheat Improvement Center (CIMMYT), Mexico; R. Wiest, Instituto Federal Sul-Rio-grandense (IFSUL), Brazil; C. A. Holbig and F. Oliveira, Universidade de Passo Fundo, Brazil; and G. Hoogenboom, University of Florida-Gainesville, USA

1 Introduction

One of the biggest accomplishments in human history has been the domestication of plants, providing a more continuous food supply and promoting the conformation of sedentary agricultural groups (Pérez-Jaramillo et al., 2016). However, since the early days of crop domestication, growers have been plagued by multitudes of pests and diseases causing hunger and social upheaval. Zadoks (2017) discussed selected historical pest and disease outbreaks in the Old World in view of their social and political consequences. The challenge persists up to present. It is estimated that crop pests and diseases are responsible for direct yield losses ranging between 20% and 40% of global agricultural productivity and regularly menace global food security (Savary et al., 2019). However, crop losses remain poorly recognized as an important driver in matters of food security, whereas plant diseases have had an enormous impact on livelihoods throughout human history (Zadocks, 2017). Crop pests and diseases impact individual farms, local commerce, national and international trade, and the global economy.

http://dx.doi.org/10.19103/AS.2019.0061.07

Dissemination of pests and diseases occurs through both natural and anthropogenic processes, facilitated by the increasing interconnectedness of the global food chain. More than half of all emerging diseases of plants are spread by an introduction (Bebber et al., 2013). Weather is the second most crucial factor (Bebber et al., 2014). Although pests are spread by human activities and aerial dispersal, prevailing climatic conditions are likely to determine their subsequent establishment and growth.

Food security depends on our ability to manage crop pests and diseases effectively. The potentially increased production and environmental risks from a changing climate are expected to make the challenge of providing sufficient food for a global population that is supposed to top 9.5 billion by 2050 which is even more difficult. Increases in human populations and demand for food, energy, and water combined with an uncertain future climate that is very likely to have higher temperatures and increased frequencies of extreme events, are sure to lead to increased food shortages unless cropping systems become more resilient to those changes (Godfray and Garnett, 2014).

Tools and techniques are needed to assist in developing strategies that can lead to higher food production and prevent crop production losses due to pest and diseases while maintaining the social, economic, and environmental sustainability. The global impacts of pests and diseases can be assessed by integrating life cycles models to crop growth models that includes damage mechanisms to simulate yield losses. Scenario development and analysis form a robust framework of such studies since the inputs of these models can be forced by global change scenarios, which include climate change scenarios. Examining the impacts of pest and disease outbreaks has a range of applications, including the design of research strategies, the evaluation, and guidance of policies, and improved plant protection management.

While the crop modeling components are continually evolving in many cases, the insect pests and pathogens are neglected or treated in a simplified fashion using conceptual models. Several groups are addressing this gap, but there are only a few instances in the literature where numerical solutions of the pests are coupled to crop models (Tonnang et al., 2017). Crop simulation models have great significance in transferring new technologies to the farmers and decision-makers, and Decision Support Systems for Agrotechnology Transfer (DSSAT) (Jones et al., 2003) has been one of the most important of them. Years of research and collaboration has turned DSSAT as a handy tool for researchers and policy-makers for decision-making and to answer what - if questions related to a cropping system (Sarkar, 2009). While the application domain has broadened, and modeling networks have expanded, DSSAT model implementations have largely remained as it was a decade ago and Fortran is still used as the programming language.

DSSAT's models are typically large constructions each containing their own implementations of very common approaches to modeling crop and soil processes. Fortran remains dominant primarily due to its legacy as the predominant language used by scientists and crop modelers (Jones et al., 2017). This reliance comes from significant past efforts spent to build those model components, which to date are still performing and functioning well, and are heavily used by many scientists as critical parts of ongoing research delivery.

The improvement and application of pest and disease models to analyze and predict yield losses including those due to climate change is still a challenge for the scientific community. Applied modeling of crop diseases and pests has mostly targeted the development of support capabilities to schedule scouting or pesticide applications. There is a need for research to both broaden the scope and evaluate the skills of pest and disease models (Donatelli et al., 2017). On the other hand, some scientific communities have the necessary knowledge to develop models that simulate the cycle of plant diseases. Therefore, there is the challenge of coupling independently developed models for different problems.

The dynamic linkage between disease and pest injuries and the host crop is created by coupling points between the pest and disease models and crop models. Coupling points are places where the values of state variables - the variables that represent the state of the physical quantities being modeled - can be exchanged with other models. The framework presented by Rabbinge and Rijsdijk (1981) and Boote et al. (1983) describes seven mechanisms of pest and disease damage on crops - that is, light stealer, leaf senescence accelerator, tissue consumer, stand reducer, photosynthetic rate reducer, turgor reducer, and assimilate sappers. The translation of these injuries into mathematical expressions offers the possibility to incorporate them into the biophysical processes simulated by crop models (Bregaglio and Donatelli, 2015).

Donatelli et al. (2017) listed three main crucial elements to be faced when a coupling point is realized: (1) suitable identification of the damage mechanisms and respective crop model outputs to be affected by the pest and disease injuries via coupling points; (2) pest and disease model outputs must be linked to crop model variables, either directly or via additional functions; and (3) synchronization of time step of the communication between the pest/disease and the crop model. In addition, other critical aspects should be addressed in integration between models. These include, for example, identifying appropriate models, specifying interactions (data conversions, variable types, etc.), and verifying the possibility of integration of source codes. Once this step is over, there is a need to choose the most appropriate integration approach and then implement the coupling between the models.

The chapter is organized as: (1) a brief overview of approaches for model coupling; (2) explaining the PEST subroutine present in the CROPGRO family of models within DSSAT; (3) using a strategy of keeping the crop model separate from disease/pest models, and coupling through Message Passing Interface (MPI) functionality; and (4) presenting a proof of concept to demonstrate the communication approach for model coupling that examines the integrations performed in crop and pest/disease simulation models.

2 Approaches to model coupling

The methodological aspects in choosing a coupling approach are extremely important for the implementation of an appropriate integration. This is due to the degree of complexity in the processes involved. The most basic way to integrate multiple model codes, called the *monolithic* approach, is to merge them into a single program. Some coupling-like techniques allow for limited interaction between models, and are collectively referred to as the *scheduled* approach. More sophisticated approaches involve frameworks designed to support model coupling. We use the term *frameworks* to refer to software systems that assist in designing software by providing a foundation upon which more complex and customized software can be built. With respect to model coupling, frameworks provide the building blocks to create coupled models. Some frameworks focus on enabling models to communicate, called *communication* frameworks, while others focus on creating models from software components, called *component* frameworks.

2.1 Monolithic approach

The monolithic approach is defined as the formulation of a computer program through a single source code, where it implements fragments of the source code of two or more models, creating a new customized model. When using this approach, a certain degree of smoothness is attained during code integration and execution, since it works with a single programming language. The monolithic approach had the advantage of being familiar to scientists, as it is not much different from the normal model writing process, that is, composing subroutines and writing source code. Another critical point is the control over all details of the model source code such as structure, data input, output formats, data types, memory allocation, among others (Bulatewicz, 2006).

2.2 Scheduled approach

Some applications are scheduled coupled when one model uses output data from another model at the given time intervals. The overlapped region(s) between the two models define the coupling interface. In some cases, there is a need for data transformation to standardize the units of measurements in both models. An application of this approach uses CSM-CROPSIM, a wheat simulation model existing in DSSAT, which simulates the growth and development of wheat in combination with a Fusarium Head Blight simulation model (Del Ponte et al., 2009). In brief, the second model uses the output from the first model as input.

2.3 Component-oriented approach

The component approach to model coupling is similar to the monolithic approach in that the result of the coupling is a single model code. However, rather than decomposing the constituent model codes into blocks of source code designed for integration into another specific model code, the scientist decomposes the model codes into software components, which are modular and reusable subroutines. With the use of components, it is possible to easily apply software engineering techniques such as testing, upgrades, comparisons, and verifications. Components are easily aggregated and regrouped into new constructions and can be reused in future compositions. An example is presented by Bregaglio and Donatelli (2015). Even using this approach, with legacy codes, there is still a need to know the codes of the models in detail. Besides, it is necessary to know the order of execution of the model in a possible conversion to components which requires a great effort of reprogramming.

2.4 Communication approach

The communication approach presents itself as the most complex, but at the same time the most used, as it provides the integration of legacy codes with more current technologies, thus increasing the life of the model (Valckle et al., 2012). The codes of the models inserted in this approach remain independent but interact through the exchange of data via messages during the execution process. The primary functions of a data exchange interface, which follows the communication-oriented approach, are the constitution of flows, the transformation of data, and sometimes can control the initialization of the model or track the overall state of the integration. The models that follow this approach can be classified by the use or not of independent applications that are intermediate between the execution and the communication models. Without the use of coupling interfaces (dependent applications) they are considered data transfer libraries, that is, custom routines for data conversion and definition of the communication mode used by the models. With the use of independent applications (with coupling interface), they have communication libraries that directly support the model-to-model interface as well as support

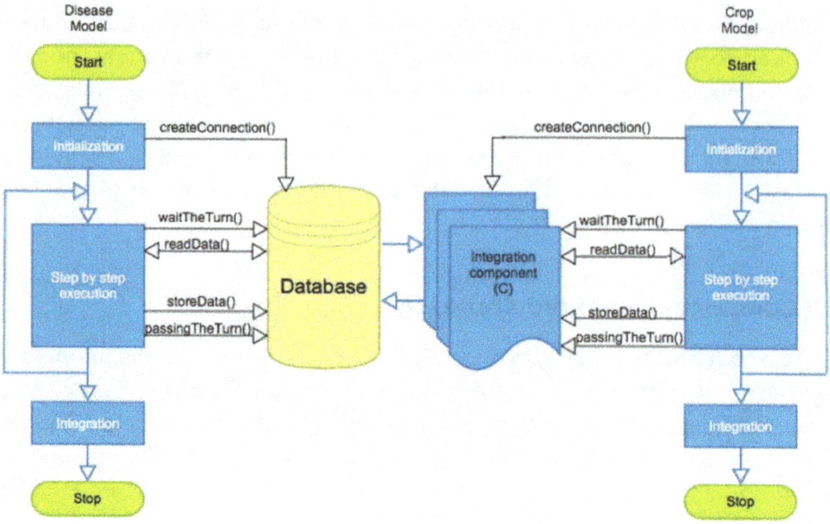

Figure 1 General coupling schema through a database management system (DBMS).

the interface model to the coupling interface. Rodrigues et al. (2012) built interfaces in the C programming language to bridge the communication between a disease simulation model written in Java and a crop model written in Fortran. Integration takes place through the exchange of data between the Fortran-C interface with the C-Java interface. Lazaretti et al. (2016) reported a communication-based approach of a crop model and a plant disease model running concurrently while exchanging data through an intermediate relational database management system (Fig. 1).

We presented a number of different frameworks that can be used for model coupling, organized into four different approaches. Clearly, the component approach is an ideal way to construct new models if components are available, but the approach is impractical for coupling existing models. The communication approach though, allows existing models to be coupled with minimal changes to the model source codes, but still makes use of existing coupling points. Since we are interested in model reuse, we will focus on the communication approach in this chapter, and refine our use of the term model coupling to refer specifically to this approach.

3 The pest and disease damage module (PEST)

Pest and disease-induced yield losses, a world-wide issue, are not addressed in most crop models. Early attempts like generic pest coupling subroutines in DSSAT-CROPGRO models allow entry of time series pest-scouting data, which in turn is used during simulation to interpolate the damage between scouting

Published by Burleigh Dodds Science Publishing Limited, 2026.

dates to predict yield loss to pest. Unlike the statistical models which cannot capture non-linearity in responses outside the boundary of data used to fit model, process-based models can efficiently achieve the underlying biological process, including process-based quantification of implications of climate change for the crop losses due to pest and diseases.

Several efforts have been done to model pest and disease reduction effects in crop growth and yield (Boote et al., 1983; Teng et al., 1998; Batchelor et al., 2000). Commonly pest damage is represented by percent difference loss in leaf mass and/or area, or disruption of plant process between treatment and control being more useful if expressed in a daily base. Boote et al. (1983) showed plant physiology aspects that can be affected by pests and how crop models can simulate it with reduction in growth and yield. Several aspects such as reduction in the stand, photosynthetic rates, light interception, cell turgor, increase in leaf senescence and respiration, and so on are caused by pests. Teng et al. (1998) presented a conceptual framework for linking pest effects on crop models belonging to the CERES and CROPGRO families using many approaches and examples with field data for many crops. The DSSAT pest and disease damage module (PEST), initially developed for the CROPGRO models (Batchelor et al., 1993), can model damage on different plant parts and tissues including leaf area reduction, assimilate loss, loss of leaves, fruits, stems, and roots (Jones et al., 2003). The module PEST of DSSAT, described in detailed by Batchelor et al. (2000), is called by the Plant module in a daily step to calculate pest damage. It has distinct rate and integration sections separated by plant growth routines, being called within the calculation section of plant growth. The PEST module of CROPGRO is structured in 11 subroutines (Table 1). In the DSSAT models observed damage from field experiments are read from input

Table 1 Subroutines of the PEST module of CROPGRO

Module	Description
PEST	The main subroutine
IPPEST	Reads simulation input information (FILEIO)
IPPARM	Reads pest data (FILEP)
IPPROG	Reads time series data (FILET)
LINDM	Interpolates pest damage linearly using observations
PESTCP	Calculates daily damage rates at each coupling point
ASMDM	Calculates assimilate damage
SEEDDM	Calculates seed and shell damage
VEGDM	Calculates leaf and stem damage
ROOTDM	Calculates root damage
OPPEST	Prepare pest damage outputs (PEST.OUT file)

files (FILET) and used in the simulations. Up to 40 pests can be simulated using a crop specific file, called the pest coefficient file (FILEP or *.PST), to apply damage to the appropriate coupling point. Up to six coupling points can be defined for each pest, in cases when the pests can damage more than one coupling point. The coupling points in the DSSAT PEST module associated with plant damage are state variables expressed in four ways: (1) daily absolute damage rate, (2) percent observed damage, (3) daily percent damage rate, and (4) daily absolute damage rate with pest competition and food preference effects. Some examples of coupling points in DSSAT are leaf area index (LAI), leaf, stem, root, and seed mass, seed and shell number, plant density, assimilates, necrotic LAI, vegetative nodes, and so on which are crop and pathogen specific (Batchelor et al., 2000).

After all growth rates, C and N rates, vegetative and reproductive senescence rates, and pest and damage rates are computed for, they are passed to a subroutine called GROW which integrates all processes and updates the values of the state variables at the end of the day.

4 Integrating a crop model and pest/disease models using Message Passing Interface (MPI)

4.1 Overview of MPI

MPI is a standardized set of libraries for parallel and high-performance computing (HPC), consisting in exchanging messages between processes. MPI has a protocol with specifications and definitions for resources optimization, defining an abstract application programming interface (API) that allows independent and compatible implementations. Due to the portability and availability of libraries for different languages such as C/C++, Fortran, and Java, MPI was quickly adopted as the standard for executing numerical software in HPC architectures (Gropp et al., 1999). Several implementations of the MPI standard are available as MPICH,[1] Intel MPI,[2] and OpenMPI.[3]

Usually, MPI has two implementation criteria: single program multiple data (SPMD) and multiple program multiple data (MIMD). In the SPMD, the same program, at independent points, runs in multiple autonomous processors simultaneously. In this criterion, tasks are split up, at runtime, and simultaneously executed on multiple processors. In MIMD, different programs can be run with different inputs in multiple autonomous processors simultaneously. It is frequently used as communication switches where applications can be written in different programming languages and easily communicated with

1 https://www.mpich.org/
2 https://software.intel.com/en-us/intel-mpi-library
3 https://www.open-mpi.org/

Published by Burleigh Dodds Science Publishing Limited, 2026.

each other, sharing information through a communicator/interface. In this way, MPI shows up as an architecture independent and efficient strategy to exchange information (Browne and Wilson, 2015). A MPI parallel technique was developed for an agroecosystem model, EPIC on global food and bioenergy studies (Kang et al., 2015).

4.2 MPI communicator for model processes

The execution using the MIMD implementation criterion allows the same approach to be used for coupling simulation models in parallel. The coupling of different simulation models requires time control, data communication, and synchronization. In this case, the development of a coupling interface allows simplifying the use and the implementation in the simulation model (Peckham et al., 2013).

The coupling interface manages the data communication and synchronization of the coupled simulation models. The purpose of the interface is to provide a set of reusable, portable standard features between different programming languages and to simplify communication of the simulation models (Peckham et al., 2013). The implementation should include methods that abstract the communication layer with MPI, so the coupling interface works as an independent module. These functions are used at the coupling point of the simulation model, enabling access to the initialization, communication, and finalization routines (Browne and Wilson, 2015; Dunlap et al., 2013).

Table 2, summarizes the main functions of the coupling interface. The *init_coupling* function initializes the MPI, enables message exchange, and check that the coupled simulation model has been started and can receive data. The *send_data* and *receive_data* functions are for data exchange between the coupled simulation models. To finalize the coupling between the models, the *finalize_coupling* function disconnects the coupled models, allowing the sequence of the execution flow. The 'getters' and 'setters' functions are to access data exchanged between models during execution. Figure 2 shows a diagrammatic representation of the models and functions.

Algorithm 1 shows pseudocode of the communication interface where the appropriate communication occurs. Pseudocode 1 represents the coupling interface operation. It is organized in three steps: initialization, model cycle, and finalization. The first step consists of initializing the simulation and coupling the models. The second step comprises the model's loop (rate, integration, and output). In this step, the logic of the model is updated and its state variables are updated. The communication functions are used to exchange information between the coupled simulation models. Finally, the termination function is called to close the connection and finalize the simulation.

Table 2 Main functions of the coupling interface

Function[a]	Description
void **init_coupling**()	MPI initialization and coupling verification.
void **send_data**(*int* destination)	Sends the stored data to the destination.
void **receive_data**(int source)	Receives the data from the source.
void **finalize_coupling**()	Finalize the coupling and close the connection with the attached models.
int **get_int**(*string* var_name)	Returns an integer value.
float **get_float**(*string* var_name)	Returns a float value.
string **get_string**(*string* var_name)	Returns a string value.
void **set_int**(*string* var_name, *int* value)	Store an integer value.
void **set_float**(*string* var_name, *float* value)	Store a float value.
void **set_string**(*string* var_name, *string* value)	Store a string value.

[a] Functions are defined by return type, function name, and parameters inside parentheses.

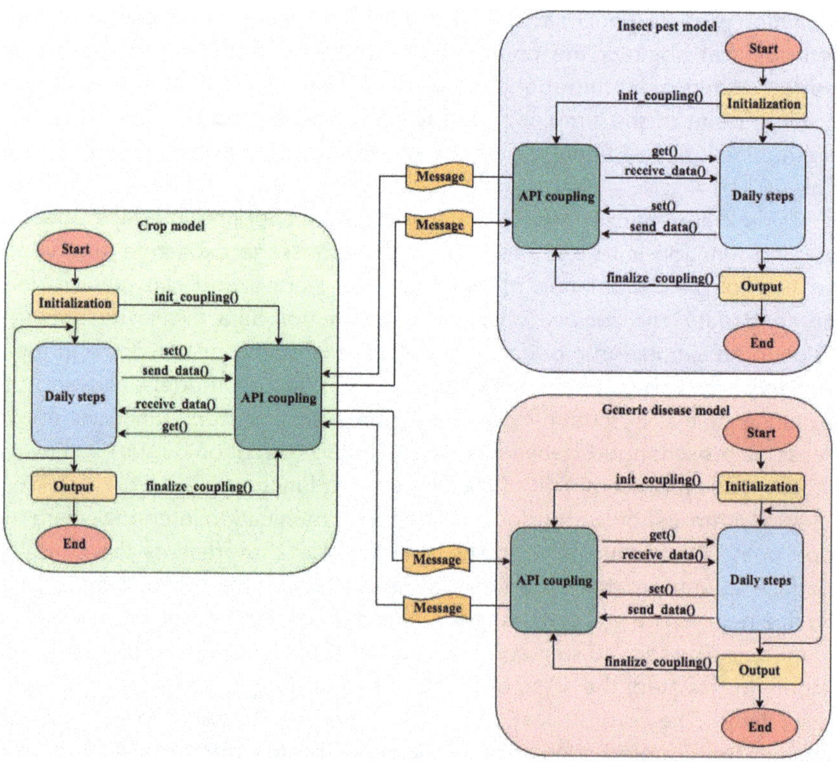

Figure 2 Diagrammatic representation of the models and functions.

4.2.1 Algorithm 1: model communication using a coupling interface

```
Set up model initialization
CALL init_coupling()
FOR daily steps:
    Rate calculations
        When the model wants to send data:
          1. Store the data using 'setters'
          2. CALL send_data()
    Integration
        When the model wants to receive data:
          1. Retrieve the data using 'getters'
          2. CALL receive_data()
    Daily output
END daily steps
Print outputs of the model
CALL finalize_coupling()
Finish the simulation
```

5 CROPGRO-soybean: insect pest and disease damage

Population growth, increasing affluence, rapid urbanization, and dietary transition are rising global demands for food and fuel products (Fan et al., 2017). Food security is decreasing in the context of the inter-linked food and economic crisis, and a number of studies (Paillard et al., 2014; Beddington et al., 2012) have indicated the need for increasing research efforts in the area of agriculture and climate change. They include the improvement of modeling capabilities to better assess the impacts on agricultural production of extreme weather events. In this context, there is also a need to integrate pests and pathogens into the climate change/food security debate. Many pests and pathogens exhibit a considerable capacity for creating, recombining, and selecting fit combinations of variants in key pathogenicity, fitness, and aggressiveness traits that there is little doubt that any new opportunities resulting from climate change will be exploited by them (Gregory et al., 2009).

As a significant source of protein for humans and livestock, the global consumption of soybean products has increased dramatically, more specifically, in Asia (Cao and Li, 2013). Hence, soybeans play a preeminent role in ensuring global food security. However, the soybean crop is exposed to a number of serious insect pests and plant diseases, which may, in turn, threaten global food security. Brazil is a prominent country in the international agricultural scene and this is due in large part to the role of soybeans in the national production of grains, placing the country as one of the main suppliers of these oilseeds (FAO, 2017).

Crop simulation models are considered important tools in the studies of interactions and changes in climate and their effects on different crops and the possible impacts of these changes on grain yield (Asseng et al., 2014). Since these models are able to make inferences about real systems, their application in agricultural systems currently represents a powerful resource for the evaluation of scenarios, management options and extrapolations of experimental results in space and time. They can also be used for an academic purpose, research assistance, support systems, and management decision management as well as strategic planning analysis and management policies (Jones et al., 2003). Here, an exploratory exercise establishes a procedure to integrate the CROPGRO-soybean model present in the DSSAT and insect pest/disease models. This procedure should enable to quantify the effects of insect pests and diseases on soybean production. The choice of DSSAT-CROPGRO-soybean simulation model relies on the fact that it has been successfully validated with many controlled field-experimental data across the world (Battisti et al., 2017) including Brazil. Most importantly, the DSSAT-CROPGRO-soybean simulation model has a PEST module which facilitates model coupling.

Here, we choose aphids to represent an insect pest. The virtual aphid, in the nymph phase as in the adult phase, can affect soybean productivity, causing direct damage. The damage is due to feeding on phloem sap (assimilate removal effect). The level of damage depends on the population density of the aphid. The population dynamics of aphids are affected by biotic factors (natural enemies such as parasitoids, predators, and pathogens) and abiotic factors (temperature, humidity, and rainfall). Temperature is one of the main factors that interfere with the development of aphid populations. This factor affects rates of development, reproduction, and survival, thus reflecting the population density on a plant (Lima et al., 2009).

Simulation systems have been used to represent a wide range of problems in the entomological area, including modeling of population growth, dispersion and migration (Parry, 2013; Toebe, 2014), life cycle (Isidoro et al., 2009; Zhou et al., 2010), and infestations (Perez and Dragicevic, 2010). In 2014, Toebe developed an agent-based model which simulates the life cycle of aphids that are considered pests to crops. In the architecture of the model, the simulation of population growth of particular aphid species considers factors that reflect the biology of the insect that needs to be parameterized.

The agents in the model are abstract representations of individuals being simulated. The environment in which the agents are embedded is delimited geographically, and agents may enter and exit this area according to their typical behavior. The environment has characteristics that include meteorological variables and host plants.

The execution of the model starts by loading the configuration settings. Upon loading the parameters, the virtual environment is assembled in the memory and the initial agent-insect population is distributed within the environment. The simulation is executed step by step, and at each step, weather data from a selected meteorological station is used as input and the routine behavior of each agent is implemented. Once the behavior of each insect is executed, the step is incremented and the process is repeated until the last step, which is the last day of the simulation. At the end of every step, the simulation results are saved in the database.

At every step of the simulation, the model considers the execution of the routine behavior of each agent in the virtual environment (Fig. 3). The first activity (development) involves the degree days accumulation in accordance with the function and development stage of the insect. It is followed by the feeding activity, which may or may not occur depending on the stage of the insect. The movement activity is also dependent on the stage of the insect and may or may not occur in apterous or winged form or both. The activity of reproduction only occurs for reproductive individuals and consists of the degree days accumulation up to a given threshold, which implies the birth of a new insect that depends on reproductive traits. Finally, the possible causes of mortality are investigated for the agent.

The effect of virtual aphid on soybean yield was simulated through CROPGRO-soybean simulation model. The coupling point in the PEST module is the variable TPSR (Daily absolute assimilate damage g[CH_2O]/m^2/d). It was empirically set that an amount of 0.0004 g[CH_2O]/d per aphid (Chander et al., 2006).

The aphid agent-based simulation model proposed by Toebe (2014) framework was used in this work. The model is generic, expandable and parameterizable and can be applied to different species of insects that infest different crops. The model was developed in the Java programming language.

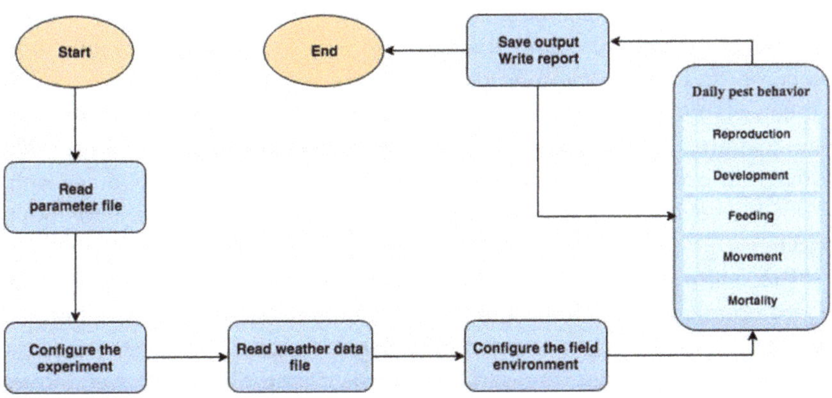

Figure 3 Fluxogram of the aphid agent-based model. Source: adapted from Toebe (2014).

The ability to add modules is useful in simulation problems because factors that were not considered in the initial design can be incorporated later.

In the case of plant disease, process-based models of the Susceptible-Exposed-Infectious-Removed (SEIR) type may represent a compelling approach for a generic modeling application. This type of model is generic even beyond the field of agriculture since the basic idea is also broadly accepted in animal and human disease epidemiology. The methods considered by this model type surely capture epidemiological processes that dictate epidemic build-up: disease transmission, the lag between infection and infectiousness of the host. Concepts and theories that exist and have been applied in a fragmented way so far can, therefore, be assembled toward an application for a generic epidemiological modeling platform (Donatelli et al., 2017).

Pavan and Fernandes (2009) developed a generic disease model, that is, the model can be parameterized to cover several diseases that occur in a given crop (Fig. 4). The model design aims a detailed representation of disease progress. The model was designed to mimic a fungal disease disease cycle. The generic disease model dynamically links to the PEST subroutine present in CROPGRO-soybean. The fungal leaf disease model was structured following the principles for coupling host and diseases dynamics introduced by Berger and Jones (1985). Disease dynamics were handled at the cohort level as proposed by Berger (1989) and applied in the development of a bean rust simulator (Berger et al., 1995). Parameters used in the model were extracted from literature (Rodrigues et al., 2012).

The number of cohorts is equal to the number of days the soybean plant has grown. Each cohort number corresponds the day the cohort has emerged. The generic model describes the disease progress on the soybean and its effects on growth and yield. Disease progress was modeled for each leaf cohort. Initially, the leaf cohort area was infection-free (healthy), but it was subjected to infection as time progressed depending on inoculum availability and environmental conditions. Disease progress of the entire plant canopy was computed by summing up disease progress on each leaf cohort. As the soybean plant naturally senesces, the total area of the cohorts matches to the entire leaf area calculated by CROPGRO-soybean. The generic disease model requires various input parameters that are produced by other modules within CROPGRO-soybean, related to crop growth, senescence, and environmental conditions.

The integration of CROPGRO-soybean and the generic disease model was via coupling points already implemented in PEST subroutine. In order to use these coupling points, the generic disease model calculates the diseased leaf area (PDLA: percent diseased leaf area %/d) and defoliation leaf damage (PLFAD: daily percent leaf area damage %). These coupling points are translated into specific model state variables, by VEGDM subroutine, like WLIDOT (daily

Published by Burleigh Dodds Science Publishing Limited, 2026.

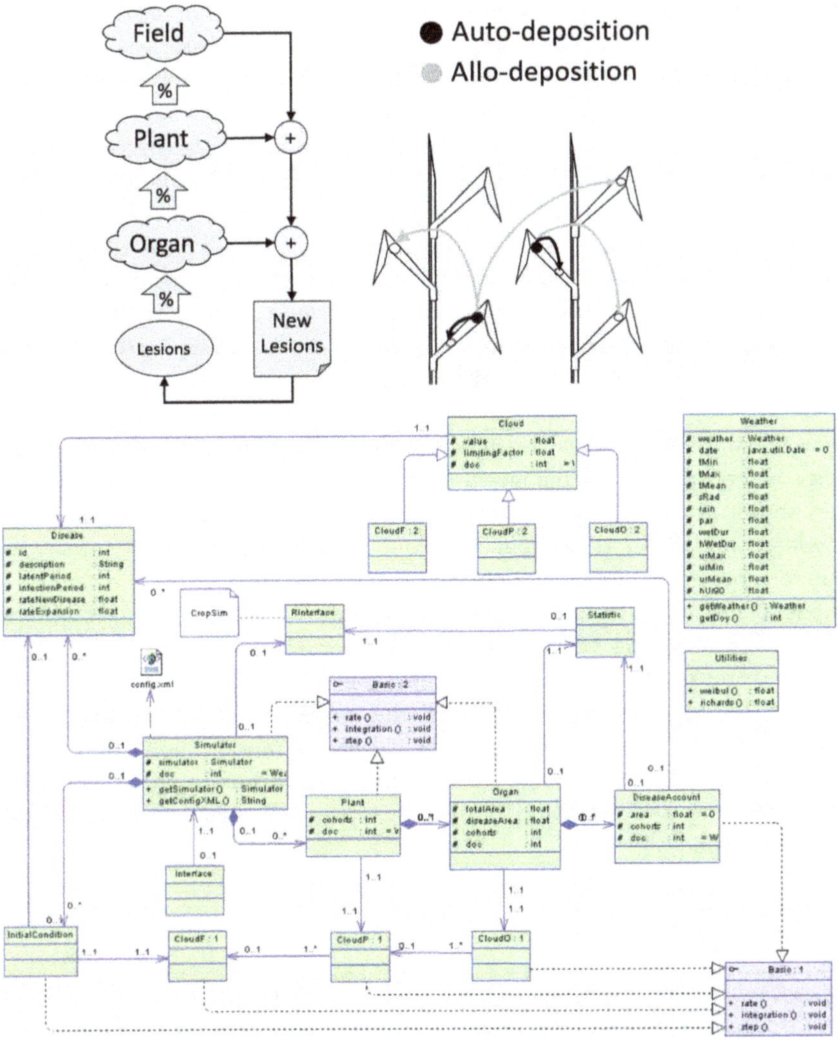

Figure 4 Diagram of the generic disease model (top image based on Willocquet and Savary, 2004).

pest or freeze damage to leaf mass g/m^2/day), LAIDOT (daily change in leaf area index due to pest damage m^2/m^2/d), and DISLA (diseased leaf area cm^2[leaf]/m^2/d) (Boote et al., 1983; Batchelor et al., 1993).

Leaf wetness is recognized as a very important determinant of plant diseases since it is fundamental for the fungal infection process. A leaf wetness duration empirical model that considers the number of hours with relative humidity above a specific threshold is the most common and easy to apply, and in this regard, several studies have been developed in many parts of the world

(Alvares et al., 2015). In our example, the number of hours of relative humidity greater than 90 was used as a proxy for leaf wetness duration.

A case study is used to explore the model coupling of a crop model, an insect pest model, and a plant disease model. The case study concerns a fictitious situation where a soybean crop is planned to be cultivated in Southern Brazil region. The growth of the soybean crop is constrained by the presence of insect pests and diseases. A hypothetical experiment that included the simulation of a generic soybean cultivar (maturation group 5.5) with and without the impact of an insect pest and/or a fungal foliar disease was prepared. The experiment was simulated over the soybean growing seasons of 2012 through 2017. Soil profile and weather data for the locality of Passo Fundo, RS, Brazil were extracted from AgroDB database (Lazzaretti et al., 2016).

The comparisons between the yield predictions and the expected yield show that the CROPGRO-soybean model was able to simulate the development and soybean yield. Both, the insect pest (aphids) and the fungal foliar disease had an impact on the soybean growth and final yield (Fig. 5). The simulated soybean yield differed between individual treatments and within each year due to weather conditions during the growing season. Overall, the simulated impact of the fungal foliar disease on the growth and final yield of soybean was greater than the impact caused by aphids.

By using the MPI_coupling interface, the CROGRO: Soybean simulation model, the agent-based model, and the generic disease model can be combined in a complex model. Most of all, without modification into the code of the models. This is made possible through the exchange of messages and also by saving and retrieving data dynamically with the coupling interface during execution.

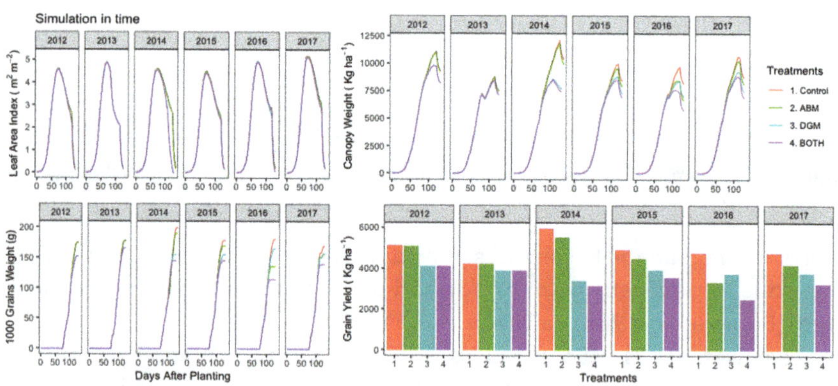

Figure 5 CROPGRO-soybean model daily simulation outputs for leaf area index, canopy weight, 1000-grain weight, and the final grain yield under no impact of pests and diseases (control), with the aphids effect (ABM), with the disease effect (DGM) and with the effect of pests and diseases (BOTH).

Published by Burleigh Dodds Science Publishing Limited, 2026.

The implementation of the MPI_coupling interface described in this chapter should be applicable for any model written in a language which has MPI bindings. These are generally directly callable from C, C++, and Fortran. Some programming languages can interface with the MPI libraries and are thus compatible with this technique. The programming languages R and Python, for example, are commonly adopted among data scientists and can interface with the MPI libraries.

The significant advantage of choosing the MPI_coupling interface is the speed of implementation. In our example, this was apparent when using the coupling points already present in PEST module of CROPGRO-soybean model.

6 Future trends and conclusion

The demand to predict the impact of pests and diseases on agricultural production systems is a critical component in the development and analysis of situations affecting producers' income and food security. We present a model coupling method which enables one to write generic and modular computational models. We show that by using this method it is possible to combine several computational models without modifying any existing code and only write new code for the coupling interface. This is a significant advantage for model development which reduces the probability of bugs and eases development, testing, and validation of computational models. The proposed model coupling approach should support system analysis including essential processes and their dynamics over an appropriate range of environmental variables. Possible applications of crop models integrated to pest and disease models include strategic decisions, such as breeding for host plant resistance in future climate scenarios, policy-making, priority-setting for research, applications for risk analysis of exotic invasive species, deployment of early warning systems, and for resource allocation. Further work should aim primarily to develop coupled models for crop model applications. It should target to serve various coupled models with flexibility, user-friendliness, and extensive coupling functions.

7 Where to look for further information

There is a Special Issue of the journal *Agricultural Systems* (volume 155) covering the foundation for the next generation of agricultural systems data, models and knowledge products. In the Introduction to this Special Issue, the authors described a vision for quickening the rate of agricultural innovation and meeting the growing global demand for food security. They also synthesize insights and formulate a strategy to advance data, models, and knowledge products that are consistent with the vision as mentioned above. In the Special Issue, there is an article on pest and disease modelling.

The Agricultural Model Intercomparison and Improvement Project - AgMIP (https://agmip.org) is a major international collaborative effort focusing on 'incorporating state-of-the-art climate products as well as crop and agricultural economic model improvements in coordinated regional and global assessments of future climate impacts'. In 2015, responding to the need to include modeling of pests and disease in agricultural assessments, AgMIP held a workshop at the University of Florida (https://conference.ifas.ufl.edu/pest/index.html).

The Community of Practice on Crop Modeling (CoPCM) is part of the CGIAR Platform for Big Data in Agriculture (https://bigdata.cgiar.org/communities-of-practice/crop-modeling/) and encompasses a wide range of quantitative applications. An example is a project 'Combining crop and disease modeling with numerical weather forecasting to inform wheat blast early warning systems in Bangladesh, Brazil, and beyond.' The CoPCM from the Big Data Platform supports the project.

8 References

Alvares, C. A., de Mattos, E. M., Sentelhas, P. C., Miranda, A. C. and Stape, J. L. 2015. Modeling temporal and spatial variability of leaf wetness duration in Brazil. *Theoretical and Applied Climatology* 120(3–4), 455–67. doi:10.1007/s00704-014-1182-3.

Asseng, S., Zhu, Y., Wang, E. and Zhang, W. 2014. Crop modeling for climate change impact and adaptation. In: *Crop Physiology: Applications for Genetic Improvement and Agronomy* (2nd edn.). Elsevier Inc., San Diego, CA, pp. 505–46. doi:10.1016/B978-0-12-417104-6.00020-0.

Batchelor, W. D., Jones, J. W., Boote, K. J. and Pinnschmidt, H. O. 1993. Extending the use of crop models to study pest damage. *Transactions of the ASAE* 36, 551–8.

Batchelor, W. D., Jones, J. W., Boote, K. J. and Porter, C. H. 2000. Pest and disease damage module. In: *DSSAT v.4.0 Documentation and Source Code Listing*. Research Report Number 2000-103. DSSAT Foundation, Gainesville, FL.

Battisti, R., Sentelhas, P. C. and Boote, K. J. 2017. Inter-comparison of performance of soybean crop simulation models and their ensemble in southern Brazil. *Field Crops Research* 200, 28–37. doi:10.1016/j.fcr.2016.10.004.

Bebber, D. P., Ramotowski, M. A. T. and Gurr, S. J. 2013. Crop pests and pathogens move polewards in a warming world. *Nature Climate Change* 3(11), 985–8. doi:10.1038/nclimate1990.

Bebber, D. P., Holmes, T. and Gurr, S. J. 2014. The global spread of crop pests and pathogens. *Global Ecology and Biogeography* 23(12), 1398–407. doi:10.1111/geb.12214.

Beddington, J., Asaduzzaman, M., Clark, M., Fernández, A., Guillou, M., Jahn, M., Erda, L., Mamo, T., Van Bo, N., Nobre, C., Scholes, R., Sharma, R., Wakhungu, J. 2012. Achieving food security in the face of climate change. CGIAR Research Program on Climate Change, Agriculture and Food Security (CCAFS), pp. 1–64.

Berger, R. D. 1989. Description and application of some general models for plant disease epidemics. In: Leonard, K. J. and Fry, W. E. (Eds), *Plant Disease Epidemiology: Genetics, Resistance and Management* (vol. 2). Kluwer Academic Publishers, Dordrecht, The Netherlands, pp. 125–49.

Berger, R. D. and Jones, J. W. 1985. A general model for disease progress with functions of variable latency and lesion expansion on growing host plants. *Phytopathology* 75(7), 792–7. doi:10.1094/Phyto-75-792.

Berger, R. D., Hau, B., Weber, G. W., Bacchi, L. M. A., Bergamin Filho, A. and Amorin, L. 1995. A simulation model to describe epidemics of rust of *Phaseolus* beans I. Development of the model and sensitivity analysis. *Phytopathology* 85(6), 715–21. doi:10.1094/Phyto-85-715.

Boote, K. J., Jones, J. W., Mishoe, J. W. and Berger, R. D. 1983. Coupling pests to growth simulators to predict yield reductions. *Phytopathology* 73(11), 1581–7. doi:10.1094/Phyto-73-1581.

Bregaglio, S. and Donatelli, M. 2015. A set of software components for the simulation of plant airborne diseases. *Environmental Modelling and Software* 72, 426–44. doi:10.1016/j.envsoft.2015.05.011.

Browne, P. A. and Wilson, S. 2015. A simple method for integrating a complex model into an ensemble data assimilation system using MPI. *Environmental Modelling and Software* 68, 122–8. doi:10.1016/j.envsoft.2015.02.003.

Bulatewicz, T. 2006. Support for model coupling: an interface-based approach. PhD thesis. Department of Computer and Information Science, University of Oregon. Available at: http://hdl.handle.net/1794/2974.

Cao, Y. and Li, D. 2013. Impact of increased demand for animal protein products in Asian countries: implications on global food security. *Animal Frontiers* 3(3), 48–55. doi:10.2527/af.2013-0024.

Chander, S., Ahuja, L. R., Peairs, F. B., Aggarwal, P. K. and Kalra, N. 2006. Modeling the effect of Russian wheat aphid, *Diuraphis noxia* (Mordvilko) and weeds in winter wheat as guide to management. *Agricultural Systems* 88(2–3), 494–513. doi:10.1016/j.agsy.2005.07.009.

Del Ponte, E. M., Fernandes, J. M. C., Pavan, W. and Baethgen, W. E. 2009. A model-based assessment of the impacts of climate variability on fusarium head blight seasonal risk in Southern Brazil. *Journal of Phytopathology* 157(11–12), 675–81. doi:10.1111/j.1439-0434.2009.01559.x.

Donatelli, M., Magarey, R. D., Bregaglio, S., Willocquet, L., Whish, J. P. M. and Savary, S. 2017. Modelling the impacts of pests and diseases on agricultural systems. *Agricultural Systems* 155, 213–24. doi:10.1016/j.agsy.2017.01.019.

Dunlap, R., Rugaber, S. and Mark, L. 2013. A feature model of coupling technologies for Earth System Models. *Computers and Geosciences* 53, 13–20. doi:10.1016/j.cageo.2011.10.002.

Fan, D., Ding, Q., Tian, Z., Sun, L. and Fischer, G. 2017. A cross-scale model coupling approach to simulate the risk-reduction effect of natural adaptation on soybean production under climate change. *Human and Ecological Risk Assessment: an International Journal* 23(3), 426–40. doi:10.1080/10807039.2016.1221308.

FAO. 2017. *FAOSTAT Production Crops*. Food and Agriculture Organization of the United Nations, Rome.

Godfray, H. C. J. and Garnett, T. 2014. Food security and sustainable intensification. *Philosophical Transactions of the Royal Society of London. Series B, Biological Sciences* 369(1639), 20120273. doi:10.1098/rstb.2012.0273.

Gregory, P. J., Johnson, S. N., Newton, A. C. and Ingram, J. S. I. 2009. Integrating pests and pathogens into the climate change/food security debate. *Journal of Experimental Botany* 60(10), 2827–38. doi:10.1093/jxb/erp080.

Gropp, W., Lusk, E. and Thakur, R. 1999. *Using MPI-2: Advanced Features of the Message-Passing Interface*. MIT Press, Cambridge, MA. ISBN 0-262-57132-3.

Isidoro, C., Fachada, N., Barata, F. and Rosa, A. 2009. Agent-based model of *Aedes aegypti* population dynamics. In: *Lecture Notes in Computer Science (including subseries Lecture Notes in Artificial Intelligence and Lecture Notes in Bioinformatics)* (vol. 5816). LNAI, 53–64. doi:10.1007/978-3-642-04686-5_5.

Jones, J. W., Hoogenboom, G., Porter, C. H., Boote, K. J., Batchelor, W. D., Hunt, L. A., Wilkens, P. W., Singh, U., Gijsman, A. J. and Ritchie, J. T. 2003. The DSSAT cropping system model. *European Journal of Agronomy* 18(3–4), 235–65. doi:10.1016/S1161-0301(02)00107-7.

Jones, J. W., Antle, J. M., Basso, B., Boote, K. J., Conant, R. T., Foster, I., Godfray, H. C. J., Herrero, M., Howitt, R. E., Janssen, S., Keating, B. A., Munoz-Carpena, R., Porter, C. H., Rosenzweig, C. and Wheeler, T. R. 2017. Brief history of agricultural systems modeling. *Agricultural Systems* 155, 240–54. doi:10.1016/j.agsy.2016.05.014.

Kang, S., Wang, D., Nichols, J. A., Schuchart, J., Kline, K. L., Wei, Y., Ricciuto, D. M., Wullschleger, S. D., Post, W. M. and Izaurralde, R. C. 2015. Development of mpi_EPIC model for global agroecosystem modeling. *Computers and Electronics in Agriculture* 111, 48–54. doi:10.1016/j.compag.2014.12.004.

Lazzaretti, A. T., Fernandes, J. M. C., Pavan, W., Toebe, J. and Wiest, R. 2016. AgroDB – integration of database management systems with crop models. In: *8th International Congress on Environmental Modelling and Software*. iEMSs, Toulouse, France, pp. 194–201.

Lima, E. A. B. F., Ferreira, C. P. and Godoy, W. A. C. 2009. Ecological modeling and pest population management: a possible and necessary connection in a changing world. *Neotropical Entomology* 38(6), 699–707. doi:10.1590/S1519-566X2009000600001.

Paillard, S., Treyer, S. and Dorin, B. 2014. *Agrimonde – Scenarios and Challenges for Feeding the World in 2050*. Springer, Dordrecht, The Netherlands, pp. 1–250. doi:10.1007/978-94-017-8745-1.

Parry, H. R. 2013. Cereal aphid movement: general principles and simulation modelling. *Movement Ecology* 1(1), 14. doi:10.1186/2051-3933-1-14.

Pavan, W. and Fernandes, J. M. C. 2009. Uso de orientação a objetos no desenvolvimento de modelos de simulação de doenças de plantas genéricos. *Revista Brasileira de Agroinformática* 9, 12–27.

Peckham, S. D., Hutton, E. W. H. and Norris, B. 2013. A component-based approach to integrated modeling in the geosciences: the design of CSDMS. *Computers and Geosciences* 53, 3–12. doi:10.1016/j.cageo.2012.04.002.

Perez, L. and Dragicevic, S. 2010. Modeling mountain pine beetle infestation with an agent-based approach at two spatial scales. *Environmental Modelling and Software* 25(2), 223–36. doi:10.1016/j.envsoft.2009.08.004.

Pérez-Jaramillo, J. E., Mendes, R. and Raaijmakers, J. M. 2016. Impact of plant domestication on rhizosphere microbiome assembly and functions. *Plant Molecular Biology* 90(6), 635–44. doi:10.1007/s11103-015-0337-7.

Rabbinge, R. and Rijsdijk, P. H. 1981. Disease and crop physiology: a modeler's point of view. In: Ayres, P. G. (Ed.), *Effects of Disease on the Physiology of the Growing Plants*. Cambridge University Press, Cambridge, UK, pp. 201–20.

Rodrigues, R. A., Pedrini, J. E., Fraisse, C. W., Fernandes, J. M. C., Justino, F. B., Heinemann, A. B., Costa, L. C. and Vale, F. X. Rd 2012. Utilization of the cropgro-soybean model to estimate yield loss caused by Asian rust in cultivars with different cycle. *Bragantia* 71(2), 308–17. doi:10.1590/S0006-87052012000200021.

Sarkar, R. 2009. Use of DSSAT to model cropping systems. *CAB Reviews: Perspectives in Agriculture, Veterinary Science, Nutrition and Natural Resources* 4(25), 1-12. doi:10.1079/PAVSNNR20094025.

Savary, S., Willocquet, L., Pethybridge, S. J., Esker, P., McRoberts, N. and Nelson, A. 2019. The global burden of pathogens and pests on major food crops. *Nature Ecology and Evolution* 3(3), 430–9. doi:10.1038/s41559-018-0793-y.

Teng, P. S., Batchelor, W. D., Pinnschmidt, H. O. and Wilkerson, G. G. 1998. Simulation of pest effects on crops using coupled pest-crop models: the potential for decision support. In: Tsuji, G. Y., Hoogenboom, G. and Thornton, P. K. (Eds), *Understanding Options for Agricultural Production*. Systems Approaches for Sustainable Agricultural Development (vol. 7). Kluwer Academic Publishers, Dordrecht, Netherlands, 221–66. doi:10.1007/978-94-017-3624-4_12.

Toebe, J. 2014. Um modelo baseado em agentes para o ciclo de vida de insetos: aplicação na interação afídeo-planta-vírus. Thesis (PhD in agronomy). Universidade de Passo Fundo.

Tonnang, H. E. Z., Hervé, B. D. B., Biber-Freudenberger, L., Salifu, D., Subramanian, S., Ngowi, V. B., Guimapi, R. Y. A., Bruce, S. Y., Kakmeni, F. M. M., Affognon, H. D., Niassy, S., Landmann, T., Ndjomatchoua, F. T., Pedro, S. A., Johansson, T., Tanga, C. M., Nana, P., Fiaboe, K. M., Mohamed, S. F., Maniani, N. K., Nedorezov, L. V., Ekesi, S. and Borgemeister, C. 2017. Advances in crop insect modelling methods–Towards a whole system approach. *Ecological Modelling* 354, 88–103. doi:10.1016/j.ecolmodel.2017.03.015.

Valcke, S., Balaji, V., Craig, A., Deluca, C., Dunlap, R., Ford, R. W., Jacob, R., Larson, J., O'Kuinghttons, R., Riley, G. D. and Vertenstein, M. 2012. Coupling technologies for earth system modelling. *Geoscientific Model Development* 5(6), 1589–96. doi:10.5194/gmd-5-1589-2012.

Willocquet, L. and Savary, S. 2004. An Epidemiological Simulation Model with Three Scales of Spatial Hierarchy. *Phytopathology* 94(8), 883-891. doi:https://doi:org/10.1094/PHYTO.2004.94.8.883.

Zadoks, J. C. 2017. On social and political effects of plant pest and disease epidemics. *Phytopathology* 107(10), 1144–8. doi:10.1094/PHYTO-10-16-0369-FI.

Zhou, Y., Arifin, S. M. N., Gentile, J., Kurtz, S. J., Davis, G. J. and Wendelberger, B. A. 2010. An agent-based model of the *Anopheles gambiae* mosquito life cycle. In: *2010 Summer Simulation Multiconference*. Society for Computer Simulation International, pp. 201–8. doi:10.1007/s11192-015-1581-y.

Part 2

Developing whole farm system, landscape and regional models

Chapter 6

Whole-farm system models in practice: diverse applications

Alison M. Laing, Cam K. McDonald, Andrew J. Ash and Diane B. Prestwidge, CSIRO Agriculture and Food, Australia; and Holger Meinke, University of Tasmania, Australia

1 Introduction

Modern agricultural production faces continued and increasing demands from consumers, governing agencies and society at large. These demands are a consequence of concerns about climate variability and change; labour scarcity and cost; volatility in the prices of both inputs and goods produced; the nexus between food production, food availability, nutrition and health (e.g. SDGs 2 and 3; https://www.un.org/sustainabledevelopment/sustainable-development-goals/); animal welfare; environmental impacts; and scarcity and competition for resources such as arable land and water (Postel, 1998; Seckler et al., 1998; Wheeler and von Braun, 2013; Jones et al., 2017a). These are global challenges that apply to high-income countries such as Australia as well as low- and middle-income countries elsewhere. Concurrently, there is an emphasis on the need to produce more food, and other agricultural products to sustain nutrition for a growing human population (Sutton et al., 2013), but to do so within sustainable development goals. In response to the challenges faced, many farmers are changing and/or adapting their management practices to mitigate risks, for example by increasing mechanisation, switching

http://dx.doi.org/10.19103/AS.2019.0061.09

to zero tillage crop establishment, or introducing new animal and/or plant breeds: these changes themselves affect production systems and may result in unforeseen, inadvertent consequences if poorly planned or implemented.

Many improvements in farm production focus on sustainably intensifying systems with the aim of producing more product (e.g. more grain, milk, meat) per unit of land, water, or energy, and/or reducing total greenhouse gas emissions or their intensity (e.g. Islam et al., 2019). Each production system is itself complex with interactions within and between system subcomponents but research into improvements has historically been conducted within the confines of agronomic disciplines, for example focussing on ways to intensify a cropping system, a dairy production system or an extensive grazing production system (e.g. Jat et al., 2016; Ash et al., 2015). Research activities in each discipline may incorporate field experiments, social science, policy engagement and simulation modelling. Models are a useful and practical method of examining, for example, the complexity, risk, variability and impact of different management practices on crop, pasture and animal production, or greenhouse gas emissions from plants or livestock (e.g. Herrero et al., 2007; van Ittersum et al., 2008; Van Wijk et al., 2009). Models complement and extend field studies, for example by infilling data not easily measured or by capturing greater diversity in management, climate and environment than is possible in the field. Farms often consist of more than one production system (Fig. 1) and there are likely to be complex interactions between the component subsystems that make up a farming system household or enterprise.

Smallholder farms in low-and middle-income countries (e.g. those in much of Asia, Africa and the Pacific) typically include multiple livestock and crop components and available labour is carefully managed to ensure the achievement of household goals (Manivong et al., 2014). For many of these

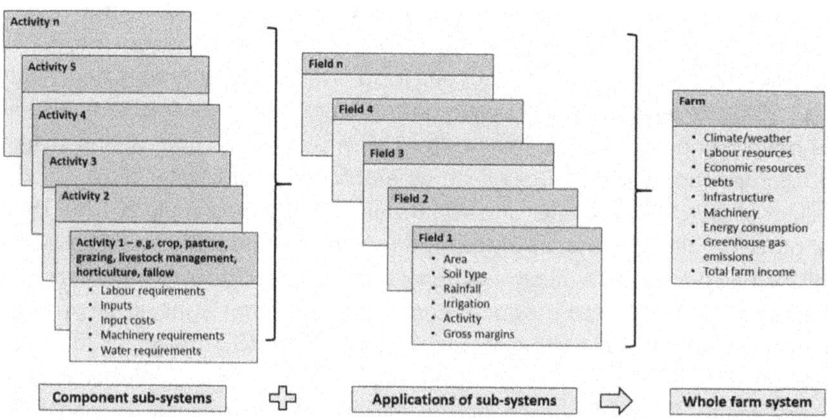

Figure 1 Complexity of components of a whole-farm system.

households, while increasing productivity and net incomes are important goals, maximising production may not be the most desirable outcome. They are usually more concerned with managing risks, as the costs of inputs (e.g. fertilisers, herbicides, high quality concentrate feeds and supplementary water) are often high, the quality and reliability of inputs is variable or poor and the sale price of farm products is low and uncertain (Newby et al., 2013). Many farmers, such as those in South East Asia (e.g. Laing et al., 2018) seek to achieve household food security by growing sufficient produce with minimum inputs (of chemicals, water and labour) so that family members can direct their labour elsewhere to higher-return activities away from the farm. In other instances farming households and businesses want to intensify their production to achieve more than the bare accomplishment of household food needs (e.g. Komarek et al., 2012; Lisson et al., 2010). Such intensification inevitably increases the risks associated with production. These risks relate to variability in the external environment (climate, economic conditions, logistics, labour availability, policy environment) as well as to the household-specific environment (soil types, financial resources, farm management capabilities). Consequently, the goal is finding the appropriate level of risk that a household or business is prepared to accept, or can sustain, on the path to intensification. Hence, *maximising* production is rarely the goal; instead producers may be interested in finding the optimal level of intensity that allows them to reduce losses in bad years, while capitalising on the good years, thereby increasing production overall.

Farm production systems in higher-income countries (e.g. Australia, New Zealand and some European countries) often have fewer subsystems and a deliberate focus on one (e.g. dairy, horticulture, beef) or two (e.g. wheat-sheep) farm enterprises. However, managing the farm subsystems to achieve defined production goals remains challenging. A high production target for individual enterprises may not result in the best economic return or environmental outcome for the farm. A whole-farm production goal that aims to achieve sustainable economic returns while minimising negative environmental consequences such as soil erosion, greenhouse gas emissions or nutrient leaching will generally be more desirable.

While the subsystem components of a farm can be highly complex, combining them into an entire farming system adds another layer of complexity resulting from interactions, feedbacks, synergies and trade-offs within and between its components. When seeking to achieve whole-farm goals (e.g. increased production, higher economic return, reduced labour requirements, lowered environmental degradation) it is necessary to account for the interactions and feedbacks between subsystem components. Individual farms differ in their biophysical conditions (soil, microclimates), resource constraints and opportunities (labour and capital), economic potential and managers' preferences for risk. Any quantitative evaluation of these interactions requires tools capable of estimating the flows of resources within the farm and its subsystems. In other words, it requires mathematically based models that are grounded in our scientific understanding of animal growth and reproduction, crop physiology, soil and environmental sciences, as well as economics and human relations. Such models are often invaluable tools in the process of undertaking nuanced, multifaceted management decisions at the farm system level.

Development of agricultural models has been ongoing for some 30–40 years (e.g. de Wit, 1958) and has provided an important predictive and assessment capability in terms of individual crops or forages (Holzworth et al., 2015) and of livestock (e.g. Johnson, 2016). However, most of these models have been developed as research tools to answer specific questions regarding a particular crop or a particular region, rather than to answer questions from end-users such as farmers or local governments (Antle et al., 2017a; Jones et al., 2017a). Many researchers combine existing models to attempt to simulate the whole farm: for example, Bell et al. (2009) combined a sheep model (GRAZPLAN) with a crop model (APSIM) to simulate mixed farming enterprises. The post-hoc combination of models does not enable a full assessment of the interactions between components of the farm system. The development of whole-farm models such as NUANCES-FARMSIM (Van Wijk et al., 2009), SEAMLESS (van Itersum et al., 2008) and IMPACT (Herrero et al., 2007) has helped to overcome some of these limitations. However, most whole-farm models lack important components of farming systems such as labour, economic resources, animal production, vulnerability to failure or are specific to a particular region (Parsons et al., 2011).

Australian scientists have led the development and application of the APSIM modelling framework for several decades (Keating et al., 2003; Holzworth et al., 2018). The APSIM framework provides a useful repository for complementary and contrasting approaches to component modelling. APSIM has been primarily conceptualised as a point source model that allows for the quantification and evolution of the components of cropping systems at a particular location. APSIM model output provided valuable input to the development of whole-farm models (e.g. Integrated Analysis Tool, McDonald et al., 2004; Lisson et al., 2010; McDonald et al., 2019). More recent developments in APSIM have led to the development of the whole-farm model APSFarm (Rodriguez et al., 2011), which is a logical extension of the individual field modelling of the APSIM model (www.apsim.info). The APSFarm model differs from APSIM in such a way that fields within a farm can transition from a crop to a fallow, or another crop, based on a set of rules related to various farm capacities, for example land, machinery, technical skills, attitude to risk. In this way various crop management strategies can be assessed at the farm scale against productivity, economic and environmental objectives, for example changes in soil fertility or in response to climate change (Rodriguez et al., 2014).

In this chapter we consider the desirable components of whole-farm models before examining how a suite of farm system models has been developed and evolved to meet changing research needs. We examine case studies of whole-farm modelling from smallholder subsistence farming systems to large scale, extensive beef enterprises using whole-farm models developed by the Australian Government's Commonwealth Scientific and Industrial Research Organisation (CSIRO) for specific applications in different research contexts. Many whole-farm models have been developed from a purely scientific/research perspective, (Antle et al., 2017a; Jones et al., 2017a) however the models we discuss below have been developed in collaboration with end-users and other stakeholders to ensure their needs are met as part of the research process (Antle et al., 2017b; McDonald et al., 2019). We review applications of these models in Indonesia, Vietnam, China and northern Australia to demonstrate a range of applications and discuss the ways in which these particular models, and farming systems models in general, are of use to farmers, researchers, policymakers and other stakeholders in modern and evolving agriculture.

2 Model selection

Model selection depends upon the capability and intended use of the model, the availability of sufficient input data and the capacity to simulate the questions to which answers are sought. For example, smallholder farm systems require examination of development technologies and sustainable intensification strategies as well information for extension services, while for commercial enterprises information on high intensity agriculture and corporate sustainability are required (Antle et al., 2017a). It is uncommon that all these will be met within a single model (and so it is important to select a model that can answer the research question(s) posed): for example while infrastructure such as good roads to major markets may be critical to a corporate enterprise, it may be of little or no interest to a primarily subsistence-based smallholder, unless it impacts on local markets.

Many models are designed to examine the risk and variability of the farming system over the medium or longer term through analysis of scenarios simulated (e.g. de Voil et al., 2009). In contrast, other approaches use models to optimise farming systems (e.g. Janssen and van Ittersum, 2007; Taylor and Adelman, 2003) and identify management strategies to achieve a pre-set goal. Some models may contain routines to optimise certain aspects of the farm enterprise (e.g. crop production, cash flow etc.). However, while such routines can identify specific management options of benefit, they can also ignore or downplay other management options which may have many benefits across the farm without necessarily being *optimal*. Optimisation routines are usually limited in the parameters that are included in the optimisation of a particular aspect of the farm, and hence present overly simplified versions of whole-farm systems, and do not adequately capture the effects of interrelationships between farm activities (Janssen and van Ittersum, 2007; McDonald et al., 2019). Models which examine risk and variability may not provide a specific 'best' management option but they may provide more detailed information about the risk and variability for the whole farm of a range of possible management options.

Many models have been designed to incorporate iterative feedback from non-modelling stakeholders including farmers and other prospective end-users in order to deliver more appropriate, feasible and realistic improved technologies and management practices (Van de Fliert and Braun, 2002; Röling, 2009; Styger et al., 2011). Farmers are experts in their farming systems and bring an intuitive understanding of likely interactions between components of their system, and the nuances of trade-offs, consequences and benefits which will need to be encapsulated into any model. Participatory engagement between farmers and researchers improves model outcomes, improves the efficiency of the process of whole-farm modelling and ensures that the management options simulated are likely to be useful, feasible and practical (Conner et al., 2015).

The complexity of a research question does not imply that the model needs to be complex. A highly complex, data-intensive model may be very accurate in its system simulation, but the output produced may be of limited value in addressing practical farming questions, or of providing answers which are applicable at scale. A model which is overly specific in its input data requirements may produce simulation results that have limited applicability outside the specific input conditions. Simulation outputs should be meaningful across farms with similar geographies and climates, of a certain production type and/or with similar production goals and challenges (i.e. not just the small number of farms on which the model was parameterised and calibrated). Outputs should also be specific enough to be meaningful, distinguishing between key factors which are of interest to the target stakeholders, such as feeding systems or proportion of cropped or grazed land on a farm.

A farm system model with highly detailed input requirements may be challenging to use in specific research environments as it takes considerable time, from both researchers and farmers, to capture and input a complex dataset. Additionally, the level of detail required for models with high-input requirements may be at a level that farmers are not comfortable disclosing, such as financial records or debt repayment plans. Models with a high degree of difficulty in setup and use are less attractive to the general research community. These models have specific applications in specialised research environments, usually by the teams who developed and/or manage them, and are often of limited value to non-specialists. Increasing model detail and complexity may improve the model's heuristic and explanatory value, but usually reduces its practical usefulness (Meinke, 2019).

Many whole-farm models, developed for specific research contexts (e.g. examining forages or crop production within a farming system), either do not capture all components of the farm in sufficient detail to properly represent the entire farm (in particular they may lack consideration of labour or economic resources), or have limited general applications outside the specific research context in which they were created (McDonald et al., 2019). Timely and affordable access to labour is a critical component of farming systems in low- and middle-income countries but it is one of the factors which is often overlooked or not adequately captured within a farming system model (Connor et al., 2015). While it is almost impossible to 'validate' a whole-farm model in the same manner as a crop growth model or animal growth model (Sterman, 2000), they require continuing evaluation against indicators determined in consultation with experts (e.g. farmers, end-users) expert in the local farm systems (Holzworth et al., 2015).

3 The development and evolution of a suite of whole-farm models: a case study

In this section we consider the general components of whole-farm models before examining how these models are developed and can evolve to meet changing needs. As indicated above, although many models of whole-farm systems exist (e.g. NUANCES-FARMSIM (Van Wijk et al., 2009); SEAMLESS (van Ittersum et al., 2008); IMPACT (Herrero et al., 2007); APSFarm (Rodriguez et al., 2011); etc.), none of them are ubiquitously used. We explore the reasons for this lack of uptake in more detail via a case study example that explores the development of three inter-related whole-farm models: the Integrated Analysis Tool (IAT, McDonald et al., 2004, 2019); the North Australian Beef Systems Analyser (NABSA, Ash et al., 2015); and the Crop Livestock Enterprise Model (CLEM, https://research.csiro.au/foodglobalsecurity/data-and-tools/models/clem/). These models have been developed by the Australian government's CSIRO to extend whole-farm modelling capacity and meet changing research needs. We discuss the key points of each model, including their heritage and limitations. We summarise the arc of model development, evolution and utility.

3.1 The Integrated Analysis Tool (IAT)

The Integrated Analysis Tool (IAT) (McDonald et al., 2004; Lisson et al., 2010) is a Microsoft® Office Excel-based spreadsheet model developed for use with smallholder farmers in low- and middle-income countries. It combines crop production data (from an external cropping system model or estimates of historical yields) with internal livestock and socio-economic modules to quantify at the farm scale the effects of different management practices on one or more components of the farm. Ruminant growth, reproduction and milk production within the IAT are based on animal-available energy and protein, which, in turn, are based on the quantity and quality of feed inputs (e.g. pastures, stored fodder, supplementary feeds etc.) while other animal classes (e.g. poultry, pigs) are accounted for more simplistically. The IAT can be used to examine the effects at the whole-farm scale of different management practices on grain production, forage and pasture production, tree and vegetable production, and ruminant and non-ruminant livestock production (McDonald et al., 2019). The IAT includes farm resources such as labour, land (including different soil types) and cash, and can be used to track domestic resource consumption (by the household or livestock), produce sold off-farm, input purchases and whole-farm economics. Users can identify periods when resources are constrained or in surplus, both across the whole farm and within key components (e.g. livestock, crops, pasture/feed, labour) of the system. The IAT was the first model to facilitate rigorous concurrent simulation of climate variability, whole-farm economics and (ruminant) livestock (Komarek et al., 2012).

A key tenet underpinning both the development and the implementation of the IAT has been participatory engagement with the farming communities for whom and with whom it is used (Lisson et al., 2010; McDonald et al., 2019). Participatory engagement is vital to successfully parameterise the model for an agricultural environment: this enables researchers to thoroughly understand the farming systems involved while modelling and quantifying the constraints and risks of current (benchmark) farm management practices (Meinke et al., 2001; Jakku and Thorburn, 2010). Additionally, farmers' adoption of improved management practices increases when the proposed new practices are relevant both to their farm goals and their household risk preferences (Connor et al., 2015). Developing alternative management practices in partnership with farmers ensures that the new options are attractive and realistic for the communities in which the research has been undertaken. Indeed, Meinke (2019) stresses that practice change is most effectively achieved via innovation platforms, where model users and developers agree on the problem domain and the methods (including the models) to be employed in solution. Such an approach empowers all participants to act within their area of expertise and respect each other's knowledge. In such instances, models become 'discussion support' rather than 'decisions support' tools (Nelson et al., 2002).

The IAT was first developed to model intensive smallholder farming systems in Indonesia (Lisson et al., 2010); it has subsequently been applied in a range of both intensive and extensive production systems in low- and middle-income countries in South and South East Asia and Africa. In China it was used to demonstrate the benefits of increasing the proportion of animal production relative to crop production to increase overall farm profitability, and to help households move from subsistence-based to market-based production (Komarek et al., 2012). In a similar way, IAT was used in Vietnam to demonstrate that planting of forages and tree legumes increased turnover and farm profitability (Parsons et al., 2011). In India it was used to assess the impact of planting forages on animal production, labour demands and methane output (Shalander et al., 2017). In Burkina Faso, Rigolot et al. (2017) used IAT in conjunction with a local innovation platform to demonstrate potential outcomes of suggested interventions. Here, the model was used to show that interventions suggested by farmers would be of little benefit to their gross margins, but those suggested by researchers could potentially have large benefits. In the process, the IAT provided a useful means of exposing farmers to expert knowledge, while helping researchers gain a better understanding of farmers' constraints. This was also the case in Ethiopia and India, where Mayberry et al. (2018) showed that better nutrition and healthcare could increase goat production substantially. Importantly, they showed that interventions that gave the highest meat production or animal turnover were not always the most profitable.

Despite the demonstrated value of the IAT in assessing alternative farm management scenarios, it is limited because the scenarios it simulates are static; once prescribed they do not change. While the IAT is effective in comparing different management scenarios, the farm management simulated within the model does not dynamically change in response to changes in external drivers (e.g. input costs, grain price etc.), or internal drivers such as labour. As well, the IAT requires input data for all submodules in order to run: this can result in information being required (and employed) in the model that is surplus to the specific requirements of the scenarios under analysis. The IAT does not directly capture the effects of pests or diseases on simulation components such as crop yields or livestock productivity; these effects can be partially captured through post-hoc reductions in the performance of factors of interest. The IAT is an example of a flexible, hierarchical model design where more detailed models of the agricultural system under investigation can be used to generate the necessary input data. For example, Mayberry et al. (2018) used IAT to create scenarios with increased minimum mortality rates, to reflect an outbreak of disease within a herd, and with reduced mortality rates to simulate the response to vaccination. These new scenarios were then compared to existing baselines.

3.2 The North Australian Beef Systems Analyser (NABSA)

The North Australian Beef Systems Analyser (NABSA) model (Hunt et al., 2014; Ash et al., 2015) is a bioeconomic model created to examine single-production enterprises such as extensive beef cattle in Australia. It enables researchers to examine the farm-level effects (e.g. on production, input requirements, herd age, or greenhouse gas emissions) of different management practices such as feeding regimes, stocking rates, breeding programmes and so on over time. The NABSA builds on the framework developed in the IAT to incorporate livestock, pasture and forage production, and land condition, and combine these with labour demands, labour availability, farm economic resources and deficits to examine production goals for larger scale beef enterprises. Unlike the IAT, the NABSA focusses primarily on extensive beef-production systems; however the complexity of the components of these systems (e.g. animal growth, reproduction, mortality and sales; pasture production; forage production) is as interconnected and inter-dependent as the subcomponents of smallholder farms, and it is this complexity that the NABSA enables the user to capture and examine.

As with the IAT model, the NABSA is not designed to optimise production systems, rather it enables sensitivity analysis through comparison of different possible scenarios of production system management. The NABSA is generally run over several decades to quantify the longer-term variability and risk of different management options.

Animal growth and reproduction within the NABSA is based on the same energy/protein module as in the IAT, however the NABSA has expanded capabilities in modelling variations in reproduction with animal age and/or climate conditions, as well as capabilities for managing stock numbers in response to drought. Pasture and forage growth and availability are input parameters provided from external pasture and/or crop production models. Key outputs from the model are productivity, gross margins, land condition (which will also influence the rate of future production on-farm) and methane production, representing the main greenhouse gas emissions from management practices on ruminant-livestock based production systems. Farm gross margins and annualised profits (or losses) are calculated based on-farm revenues from animal sales less production costs; these can include longstanding farm financial commitments such as debt, interest payments and depreciation of large purchases.

Modelling using the NABSA commences with the establishment of baseline simulations to accurately reflect geographic location and climate, current herd structure and management and ensure input data from (native) pasture and forage crop simulations are accurate. Whole-farm economics, which may include components such as labour requirements and costs, irrigation infrastructure and costs, debt and interest and so on are captured, as are the production goals of the baseline farm simulations. Alternative management scenarios are then developed and run using the NABSA, with output compared to the baseline.

The NABSA does not directly include the effect of pests or diseases, either on livestock or crops, although their impacts on-farm production and their economic cost can be factored into simulations. In a similar manner to the

IAT, scenarios examined in the NABSA are static and cannot be altered during a simulation in response to external factors such as climate or economics. Instead, different static scenarios (representing different static climatic or economic environments) are compared against each other and against a baseline case.

3.3 The Crop Livestock Enterprise Model (CLEM)

The Crop Livestock Enterprise Model (CLEM) is a bioeconomic model and a new iteration of whole-farm modelling, incorporating and building on many of the features of both the IAT and the NABSA from which it was developed. It has been designed for use in diverse agricultural production systems with varying levels of intensification, land use and technology. It is intended to guide whole-of-farm decision-making by integrating livestock, pasture and crop production with labour and economic resources. The CLEM is capable of simulating the scenarios of both the IAT and the NABSA models and overcomes the more specific focus of each. As with the IAT and the NABSA, the CLEM does not directly simulate plant production: these data are imported (e.g. from pasture and crop growth models) and are used in the CLEM to examine scenarios of different management practices relating to livestock production, labour availability and demand and the economics of the whole farm and of system components.

Limitations in the IAT and the NABSA led to developments in the CLEM: for example, the clunky, Excel-based platform of the IAT and the NABSA, which has a poor user interface and requires complex inputs was replaced by a tree framework consistent with the APSIM suite of models: CLEM is considerably more user-friendly than its predecessor models. This tree structure adds flexibility to the model; modules which are not relevant to a particular research question do not need to be included in the simulation, making whole-farm simulation a faster process and, in data-limited environments, at risk of fewer errors or potentially erroneous assumptions.

As with previous models, the use of participatory engagement and iterative consultation with research partners and end-users such as farmers and other stakeholders is an important part of modelling with the CLEM.

The CLEM is currently being applied to case studies in Australia and internationally in Africa, South Asia and South East Asia. Case studies are underway and thus not presented here, however current research includes examining: (1) the long-term farming system production risk and variability under different allocations of labour, crop and livestock resources for smallholder farmers in India; (2) modelling strategies for farmers to transition from traditional mixed smallholder systems relying on draught cattle to modern, mechanised farms in Myanmar; and (3) mitigating the effects of climate change in Australian beef and wheat-sheep systems. The CLEM (unlike the IAT and the NABSA) is widely available online to researchers (www.apsim.info/clem).

4 Whole-farm modelling in smallholder systems: IAT

The IAT is of relatively low complexity and operates on a monthly time step and while it can be run over the short term (1-2 years) its strength lies in medium (5-10 years) to long (15+ years) term simulations where it is used to quantify the variability and risk of different farm management options by comparing the performance of alternative management scenarios to a baseline, generally representing current farm management practices. The IAT is designed for scenario analyses, where several alternative management practices are compared to the baseline practice, in order to examine differences between scenarios and identify the benefits and limitations of each for a particular farming system. As part of these sensitivity analyses, key parameters can be systematically altered to examine their impacts on the production benefits and costs, and the longer-term variability of the farm system.

Modelling using the IAT commences with understanding the current farm management practices by benchmarking the existing crop, forage and pasture production practices; livestock management including husbandry, feeding and reproductive regimes; labour requirements for plant and animal activities and off-farm income; labour availability; and farm economic resources or constraints. Once benchmarked, the baseline model is run over several years so that the medium- to long-term performance can be used to provide an understanding of the risk and variability in the existing system. Improved management practices, where possible identified in partnership with farmers and other stakeholders, are then simulated and their risk profiles compared to that of the baseline in terms of their costs and benefits. Proposed improvements to management practices can be compared directly to the baseline and, as well, the effect of combining smaller management changes can be assessed. Often there are many areas (e.g. improved feedstock, improved livestock breeding) which can both singly and cumulatively benefit the smallholder household. In the summaries below cumulative benefits are highlighted to demonstrate the multifaceted analyses of inter-related components of the farm systems which have been examined.

4.1 Example 1: rice-cattle production system in Indonesia

The IAT was used in eastern Indonesia (McDonald et al., 2004; Lisson et al., 2010) to examine the potential benefits and costs for smallholder farmers of improving a representative current enterprise against a series of improved management options focussing on crop residues and forages for livestock, developed in collaboration with farming communities in eastern Indonesia. The baseline system comprised two adults (who provided the majority of farm labour) and two children; the family grew rainfed rice and groundnut crops annually, primarily for their own consumption, and managed two cows which were fed a combination of crop residues and cut and carry forages collected daily. Cash income was primarily from the sale of young cattle and supplemented by the sale of any surplus grain. The options proposed to improve whole-farm performance were: (1) increasing the quality and longevity of crop residues; (2) increasing the land area over which forages were grown; (3) increasing the number of cattle; and (4) optimising reproduction by introducing seasonal mating. These options were simulated cumulatively (each built on the previous) over a five-year simulation period.

As described in Lisson et al. (2010), relative to the baseline, increasing the quality and storage potential of crop residues lowered the farm fodder deficit and increased cattle sales and the farm cash balance to approximately USD $1554[1] (from a base of USD $989) over the five years, but labour remained in deficit. Increasing the area planted to forages created a fodder surplus and resulted in increased cattle numbers and cash balance (to USD $1625), and alleviated the labour deficit (on-farm production of forages reduced the labour required to feed animals relative to that required to source off-farm forages). Increasing the number of cattle on the farm eliminated the fodder surplus but considerably increased the number of cattle sold and hence increased the cash balance (to USD $2684) without causing a labour deficit. Introducing seasonal mating increased the number of cattle sold further, resulted in a fodder deficit, but maintained a sufficiency of labour within the household and again increased the cash balance (to USD $3037). The suite of IAT simulations demonstrated to the stakeholders involved in the research that there were plausible farm-level benefits for farmers who took up some or all of the proposed management changes. Also, the simulations highlighted trade-offs in the different management options; for example, they quantified the additional feed requirements of increasing livestock numbers. This knowledge better prepared farmers for the likely outcomes of implementing the improved management options. Across 30 participant households a total of 157 improved management options were taken up, often with individual interpretations of their specific on-farm implementation, and with long-lasting success (Lisson et al., 2010).

4.2 Example 2: crop-small ruminant production system in China

The IAT was applied in north-central China to quantify the effects and trade-offs of different strategies to increase net farm income by increasing fodder (lucerne)

1 Using a conversion rate of 1 US dollar to 14,157 Indonesian rupiahs.

and livestock production (Komarek et al., 2012). Over a 40-year simulation period Komarek et al. (2012) compared three baseline systems, representing farming systems which (1) had low land to labour ratios (these farmers relied on off-farm income); (2) were livestock focussed and relied on a combination of off-farm income and livestock for security; and (3) were managed by subsistence farmers who had very little cash reserves. The farms grew wheat and maize crops and produced pen-fed small ruminants (sheep or goats). For each of these farming systems, two new, cumulative, farm management options were also simulated. The first potential management practice replaced some of the wheat and maize crops grown with lucerne to increase on-farm fodder quantity and quality. The second option was to intensify (double) livestock numbers while maintaining the additional fodder grown and feed-to-animal ratios as in the first scenario.

The IAT output suggested that replacing grain crops with animal fodder increased the likelihood of needing to purchase grain for human consumption. However, livestock intensification facilitated by the additional farm-grown lucerne increased total farm income and thus, for some of the farming systems examined, it was financially advantageous to increase net farm income from livestock production. Additionally, intensifying livestock production generally reduced income variability without reducing average farm incomes. Smallholder farmers in this study were constrained more by land than by labour; off-farm earnings were the key to farm and household economic security. Options which increased on-farm income while minimising on-farm labour requirements were therefore important. Livestock (rather than crop) intensification proved to be a suitable strategy. However, the intensification options were not attractive to all farming households: participatory engagement identified that replacing farm-grown grain with market-bought resources had negative cultural implications as farming households, in particular more conservative households, preferred to meet their grain requirements on-farm. This case study demonstrates how a whole-farm model can be used to engage with farmers to generate new insights for all participants, including the scientists, involved. Such engagement is the hallmark of co-innovation, which is at the centre of most successful innovation platforms.

4.3 Example 3: goat production in India and Ethiopia

The IAT was applied on smallholder farms in the arid zone in India and in two grazing zones (pastoral lowlands and highland rainfall deficient) in Ethiopia to examine the effects of different management options to improve on-farm goat production, income and profitability (Mayberry et al., 2018). This research is an example of the IAT model used with national and local datasets and published literature, and without the participation of local farmers as the goal was to generate insights that could help prioritise research and development investments by donors, development agencies and governments.

Mayberry et al. (2018) used the IAT model to explore three management options to improve goat production: (1) improving goat nutrition through increasing the quality and quantity of feed; (2) reducing goat mortality and improving herd health through improved disease control; and (3) in Ethiopia only, replacing native animals with improved goat breeds which have both higher liveweight potential and sale prices but also higher feed requirements and production costs. The effects of these management options were considered separately and in combination.

The IAT simulations showed that an integrated approach to increasing production across the whole farm was likely to bring more, and larger, benefits than the adoption of single strategies. Not all management strategies examined were as beneficial as they may have appeared at inception: for example, high input costs may outweigh and negate potential benefits of a strategy.

5 Whole-farm modelling for a single enterprise farm business: NABSA

Ash et al. (2015) used the NABSA to examine the potential production and financial effects of introducing improved management strategies to extensive beef-production enterprises across a range of environments and production systems in northern Australia. They considered specific strategies to improve production: improving animal reproduction through improved genetics; improving animal growth efficiency through improved genetics; improving animal digestive ability through rumen modification; supplementing existing native pastures with cost-effective protein sources; introducing improved pastures; and combining the first four strategies to examine the potential for cumulative benefits of a combined improvement strategy compared to stepwise improvement strategies. Ash et al. (2015) examined three case study regions and demonstrated that, while there are many changes to farm management that are likely to improve whole-farm performance, increasing the protein in animal feed is particularly beneficial. Concurrent improvements in many components of farm management resulted in the greatest increases in livestock production and whole-farm profitability. The research also quantified environmental changes, in terms of methane emissions, of the altered management practices relative to baseline simulations.

The NABSA has also been used to examine the effects of irrigation management practices on livestock production in northern Australia. Monjardino et al. (2015) examined the economics of irrigating forages in the Flinders catchment in north-western Queensland during the dry season when unimproved native pastures are typically low quality and of limited availability, and feed shortages occur. The authors reported benefits to whole-farm production (e.g. liveweight gain and animal turnoff rates) from dry season irrigation, however they noted significant costs in the installation of appropriate irrigation infrastructure which severely restrict the overall potential benefits of any irrigation practices without external support for producers. However, Monjardino et al. (2014) examined the effect of integrating irrigated forages into beef cattle enterprises in the Gilbert catchment in north Queensland. This study included a range of forage and irrigation regimes on a cattle breeding enterprise and concluded that investing in irrigation development to improve the profitability of the beef-production enterprise was unlikely to be feasible. While increased irrigation improved the beef-production system, primarily by increasing stocking rates and animal turnoff, the significant costs of installing and operating an irrigation system outweighed the resulting production benefits. MacLeod et al. (2018) used the NABSA to examine whether small scale (mosaic) irrigation schemes could be used to improve the productivity of beef enterprises in five locations across Queensland and the Northern Territory, particularly in comparison to improving pastures, or on-farm infrastructure (e.g. water or fencing). They concluded that there is potential for mosaic irrigation practices to result in increased beef cattle production but that here as well the overall farm economics of increased irrigation management are less obvious, largely due to high infrastructure costs and the potential for other management interventions such as pasture or field infrastructure improvement to bring production benefits with lower overhead costs.

6 Future trends

Reviews (e.g. Kotir et al., 2019; Antle et al., 2017b; Jones et al., 2017b) indicate support for ongoing farming-system level approaches that integrate the biophysical components of farms (e.g. livestock, crop and forage production) with farm-level economics and labour requirements and constraints, as well as with the broader issues of global food security, and economic and environmental sustainability. Jones et al. (2017b) identify the potential for whole-farm models to be integrated with remote sensing data to allow potential production and management strategies to be assessed at the regional scale. This information could then be used to identify areas of potential industry expansion (e.g. meat works, grain storage) or for infrastructure planning. To be useful for decision-making, models need to contain the right level of detail. Generally the usefulness of a model for decision-making declines with increasing complexity, largely due to over-parameterisation and multiplication of uncertainties associated with parameter values.

McDonald et al. (2019) and Kotir et al. (2019) both note that the IAT model performs well against the independent metrics identified by Antle et al. (2017b). It is likely that the applications of farm systems models in general will broaden:

their applicability and use as education and research tools is on the rise (A. Smith and D. McGill, both University of Melbourne, pers comm). In other contexts, farm systems models are effective tools to discern and distil actionable knowledge from the ever-increasing amounts of data which are readily available to farmers, farm managers, researchers, policymakers and other stakeholders. As Meinke (2019) points out, the heuristic value of models is still under-exploited.

Farm system models have evolved from simple, spreadsheet-based tools which require complex inputs (e.g. the IAT and the NABSA) to modular, tree-based structures (e.g. the CLEM) which provide sophisticated simulations of the whole farm at scales from subsistence smallholder systems to large single enterprise farm businesses.

7 Conclusion

The most useful whole-farm models are adaptable, capable of capturing a sophisticated level of detail about farm management, biophysical conditions and economics and are applicable across a range of research environments to facilitate the consideration of a broad suite of research questions. They are also scalable to suit the topic of investigation. Examples of important trade-offs to be considered by an all-inclusive whole-farm system model include: crop and livestock production; food-energy-water usage; labour requirements, use and trade-offs against other activities (e.g. children's education, higher-value off-farm income); and the effects of increasing mechanisation for both landed and landless farmers.

Models are useful because they reduce real-life complexity to a set of key driving forces of the systems of interest. It is this simplification in whole-farm system models that makes them suitable analytical tools. Hence, most farm system models do not include all the components of a farm. When a targeted selection of critical model components is combined with appropriate research questions and model parameterisation (often developed in partnership with farmers and farm managers), this is a strength, rather than a weakness. Farm system models are tools that facilitate participatory research and engagement and apply at all socio-economic settings in both emerging-economy countries as well as middle- and higher-income countries. The successful use of farm modelling tools requires selection of the most appropriate model, as well as framing the correct questions to be addressed, and applying the model in a participatory manner that facilitates co-learning between researchers and end-users.

8 Where to look for further information

The study of whole-farm modelling is rapidly evolving in the context both of subsistence smallholder farms and of commercial production enterprises. As models evolve and become more sophisticated their ability to capture nuances of whole-farm systems increases. Concurrently, as scientists become aware of the complex, multi faceted nature of whole-farm modelling the scope of research undertaken broadens. One simple example of this is the CLEM, which is currently being used to examine the effect at the water catchment scale of many smallholders irrigating their farming systems in the Nepal Terai and to inform government stakeholders to enable them to make more knowledgeable policy decisions. Policies which subsidise irrigation may increase individual farm prosperity but also result in a tragedy of the commons, whereby farming across the whole catchment is at risk. This is a new and innovative application of the CLEM, which was only recently developed as a research need was identified.

The modelling community publishes up-to-date whole-farm research in leading agronomic journals and it is in these where future investigations on the topic will be found.

The Commonwealth Scientific and Industrial Research Organisation (CSIRO) is funded by the Australian government to conduct scientific research within Australia and internationally. Within CSIRO, Agriculture & Food delivers world-leading research examining trade-offs within farming systems at both small and large farm scales. The University of Tasmania hosts the Tasmanian Institute of Agriculture, which supports sustainability in the agriculture and food sectors and promotes critical thinking, and impactful transdisciplinary research.

9 References

Antle, J. M., Jones, J. W. and Rosenweig, C. E. 2017a. Next generation agricultural system data, models and knowledge products: introduction. *Agricultural Systems* 155, 186-90. doi:10.1016/j.agsy.2016.09.003.

Antle, J. M., Basso, B., Conant, R. T., Godfray, H. C. J., Jones, J. W., Herrero, M., Howitt, R. E., Keating, B. A., Munoz-Carpena, R., Rosenweig, C. E., Tittonell, P. and Wheeler, T. R. 2017b. Towards a new generation of agricultural system data, models and knowledge products: design and improvement. *Agricultural Systems* 155, 255-68. doi:10.1016/j.agsy.2016.10.002.

Ash, A. J., Hunt, L., McDonald, C. K., Scanlan, J. C., Bell, L. W., Cowley, R., Watson, I., McIvor, J. and MacLeod, N. D. 2015. Boosting the productivity and profitability of northern Australian beef enterprises: exploring innovation options using simulation modelling and systems analysis. *Agricultural Systems* 139, 50-65. doi:10.1016/j.agsy.2015.06.001.

Bell, L. W., Hargreaves, J. N. G., Lawes, R. A. and Robertson, M. J. 2009. Sacrificial grazing of wheat crops: identifying tactics and opportunities in Western Australia's grainbelt using simulation approaches. *Animal Production Science* 49(10), 797-806. doi:10.1071/AN09014.

Connor, D. J., van Rees, H. and Carberry, P. S. 2015. Impact of systems modelling on agronomic research and adoption of new practices in smallholder agriculture. *Journal of Integrative Agriculture* 14(8), 1478-89. doi:10.1016/S2095-3119(15)61069-3.

de Voil, P. G., Rodriguez, D., Power, B. and Rossing, W. A. H. 2009. Simulation of whole farm management decisions. In: Anderssen, R. S., Braddock, R. D. and Newham, L. T. H. (Eds), *18th World IMACS Congress and MODSIM09 International Congress on Modelling and Simulation*, Cairns, Australia, 2009. Modelling and Simulation Society of Australia and New Zealand and International Associated for Mathematics and Computers in Simulation, pp. 624–48.

de Wit, C. T. 1958. *Transpiration and Crop Yields*. Volume 64 of Agricultural research report/Netherlands, Volume 59 of Mededeling (Instituut voor Biologisch en Scheikundig Onderzoek va Landbouwgewasses), Verslagen van landbouwkundige onderzoekingen. Institute of Biological and Chemical Research on Field Crops and Herbage, The Netherlands.

Herrero, M., Gonzales-Estrada, E., Thornton, P. K., Quiros, C., Waithaka, M. M., Ruiz, R. and Hoogenboom, G. 2007. IMPACT: generic household-level databases and diagnostic tools for integrated crop-livestock systems and analysis. *Agricultural Systems* 92(1-3), 240–65. doi:10.1016/j.agsy.2006.03.008.

Holzworth, D. P., Snow, V., Janssen, S., Athanasiadis, I. N., Donatelli, M., Hoogenboom, G., White, J. W. and Thorburn, P. 2015. Agricultural production systems modelling and software: current status and future prospects. *Environmental Modelling and Software* 72, 276–86. doi:10.1016/j.envsoft.2014.12.013.

Holzworth, D. P., Huth, N. I., Fainges, J., Brown, H., Zurcher, E., Cichota, R., Verrall, S., Herrmann, N. I., Zheng, B. and Snow, V. 2018. APSIM Next Generation: overcoming challenges in modernising a farming systems model. *Environmental Modelling and Software* 103, 43–51. doi:10.1016/j.envsoft.2018.02.002.

Hunt, L., Ash, A. J., MacLeod, N. D., McDonald, C. K., Scanlan, J. C., Bell, L. W., Cowley, R., Watson, I. and McIvor, J. 2014. Research opportunities for sustainable productivity improvement in the northern beef industry: a scoping study. Final report for MLA. Meat and Livestock Australia, Sydney, NSW, Australia, 210pp.

Islam, S. M., Gathala, M. K., Tiwari, T. P., Timsina, J., Laing, A. M., Maharjan, S., Chowdhury, A. K., Bhattacharya, P. M., Dhar, T., Mitra, B., Kumar, S., Srivastwa, P. K., Dutta, S. K., Shrestha, R., Manandhar, S., Sherestha, S. R., Paneru, P., Siddquie, N.-E.-A., Hossain, A., Islam, R., Ghosh, A. K., Rahman, M. A., Kumar, U., Rao, K. K. and Gérard, B. 2019. Conservation Agriculture based sustainable intensification: increasing yields and water productivity for smallholders of the Eastern Gangetic Plains. *Field Crops Research* 238, 1–17. doi:10.1016/j.fcr.2019.04.005.

Jakku, E. and Thorburn, P. J. 2010. A conceptual framework for guiding the participatory development of agricultural decision support systems. *Agricultural Systems* 103(9), 675–82. doi:10.1016/j.agsy.2010.08.007.

Janssen, S. and van Ittersum, M. K. 2007. Assessing farm innovations and responses to policies: a review of bio-economic farm models. *Agricultural Systems* 94(3), 622–36. doi:10.1016/j.agsy.2007.03.001.

Jat, M. L., Dagar, J. C., Sapkota, T. B., Singh, Y., Govaerts, B., Ridaura, S. L., Saharawat, Y. S., Sharma, R. K., Tetarwal, J. P., Jat, R. K., Hobbs, H. and Stirling, C. 2016. Climate Change and Agriculture: adaptation strategies and mitigation opportunities for food security in South Asia and Latin America. *Advances in Agronomy* 137, 127–235. doi:10.1016/bs.agron.2015.12.005.

Johnson, I. R. 2016. *DairyMod and the SGS Pasture Model: a Mathematical Description of the Biophysical Model Structure*. IMJ Consultants, Dorrigo, NSW, Australia.

Jones, J. W., Antle, J. M., Basso, B., Boote, K. J., Conant, R. T., Foster, I., Godfray, H. C. J., Herrero, M., Howitt, R. E., Jansen, S., Keating, B. A., Munoz-Carpena, R., Porter, C. H., Rosenweig, C. E. and Wheeler, T. R. 2017a. Brief history of agricultural systems modelling. *Agricultural Systems* 155, 240–54. doi:10.1016/j.agsy.2016.05.014.

Jones, J. W., Antle, J. M., Basso, B., Boote, K. J., Conant, R. T., Foster, I., Godfray, H. C. J., Herrero, M., Howitt, R. E., Jansen, S., Keating, B. A., Munoz-Carpena, R., Porter, C. H., Rosenweig, C. E. and Wheeler, T. R. 2017b. Towards a new generation of agricultural system data, models and knowledge products: state of agricultural system science. *Agricultural Systems* 155, 269–88.

Keating, B. A., Carberry, P. S., Hammer, G. L., Probert, M. E., Robertson, M. J., Holzworth, D., Huth, N. I., Hargreaves, J. N. G., Meinke, H., Hochman, Z., McLean, G., Verburg, K., Snow, V., Dimes, J. P., Silburn, M., Wang, E., Brown, S., Bristow, K. L., Asseng, S., Chapman, S., McCown, R. L., Freebairn, D. M. and Smith, C. J. 2003. An overview of APSIM, a model designed for farming systems simulation. *European Journal of Agronomy* 18(3–4), 267–88. doi:10.1016/S1161-0301(02)00108-9.

Komarek, A. M., McDonald, C. K., Bell, L. W., Whish, J. P. M., Robertson, M. J., MacLeod, N. D. and Bellotti, W. D. 2012. Whole-farm effects of livestock intensification in smallholder systems in Gansu, China. *Agricultural Systems* 109, 16–24. doi:10.1016/j.agsy.2012.02.001.

Kotir, J. H., Bell, L. W. and Kirkegaard, J. A. 2019. Diverse bio-economic whole-farm models differ in capabilities to represent critical farm attributes: a review. *Agricultural Systems* (submitted).

Laing, A. M., Roth, C. H., Chialue, L., Gaydon, D. S., Grünbühel, C. M., Inthavong, T., Phengvichith, V., Schiller, J., Sipaseuth, Thiravong, K. and Williams, L. J. 2018. Mechanised dry seeding is an adaptation strategy for managing climate risks and reducing labour costs in rainfed rice production in lowland Lao PDR. *Field Crops Research* 225, 32–46. doi:10.1016/j.fcr.2018.05.020.

Lisson, S., MacLeod, N., McDonald, C., Corfield, J., Pengelly, B., Wirajaswadi, L., Rahman, R., Bahar, S., Padjung, R., Razak, N., Puspadi, K., Dahlanuddin, Sutaryono, Y., Saenong, S., Panjaitan, T., Hadiawati, L., Ash, A. and Brennan, L. 2010. A participatory, farming systems approach to improving Bali cattle production in the smallholder crop-livestock systems of eastern Indonesia. *Agricultural Systems* 103(7), 486–97. doi:10.1016/j.agsy.2010.05.002.

MacLeod, N. D., Mayberry, D. E., Revell, C., Bell, L. W. and Prestwidge, D. B. 2018. An exploratory analysis of the scope for dispersed small-scale irrigation developments to enhance the productivity of northern beef cattle enterprises. *The Rangeland Journal* 40(4), 381–99. doi:10.1071/RJ18026.

Manivong, V., Cramb, R. and Newby, J. C. 2014. Rice and remittances: crop intensification versus labour migration in Southern Laos. *Human Ecology* 42(3), 367–79. doi:10.1007/s10745-014-9656-6.

Mayberry, D., Ash, A. J., Prestwidge, D. B. and Herrero, M. 2018. Closing yield gaps in smallholder goat production systems in Ethiopia and India. *Livestock Science* 214, 238–44. doi:10.1016/j.livsci.2018.06.015.

McDonald, C. K., MacLeod, N., Lisson, S., Ash, A. J., Pengelly, B., Brennan, L., Corfield, J., Wirajaswadi, L., Panjaitan, T., Saenong, S., Sutaryono, Y., Padjung, R., Rahman, R. and Bahah, S. 2004. Improving Bali cattle production in mixed crop-livestock systems in eastern Indonesia using an integrated modelling approach. In: Wong, H. K., Liang, J. B., Jelan, Z. A. and Ho, Y. W. (Eds), *New Dimensions and Challenges for Sustainable*

Livestock. Proceedings of the 11th Animal Science Congress, Kuala Lumpur (vol II), pp. 116–9.

McDonald, C. K., MacLeod, N. D., Lisson, S. and Corfield, J. P. 2019. The Integrated Analysis Tool (IAT) - A model for the evaluation of crop-livestock and socio-economic interventions in smallholder farming systems. Agricultural Systems 176, 102659. DOI: 10.1016/j.agsy.2019.102659.

Meinke, H. 2019. The role of modelling and systems thinking in contemporary agriculture. In: Accorsi, R. and Manzini, R. (Eds), *Sustainable Food Supply Chains: Planning, Design, and Control through Interdisciplinary Methodologies*. Elsevier, San Diego, CA, p. 394.

Meinke, H., Baethgen, W. E., Carberry, P. S., Donatelli, M., Hammer, G. L., Selvaraju, R. and Stöckle, C. O. 2001. Increasing profits and reducing risks in crop production using participatory systems simulation approaches. *Agricultural Systems* 70(2–3), 493–513. doi:10.1016/S0308-521X(01)00057-9.

Monjardino, M., MacLeod, N. D. and Prestwidge, D. B. 2014. Does it pay to integrate irrigated forages in a beef cattle breeding operation in north Queensland? *Proceedings of the 58th AARES Annual Conference*, NSW, 12pp.

Monjardino, M., MacLeod, N. D., McKellar, L. and Prestwidge, D. B. 2015. Economic evaluation of irrigated forage production in a beef cattle operation in the semi-arid tropics of northern Australia. *Agricultural Systems* 139, 122–43. doi:10.1016/j.agsy.2015.07.004.

Nelson, R. A., Holzworth, D. P., Hammer, G. L. and Hayman, P. T. 2002. Infusing the use of seasonal climate forecasting into crop management practice in North East Australia using discussion support software. *Agricultural Systems* 74(3), 393–414. doi:10.1016/S0308-521X(02)00047-1.

Newby, J. C., Manivong, V. and Cramb, R. A. 2013. Intensification of lowland rice-based farming systems in Laos in the context of diversified rural livelihoods. Contributed paper for the *57th AARES Annual Conference*, Sydney, Australia.

Parsons, D. J., McDonald, C. K., Ba, N. X., Tuan, D. T., Lisson, S. N., Corfield, J., Phung, L. D., Quan, N. H., Van, N. H., Ngoan, L. D. and Lane, P. 2011. Improving cattle profitability in mixed crop-livestock systems in south central coast Vietnam using an integrated modelling approach. *Proceedings of the 5th World Congress of Conservation Agriculture and 3rd Farming Systems Design Conference*, Brisbane, pp. 242–5.

Postel, S. L. 1998. Water for food production: will there be enough in 2025? *BioScience* 48(8), 629–37. doi:10.2307/1313422.

Rigolot, C., de Voil, P., Douxchamps, S., Prestwidge, D., Van Wijk, M., Thornton, P. K., Rodriguez, D., Henderson, B., Medina, D. and Herrero, M. 2017. Interactions between intervention packages, climatic risk, climate change and food security in mixed crop-livestock systems in Burkina Faso. *Agricultural Systems* 151, 217–24. doi:10.1016/j.agsy.2015.12.017.

Rodriguez, D., deVoil, P., Power, B., Cox, H., Crimp, S. and Meinke, H. 2011. The intrinsic plasticity of farm businesses and their resilience to change. An Australian example. *Field Crops Research* 124(2), 157–70. doi:10.1016/j.fcr.2011.02.012.

Rodriguez, D., Cox, H., deVoil, P. and Power, B. 2014. A participatory whole farm modelling approach to understand impacts and increase preparedness to climate change in Australia. *Agricultural Systems* 126, 50–61. doi:10.1016/j.agsy.2013.04.003.

Röling, N. 2009. Pathways for impact: scientists' different perspectives on agricultural innovation. *International Journal of Agricultural Sustainability* 7(2), 83–94. doi:10.3763/ijas.2009.0043.

Seckler, D., Amarasinghe, U., Molden, D., de Silva, R. and Barker, R. 1998. World water demand and supply, 1990 to 2025: senarios and issues. Research report 19. International Water Management Institute, Colombo, Sri Lanka.

Shalander, K., Sravya, M., Pramanik, S., Dakshina, M. K., Balaji, N. B., Samuel, J., Prestwidge, D. and Whitbread, A. 2017. Potential for enhancing farmer income in semi-arid Telangana: a multi-model systems approach. *Agricultural Economics Research Review* 30, 300.

Sterman, J. D. 2000. *Business Dynamics: Systems Thinking and Modelling for a Complex World*. McGraw Hill, Boston.

Styger, E., Aboubacrine, G., Attaher, M. A. and Uphoff, N. 2011. The system of rice intensification as a sustainable innovation: introducing, adapting and scaling up a system of rice intensification practices in the Timbuktu region of Mali. *International Journal of Agricultural Sustainability* 9(1), 67–75. doi:10.3763/ijas.2010.0549.

Sutton, M. A., Bleeker, A., Howard, C. M., Bekunda, M., Grizzett, B., de Vries, W., van Grinsven, H. J. M., Abrol, Y. P., Adhya, T. K., Billen, G., Davidson, E. A., Datta, A., Diaz, R., Erisman, J. W., Liu, X. J., Oenema, O., Palm, C., Raghuram, N., Reis, S., Scholz, R. W., Sims, T., Westhoek, H. and Zhang, F. S. 2013. *Our Nutrient World: the Challenge to Produce More Food and Energy with Less Pollution. Global Overview of Nutrient Management*. Centre for Ecology and Hydrology, Edinburgh, UK, 128pp.

Taylor, J. E. and Adelman, I. 2003. Agricultural household models: genesis, evolution and extensions. *Review of Economics of the Household* 1(1/2), 33–58. doi:10.1023/A:1021847430758.

Van de Fliert, E. and Braun, A. R. 2002. Conceptualizing integrative, farmer participatory research for sustainable agriculture: from opportunities to impact. *Agriculture and Human Values* 19, 25–38.

Van Ittersum, M. K., Ewert, F., Heckelei, T., Wery, J., Olsson, J. A., Anderson, E., Bezlepkina, I., Brouwer, F., Donatelli, M., Flichman, G., Olsson, L., Rizzoli, A. E., van der Wal, T., Wien, J. E. and Wolf, J. 2008. Integrated assessment of agricultural systems: a component-based framework for the European Union (SEAMLESS). *Agricultural Systems* 96(1–3), 150–65. doi:10.1016/j.agsy.2007.07.009.

Van Wijk, M. T., Tittonell, P., Ruffino, M. C., Herrero, M., Pacini, C., de Ridder, Nd and Giller, K. E. 2009. Identifying key entry points for strategic management of smallholder farming systems in sub-Saharan Africa using the dynamic farm scale simulation model NUANCES-FARMSIM. *Agricultural Systems* 102(1–3), 89–101. doi:10.1016/j.agsy.2009.07.004.

Wheeler, T. and von Braun, J. 2013. Climate change impacts on global food security. *Science* 341(6145), 508–13. doi:10.1126/science.1239402.

Chapter 7

The DSSAT crop modeling ecosystem

Gerrit Hoogenboom, Cheryl H. Porter, Kenneth J. Boote and Vakhtang Shelia, University of Florida, USA; Paul W. Wilkens[1] and Upendra Singh, International Fertilizer Development Center, USA; Jeffrey W. White, USDA-ARS, USA; Senthold Asseng, University of Florida, USA; Jon I. Lizaso, Universidad Politécnica de Madrid, Spain; L. Patricia Moreno, University of Florida, USA; Willingthon Pavan, Universidade de Passo Fundo, Brazil; Richard Ogoshi, University of Hawaii, USA; L. Anthony Hunt, University of Guelph, Canada; Gordon Y. Tsuji, University of Hawaii, USA; and James W. Jones, University of Florida, USA

1 Introduction

Traditionally, research for agricultural development and improvement is based on small plot experiments that are conducted for multiple years on a research station

1 This chapter is written in memory of Paul Wilkens, who passed away November 27, 2017, due to brain cancer. Paul was a key member of the DSSAT Development Team and his many contributions to the DSSAT community and his low-key humor are greatly missed.

http://dx.doi.org/10.19103/AS.2019.0061.10

and, on occasion, in multiple locations. The outcomes of these experiments are then transmitted in the form of recommendations to farmers through state-wide and county-based extension services. Although this approach works well for the United States and Europe where farms are normally well managed with respect to fertilizer, irrigation inputs, and pests and diseases, in some countries funding and resource challenges make this approach less practical. In the early 1980s, the United States Agency for International Development (USAID) made a bold step to support a project that was based on systems analysis of agricultural production to address food security in developing countries. This project for improving agricultural production, called the International Benchmark Sites Network for Agrotechnology Transfer (IBSNAT), was developed by Goro Uehara, a soil physicist and professor at the University of Hawaii. Uehara's previous work with the Benchmark Soils Project, which was also funded by USAID, showed that research only on soils cannot address food security in developing countries. The IBSNAT Project was funded from 1982 through 1992. In subsequent years, USAID and other funding agencies have not been as supportive of funding for basic model development and improvement as for providing funding for model applications.

The systems analysis approach of the IBSNAT Project was based on biophysical models that predict crop growth, development, and yield using daily weather data, local soil conditions, crop management, and genetics as input. At the start of the IBSNAT Project, crop modeling teams from the University of Florida (SOYGRO and PNUTGRO models) and from USDA-ARS in Temple, Texas (CERES-Maize and CERES-Wheat models) were invited to collaborate with scientists from the University of Puerto Rico, the University of Edinburgh (Scotland), the University of Guelph (Canada), and the International Fertilizer Development Center in Muscle Shoals, Alabama (Wilkerson et al., 1983; Boote et al., 1986; Ritchie et al., 1985; Jones and Kiniry, 1986). The early versions of these crop models were based on nonuniform and nonstandard input and output files, making it challenging for users to apply models for different crops to the same farming system. Therefore, the Minimum Data Set (MDS) system was developed to standardize the inputs required for these crop models as well as the file formats used (ICRISAT, 1984). This standardization facilitated the development of data utility programs for processing weather, soil, management input data, and experimental observation files, as well as tools for application and display of output data for the models, forming the basis for the Decision Support System for Agrotechnology Transfer (DSSAT) software. For further details of IBSNAT activities and outcomes, see IBSNAT (1993), Uehara and Tsuji (1998), and Jones et al. (2017).

2 The DSSAT ecosystem

The combination of different models, tools, utilities, and applications requires the development of a unique interface that provides easy access for a user who

Published by Burleigh Dodds Science Publishing Limited, 2026.

may not be familiar with crop models in general, especially with the challenges of formatting input and output files. Jim Jones conceptualized the design of DSSAT to be an integrated crop modeling platform (Jones et al., 1998). DSSAT provides tools to assist a user to prepare the different input files that are needed for running a model, to define the experiments and treatments or scenarios a user wants to simulate, and to conduct an analysis of crop model outputs from the simulations, including a comparison with observed data for model evaluation and strategic analyses for model scenarios (Fig. 1). In order to facilitate the interaction between the crop models, the data tools, the utilities, and the application programs, a very strict protocol is required for the file naming convention, specific file formats, and system settings that define the location and names of the model input and output files. This approach was first presented to DSSAT users in DSSAT Version 2.1 (IBSNAT, 1989) and DSSAT Version 3.0 (IBSNAT, 1993) at the end of the IBSNAT Project. The original design and concept are still viable in the most current version of DSSAT Version 4.7.5 (Hoogenboom et al., 2019) and in a proposed future implementation in jDSSAT (Resenes et al., 2019).

Over time, both the file formats and the file naming conventions have changed, but the approach is still the same. The same flat ASCII file structures

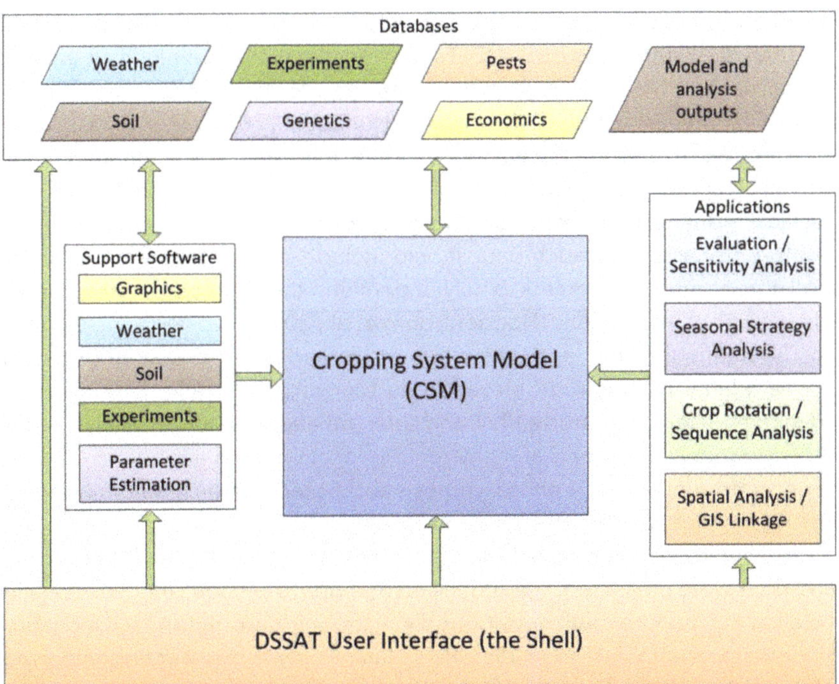

Figure 1 The DSSAT crop modeling ecosystem.

Published by Burleigh Dodds Science Publishing Limited, 2026.

are used to provide ultimate portability, something learned after using early proprietary data base software that was no longer supported. As a result, a user can easily simulate crop growth, development, and yield for different crops for the same field while making only minor changes to the input files that are crop specific, such as variety selection, planting date, or plant density. The strict protocol also has allowed programmers to develop new tools and utilities that can be easily incorporated into DSSAT based on standard input and output formats. Examples include the graphics program EasyGrapher (Yang et al., 2014b), the genetic-specific parameter optimizer GLUE (He et al., 2010), and a platform-independent DSSAT shell (Resenes et al., 2019). The only challenge has been the use of two digits to represent a year, which will be resolved in the next release of DSSAT with the introduction of ICASA Version 2.0 file formats that were defined several years ago (White et al., 2013).

3 Minimum Data Set (MDS) for crop modeling

In order to run a crop simulation model, a minimum set of input data is required. The challenge is to define a MDS that is relatively easy to collect by crop model users and one that also provides reasonable simulation results. Unfortunately, the larger crop modeling community has never been able to come to an agreement on a standard definition for MDS (Hunt et al., 1994). One of the outcomes and successes of the IBSNAT Project was the definition of an MDS that was acceptable to all crop model developers for the CERES-Maize, CERES-Wheat, SOYGRO, and PNUTGRO models (Hunt et al., 2001). This MDS includes daily weather data, soil surface and soil profile information, crop management, and initial conditions at the start of the simulation. Although the MDS is specifically defined for crop model applications, the IBSNAT community also tried to emphasize that such data should include basic information collected for all agronomic experiments to fully understand the Genotype * Environment * Management interactions (Hoogenboom et al., 2012).

The minimum weather data include the metadata for the weather station, especially latitude, longitude, elevation, and sensor height, and daily maximum and minimum temperature, rainfall, and solar radiation. Although solar radiation is not commonly measured at many remote locations, it is a required input for the accurate simulation of photosynthesis and potential transpiration using the Priestley-Taylor equation (Priestley and Taylor, 1972).

The minimum soil data include the metadata for the location where the soil conditions were measured: soil surface color, slope, drainage, and permeability, as well as soil texture, bulk density, and soil organic carbon for each individual soil horizon. The DSSAT crop models simulate only a one-dimensional water balance with vertical flow to meet the requirements for relatively simple inputs for model users, especially for applications.

The crop management data include the crop and cultivar selection, planting date, plant density, row spacing, sowing depth, irrigation, and fertilizer inputs. For irrigated treatments and scenarios, the dates, amounts, and the type of irrigation system must be defined; for fertilized treatments and scenarios, the dates, amounts, and types of inorganic fertilizer must be defined, as well as depth of incorporation. For organic fertilizers using plant or animal material, the type and composition of the organic fertilizers have to be defined. If a crop, such as rice, tomatoes, or other vegetables, is transplanted, the initial weight of the transplant material, age, and the temperature of the nursery have to be defined. For potatoes, the weight of the seed-potato is an input, for cassava the weight and length of the stick and orientation of planting are defined, and for sugarcane the initial cane is defined. Although these inputs might seem complex, proper recording for all management activities will capture most of this information. For a few crops, including potatoes and cassava, the harvest date must be defined as well.

Boundary or initial conditions at the start of the simulation are also very important, especially for the soil environment, requiring initial soil moisture, nitrate, and ammonia for each horizon or soil layer, as well as the aboveground biomass residue and roots of the previous crop and their composition. Although these conditions can be challenging to measure unless equipment and personnel are available, they can be estimated using the tools and utilities that are provided with DSSAT.

The previously listed input data for weather, soil, crop management, and initial conditions are the MDS required for running the model. For model calibration, evaluation, and improvement, crop and soil measurements are required so that comparisons can be made between simulated and observed data. Depending on the research goals and objectives, measurements can include yield and yield components, detailed crop phenology, crop growth analysis, and soil profile measurements such as soil moisture, nitrate, and ammonia, organic carbon, and other information (Hoogenboom et al., 2012). The number of measurements needed should be based on the model application rather than requiring a researcher to collect as much data as possible. For example, variety trial data that are collected for multiple locations and multiple years for the same cultivars or hybrids can be very useful, but normally in these trials only yield, some yield components, and phenological events are recorded. Recently, there have been some discussions about a classification of experimental data sets for crop modeling (Boote et al., 2015; Kersebaum et al., 2015).

4 Input data tools

Most researchers have their own individual standard methodology for recording experimental data in field books, spreadsheets, and other electronic media. These individual differences make it somewhat challenging to convert the measured data into a format that can be directly applied in a crop modeling system. DSSAT, therefore, provides specific tools for entering weather, soil, crop management, and observational data.

4.1 XBUILD

XBuild is the tool for entering crop management data that are stored in a crop management file (Fig. 2). The tool is designed so that the user first enters information that defines the field, especially the weather station and soil profile that are associated with that experiment, followed by the crop and cultivar selection, and planting information. Initial conditions are defined in the Environmental Section of the tool. The user can enter different levels for each management scenario, such as multiple cultivars or hybrids, different planting dates, and different input levels and application dates for irrigation and fertilizer. Following the entry of all specific information, the user then defines the specifics for each individual treatment, including field location, crop and cultivar, planting details, initial conditions, and the appropriate irrigation and fertilizer level, similar to the way a researcher defines a treatment for an agronomic experiment.

4.2 WeatherMan

WeatherMan allows for the entry and formatting of weather data into DSSAT weather files (Fig. 2). A user can import weather data preferably from spreadsheets, but WeatherMan can also handle other formats including CSV and ASCII text files. Once the data have been imported into WeatherMan,

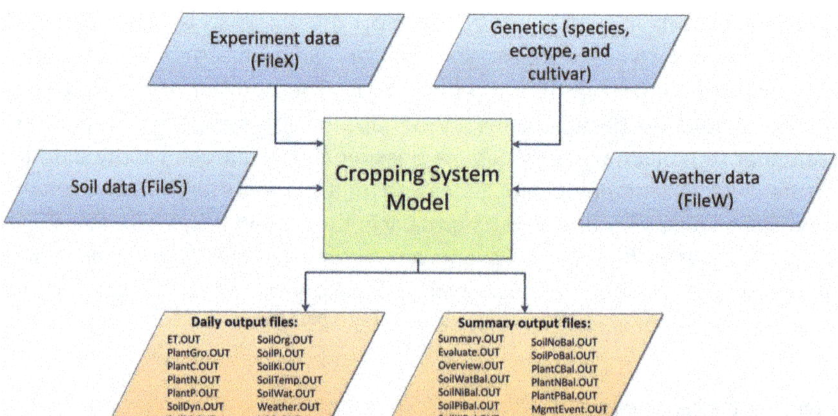

Figure 2 The DSSAT Input and Output file system.

Published by Burleigh Dodds Science Publishing Limited, 2026.

quality control procedures can be applied to identify extreme values or extreme changes in values for two consecutive days and any missing values. A new internal database is created in WeatherMan with what is referred to as the 'corrected' data. The final procedure is to export the data back into DSSAT format weather files.

4.3 SBUILD

The soil water balance simulation in DSSAT is based on the tipping bucket approach with three key soil moisture variables, including Saturated Water Content (SAT), Drained Upper Limit (DUL), and the Lower Limit (LL) of plant extractable water. Although there are procedures for measuring these, they are not very common and require a significant amount of experimental resources. The SBUILD program of DSSAT allows a user to enter soil surface information, including soil color, slope, permeability, and drainage characteristics, and soil texture, bulk density, and organic carbon for each soil horizon. SBUILD then uses internal pedotransfer functions to calculate SAT, DUL, and LL for each soil horizon or layer, and it saves the information for that particular soil profile in the soil input file (Fig. 2).

4.4 ATCreate

Measurement data for model evaluation can be differentiated into two types in DSSAT. The first type is referred to as the summary data, and includes the key phenological stages, yield and yield components at final harvest, and other measurements that can be obtained at critical stages, such as maximum leaf area index (LAI) or grain nitrogen concentration. The summary data are stored in FileA as a single line per treatment. The second type of measurement data is referred to as time series data for growth analysis, soil moisture content, soil nitrogen measurements, and other relevant data that can be used for model evaluation. The time series data are stored in FileT and organized by treatment and then observation date. There is also the ability to store the observations for the individual replicates. The ATCreate program allows users to enter observations either manually or by importing a spreadsheet or text file, thereby creating the FileA and FileT for each experiment. It is important to select the appropriate header for each column of data so that the other programs within DSSAT are able to recognize the observed variables. The file called DATA.CDE holds the names of these variables (short name, long name, and units) so that header names are shared and are also readable by the graphics program and other programs in DSSAT.

5 The Cropping System Model (CSM)

The main engine of the DSSAT ecosystem is the Cropping System Model (CSM; Fig. 3). For most users, the model is run through the DSSAT Interface, but for power users, it can also be run through a command line interface on iOS, Linux, and Unix platforms. The original crop models in the first version of DSSAT were CERES-Maize, CERES-Wheat, SOYGRO, and PNUTGRO. These original models morphed over time from many independent models to a single agricultural systems model that encompasses all the original crop models as individual crop modules (Jones et al., 2003).

Development of models for new crops was initially based on creating new, stand-alone models, such as the model for dry beans BEANGRO, which was developed based on SOYGRO (Hoogenboom et al., 1994). In the early 1990s, the DSSAT developers realized that code modifications were often made redundantly for the separate SOYGRO (Jones et al., 1987), PNUTGRO (Boote et al., 1987), and BEANGRO (Hoogenboom et al., 1990) models. Therefore, we pulled all crop-specific parameters and relationships out of the FORTRAN code and placed them into external species files (per crop), thus allowing a single generic executable CROPGRO code to represent three crop species

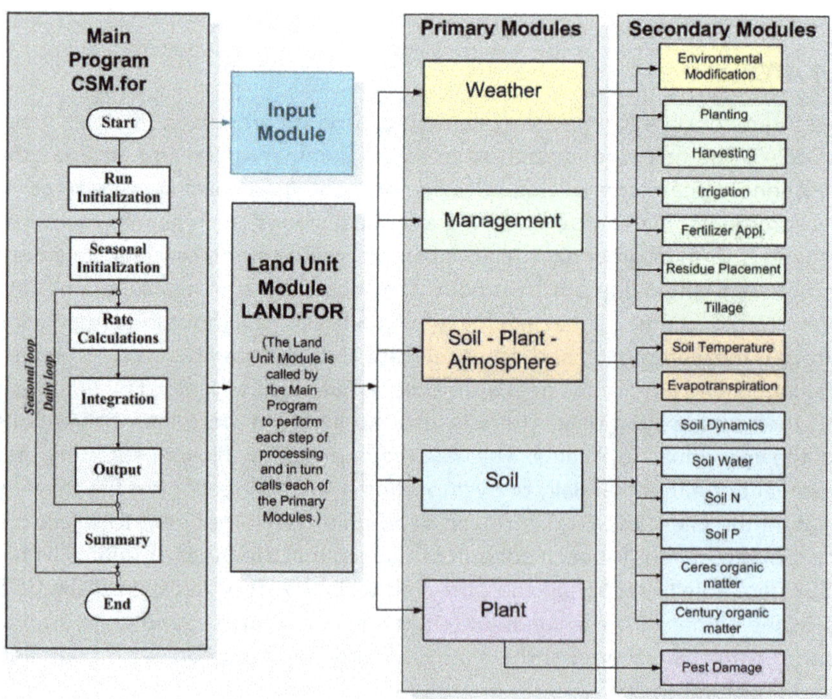

Figure 3 The structure of the Cropping System Model.

Published by Burleigh Dodds Science Publishing Limited, 2026.

(Hoogenboom et al., 1991, 1992). These improvements were part of the DSSAT v3.5, a stable, well-used software system (Hoogenboom et al., 1999). This template approach allowed later adaptations of species files to represent many other crop species using the same source code.

During the 1990s, code improvements were made to the CROPGRO model to add mechanistic leaf-level photosynthesis coupled with hedgerow light interception to simulate leaf-to-canopy assimilation running on an hourly basis for sunlit and shaded leaf classes (Boote and Pickering, 1994; Pickering et al., 1995; Boote et al., 1998). The leaf-level photosynthesis captures the rubisco kinetics of Farquhar and von Caemmerer (1982) and mechanistically simulates photosynthesis response to CO_2, O_2, temperature, solar radiation, LAI, and leaf state (specific leaf mass and leaf N concentration). The simulated response to CO_2 is thus an outcome of this rubisco kinetics, rather than resulting from an externally prescribed CO_2 response curve often used in other models. The simulated leaf-level and canopy-level photosynthesis were tested against observed data and shown to be accurate by Alagarswamy et al. (2006). Since that time, this hourly leaf-level photosynthesis method has been the default for the CROPGRO model crops, in place of the older, but still available, daily canopy photosynthesis option. The CERES-based models in DSSAT are based on radiation-use efficiency (RUE), and they use an externally prescribed CO_2 response modifier based on observed CO_2 response data. See Boote et al. (2010) for a discussion of the CO_2 response curves for C3 and C4 crops simulated by the CERES-based models in DSSAT and for an evaluation of CERES and CROPGRO version crops against metadata on observed CO_2 response. A simple ozone impact routine was recently introduced in one of the wheat models in DSSAT (Guarin et al., 2019).

In the decade from 2000 to 2010, the DSSAT-CSM was created (Jones et al., 2003). This single executable program was able to simulate all the crop models, including the CERES models (Ritchie et al., 1998) and the CROPGRO models (Boote et al., 1998), that until this point were available only as individual models. With CSM, each crop module shares the same routines for the simulation of soil water dynamics, soil N dynamics, soil C dynamics, management operations, and daily weather processes. All input and output data use the same structure, naming conventions, and formats (see scheme in Fig. 2). This consolidation of the soil processes enabled the simulation of a true crop rotation, that is, a sequence of different crops grown in rotation in which the shared soil water, N, and C balances are run in a continuous simulation allowing carry-over of soil water, N, soil C from one cropping season to the next. During this same time period, the daily DSSAT-CENTURY soil C module was developed (Gijsman et al., 2002), providing simulation of surface residue decomposition and in-season contribution of senesced plant components, which are very important for long-term simulations of crop rotations and perennial systems.

5.1 Additional Crops in DSSAT

There are two methods for adding new crop modules into DSSAT. The first and easier approach uses the CROPGRO template and data from field experiments, journal articles, non-refereed publications and reports, and variety trials to calibrate the genetic parameters which control the growth and development characteristics of the new crop. This approach does not require any modification of the existing model software and computer code. The second approach is to add a completely new crop module into the CSM code, such as when growth or phenological characteristics of a new crop are very different from those described in the CROPGRO template. In this second case, both model coding and calibration of parameters are required. As an example, the CERES-Sugarbeet model was one of the most recent modules added to CSM (Anar et al., 2019).

During the past 20 years, a number of additional crops have been added to those originally available in the 1998 DSSAT v3.5 release. Figure 4 shows the crop models available in DSSAT v4.7, including a few models that are currently under development and will be released in a future version of DSSAT. Crops added since v3.5 that use the CROPGRO template include chickpea (Hoogenboom et al., 1997), tomato (Scholberg et al., 1997; Boote et al., 2012), cowpea (Boote, 1998, unpublished), mucuna or velvet bean (Hartkamp et al., 2002), faba bean (Boote et al., 2002), cotton (Boote, 2010, unpublished; Pathak et al., 2007), pigeon pea (Alderman et al., 2015), safflower (Singh et al., 2015), canola (Deligios et al., 2013), sunflower (Boote, 2014, unpublished), green bean (Boote, 2009, unpublished), cabbage, and pepper. All the CROPGRO type crops share the same source code, but they are facilitated by different species, ecotype, and cultivar files. The adaptation process for new CROPGRO template crops, as described for faba bean by Boote et al. (2002) and for pigeon pea by Alderman et al. (2015), makes use of available literature information for cardinal

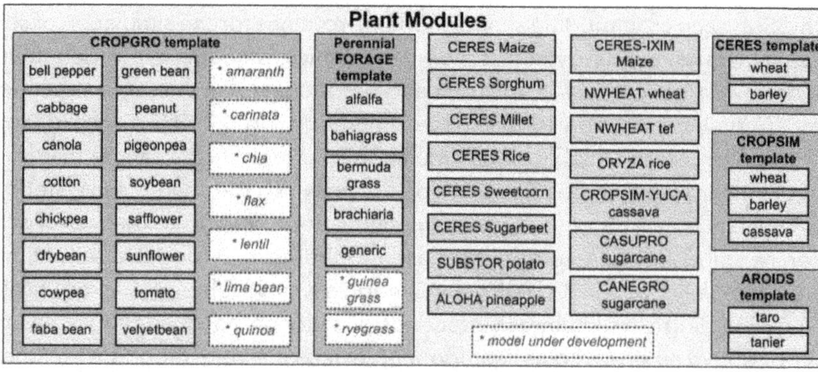

Figure 4 Plant modules of the Cropping System Model.

Published by Burleigh Dodds Science Publishing Limited, 2026.

temperatures, composition, and so forth, along with observed time series growth analysis (LAI, total crop mass, reproductive mass) and subsequent inverse model optimization of parameters (in the species file). Crop models under development using the CROPGRO template include chia (Mack et al., 2019, paper submitted), quinoa, and carinata (Boote et al., 2019, in progress).

Sweet corn (Lizaso et al., 2007) and sugarbeet (Anar et al., 2019) were added as new crop modules following the style of CERES models. Other models in CSM were adapted from an existing model to use the modular format of CSM (Jones et al., 2003), such as CANEGRO sugarcane (Singels et al., 2008; Jones and Singels, 2018), ALOHA pineapple (Zhang et al., 1997), and NWheat (Asseng et al., 2000). NWheat was also used as a template for a new tef model (Paff and Asseng, 2019). The CROPSIM model (Hunt and Pararajasingham, 1995) was added to DSSAT-CSM as a template model for wheat, barley, and cassava. The CROPSIM template was also used to develop a new crop model specific for cassava called YUCA. The perennial forage model (Rymph, 2004; Pequeno et al., 2018) is based on the CROPGRO model, but it differs enough that it is a separate model. It is also a template model, allowing simulation of *brachiaria* and *cynodon* (Pequeno et al., 2014), and alfalfa (*Medicago sativa*; Malik et al., 2018). The SIMPLE modeling approach by Zhao et al. (2019) will also be included for the development of models for crops for which limited data are available.

6 Water balance processes

All the DSSAT models share the same soil water balance subroutine. On a daily basis, the soil water balance is computed by adding irrigation and rainfall and subtracting surface runoff, drainage, plant transpiration, and soil evaporation. Within a soil column, soil water is redistributed by vertical drainage, capillary rise, and tillage. Rainfall is supplied as a user input in weather files. Irrigation is specified in the experimental details input file which supplies information about the type of irrigation, the efficiency of water supply, and the amount of irrigation applied. Partitioning of rainfall to infiltration and surface runoff is based on the SCS curve number approach (Ritchie, 1998).

Drainage of soil water follows the tipping bucket approach for layered soils with only one-dimensional flow (Ritchie, 1998). Successive soil layers are defined by the LL (wilting point), DUL, and saturated volumetric soil water content. Downward water movement within the soil depends on a soil drainage factor (fraction per day), limited by the saturated hydraulic conductivity of soil layers.

Actual evapotranspiration (ET) depends on total ETo demand using either of two options: Priestley-Taylor (Priestley and Taylor, 1972), based on standard weather data input, or FAO-56 (Allen et al., 1998), which additionally requires wind speed and relative humidity as input data. After it is calculated, ETo is partitioned to the potential transpiration of the crop canopy (Ep) or potential evaporation of the soil (Es) as a function of the LAI and an energy extinction coefficient (Kep). Kep differs for each crop in CROPGRO, but it is more complex for the CERES crops where a 'mixed' function of extinction of photosynthetically active radiation is used. The actual soil evaporation depends on the potential Es and the soil water content, using either the older Stage 1 (square root of time method) or the Suleiman-Ritchie method (Ritchie et al., 2009). Actual transpiration of the crop is the minimum of the potential Ep or the root water uptake. Potential root water uptake from successive layers follows the approach described by Ritchie (1998), and it is dependent on root length density and the fraction of available soil water content in each layer. Total root water uptake is then integrated over all layers, and transpiration is reduced if potential root water uptake is less than potential Ep. The daily photo-assimilation is reduced as a function of actual transpiration (root uptake) over potential Ep, using a drought stress factor called SWFAC. Expansive processes are reduced somewhat sooner by a similar factor called TURFAC. See Boote et al. (2009) for a review of water balance, ET, and simulation of water stress effects in the CROPGRO model.

7 Nitrogen (N) balance processes

Soil nitrogen dynamics in CSM (Godwin and Singh, 1998) are handled in the soil inorganic N module and in two soil organic matter modules. In the inorganic N module, a mass balance accounts for all additions of inorganic N to the soil, all processes transforming N from one type to another, and all removals of inorganic N from the soil column. Additions of inorganic N are from fertilizer applications and from mineralized N resulting from decomposition of organic matter. Fertilizer applications are defined in the experimental details file and include the date applied, fertilizer type, amount of N applied, application method, and the depth and percentage of incorporation into the soil.

Daily transformations of nitrate, ammonium, and urea are computed based on process rates of nitrification, denitrification, ammonia volatilization, and urea hydrolysis. Removals of inorganic N from the system are based on plant uptake, immobilization due to decomposing organic matter, leaching, and N gas losses due to ammonia volatilization, denitrification, and nitrification. Gaseous emissions of N_2O, NO, and CO_2 are computed based on organic matter decomposition, nitrification, and denitrification processes. N gas emission algorithms are based on the DayCent model (Del Grosso et al., 2001). For flooded rice systems, in addition to the processes listed previously, the model simulates the chemical and biological processes occurring in the floodwater. These processes are discussed in more detail in Chapter 3 of this book.

Two options are available in DSSAT for computation of soil organic matter dynamics: the original CERES-based module (Godwin and Singh, 1998) and the CENTURY-based module (Gijsman et al., 2002). The main difference is the inclusion of surface fresh organic matter and three pools of soil organic matter in the Century model. The more complex CENTURY model allows more control over initialization of stable C pools and, therefore, overall decomposition dynamics, but it also requires additional input data which are difficult to obtain. These routines interact with the inorganic N and P modules by computing transformation of organic N and P into inorganic forms as a product of mineralization. Conversely, immobilization can remove inorganic N and P from the soil and reduce plant-available nutrients.

7.1 Plant nitrogen processes

The details of modeling plant N uptake in CSM vary between the individual crop modules listed in Fig. 4, but for all crop modules, N uptake is computed as the minimum of N demand and N supply. The potential N supply from the soil profile is a function of rooting density, nitrate and ammonium concentrations, and soil water in each soil layer (Godwin and Singh, 1998). Soil N supply is influenced by environmental factors such as soil temperature, soil moisture, soil pH, and management of N fertilizers and organic amendments. Root morphology, root architecture, and root length density may limit the ability of the crop to access the N supply.

Crop N demand differs with the growth stage of a crop, with higher critical N concentrations, and, therefore, higher N demand during early crop growth and development. N demand is driven by plant growth rate, growth stage, and tissue N status as a function of the growth stage (Godwin and Singh, 1998). Total crop N demand is the summation of all deficiency demands from various plant organs plus the demand by new growth. For legumes, simulated with the CROPGRO model, when the supply of N is less than the demand, carbohydrates are metabolized to meet the crop N demand via N-fixation (Boote et al., 1998). The N-fixation rate is influenced by soil temperature, soil water deficit, soil aeration, and plant reproductive age (Boote et al., 2008). For all other crops, when the N supply is less than N demand, vegetative tissues are grown at lower N concentrations. If this condition persists, N deficiency symptoms arise, resulting in a reduction in LAI, reduced photosynthesis (growth and yield reduction), and accelerated senescence.

8 Inorganic soil phosphorus (P) processes

The CSM inorganic soil P module maintains state variables for labile, active, and stable forms of phosphorus. Transformation between the pools assumes first-order kinetics with rate constants computed based on soil chemical and physical properties. Additions to the system are from fertilizer application and mineralization due to decomposition of organic matter. Removal of P from the system is from plant uptake and immobilization of P by microbes to meet decomposition demand. Tillage events will redistribute soil P in the layers affected.

Computation of plant-available P assumes that soil P is relatively immobile and that only soluble P in close proximity to roots is available for uptake. The soil column is partitioned into root and non-root volume zones using a species-dependent root radius parameter and dynamically varying root length density to define the root zone volume for P uptake. Pools of labile, active, and stable P are maintained separately for root and non-root soil zones. As roots proliferate, mass of soil and nutrients are transferred from non-root to root soil zones, making more P available to the plant with higher root density. Soluble P is a proportion of labile P calculated daily and dependent on soil water content, labile P, and soil texture in each layer. This soluble P in the root zone is the daily P supply available for potential root uptake.

Soil P initialization is critical to a successful simulation of P processes, but data are often difficult or expensive to obtain. Labile P is computed from the measured extractable P using an expert system that depends on the laboratory extraction method used and soil characteristics.

8.1 Plant P processes

Modeling P demand is similar to that of crop N demand. Each day demand for each plant part is calculated as the amount of P required to bring tissue concentration up to a stage-dependent optimal concentration, plus the demand for new growth. This demand can be met through the soil supply and by mobilization of P to grain from vegetative, pod, or root tissue. P uptake is defined as the minimum of supply and demand. The amount of P taken up by roots may be further limited by a species-dependent minimum vegetative N:P ratio, limiting P uptake with low vegetative N concentrations. When supply falls short of demand, P stresses occur affecting rates of photosynthesis, vegetative and reproductive growth, and senescence.

9 Modeling genetics in DSSAT

For the CERES-style crops, species genetic attributes are present in the source code (as allometric relationships of partitioning to growth stage) as well as in the species, ecotype, and cultivar files. The genetic attributes of the CROPGRO-style crops are contained in the species, ecotype, and cultivar files. For CROPGRO-style crops with its single generic source code, the species file contains all

Published by Burleigh Dodds Science Publishing Limited, 2026.

parameters and parameterized relationships for sensitivity of processes (leaf appearance rate, rate of reproductive progress, photosynthesis, respiration, leaf area expansion, protein mobilization, pod addition, and seed growth rate) to temperature, along with compositions, N effects on photosynthesis, and many other parameters. The species file and the ecotype file are reserved for the model developers, and model users should only modify the cultivar file to mimic different cultivars within a crop species. For example, the cultivar file contains critical photoperiod parameters, photothermal durations (or heat units) required to reach given growth stages, along with other traits affecting photosynthesis, determinacy, leaf appearance rate, seed size, seed fill duration, and seed composition. The number of cultivar coefficients varies, depending on the crop module that is being used. For instance, the CERES-Maize model includes six cultivar coefficients, while the CROPGRO model includes 18 cultivar coefficients.

9.1 Estimating genotype-specific parameters

The DSSAT modeling system defines genotype-specific inputs, normally referred to as the Genotype-Specific Parameters (GSPs), thus allowing a user to define differences among cultivars, varieties, hybrids, clones, and other seed material. Although the user has a lot of flexibility in evaluating different local management scenarios with respect to genotypic performance, there are also challenges. As model developers, the DSSAT group is unable to provide local cultivar-specific parameters beyond those with specific experiments included in DSSAT, which means that a model has to be calibrated first for local genetics, requiring some of the critical observations associated with the MDS described previously. Once the crop management and observational data have been entered, the specific cultivar then has to be calibrated, either manually or using optimization tools. The ultimate goal is to minimize the error between simulated and observed phenological dates, yield, and yield components. Within the DSSAT ecosystem, there are two tools that can be used for crop cultivar calibration, the *GLUE* tool and the *GENCALC* tool. In addition, the sensitivity analysis utility can also be used to improve the value of one or more cultivar coefficients by setting the range and increment for a particular cultivar coefficient and by comparing simulated with observed data. A comparison of the performance of these two tools for rice was conducted by Buddhaboon et al. (2018).

9.2 General Likelihood Uncertainty Estimation (GLUE)

The General Likelihood Uncertainty Estimation (GLUE) is a statistical approach that results in multiple sets of parameter values that are equally as likely as the final solution. This approach was first introduced by Beven and Binley (1992) for modeling hydrological processes. The initial evaluation of GLUE for DSSAT was made for the CSM-CERES-Sweetcorn model by He et al. (2009, 2010), which required defining the means and variances of all cultivar parameters for sweetcorn based on the existing cultivar database of DSSAT. The approach was successful and provided in DSSAT as a new tool for estimating GSPs.

To estimate the most likely values for the GSPs for a new cultivar, a user first has to provide the required input files associated with weather, soil, and crop management, and basic observations, especially for phenology, yield, and yield components. Although a user can estimate the GSPs for only one treatment, the results will normally not be very robust. Therefore, we recommend that at least two non-stressed treatments from different environments representing either different locations, planting dates, or years be used. Once the model runs properly, the GSPs can be estimated with GLUE, first for phenological GSPs, and then for the yield and yield component GSPs. For most of the crops in DSSAT, the means and variances for the GSPs are provided in an input file that is used by GLUE for estimating the uncertainty. The ultimate outcome of GLUE is a list of the most likely value for each individual GSP that is being estimated.

9.3 Genetic Coefficient Calculator (GENCALC)

The Genetic Coefficient Calculator or GENCALC uses a rule-based approach to determine the value for one or more GSPs (Hunt et al., 1993). In the input file for GENCALC, one or more GSPs are associated with one particular plant trait, as described in Section 9.2. During the calibration process, these GSPs are varied to minimize the error between the simulated and observed trait. GENCALC normally optimizes the phenological GSPs first, followed by growth, and then yield components and yield for final optimization. Most crop models do not have a specific GSP that controls only yield and most GSPs affect multiple yield components. GENCALC should be used by more advanced DSSAT users who are familiar with the GSPs of a particular model and are comfortable editing the GENCALC rules file (Anothai et al., 2008).

10 Model analysis utilities for performance evaluation

For performance evaluation of the model with experimental data, visualization tools that not only provide a visual comparison between simulated and observed data, but also statistical analysis, are critical (Yang et al., 2014a). The main tool in DSSAT for visualization and comparison of model simulations with observed data is GBuild. Another tool is EasyGrapher, originally developed by scientists associated with Agriculture and Agri-Food Canada (Yang et al., 2014b).

Published by Burleigh Dodds Science Publishing Limited, 2026.

10.1 GBuild

GBuild is an analysis utility for visualization of simulated and experimental data (Uryasev et al., 2004). It gives a user the ability to easily plot graphs that are routinely used during crop model development and evaluation. The basic design of GBuild is based on a set of codes that are headers for each column of data that represent different variables. The file selection in GBuild allows a user to select one or more output files for plotting as well as any combination of the variables and runs/treatments, and then proceed to display the graph. The graphic-type selection options provide different views of the simulated results and include time series, for example displaying the simulated data as a function of date or days after planting (DAP), and simulated data versus experimental data. In order to compare the simulated results with observations from experiments, GBuild includes statistics for time series data with emphasis on the d-statistics (Willmott et al., 1985) and Root Mean Square Error (RMSE) and regression statistics for phenological and end-of-season data, such as flowering and maturity dates and yield and yield components. The graphic output of simulated and observed data can be visualized, printed, and exported into an Excel spread sheet with the statistics or exported to a text file with data only.

10.2 Sensitivity Analysis Tool

In addition to evaluating the model with real-world data, it is also important to understand the response of the model to one specific input, such as weather data, cultivars or hybrids, soil data, and values for individual GSPs. This approach, in which all inputs are kept constant except for one input or parameter, is called sensitivity analysis. A recently developed tool now available in DSSAT called SensitivityAnalysis enables the user to evaluate the model sensitivity to changes of cultivars, single GSPs, soil profiles, weather inputs for different location or year, plant and row spacing, and various other options. Variables that have numeric values, such as planting date, can be varied using a starting value, an increment value, and the number of iterations. The program automatically creates a new experimental file ready to run, with the selected sensitivity input variation. Following the simulations, the linked GBuild graphics program allows for a visual analysis of simulation results and associated statistics.

11 Application programs

Crop simulation models embedded in decision support systems are very powerful tools for scenario analyses. So far, this chapter has provided an overview of the structure and science of the crop simulation models that are included in DSSAT, as well as the tools and utilities for weather, soil, experimental and observational data entry, and crop model calibration and evaluation. Once a crop model has been calibrated, the most important and useful aspects are associated with the applications. DSSAT, therefore, includes several application programs. The seasonal analysis program is used for single-season scenario evaluations that account for both weather and economic uncertainties. The sequence analysis program is used for the analysis of crop rotation, and, in addition to weather and economic uncertainties, takes into account the effects of long-term cropping systems including changes in the soil system with respect to soil water, carbon, nitrogen, and other nutrient components (Fig. 5).

11.1 Seasonal analysis

The *'Seasonal Analysis'* application allows a user to explore the effects of weather variability and to evaluate the uncertainty and risk factors associated with various management and genetic inputs (Thornton and Hoogenboom, 1994). The DSSAT-CSM integrates the interaction of weather, soil, management, and genetic factors, enabling a user to simulate many hypothetical scenarios quickly

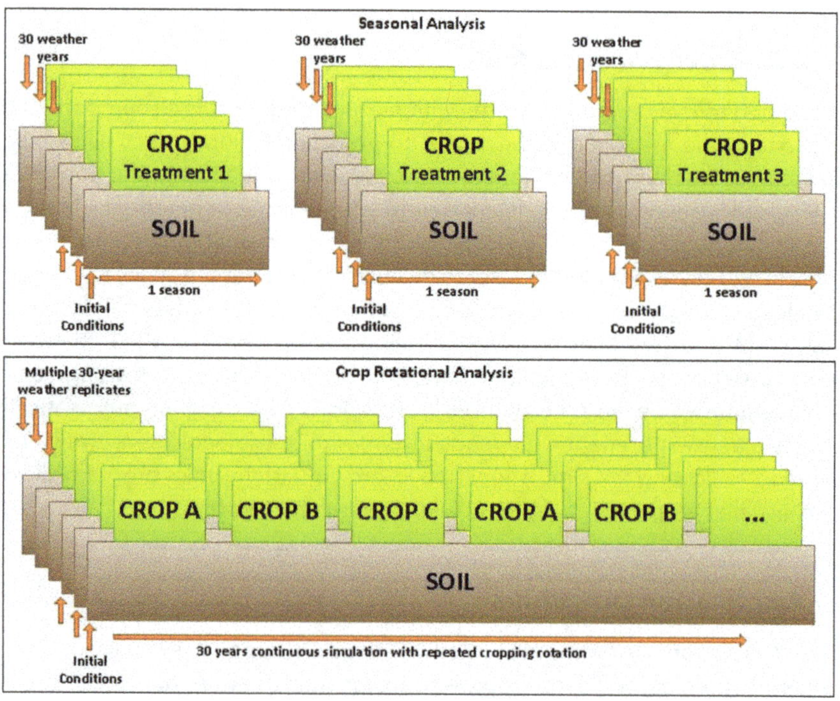

Figure 5 DSSAT scheme for the simulation of seasonal analysis scenarios (top) and crop rotational analysis scenarios (bottom).

Published by Burleigh Dodds Science Publishing Limited, 2026.

and efficiently using long-term historical weather records or stochastically generated weather data. The model simulates a distribution of yields and other outputs, converting uncertainty in weather into uncertainty in yield for the specified management scenarios.

The seasonal analysis application works on a field scale and emphasizes weather uncertainty. Economic risks can also be estimated using costs of inputs and prices of products, including the variability associated with these costs and prices. The application can be used to select optimal crop and variety, planting options, irrigation options, application of fertilizer and other agrochemical inputs, marketing options, insurance risks, policy advisement, and investments in equipment, technology, and diversification of land use. Typically, one season of simulation per scenario per weather year is conducted (Fig. 5). For each scenario, a number of weather years are simulated, with re-initialization of soil variables done at the beginning of each simulation so that the results reflect the variation in model outputs due to interannual weather variability.

A graphical interface allows a user to explore distributions of outcomes for variables including crop yields, farm profits, and environmental factors such as nitrogen leaching and irrigation requirements. The application includes graphical options such as box plots, cumulative function plots, and mean-variance plots for biophysical and economic variables (Fig. 6). This feature allows the user to optimize management practices that will benefit the farmer and to select best management practices relative to maximum profit, minimum risk of low profit or yield, minimum degradation of the environment, or other criteria.

11.2 Crop rotation analysis

Crop rotation analysis (or sequence analysis) application allows a user to produce long-term simulations of a given cropping system for predictions of farming system sustainability such as soil carbon loss, soil fertility degradation, decreasing yields, and increased greenhouse gas emissions (Thornton et al., 1995). Users can explore the sustainability of various options over a long period of time and optimize options for managing the land to sustain productivity, soil health, and natural resources (Tsuji et al., 1998). Soil organic matter is related to crop nutrient availability and thus to yield, income, and food security. Organic matter also improves water- and nutrient-use efficiency and reduces losses and environmental pollution. The soil also provides a sink for atmospheric C, a potentially important climate change mitigation mechanism referred to as soil carbon sequestration.

Crop rotation analysis in DSSAT is generally used to explore cropping system options based on a pattern of crops planted in sequence. Figure 5 illustrates the process and shows how this analysis differs from the seasonal

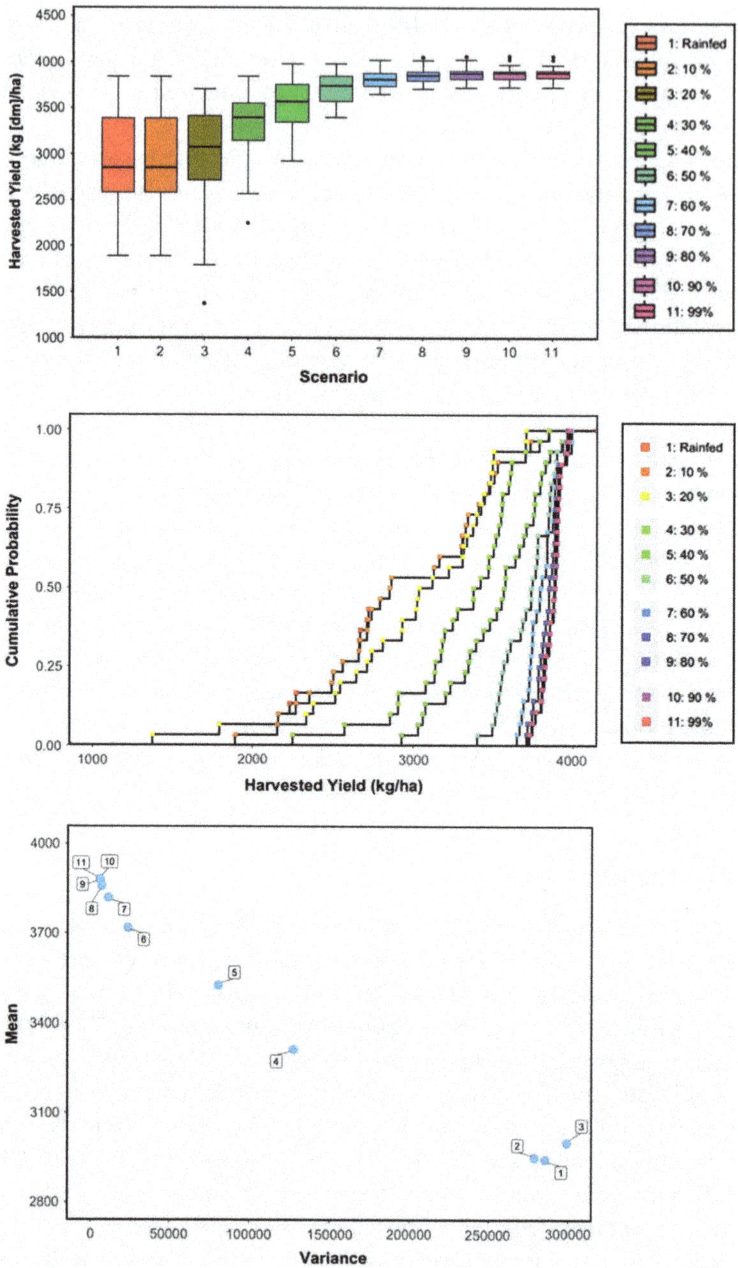

Figure 6 Box and whisker (top), cumulative probability (center), and mean-variance (bottom) plots a for scenario analysis of the impact of irrigation management on soybean yield in Gainesville, Florida, USA. The irrigation scenarios ranged from rainfed (scenario 1), a 10% threshold of extractable water (scenario 2), to a 99% threshold of extractable soil water (scenario 11) at 10% intervals for the top 30 cm of the soil profile.

Published by Burleigh Dodds Science Publishing Limited, 2026.

analysis. In the crop rotation analysis, soil variables are initialized once at the beginning of a long-term, continuous simulation. A crop sequence kernel or pattern is defined. In the example in Fig. 5, the kernel is represented by Crop A, followed by Crop B, followed by Crop C. The kernel is repeated as many times as the user specifies for a given simulation. If the effects of weather variability are important to the analysis, weather data can be generated to allow multiple realizations of daily weather data associated with the climatology being analyzed. In the example of Fig. 5, 30 realizations of a 30-year period of simulation are generated, allowing a distribution of possible outcomes for each variable predicted and each year in the time series.

Dynamics of soil organic carbon are of prime importance in these long-term simulations. In a system with poor organic carbon management (e.g. low inputs and removal of all crop residues from the system), soil organic carbon can be rapidly depleted, especially in the tropics. In sequence simulations, soil organic carbon is 'carried over' from one season to the next and these long-term soil fertility dynamics can be analyzed. Initialization of soil organic carbon state variables is often difficult, but it is very important to the predictive accuracy of the model. Often data on soil organic carbon composition are not available. Methods have been developed to estimate the amount of stable, intermediate, and microbial soil carbon present in the system (Basso et al., 2011; Porter et al., 2010).

11.3 Spatial analysis

The crop models in DSSAT are point-based models, in that the inputs are based on site-specific information such as the weather data from a local weather station, the soil data from a local profile at the experimental site, and crop management for a plot or field. For many applications, there is significant interest in understanding the variability across space for crop growth and development. Therefore, the models can be operated at the spatial scale, providing all input data at a spatial level, either for a polygon in which the inputs are considered the same or for a grid that is evenly distributed across an area. One of the weaknesses of the current system is that the models do not allow for interaction across space. However, the strength of the CSM crop model is that it can simulate at a spatial scale as small as 1 m or less for precision agriculture to 1 arc-degree for global simulations. The current DSSAT software does not include a specific tool for preparation of input files and visualization of output files. However, the underlying crop model and associated input and output files with GPS coordinates can be easily integrated into other systems if the DSSAT file naming convention and system structure are maintained.

There have been many approaches for integration of the DSSAT crop models with various Geographical Information Systems (GIS) and spatial databases, starting with the Agricultural and Environmental Geographic Information System (AEGIS; Lal et al., 1993; Luyten et al., 1994) and AEGIS/WIN (Engel et al., 1997). Most of the spatial applications have been conducted external to the DSSAT Windows Shell due to the complexity of the GIS systems and there are various approaches to coupling or linking crop models with GIS (Hartkamp et al., 1999; Thorp et al., 2008). Due to the rapid changes in GIS technology and software, as well as costs associated with some of the GIS systems, recent spatial applications have concentrated on using scripting languages for pre- and post-processing of the input data as well as for visualization. One example is MINK for global gridded simulations, developed by the International Food Policy Research Institute (IFPRI; Robertson, 2017). Another example is pDSSAT that has been developed for global gridded climate change applications in agriculture (Elliott et al., 2014). One of the limitations of these systems is that all gridded spatial inputs have to be referenced to the same grid. A recent development is used for spatial simulations based on present coordinates, allowing for flexible input data with spatial different resolutions for crop mask, weather, soil, and crop management. This tool, called DSSAT-pythia, can be run on any platform, including Linux, Windows, and iOS. It does not require any GIS system for data preparation, and it can use open-source display systems for thematic mapping of crop model outputs.

12 Example applications

The current DSSAT ecosystem includes at least one real-world experiment per crop that was used either for model development, calibration, or evaluation. As model developers, we feel that it is important to show the performance of the model when making it available to the DSSAT modeling community. DSSAT as an application is used extensively for a range of applications from gene-based modeling for plant breeding to climate change impact assessment across the globe for policy decisions. An initial overview of the range of applications was presented by Jones et al. (2003), and published applications of DSSAT and CSM have increased exponentially during the past 15 years. Rather than providing a detailed literature review, we provide a few illustrative case studies here to demonstrate the approach that is normally used for developing a specific application.

12.1 Interaction of nitrogen and water management on performance of maize

An experiment was conducted at the University of Florida in 1982 to study the interaction of nitrogen and irrigation management on maize (Bennett et al., 1985). The experiment included three levels of irrigation, that is, rainfed, stress during early growth (vegetative stress), and irrigated, and two levels of nitrogen fertilizer, that is, 116 kg N/ha in three applications and 401 kg N/ha in six applications, for a total of six treatments. The maize hybrid McCurdy 84aa was planted on February 26, 1982, at a plant density of 7.2 plants/m^2. The crop was well managed; phenology was observed nondestructively and growth analysis samples were taken on a regular basis. The CSM-CERES-Maize model was calibrated for the non-stressed treatment, but over time some of the other treatments were used for evaluating the response to water and nitrogen. Because of the sandy soils, there was a strong difference between the irrigated and rainfed treatments (Fig. 7). The number of days to flowering and

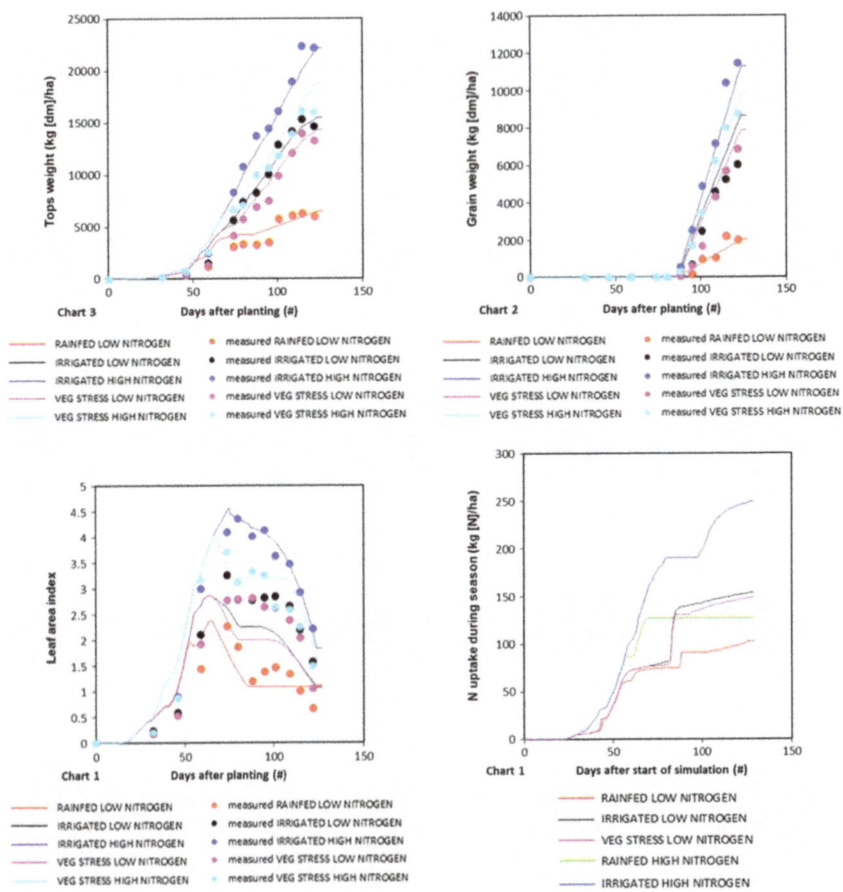

Figure 7 Simulated and measured tops weight (left top), grain weight (right top), leaf area index (bottom left), and N uptake (bottom right) for a maize experiment conducted in Gainesville, Florida, USA, in 1982. The experiment consists of three irrigation levels (rainfed, vegetative stress, and irrigated) and two nitrogen levels (low and high) for a total of six treatments.

physiological maturity were predicted well (RMSE for flowering duration: 1 day; RMSE for maturity duration: 1 day), and yield and yield components were also reasonable (RMSE for Tops Weight: 1280 kg/ha; RMSE for yield: 1087 kg/ha). No observations were available for the soil processes, but the model showed a clear response across the six treatments. Nitrogen uptake was highest for the irrigated-high nitrogen treatment and lowest for the rainfed-low nitrogen treatment (Fig. 7). Mineralization was slightly lower for the rainfed treatments compared to the irrigated treatments, but the differences were small. Nitrogen leaching was highest for the irrigated-high nitrogen treatment and lowest for the rainfed-low nitrogen treatment (Fig. 8). Inorganic nitrogen remaining

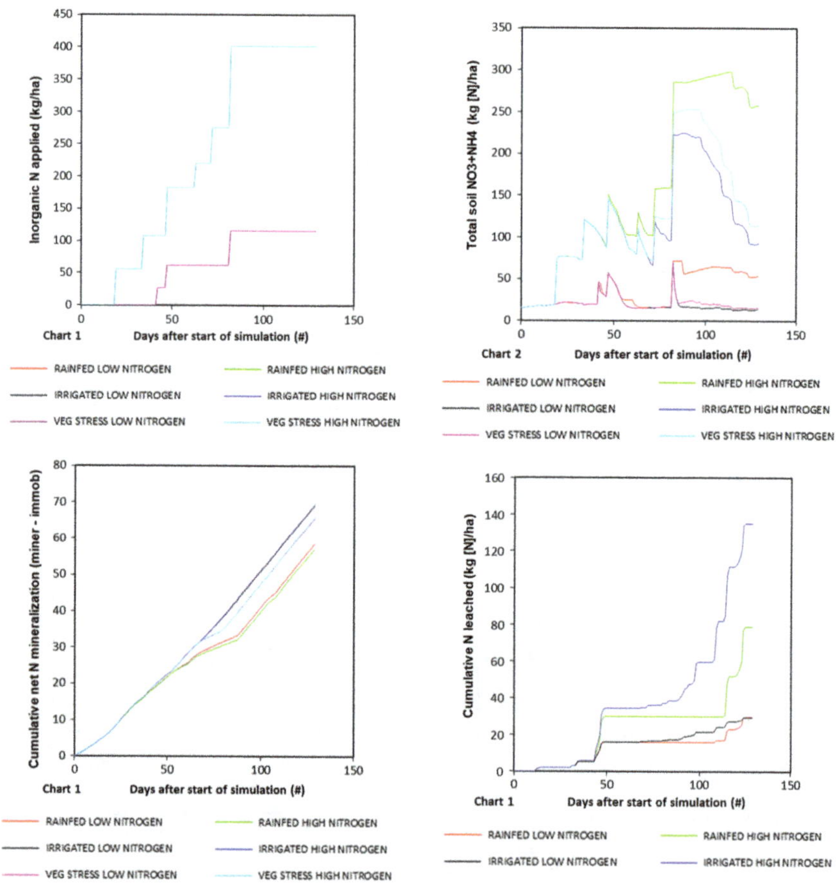

Figure 8 Simulated inorganic N applied (left top), total inorganic N in the soil profile (right top), cumulative N mineralized (bottom left) and N leached (bottom right) for a maize experiment conducted in Gainesville, Florida, USA, in 1982. The experiment consisted of three irrigation levels (rainfed, vegetative stress, and irrigated) and two nitrogen levels (low and high) for a total of six treatments.

Published by Burleigh Dodds Science Publishing Limited, 2026.

at the end of the growing season was highest for the rainfed-high nitrogen treatment, with most of the applied nitrogen remaining in the soil, and lowest for the irrigated and vegetative stress-low nitrogen treatment (Fig. 8). Overall, the model was able to simulate a close interaction between nitrogen and irrigation management and the impact was not only on yield, but also on the environment.

12.2 Impact of irrigation management on performance of soybean

An experiment was conducted at the University of Florida in 1978 to study the impact of irrigation on soybean growth, development, and yield (Wilkerson et al., 1983). The experiment included two treatments, that is, rainfed and irrigated with 21 irrigation applications for a total of 206 mm of supplemental irrigation. Soybean is a nitrogen-fixing crop, so no nitrogen fertilizer was applied. The cultivar Bragg was planted June 29, 1978, at a plant density of 29.9 plants/m^2. The crop was well managed; phenology was observed nondestructively and growth analysis samples were taken on a regular basis. The CSM-CROPGRO-Soybean model had been calibrated for the Bragg cultivar grown in the treatments for this experiment, as well other prior experiments conducted on this cultivar at the same location. The number of days to flowering and physiological maturity were predicted well, and yield and yield components were simulated very well (RMSE for Tops Weight: 218 kg/ha; RMSE for yield: 157 kg/ha). Early during the growing season, there was no treatment effect on aboveground biomass and LAI (Fig. 9) due to the high amount of rainfall received during this period (Fig. 10). Around 60 DAP, rainfall ended dramatically, with a decrease in extractable soil moisture (Fig. 10). This resulted in severe drought stress for the rainfed treatment (Fig. 10), reducing dry weight gain in tops and seed, and accelerating LAI senescence for the rainfed treatment compared to the irrigated treatment (Fig. 9).

Following model evaluation, the same experiment was set up as a strategy analysis scenario, using the same input conditions for each year and long-term weather data. Because 30 years of continuous historical weather data were not available, the internal weather generator WGEN was used. However, for the irrigation management, the automatic irrigation option was selected with different threshold values to determine when to irrigate. When the soil moisture content in the top 30 cm of the soil profile drops below this threshold value, an irrigation event is triggered by the model. The irrigation thresholds ranged from 10% to 99% (remaining soil water at which to irrigate) for a total of ten irrigation scenarios and one rainfed scenario. Final results can be analyzed either as box and whisker plots, cumulative probability graphs, or a mean-variance graph (Fig. 6). Depending on the analysis question, these graphs provide different functionalities based on the overall objectives. A researcher

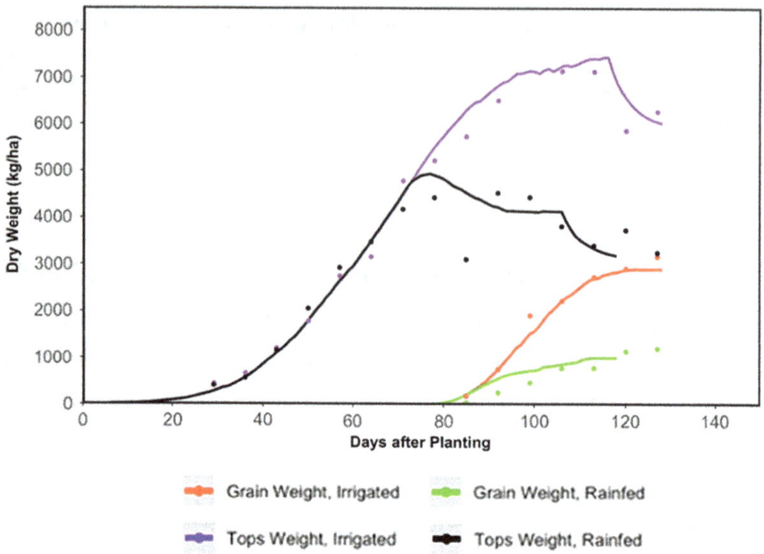

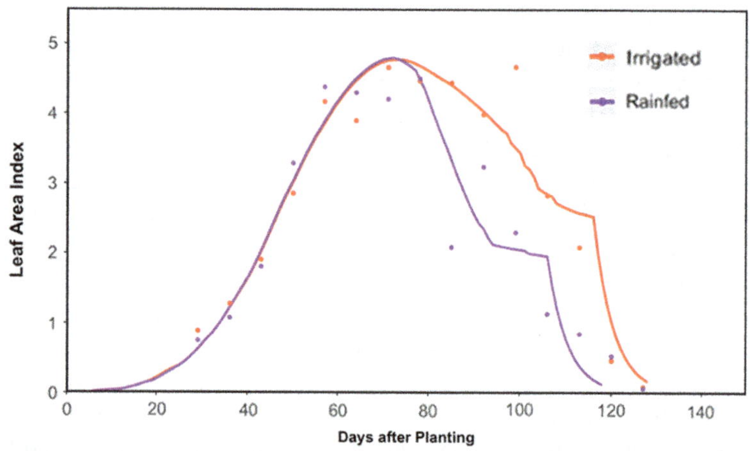

Figure 9 Simulated and measured tops and grain weight (left) and leaf area index (right) for a soybean experiment conducted in Gainesville, Florida, USA, in 1978. The two treatments were irrigated and rainfed.

might be interested in maximizing yield, maximizing water-use efficiency, or minimizing the impact on the environment or water use for irrigation.

The results show clearly that yield increased with an increase in the threshold value, while the variance and variability were reduced (Fig. 6). However, the amount of supplemental irrigation required also increased to over 300 mm for the highest threshold value with more than 30 irrigation applications (Fig. 11). In contrast, the amount of water applied upto a

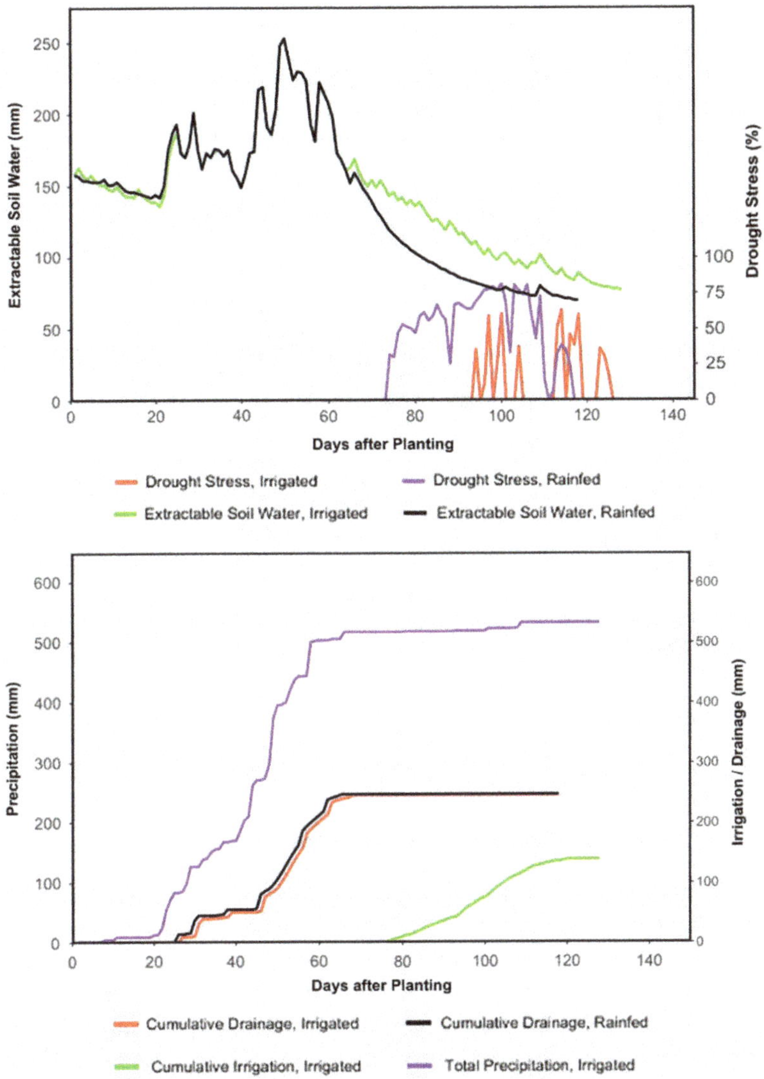

Figure 10 Simulated extractable water and drought stress (left) and cumulative irrigation, drainage and precipitation (right) for a soybean experiment conducted in Gainesville, Florida, USA, in 1978. The two treatments were irrigated and rainfed.

threshold value of 50% was less than 100 mm. The scenario with a threshold value of 50% showed the best water-use efficiency taking into consideration the associated uncertainty as well as yield (Fig. 6) and total water use (Fig. 11). If ground or surface water are limited due to governmental restrictions, water rights, or a drought, the model can be used to help determine the best

Published by Burleigh Dodds Science Publishing Limited, 2026.

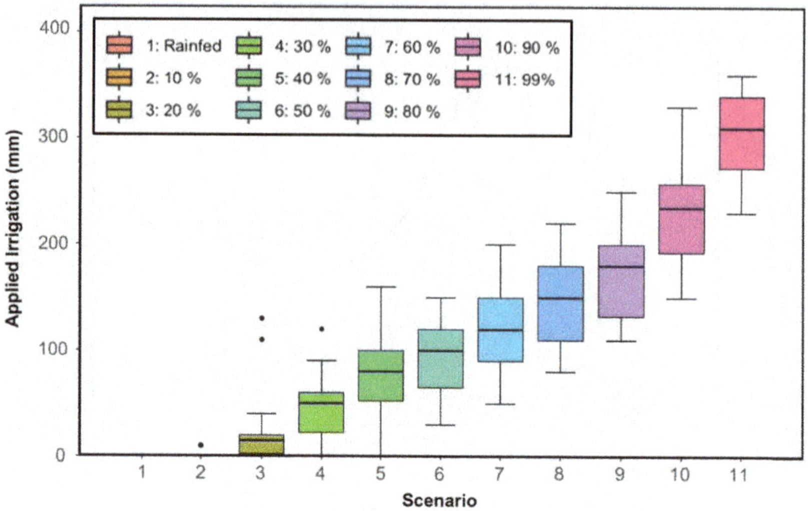

Figure 11 Total irrigation applied for a scenario analysis of the impact of irrigation management on soybean yield in Gainesville, Florida, USA. The irrigation scenarios ranged from rainfed (scenario 1), a 10% threshold of extractable water (scenario 2) to a 99% threshold of extractable soil water (scenario 11) at 10% intervals for the top 30 cm of the soil profile.

scenario that maximizes yield while at the same time optimizing water use for irrigation.

12.3 Residual soil moisture for crop rotations

In many regions around the world, crop production is restricted to the rainy season when sufficient moisture is available to grow a crop. However, in many instances, some soil moisture might be remaining at the end of the rainy season to allow for a second crop that requires less water. In India, soybean has become a dominant crop as a source for cooking oil, but it has to be grown during the rainy season due to the crop's water requirements. Chickpea is an important pulse crop that requires a lot less water. A soybean-chickpea rotation, therefore, has become quite common (Singh et al., 1999a,b). We defined a scenario to simulate this crop rotation using the different components described previously (Fig. 5). The location was Hyderabad, India, using a local soil. The soybean variety PK-472 (maturity group 8) was planted on June 25 and the chickpea crop was planted immediately following harvest of the soybean crop. The period between the harvest of the chickpea crop and planting of the soybean crop was considered to be a fallow period. For long-term weather, we used the WGEN weather generator, starting in 1978 and ending in 2020 for a duration of 42 years. Please note that these weather years are hypothetical years and do not represent the real weather conditions. The weather sequence from 1978 through 2020 was repeated 30 times.

Total seasonal precipitation for soybean ranged from 500 mm to 800 mm with some outliers, while seasonal rainfall for chickpea was less than 80 mm (Fig. 12). The median for plant extractable soil moisture at final harvest of

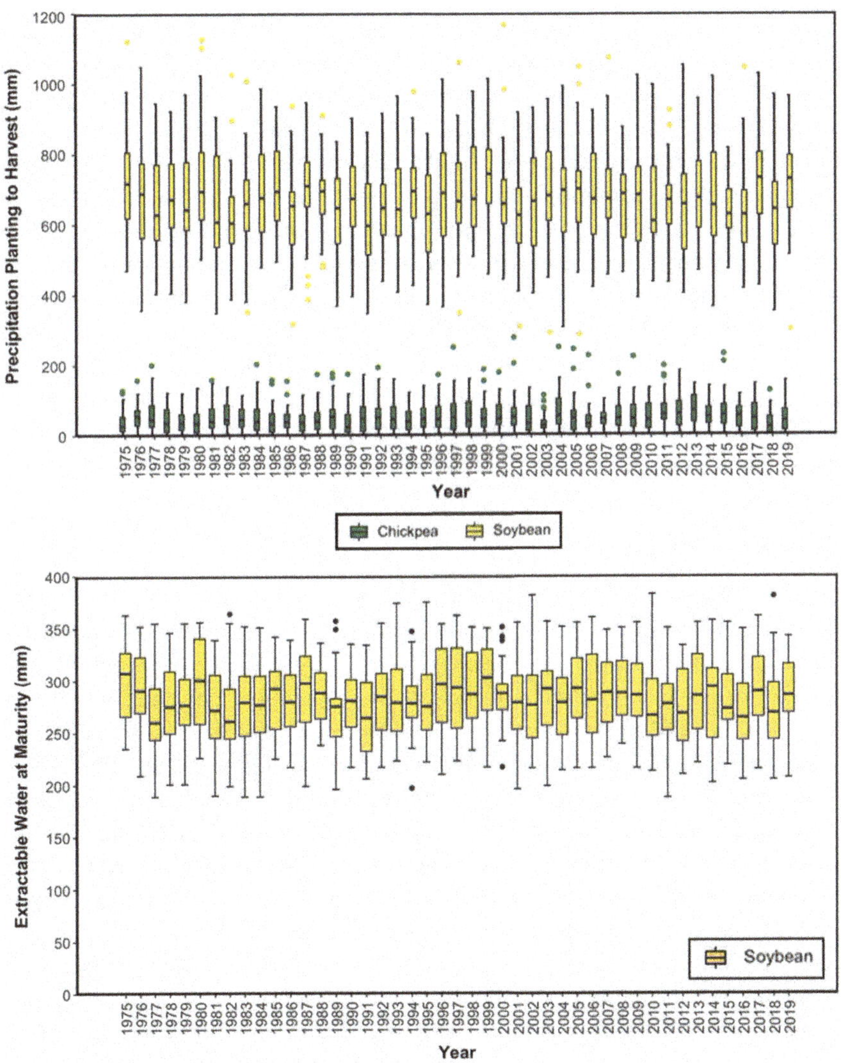

Figure 12 Seasonal precipitation from planting to harvest (top) and extractable soil moisture at soybean harvest (bottom) for a soybean-chickpea rotation in Hyderabad, India. Soybean was planted on June 25 and chickpea was planted immediately following the soybean harvest.

Published by Burleigh Dodds Science Publishing Limited, 2026.

soybean ranged from 250 mm to 310 mm and was rather variable (Fig. 12). Yield for soybean was fairly constant, varying between 2.8 t/ha and 3.3 t/ha, while chickpea yield was much more variable, ranging between 0 t/ha (crop failure) and 0.4 t/ha (Fig. 13). Water-use efficiency based on total yield over precipitation was fairly constant for soybean ranging between 4 kg/mm and 5 kg/mm (Fig. 13), while for chickpea it ranged from 0 kg/mm to 8 kg/mm, with the outliers not even shown in the analysis (Fig. 13).

For a proper cropping systems analysis, one should consider not only a single growing season, as discussed previously, but also other crops that might be grown or the fallow period between harvest of one crop and planting of the following crop. The physical, chemical, and biological processes in the soil are continuous, and, therefore, they should be simulated. In the example for the soybean-chickpea rotation, we analyzed the impact of residual soil moisture of the rainy season for growing a second crop during the dry season. Boote (Chapter 17) analyzed the impact of a continuous corn sequence/rotation on yield for the same Hyderabad location, showing a decrease in yield over time for an unfertilized crop associated with soil C decline, whereas an N-fertilized maize treatment did not decline in yield. Basso (Chapter 8) analyzed long-term soil C dynamics for a crop rotation. Overall, simulation models can be very powerful tools to analyze long-term crop rotations, especially when good experimental data for long-term crop rotations in general are lacking.

12.4 Yield forecasting

The application of crop simulation models for in-season yield forecasting has been of interest to many in the agriculture community, but so far it has been challenging due to the complexity and access to current weather data, weather forecasts, and climate outlooks (Georgiev and Hoogenboom, 1999; Hoogenboom, 2000). A new tool was recently developed external to DSSAT but based on the DSSAT crop model engine and associated input files and DSSAT system setup. The Climate Change, Agriculture and Food Security (CCAFS) Regional Agricultural Forecasting Toolbox (CRAFT) is a Windows desktop application that provides relatively easy access to gridded crop modeling and yield forecasting along with risk analysis and climate change impact assessments at spatial resolutions of 5 arc-minutes (0.083° or 10 km) and 30 arc-minutes (0.5° or 50 km; Shelia et al., 2019). The input data are prepared based on the GIS shape files of the region, weather files, and soil profile(s) data for each grid cell, masked data for crop and management, including organic and inorganic fertilizer, irrigation, and other inputs, and then uploaded into a MySQL database.

The toolbox can generate and conduct multiple simulation scenarios, maps, statistics, and interactive visualizations for a region and for each grid cell

Published by Burleigh Dodds Science Publishing Limited, 2026.

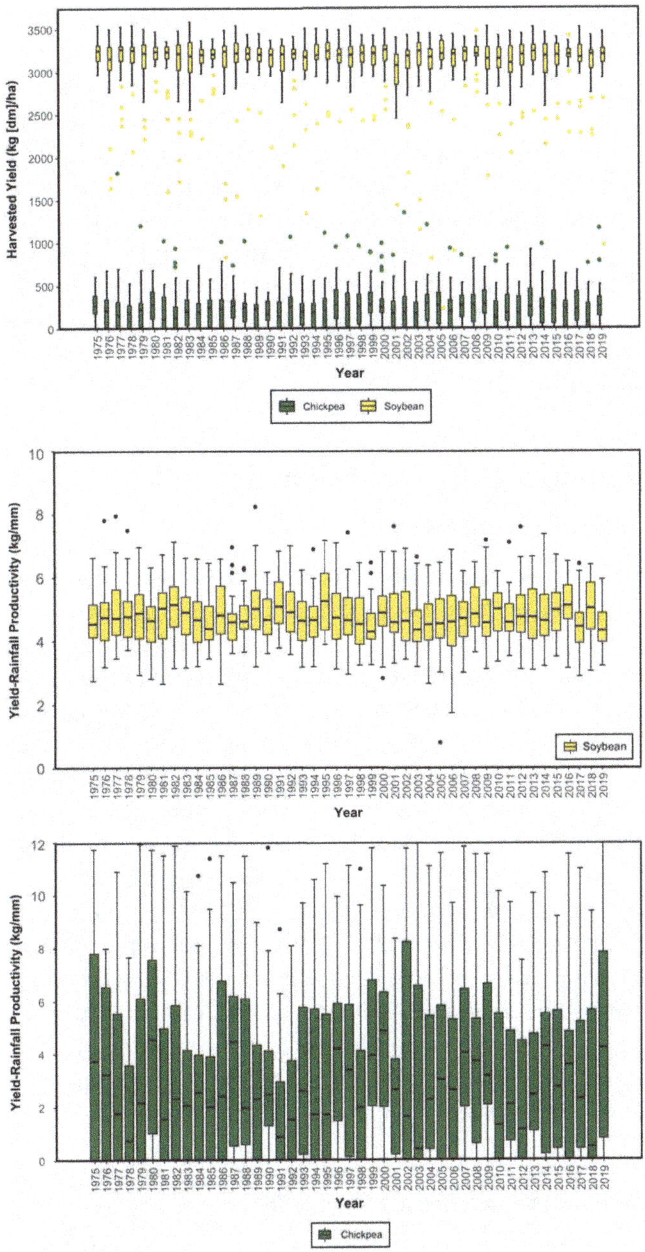

Figure 13 Final yield for soybean and chickpea (top) and water-use efficiency for soybean (center) and chickpea (bottom) for a soybean-chickpea rotation in Hyderabad, India. Soybean was planted on June 25 and chickpea was planted immediately following the soybean harvest.

of the selected region. The core of CRAFT is the crop engine that can run the crop simulation models not only for DSSAT, but also for other crop models such as APSIM and SARRA-H based on the Agricultural Model Intercomparison and Improvement Project (AgMIP) data tools (www.agmip.org). Gridded simulations can be conducted for any region for up to three different spatial scales such as a country, a state/province, or a district. CRAFT also has the embedded Climate Predictability Tool (CPT) for obtaining probabilistic seasonal climate forecasts (Mason and Tippett, 2016), and it uses a statistical approach to integrate the seasonal climate forecast with the crop yield forecast (Hansen et al., 2004, 2006; Mishra et al., 2008). Spatial aggregation from yield to production can be performed to account for the heterogeneity of environment and management of spatial data sets.

13 Developing a global modeling network

13.1 Software development

One of the unique aspects of the original IBSNAT Project was its participatory approach that encouraged scientists from different disciplines and different organizations to collaborate in support of the development of one unique DSSAT ecosystem. This network of model developers has expanded since the end of the IBSNAT Project in 1993 to include many scientists from Brazil, China, South Korea, France, Nigeria, South Africa, Kenya, and many other countries across the globe. The mutual support and collaboration are facilitated using an Open-Source approach, in which the source code of the crop simulation models along with the tools and utility programs are freely available. Currently, the source code for the CSM can be obtained from GitHub (https://github.com/DSSAT) upon request. We are planning to make CSM completely Open Source using the 3-Clause BSD License (BSD-3-Clause) once a few minor intellectual property rights have been resolved. Since 2014, the DSSAT Development Team has also facilitated biannual DSSAT Development Sprints, which are hackathons for crop modelers. During this 1-week, hands-on 'working' workshop, the emphasis is on crop model improvement and the advancement of application programs, tools, and utilities.

13.2 Software distribution and website

Previously, the DSSAT software had been sold to cover some of the administrative and development costs. In 2011, the distribution of DSSAT was changed to a free download system from the DSSAT portal (www.DSSAT.net) and since then the interest in DSSAT has exponentially increased. There have been 6630 downloads of DSSAT Version 4.5, 6500 downloads of DSSAT Version 4.6, and over 7100 downloads of DSSAT Version 4.7 from November 2017 through September 2019. For dissemination of information to the DSSAT user network, a DSSAT Listserv is used, and currently there are over 15000 unique e-mail addresses. User support and documentation are also provided through the DSSAT portal (www.DSSAT.net).

13.3 Training

Because DSSAT is a comprehensive software program, it requires training of agricultural scientists who traditionally specialize in single disciplines and who may not be very familiar with the systems approach that encompasses multiple disciplines. The first workshop sponsored by the IBSNAT Project was held in Venezuela in 1984, followed by a more extensive crop modeling training workshop at the University of Florida in 1985. Annual crop modeling training workshops have been held in the United States since then, initially as 2-week workshops and then condensed to intensive, 6-day training programs. Since 2002, these workshops have been hosted at the University of Georgia's Griffin campus, with an average of 50 international participants from universities, governmental organizations, and private industries. The rapid expansion of the DSSAT user network has also resulted in requests for international training programs across the globe, and recent workshops were held in Argentina, Australia, Indonesia, Nigeria, the Philippines, South Africa, Tanzania, Thailand, Tunisia, and Vietnam. In addition, some of our expert users are now DSSSAT trainers, facilitating workshops in Pakistan, Indonesia, Turkey, Brazil, China, and other countries. Ideal capacity building requires multiple workshops, starting with the basics of crop modeling and data requirements, followed by data collection for model evaluation, and finishing with model applications, the most critical part of crop modeling and decision support (Kihara et al., 2012).

14 The future of DSSAT

14.1 Collaboration

One of the challenges of maintaining and developing scientific software platforms in agriculture is the limited availability of resources for software development. Most grant agencies, both domestic and international, are not as interested in advancing scientific models as in applications. In addition, many agricultural scientists are not good programmers, especially in computer languages such as Fortran, Delphi, and Visual Basic that are currently used in DSSAT. Since the start of the development of DSSAT and the crop models, the emphasis has been on scientists developing the crop models, rather than relying on professional programmers who develop the code based on input provided by the scientists. However, this development model is not sustainable, especially when financial resources are limited. Therefore, collaboration is required among programmers who can code in scientifically sound code and scientists with state-of-the-art expertise. With the migration to Open Source and the sharing of the source code, the DSSAT Development group hopes to expand the community of those interested in advancing and improving the DSSAT ecosystem, including the models as well as the tools and utility programs. The DSSAT Development Sprints are part of this collaboration and so far the sprints have resulted in crop model improvement for irrigation management (Lopez et al., 2017), a new tool for soil data retrieval from the internet (Kim et al., 2018), and a new modeling engine for different operating systems (Resenes et al., 2019).

14.2 Mixed languages

The current source code of the CSM is Fortran. Although Fortran is computationally very efficient, it is not commonly taught in computer science courses. We are, therefore, evaluating a mixed-language approach, in which different programming languages can be combined into one that can be executable. The initial application for alternate languages will focus on the Input and Output file system of CSM, referred to as flexible I/O, to handle the Y2K issue with the current DSSAT input files and to provide a mechanism that facilitates adding or removing new parameters to the cultivar, ecotype, or species files.

14.3 Insect pests, diseases, and weed modeling

One weakness of the DSSAT ecosystem and many other crop modeling systems, is the limited capability for handling the impact of biotic stresses caused by insect pests, diseases, and weeds. DSSAT currently has a static system that allows a user to define biotic stressors based on field damage observations. However, there is no coupling with dynamic pest and disease models. We are currently evaluating using Docker containers and images that allow for the coupling of two or more models in order to provide opportunities for running multiple instances of two models in parallel.

14.4 Gene-based modeling

One of the most challenging aspects of the DSSAT crop models are the GSPs, which, for all models, must be estimated for local cultivars and hybrids prior to any real-world application. There have been efforts to bridge the gap between biotechnology, genetics, plant breeding, and crop modeling using either genes or quantitative trait loci (QTLs). The first, simple, gene-based model, GENEGRO, was developed by White and Hoogenboom (1996) more than 20 years ago, linking several genes to the GSPs of the dry bean model, BEANGRO. Predictions for phenology were as accurate as the original model, while final yield and biomass predictions were more challenging. A similar gene-based approach was applied for soybean by Messina et al. (2006). More recent developments are based on QTLs that are directly or indirectly linked to GSPs or plant traits (Wallach et al., 2018). As the cost of mapping QTLs/genes becomes cheaper, it is expected that rapid advances can be made in this area. A future model would have QTLs/genes linked to one or more growth and development processes via modules that would allow for the input of gene maps directly into crop models (Hoogenboom and White, 2003; White and Hoogenboom, 2003). If successful, this improvement would then reduce the requirements for calibration of a new cultivar for local conditions, assuming that QTL knowledge is public and proper phenotyping of QTL actions has been done (Hoogenboom et al., 2004).

15 Summary

The current agricultural production system is challenged with weather and climate extremes and variability and economic risks. There is pressure to grow more and healthy food using sustainable practices. At the same time, technology is rapidly improving with new sensor technologies, the Internet of Things, edge computing, and remote sensing. The amount of data that are being collected for agricultural production system is exponentially expanding, providing opportunities for data analytics for strategic and actionable decisions. The DSSAT ecosystem can play a major role in helping to understand the interaction between Genotype, Environment, and Management (G * E * M) and to provide alternative management options that increase crop yield and quality, optimize resource use, and minimize environmental impact for long-term sustainable agricultural production.

The DSSAT crop modeling ecosystem is one of the oldest and most widely used crop modeling platforms across the world. The success of DSSAT is based on the inclusiveness and participatory approach that has been used since the original development of the CERES and CROPGRO family of models and the emphasis on sharing data and model code. DSSAT is not just a software program, but an ***ecosystem*** of:

- Crop model users;
- Crop model trainers;
- Crop model developers;
- Models for the most important food, feed, fiber, and fuel crops;

- Tools and utilities for data preparation;
- Minimum data for model calibration and evaluation; and
- Application programs for assessing real-world problems.

16 Acknowledgements

The authors appreciate the interest of the DSSAT Users' community in the use of the DSSAT-CSM for a very wide range of applications as shown by the large number of download requests for DSSAT and modeling publications based on DSSAT. Your encouragement motivates us to continuously improve DSSAT! The authors would like to thank Dr. Carol Wilkerson for providing editorial support and Mr. Alwin Hopf for creating the simulation graphics.

17 References

Alagarswamy, G., Boote, K. J., Allen Jr., L. H. and Jones, J. W. 2006. Evaluating the CROPGRO-Soybean model ability to simulate photosynthesis response to carbon dioxide levels. *Agronomy Journal* 98(1), 34–42. doi:10.2134/agronj2004-0298.

Alderman, P. D., Boote, K. J., Jones, J. W. and Bhatia, V. S. 2015. Adapting the CSM-CROPGRO model for pigeonpea using sequential parameter estimation. *Field Crops Research* 181, 1–15. doi:10.1016/j.fcr.2015.05.024.

Allen, R. G., Pereira, L. S., Raes, D. and Smith, M. 1998. Crop evapotranspiration. Guidelines for computing crop water requirements. FAO Irrigation and drainage paper no 56. FAO, Rome, Italy.

Anar, M. J., Lin, Z., Hoogenboom, G., Shelia, V., Batchelor, W. D., Teboh, J. M., Ostlie, M., Schatz, B. G. and Khan, M. 2019. Modeling growth, development and yield of sugarbeet using DSSAT. *Agricultural Systems* 169(1), 58–70. doi:10.1016/j.agsy.2018.11.010.

Anothai, J., Patanothai, A., Jogloy, S., Pannangpetch, K., Boote, K. J. and Hoogenboom, G. 2008. A sequential approach for determining the cultivar coefficients of peanut lines using end-of-season data of crop performance trials. *Field Crops Research* 108(2), 169–78. doi:10.1016/j.fcr.2008.04.012.

Asseng, S., van Keulen, H. and Stol, W. 2000. Performance and application of the APSIM Nwheat model in the Netherlands. *European Journal of Agronomy* 12(1), 37–54.

Basso, B., Gargiulo, O., Paustian, K., Robertson, G. P., Porter, C., Grace, P. R. and Jones, J. W. 2011. Procedures for initializing soil organic carbon pools in the DSSAT-CENTURY model for agricultural systems. *Soil Science Society of America Journal* 75(1), 69–78. doi:10.2136/sssaj2010.0115.

Bennett, J. M., Jones, J. W., Zur, B. and Hammond, L. C. 1985. Interactive effects of nitrogen and water stresses on water relations of field-grown corn leaves. *Agronomy Journal* 78(2), 273–80.

Beven, K. and Binley, A. 1992. The future of distributed models: model calibration and uncertainty prediction. *Hydrological Processes* 6(3), 279–98. doi:10.1002/hyp.3360060305.

Boote, K. J. and Pickering, N. B. 1994. Modeling photosynthesis of row crop canopies. *HortScience* 29(12), 1423–34. doi:10.21273/HORTSCI.29.12.1423.

Boote, K. J., Jones, J. W., Mishoe, J. W. and Wilkerson, G. G. 1986. Modeling growth and yield of groundnut. In: *Agrometeorology of Groundnut*. Proc. Int. Symp., 21–26 August 1985, ICRISAT Sahelian Center, Niamey, Niger. ICRISAT, Patancheru, India, pp. 243–54.

Boote, K. J., Jones, J. W., Hoogenboom, G., Wilkerson, G. G. and Jagtap, S. S. 1987. *PNUTGRO V1.0. Peanut Crop Growth Simulation Model. User's Guide*. Department of Agronomy and Department of Agricultural Engineering, University of Florida, Gainesville, FL, 48pp.

Boote, K. J., Jones, J. W., Hoogenboom, G. and Pickering, N. B. 1998. The CROPGRO model for grain legumes. In: Tsuji, G. Y., Hoogenboom, G. and Thornton, P. K. (Eds), *Understanding Options for Agricultural Production*. Kluwer Academic Publishers, Dordrecht, pp. 99-128.

Boote, K. J., Mínguez, M. I. and Sau, F. 2002. Adapting the CROPGRO legume model to simulate growth of faba bean. *Agronomy Journal* 94(4), 743–56. doi:10.2134/agronj2002.0743.

Boote, K. J., Hoogenboom, G., Jones, J. W. and Ingram, K. T. 2008. Modeling N-fixation and its relationship to N uptake in the CROPGRO model. In: Ma, L., Ahuja, L. and Bruulsema, T. (Eds), *Quantifying and Understanding Plant Nitrogen Uptake for Systems Modeling*. Taylor & Francis Group LLC, Boca Raton, FL.

Boote, K. J., Sau, F., Hoogenboom, G. and Jones, J. W. 2009. Experience with water balance, evapotranspiration, and prediction of water stress effects in the CROPGRO model. In: Ahuja, L. R., Reddy, V. R., Saseendran, S. A. and Yu, Q. (Eds), *Response of Crops to Limited Water: Modeling Water Stress Effects on Plant Growth Processes, Volume 1 of Advances in Agricultural Systems Modeling*. ASA-CSSA-SSSA, Madison, WI.

Boote, K. J., Allen Jr., L. H., Prasad, P. V. V. and Jones, J. W. 2010. Testing effects of climate change in crop models. In: Hillel, D. and Rosenzweig, C. (Eds), *Handbook of Climate Change and Agroecosystems*. Imperial College Press, London UK.

Boote, K. J., Rybak, M. R., Scholberg, J. M. S. and Jones, J. W. 2012. Improving the CROPGRO-Tomato model for predicting growth and yield response to temperature. *HortScience* 47(8), 1038–49. doi:10.21273/HORTSCI.47.8.1038.

Boote, K. J., Porter, C., Jones, J. W., Thorburn, P. J., Kersebaum, K. C., Hoogenboom, G., White, J. W. and Hatfield, J. L. 2015. Sentinel site data for crop model improvement–definition and characterization. In: *Improving Crop Modeling Tools to Assess Climate Change Effects on Crop Response. Advances in Agricultural Systems Modeling 07*. American Society of Agronomy, Crop Science Society of America, and Soil Science Society of America, Madison, WI.

Buddhaboon, C., Jintrawet, A. and Hoogenboom, G. 2018. Methodology to estimate rice genetic coefficients for the CSM-CERES-Rice model using GENCALC and GLUE genetic coefficient estimators. *The Journal of Agricultural Science* 156(4), 482–92. doi:10.1017/S0021859618000527.

Del Grosso, S. J., Parton, W. J., Mosier, A. R., Hartman, M. D., Brenner, J., Ojima, D. S. and Schimel, D. S. 2001. Simulated interaction of carbon dynamics and nitrogen trace gas fluxes using the DAYCENT model. In: Schaffer, M., Ma, L. and Hansen, S. (Eds), *Modeling Carbon and Nitrogen Dynamics for Soil Management*. CRC Press, Boca Raton, FL, pp. 303–32.

Deligios, P. A., Farci, R., Sulas, L., Hoogenboom, G. and Ledda, L. 2013. Predicting growth and yield of winter rapeseed in a Mediterranean environment: model adaptation at a field scale. *Field Crops Research* 144, 100–12. doi:10.1016/j.fcr.2013.01.017.

Elliott, J., Kelly, D., Chryssanthacopoulos, J., Glotter, M., Jhunjhnuwala, K., Best, N., Wilde, M. and Foster, I. 2014. The parallel system for integrating impact models and sectors (pSIMS). *Environmental Modelling and Software* 62, 509–16. doi:10.1016/j.envsoft.2014.04.008.

Engel, T., Hoogenboom, G., Jones, J. W. and Wilkens, P. W. 1997. AEGIS/WIN - a computer program for the application of crop simulation models across geographic areas. *Agronomy Journal* 89(6), 919–28. doi:10.2134/agronj1997.00021962008900060012x.

Farquhar, G. D. and von Caemmerer, S. 1982. Modelling of photosynthetic response to environmental conditions. In: Lange, O. L., Nobel, P. S., Osmond, C. B. and Ziegler, H. (Eds), *Physiological Plant Ecology II. Encyclopedia of Plant Physiology (New Series)* (vol. 12/B). Springer-Verlag, Berlin, Germany, pp. 549-87.

Georgiev, G. A. and Hoogenboom, G. 1999. Near real-time agricultural simulations on the web. *Simulation* 73(1), 22–8. doi:10.1177/003754979907300104.

Gijsman, A. J., Hoogenboom, G., Parton, W. J. and Kerridge, P. C. 2002. Modifying DSSAT crop models for low-input agricultural systems using a soil organic matter-residue module from CENTURY. *Agronomy Journal* 94(3), 462–74. doi:10.2134/agronj2002.4620.

Godwin, D. C. and Singh, U. 1998. Nitrogen balance and crop response to nitrogen in upland and lowland cropping systems. In: Tsuji, G. Y., Hoogenboom, G. and Thornton, P. K. (Eds), *Understanding Options for Agricultural Production*. Kluwer Academic Publishers, Dordrecht, the Netherlands, pp. 55–77.

Guarin, J. R., Kassie, B., Mashaheet, A. M., Burkey, K. and Asseng, S. 2019. Modeling the effects of tropospheric ozone on wheat growth and yield. *European Journal of Agronomy* 105, 13–23. doi:10.1016/j.eja.2019.02.004.

Hansen, J. W., Potgieter, A. and Tippet, M. K. 2004. Using a general circulation model to forecast regional wheat yields in Northeast Australia. *Agricultural and Forest Meteorology* 127(1–2), 77–92. doi:10.1016/j.agrformet.2004.07.005.

Hansen, J. W., Challinor, A., Ines, A., Wheeler, T. and Moron, V. 2006. Translating climate forecasts into agricultural terms: advances and challenges. *Climate Research* 33, 27–41. doi:10.3354/cr033027.

Hartkamp, A. D., White, J. W. and Hoogenboom, G. 1999. Interfacing geographic information systems with agronomic modeling: a review. *Agronomy Journal* 91(5), 762–72. doi:10.2134/agronj1999.915761x.

Hartkamp, A. D., Hoogenboom, G. and White, J. W. 2002. Adaptation of the CROPGRO growth model to velvet bean (*Mucuna pruriens*): I. Model development. *Field Crops Research* 78(1), 9–25. doi:10.1016/S0378-4290(02)00091-6.

He, J., Dukes, M. D., Jones, J. W. and Graham, W. D. 2009. Applying GLUE for estimating CERES-Maize genetic and soil parameters for sweet corn production. *Transactions of the ASABE* 52(6), 1907–21.

He, J., Jones, J. W., Graham, W. D. and Dukes, M. D. 2010. Influence of likelihood function choice for estimating crop model parameters using the generalized likelihood uncertainty estimation method. *Agricultural Systems* 103(5), 256–64. doi:10.1016/j.agsy.2010.01.006.

Hoogenboom, G. 2000. Contribution of agrometeorology to the simulation of crop production and its applications. *Agricultural and Forest Meteorology* 103(1–2), 137–57. doi:10.1016/S0168-1923(00)00108-8.

Hoogenboom, G. and White, J. W. 2003. Improving physiological assumptions of simulation models by using gene-based approaches. *Agronomy Journal* 95(1), 82–9. doi:10.2134/agronj2003.0082.

Hoogenboom, G., White, J. W., Jones, J. W. and Boote, K. J. 1990. *BEANGRO V1.00. Dry Bean Crop Growth Simulation Model. User's Guide*. Florida Agricultural Experiment Station Journal No. N-00379. University of Florida, Gainesville, FL, 120pp.

Hoogenboom, G., Jones, J. W. and Boote, K. J. 1991. *Predicting Growth and Development of Grain Legumes with a Generic Model*. ASAE Paper 91-4501. American Society of Agricultural Engineers, St. Joseph, MI.

Hoogenboom, G., Jones, J. W. and Boote, K. J. 1992. Modeling growth, development and yield of grain legumes using SOYGRO, PNUTGRO, and BEANGRO: a review. *Transactions of the ASAE* 35(6), 2043–56.

Hoogenboom, G., White, J. W., Jones, J. W. and Boote, K. J. 1994. BEANGRO, a process oriented dry bean model with a versatile user interface. *Agronomy Journal* 86(1), 182–90. doi:10.2134/agronj1994.00021962008600010032x.

Hoogenboom, G., Singh, P., Alagarswamy, G. and Virmani, S. M. 1997. *Prediction of Yield and Water Use for Chickpea-Based Cropping Systems*. Resource Management Program, International Crops Research Institute for the Semi-Arid Tropics (ICRISAT), Patancheru, Andhra Pradesh, India, 17pp.

Hoogenboom, G., Wilkens, P. W., Thornton, P. K., Jones, J. W., Hunt, L. A. and Imamura, D. T. 1999. Decision support system for agrotechnology transfer v3.5. In: Hoogenboom, G., Wilkens, P. W. and Tsuji, G. Y. (Eds), *DSSAT Version 3* (vol. 4). University of Hawaii, Honolulu, HI, pp. 1–36.

Hoogenboom, G., White, J. W. and Messina, C. D. 2004. From genome to crop: integration through simulation modeling. *Field Crops Research* 90(1), 145–63. doi:10.1016/j.fcr.2004.07.014.

Hoogenboom, G., Jones, J. W., Traore, P. C. S. and Boote, K. J. 2012. Experiments and data for model evaluation and application. In: Kihara, J., Fatondji, D., Jones, J. W., Hoogenboom, G., Tabo, R. and Bationo, A. (Eds), *Improving Soil Fertility Recommendations in Africa Using the Decision Support Systems for Agrotechnology Transfers (DSSAT)*. Springer, Dordrecht, the Netherlands, pp. 9–18.

Hoogenboom, G., Porter, C. H., Shelia, V., Boote, K. J., Singh, U., White, J. W., Hunt, L. A., Ogoshi, R., Lizaso, J. I., Koo, J., Asseng, S., Singels, A., Moreno, L. P. and Jones, J. W. 2019. *Decision Support System for Agrotechnology Transfer (DSSAT) Version 4.7*. DSSAT Foundation, Gainesville, FL. Available at: www.DSSAT.net.

Hunt, L. A. and Pararajasingham, S. 1995. CROPSIM–WHEAT: a model describing the growth and development of wheat. *Canadian Journal of Plant Science* 75(3), 619–32. doi:10.4141/cjps95-107.

Hunt, L. A., White, J. W. and Hoogenboom, G. 2001. Agronomic data: advances in documentation and protocols for exchange and use. *Agricultural Systems* 70(2–3), 477–92. doi:10.1016/S0308-521X(01)00056-7.

Hunt, L. A., Pararajasingham, S., Jones, J. W., Hoogenboom, G., Imamura, D. T. and Ogoshi, R. M. 1993. GENCALC - software to facilitate the use of crop models for analyzing field experiments. *Agronomy Journal* 85(5), 1090–4. doi:10.2134/agronj1993.00021962008500050025x.

Hunt, L. A., Jones, J. W., Hoogenboom, G., Godwin, D. C., Singh, U., Pickering, N., Thornton, P. K., Boote, K. J. and Ritchie, J. T. 1994. General input and output files structures for crop simulation models. In: Uhlir, P. F. and Carter, G. C. (Eds), *Crop Modeling and Related Environmental Data. A Focus on Applications for Arid and Semiarid Regions in Developing Countries*. CODATA, Paris, France, pp. 35–73 (ISBN 1-884893-00-7).

International Benchmark Sites Network for Agrotechnology Transfer (IBSNAT). 1989. *The Decision Support System for Agrotechnology Transfer Version 2.1 (DSSAT v2.1) User's Guide*. Department of Agronomy and Soil Science, College of Tropical Agriculture and Human Resources, University of Hawaii, Honolulu, HI.

International Benchmark Sites Network for Agrotechnology Transfer (IBSNAT). 1993. *The IBSNAT Decade*. Department of Agronomy and Soil Science, College of Tropical Agriculture and Human Resources, University of Hawaii, Honolulu, HI, p. 178.

International Benchmark Sites Network for Agrotechnology Transfer (IBSNAT). 1994. *The Decision Support System for Agrotechnology Transfer Version 3.0 (DSSAT v3) User's Guide*. Department of Agronomy and Soil Science, College of Tropical Agriculture and Human Resources, University of Hawaii, Honolulu, HI.

International Crops Research Institute for the Semi-Arid Tropics (ICRISAT). 1984. *Proceedings of the International Symposium on Minimum Data Sets for Agrotechnology Transfer*. ICRISAT Center, Patancheru, India.

Jones, C. A. and Kiniry, J. R. (Eds). 1986. *CERES-Maize: A Simulation Model of Maize Growth and Development. Texas A & M*. University Press, College Station, TX.

Jones, M. R. and Singels, A. 2018. Refining the Canegro model for improved simulation of climate change impacts on sugarcane. *European Journal of Agronomy* 100, 76–86. doi:10.1016/j.eja.2017.12.009.

Jones, J. W., Boote, K. J., Jagtap, S. S., Hoogenboom, G. and Wilkerson, G. G. 1987. *SOYGRO V5.4. Soybean Crop Growth Simulation Model. User's Guide*. Department of Agricultural Engineering and Department of Agronomy, University of Florida, Gainesville, FL, 50pp.

Jones, J. W., Tsuji, G. Y., Hoogenboom, G., Hunt, L. A., Thornton, P. K., Wilkens, P. W., Imamura, D. T., Bowen, W. T. and Singh, U. 1998. Decision support system for agrotechnology transfer: DSSAT v3. In: Tsuji, G. Y., Hoogenboom, G. and Thornton, P. K. (Eds), *Understanding Options for Agricultural Production. Systems Approaches for Sustainable Agricultural Development* (vol. 7). Springer, Dordrecht, the Netherlands.

Jones, J. W., Hoogenboom, G., Porter, C. H., Boote, K. J., Batchelor, W. D., Hunt, L. A., Wilkens, P. W., Singh, U., Gijsman, A. J. and Ritchie, J. T. 2003. The DSSAT cropping system model. *European Journal of Agronomy* 18(3–4), 235–65. doi:10.1016/S1161-0301(02)00107-7.

Jones, J. W., Antle, J. M., Basso, B. O., Boote, K. J., Conant, R. T., Foster, I., Godfray, H. C. J., Herrero, M., Howitt, R. E., Janssen, S., Keating, B. A., Munoz-Carpena, R., Porter, C. H., Rosenzweig, C. and Wheeler, T. R. 2017. Brief history of agricultural systems modeling. *Agricultural Systems* 155, 240–54. doi:10.1016/j.agsy.2016.05.014.

Kersebaum, K. C., Boote, K. J., Jorgenson, J. S., Nendel, C., Bindi, M., Fruehauf, C., Gaiser, T., Hoogenboom, G., Kollas, C., Olesen, J. E., Rotter, R. P., Ruget, F., Thorburn, P. J., Trnka, M. and Wegehenkel, M. 2015. Analysis and classification of data sets for calibration and validation of agro-ecosystem models. *Environmental Modelling and Software* 72, 402–17. doi:10.1016/j.envsoft.2015.05.009.

Kihara, J., Fatondji, D., Jones, J. W., Hoogenboom, G., Tabo, R. and Bationo, A. (Eds). 2012. *Improving Soil Fertility Recommendations in Africa Using the Decision Support for Agrotechnology Transfer (DSSAT)*. Springer, Dordrecht, the Netherlands, 195pp. ISBN: 978-94-007-2959-9.

Kim, K. S., Yoo, B. H., Shelia, V., Porter, C. H. and Hoogenboom, G. 2018. START: a data preparation tool for crop simulation models using web-based soil

Published by Burleigh Dodds Science Publishing Limited, 2026.

databases. *Computers and Electronics in Agriculture* 154, 256-64. doi:10.1016/j.compag.2018.08.023.

Lal, H., Hoogenboom, G., Calixte, J.-P., Jones, J. W. and Beinroth, F. H. 1993. Using crop simulation models and GIS for regional productivity analysis. *Transactions of the ASAE* 36(1), 175-84.

Lizaso, J. I., Boote, K. J., Cherr, C. M., Scholberg, J. M. S., Casanova, J. J., Judge, J., Jones, J. W. and Hoogenboom, G. 2007. Developing a sweet corn simulation model to predict fresh market yield and quality of ears. *Journal of the American Society for Horticultural Science* 132(3), 415-22. doi:10.21273/JASHS.132.3.415.

Lopez, J. R., Winter, J. M., Elliott, J., Ruane, A. C., Porter, C. H. and Hoogenboom, G. 2017. Integrating growth stage deficit irrigation into a process-based crop model. *Agricultural and Forest Meteorology* 243(1), 84-92. doi:10.1016/j.agrformet.2017.05.001.

Luyten, J. C., Jones, J. W., Calixte, J. P., Hoogenboom, G. and Negabhan, B. 1994. *AEGIS+. Agricultural and Environmental Geographic Information System Plus Version 2.0. User's and Developer's Manual*. Research Report AGE No. 94-1. Agricultural Engineering Department, University of Florida, Gainesville, FL, 107pp.

Malik, W., Boote, K. J., Hoogenboom, G., Cavero, J. and Dechmi, F. 2018. Adapting the CROPGRO model to simulate alfalfa growth and yield. *Agronomy Journal* 110(5), 1777-90. doi:10.2134/agronj2017.12.0680.

Mason, S. J. and Tippett, M. K. 2016. *Climate Predictability Tool Version 15.3*. Columbia University Academic Commons, New York, NY. doi:10.7916/D8668DCW.

Messina, C. D., Jones, J. W., Boote, K. J. and Vallejos, C. E. 2006. A gene-based model to simulate soybean development and yield responses to environment. *Crop Science* 46(1), 456-66. doi:10.2135/cropsci2005.04-0372.

Mishra, A., Hansen, J. W., Dingkuhn, M., Baron, C., Traore, S. B., Ndiaye, O. and Ward, M. N. 2008. Sorghum yield prediction from seasonal rainfall forecasts in Burkina Faso. *Agricultural and Forest Meteorology* 148(11), 1798-814. doi:10.1016/j.agrformet.2008.06.007.

Paff, K. and Asseng, S. 2019. Comparing the effects of growing conditions on simulated Ethiopian tef and wheat yields. *Agricultural and Forest Meteorology* 266-267, 208-20. doi:10.1016/j.agrformet.2018.12.010.

Pathak, T. B., Fraisse, C. W., Jones, J. W., Messina, C. D. and Hoogenboom, G. 2007. Use of global sensitivity analysis for CROPGRO cotton model development. *Transactions of the ASAE* 50(6), 2295-302.

Pequeno, D. N. L., Pedreira, C. G. S. and Boote, K. J. 2014. Simulating forage production of Marandu palisade grass (*Brachiaria brizantha*) with the CROPGRO-Perennial Forage model. *Crop and Pasture Science* 65(12), 1335-48. doi:10.1071/CP14058.

Pequeno, D. N. L., Pedreira, C. G. S., Boote, K. J., Alderman, P. D. and Faria, A. F. G. 2018. Species-genotypic parameters of the CROPGRO Perennial Forage Model: implications for comparison of three tropical pasture grasses. *Grass and Forage Science* 73(2), 440-55. doi:10.1111/gfs.12329.

Pickering, N. B., Jones, J. W. and Boote, K. J. 1995. Adapting SOYGRO V5.42 for prediction under climate change conditions. In: Rosenzweig, C., Jones, J. W. and Allen Jr., L. H. (Eds), *Climate Change and Agriculture: Analysis of Potential International Impacts*. ASA Spec. Pub. No. 59. ASA-CSSA-SSSA, Madison, WI, pp. 77-98.

Porter, C. H., Jones, J. W., Adiku, S., Gijsman, A. J., Gargiulo, O. and Naab, J. B. 2010. Modeling organic carbon and carbon-mediated soil processes in DSSAT v4.5.

International Journal of Operational Research 10(3), 247–78. doi:10.1007/s12351-009-0059-1.

Priestley, C. H. B. and Taylor, R. J. 1972. On the assessment of surface heat and evaporation using large scale parameters. *Monthly Weather Review* 100(2), 81–92. doi:10.1175/1520-0493(1972)100<0081:OTAOSH>2.3.CO;2.

Resenes, J., Pavan, W., Holbig, C., Fernandes, J. M. and Hoogenboom, G. 2019. jDSSAT: a JavaScript modules for DSSAT-CSM integration. *SoftwareX* 10. doi:10.1016/j.softx.2019.100271.

Ritchie, J. T. 1998. Soil water balance and plant water stress. In: Tsuji, G. Y., Hoogenboom, G. and Thornton, P. K. (Eds), *Understanding Options for Agricultural Production*. Kluwer Academic Publishers, Dordrecht, the Netherlands, pp. 41–54.

Ritchie, J. T., Godwin, D. C. and Otter-Nacke, S. 1985. *CERES-Wheat: A User-oriented wheat Yield Model*. Preliminary Documentation. AGRISTARS Publication No. YM-U3-04442-JSC-18892. Michigan State University, East Lansing, MI.

Ritchie, J. T., Singh, U., Godwin, D. C. and Bowen, W. T. 1998. Cereal growth, development and yield. In: Tsuji, G. Y., Hoogenboom, G. and Thornton, P. K. (Eds), *Understanding Options for Agricultural Production*. Kluwer Academic, Dordrecht, pp. 79–98.

Ritchie, J. T., Porter, C. H., Judge, J., Jones, J. W. and Suleiman, A. A. 2009. Extension of an existing model for soil water evaporation and redistribution under high water content conditions. *Soil Science Society of America Journal* 73(3), 792–801. doi:10.2136/sssaj2007.0325.

Robertson, R. D. 2017. *Mink: Details of a Global Gridded Crop Modeling System*. International Food Policy Research Institute (IFPRI), Washington DC. Available at: http://ebrary.ifpri.org/cdm/ref/collection/p15738coll2/id/131406.

Rymph, S. J. 2004. Modeling growth and composition of perennial tropical forage grasses. Ph.D. dissertation. University of Florida, Gainesville, FL.

Scholberg, J. M. S., Boote, K. J., Jones, J. W. and McNeal, B. L. 1997. Adaptation of the CROPGRO model to simulate the growth of field-grown tomato. In: Kropff, M. J., Teng, P. S., Aggarwal, P. K., Bouman, J., Bouman, B. A. M., Jones, J. W. and van Laar, H. H. (Eds), *Systems Approaches for Sustainable Agricultural Development: Applications of Systems Approaches at the Field Level*. Kluwer Academic Publishers, Dordrecht, the Netherlands, pp. 133–51.

Shelia, V., Hansen, J., Sharda, V., Porter, C., Aggarwal, P., Wilkerson, C. J. and Hoogenboom, G. 2019. A multi-scale and multi-model gridded framework for forecasting crop production, risk analysis, and climate change impact studies. *Environmental Modelling and Software* 115, 144–54. doi:10.1016/j.envsoft.2019.02.006.

Singels, A., Jones, M. and van den Berg, M. 2008. *DSSAT v4.5 Canegro Sugarcane Plant Module: Scientific Documentation*. SASRI, Mount Edgecombe, South Africa, p. 34. Available at: https://sasri.sasa.org.za/agronomy/icsm/documents/DSSAT%20Canegro%20SCIENTIFIC%20documentation_20081215.pdf.

Singh, P., Alagarswamy, G., Hoogenboom, G., Pathak, P., Wani, S. P. and Virmani, S. M. 1999a. Soybean-chickpea rotation on Vertic Inceptisols: 2. Long-term simulation of water balance and crop yields. *Field Crops Research* 63(3), 225–36. doi:10.1016/S0378-4290(99)00038-6.

Singh, P., Alagarswamy, G., Pathak, P., Wani, S. P., Hoogenboom, G. and Virmani, S. M. 1999b. Soybean-chickpea rotation on Vertic Inceptisols: 1. Effect of soil depth and landform on light interception, water balance and crop yields. *Field Crops Research* 63(3), 211–24. doi:10.1016/S0378-4290(99)00037-4.

Singh, S., Boote, K. J., Angadi, S. V., Grover, K., Begna, S. and Auld, D. 2015. Adapting the CROPGRO model to simulate growth and yield of spring safflower in semiarid conditions. *Agronomy Journal* 108(1), 64–72. doi:10.2134/agronj15.0272.

Thornton, P. K. and Hoogenboom, G. 1994. A computer program to analyze single-season crop model outputs. *Agronomy Journal* 86(5), 860–8. doi:10.2134/agronj1994.00021962008600050020x.

Thornton, P. K., Hoogenboom, G., Wilkens, P. W. and Bowen, W. T. 1995. A computer program to analyze multiple-season crop model outputs. *Agronomy Journal* 87(1), 131–6. doi:10.2134/agronj1995.00021962008700010023x.

Thorp, K. R., DeJonge, K. C., Kaleita, A. L., Batchelor, W. D. and Paz, J. O. 2008. Methodology for the use of DSSAT models for precision agriculture decision support. *Computers and Electronics in Agriculture* 64(2), 276–85. doi:10.1016/j.compag.2008.05.022.

Tsuji, G. Y., Hoogenboom, G. and Thornton, P. K. (Eds). 1998. *Understanding Options for Agricultural Production*. Kluwer Academic Publishers, Dordrecht, the Netherlands.

Uehara, G. and Tsuji, G. Y. 1998. Overview of IBSNAT. In: Tsuji, G. Y., Hoogenboom, G. and Thornton, P. K. (Eds), *Understanding Options for Agricultural Production. Systems Approaches for Sustainable Agricultural Development* (vol. 7). Springer, Dordrecht, the Netherlands.

Uryasev, O., Jones, J. W. and Hoogenboom, G. 2004. Graphical Display Program (GBuild). Chapter 6. In: Wilkens, P. W., Hoogenboom, G., Porter, C. H., Jones, J. W. and Uryasev. O. (Eds), *DSSAT v.4. Vol 2. Data Management and Analysis Tools*. International Consortium for Agricultural Systems Applications University of Hawaii, Honolulu, Hawaii, pp. 155–74.

Wallach, D., Hwang, C., Correll, M. J., Jones, J. W., Boote, K. J., Hoogenboom, G., Gezan, S., Bhaktae, M. and Vallejos, C. E. 2018. A dynamic model with QTL covariables for predicting flowering time of common bean (*Phaseolus vulgaris*) genotypes. *European Journal of Agronomy* 101(1), 200–9. doi:10.1016/j.eja.2018.10.003.

White, J. W. and Hoogenboom, G. 1996. Simulating effects of genes for physiological traits in a process-oriented crop model. *Agronomy Journal* 88(3), 416–22. doi:10.2134/agronj1996.00021962008800030009x.

White, J. W. and Hoogenboom, G. 2003. Gene-based approaches to crop simulation: past experiences and future opportunities. *Agronomy Journal* 95(1), 52–64. doi:10.2134/agronj2003.0052.

White, J. W., Hunt, L. A., Boote, K. J., Jones, J. W., Koo, J., Kim, S., Porter, C. H., Wilkens, P. W. and Hoogenboom, G. 2013. Integrated description of agricultural field experiments and production: the ICASA version 2.0 data standards. *Computers and Electronics in Agriculture* 96, 1–12.

Wilkerson, G. G., Jones, J. W., Boote, K. J., Ingram, K. T. and Mishoe, J. W. 1983. Modeling soybean growth for crop management. *Transactions of ASAE* 26, 63–73.

Willmott, C. J., Ackleson, S. G., Davis, R. E., Feddema, J. J., Klink, K. M., Legates, D. R., O'Donnell, J. and Rowe, C. M. 1985. Statistics for the evaluation and comparison of models. *Journal of Geophysical Research* 90(C5), 8995-9005. doi:10.1029/JC090iC05p08995.

Yang, J. M., Yang, J. Y., Liu, S. and Hoogenboom, G. 2014a. An evaluation of the statistical methods for testing the performance of a crop simulation model with observed data. *Agricultural Systems* 127(1), 81-9.

Yang, J. Y., Drury, C. F., Yang, J. M., Li, Z. T. and Hoogenboom, G. 2014b. EasyGrapher: software for data visualization and statistical evaluation of DSSAT cropping system model and the CANB model. *International Journal of Computer Theory and Engineering* 6(3), 210–4. doi:10.7763/IJCTE.2014.V6.864.

Zhang, J., Bartholomew, D. P. and Malezieux, E. 1997. ALOHA-Pineapple V.2.1: a computer model to predict the growth, development and yield of pineapple. *Acta Horticulturae* 425, 287–96.

Zhao, C., Liu, B., Xiao, L., Hoogenboom, G., Boote, K. J., Kassie, B. T., Pavan, W., Shelia, V., Kim, K. S., Hernandez-Ochoa, I. M., Wallach, D., Porter, C. H., Stockle, C. O., Zhu, Y. and Asseng, S. 2019. A SIMPLE crop model. *European Journal of Agronomy* 104(1), 97–106. doi:10.1016/j.eja.2019.01.009.

Published by Burleigh Dodds Science Publishing Limited, 2026.

Chapter 8

Modeling crop rotations: capturing short- and long-term feedbacks for sustainability and soil health

B. Basso and R. A. Martinez-Feria, Michigan State University, USA; and B. Dumont, University of Liege, Belgium

1 Introduction

Crop rotations are cyclic sequences of crop-plant species grown on the same parcel of land. This contrasts with intercropping (multiple species grown simultaneously), or monocropping (growing a single species continuously). Throughout most of human history, rotating crops was one of the few tools farmers had at their disposal to sustain soil fertility, combat pests, and improve yields (Bullock, 1992; Karlen et al., 1994; Leighty, 1938). Since the 1950s, the use of synthetic fertilizers and biocides to manage fertility and pests has allowed monocropping on a more prominent scale. However, crop rotations are still used extensively worldwide (Alhameid et al., 2017). The reasons for the current widespread adoption are very much the same benefits that early farmers identified. Well-designed rotations can (1) effectively break growth cycles of weeds (Liebman and Dyck, 1993; Stevenson and van Kessel, 1996), insects (Miller et al., 2006), and diseases (Curl, 1963; Médiène et al., 2011); (2) improve soil fertility and health by alternating plant nutrient requirements, as well as the quality and quantity of crop residue inputs into the soil (Bennett et al., 2012; Kay, 1990; Reeves, 1994); (3) allow for more flexibility in logistics, as distinct crop growth cycles spread out the allocation of labor (Zentner et al., 2002); and (4) manage risks related to market fluctuations (Zentner et al., 2001).

http://dx.doi.org/10.19103/AS.2019.0061.11

Given their large footprint and important role in crop production, examining the biophysical dynamics resulting from growing diverse crops in sequence has been a common topic of research in the agronomy and agroecology fields. Alternating among crop species fundamentally changes the cycling of water and nutrients and their distribution in the soil profile (Fletcher et al., 2011; Hirsh and Weil, 2019; Ryan et al., 2009). Additionally, differing quantities and qualities of aboveground and belowground crop residues influence the amount and timing of plant nutrient availability, water infiltration and runoff, soil temperature and long-term soil carbon storage, and potential risk of nutrient leaching (Blanco-Canqui and Lal, 2009; Poffenbarger et al., 2017; Puntel et al., 2016). These 'legacy', 'carry-over', or 'rotation' effects are often observed as increased crop yields or greater resilience to environmental stress (Reeves, 1994; Ryan et al., 2008). Because of these complex feedbacks, legacy effects and their interactions with weather are generally not well understood, and their ultimate influence on yields and environmental quality is difficult to predict.

Crop simulation models offer a way of parsing through this complexity. Their explicit representation of the fundamental processes driving crop growth and development, as well as water, carbon, and nutrient cycles, have made them popular tools for evaluating many aspects of cropping systems. These models use information on soils, weather, crop cultivars, and management as inputs in mathematical algorithms to calculate changes in the system state across various temporal scales (Wallach et al., 2014). Many of these have the capability to simulate crop rotations, and their implementation in this context has been useful to examine differences in water and nutrient flows among crop sequences and phases (Dietzel et al., 2016; Martinez-Feria et al., 2018; Post et al., 2007; Salado-Navarro and Sinclair, 2009), estimate environmental nitrogen (N) losses over long periods (Basso et al., 2016; Gillette et al., 2018; Kovács et al., 1995; Martinez-Feria et al., 2016), predict changes in soil carbon storage (Basso et al., 2015, 2018; Berntsen et al., 2007; Hlavinka et al., 2014; Jarecki et al., 2018), and assess the economic impacts of management (Puntel et al., 2016; Araya et al., 2017; Nielsen et al., 2009).

Methodologically speaking, two general approaches exist when simulating crop growth cycles and, by extension, crop rotations (Basso et al., 2015). The first and most widely used approach is to simulate single crop phases independently using the same initial soil state conditions (e.g. water, organic carbon, or nutrient levels) for every simulation cycle. This approach, often referred to as the seasonal 'reset' mode, is a holdover from when crop models were only capable of simulating single crop growing seasons. There are obvious limitations with this approach, namely that the legacy effects on soils, such as the ones discussed above, are ignored (Ewert et al., 2015; Teixeira et al., 2015). Despite this, running models under reset mode has been standard practice in impact assessments such as those for climate change (Kollas et al.,

2015; White et al., 2011). Most models today are now capable of running continuous simulations over multiple crop rotation phases in what is known as 'sequential' mode, and several studies have discussed its advantages, both in terms of predicting yield (Kollas et al., 2015; Teixeira et al., 2015, 2018) and soil carbon storage (Basso et al., 2015, 2018). Under this approach, the legacy effects from one phase of the rotation to the next emerge from the simulation process itself; thus the user is only required to provide initial conditions for the first year of the simulation.

Although running crop models under either reset or sequential mode each have their limitations and advantages, they also provide distinct insights on the behavior of the system. To illustrate and discuss this point, in this chapter we focus on two case studies that evaluate short- and long-term outcomes of distinct crop rotations. In the first case study, we use APSIM (Agricultural Production Systems sIMulator; Holzworth et al., 2014) run with reset mode to examine short-term weather and legacy impacts on hydrological nitrate (NO_3) losses and evaluate mitigation strategies in distinct crop rotation phases. The second case study uses system approach to land use sustainability (SALUS) (Basso and Ritchie, 2015) run on sequential mode to identify adaptations to crop rotations for maintaining long-term productivity and soil health under climate change.

2 Reset mode crop models: the example of mitigating nitrate loss from corn-based crop rotations

Loss of NO_3 from corn (*Zea mays* L.) and soybean (*Glicyne max* L [Merr.]) cropland into surface waters is one of the major environmental impacts of crop production in the Midwest US (David et al., 2010; Robertson and Vitousek, 2009). While many studies attribute the loss of NO_3 to the overuse of nitrogen (N) fertilizer (Cassman et al., 2002; Zhang et al., 2015), the release of native soil N can also contribute substantially to NO_3 loss (Bowles et al., 2018; Martinez-Feria et al., 2018). Without a strong sink (e.g. plant growth) to retain N during the extensive fallow periods (October to May), NO_3 from fertilizer or organic matter mineralization sources builds up in soils and is transported into subsurface drainage systems during heavy rains (Randall and Mulla, 2001). With spring rainfall increasing in many areas of the Midwest (Melillo et al., 2014), the NO_3 loss problem will continue to worsen (Bowles et al., 2018).

From a crop management perspective, two strategies are often cited as having great potential to mitigate NO_3 losses into subsurface drainage (Christianson et al., 2018): (1) applying N fertilizer in season rather than before planting and (2) growing cover crops during the fallow period. In this case study, we use APSIM to evaluate the NO_3 loss reduction effectiveness of these practices across ranges of weather and legacy conditions, aiming to identify scenarios where these practices may be the most effective.

2.1 Sites and data sources

Soil and weather data from four long-term experimental field sites located in Iowa were used to configure and drive the simulation model (Table 1). The Ames and Nashua sites have been described in detail in previous studies (Dietzel et al., 2016; Martinez-Feria et al., 2018), while the information for Gilmore and Crawfordsville were extracted from the Sustainable Corn CAP Research Database (Abendroth et al., 2017). Soil information for each site was obtained from the SSURGO database (Soil Survey Staff). The soils in these sites are deep, fertile, and artificially drained using subsurface drain tubes. Daily weather (1987–2016) for all sites was retrieved from the Daymet dataset (Thornton et al., 2018) using the single pixel extraction tool (downscaled to 1 km × 1 km resolution).

2.2 The APSIM model

APSIM is an open-source cropping systems simulation platform with interconnected crop, hydrological and nitrogen cycling process-based models. Using daily weather and user-defined soil and management information, the model calculates many soil-plant-atmosphere variables, including crop growth processes, soil water, soil temperature, and N and C cycling. For in-depth descriptions of APSIM, we refer the reader to Holzworth et al. (2014).

2.3 Model configuration

APSIM (version 7.8) was configured using the information on soil, drainage specifications, and management available for each site (Table 1). When configuring the simulations, we used the following APSIM modules: *maize* (corn), *soybean, wheat* (for rye cover crop), *SWIM3* (soil hydrology), and *soilN* (C and N cycling). The corn and soybean APSIM cultivars used have been calibrated to broadly characterize locally adapted commercial genotypes in the region (Archontoulis et al., 2014a,b). We selected maturity groups appropriate for each site based on the management records available. The *wheat* module was calibrated following Dietzel et al. (2016) and Martinez-Feria et al. (2016) and was used to simulate the rye cover crop.

Before conducting simulation experiments, we ran the model for a 'spin-up' period. This was to remove the confounding effects of buildup or decline in soil organic carbon (SOC) humic or microbial pools (Puntel et al., 2016). In this case, a maize-soybean rotation was simulated sequentially for 15 years at each

Table 1 Summary of the sites used for simulation modeling

	Location		Soil			N fertilizer rate[d] (kg N ha^{-1})		Subsurface drainage	
Site	Lat.	Long.	Texture	OC (%)	PAWC (mm)	Corn after corn	Corn after soybean	Depth (cm)	Spacing (m)
Ames[a,b]	41.92	−93.75	Silty clay loam	2.0	132	211	175	110	13.5
Gilmore[c]	42.75	−94.50	Silty clay loam	2.2	123	211	157	110	7.6
Nashua[a]	42.93	−92.57	Loam	1.3	124	211	157	120	28.5
Crawfordsville[c]	41.19	−91.48	Silty clay loam	1.6	110	228	171	122	18.3

Source: [a] Martinez-Feria et al. (2018); [b] Dietzel et al. (2016); [c] Abendroth et al. (2017); [d] Based on university recommendations (N rate calculator: http://cnrc.agron.iastate.edu/) and Sawyer et al. (2006).
OC = organic carbon; PAWC = plant available water content (0–1 m).

site. Initial values for soil nitrate and moisture, and above- and below-ground residue amount and C:N were also derived from this step. To avoid introducing bias from a given set of conditions experienced during the last year of the spin-up, we used the average value of these variables at harvest for the last five simulated years for each crop. The values for initial conditions, as well as further details on model configuration, testing, and performance against measured data, are provided by Martinez-Feria (2018).

2.4 Simulation experiments

The simulation experiments were designed to quantify the impact of weather and management factors on cumulative annual NO_3 loads (kg N ha^{-1}) in subsurface drains for cropland under a corn-corn-soybean rotation at the long-term sites. Each of the phases of the rotation was simulated with 30 years of historical weather (1987–2016). To decouple the effect of weather-year from the legacy rotation effects, the soil states (i.e. moisture and N levels) were reset every year on 20 October. This means that simulation accounted for the period from 20 October to 19 October of the following year, roughly representing a harvest-to-harvest cycle.

As default initial values of soil N, we use the average soil NO_3 content on 19 October in spin-up runs. This was then increased and decreased by a factor of 0.5, which provided a low, average, and high value. Similarly, the initial water content in the profile was altered by initializing the water table 12 cm above, 12 cm below, and at the depth of subsurface drains, providing a shallow, deep, and average water table level, respectively.

In addition to these legacy initialization factors, we evaluated the impact of two management practices in each of the rotation phases. In the corn phases, we simulated two N fertilizer application timing treatments: at planting and in-season split (with 50% applied at planting and 50% applied 40 days later). N fertilizer rates followed university recommendations (http://cnrc.agron.iastate.edu/; Sawyer et al., 2006; Table 1). In the soybean phase, we also evaluated the impact of planting time, with early (late April) and late (late May) planting treatments. Finally, we also evaluated the inclusion of winter rye cover crop in all crop phases. Cover crops in each simulation cycle were planted every year on 1 November and terminated in the spring 7 days before planting of the main crop. The combination of 4 sites, 3 rotation phases, 9 sets of initial conditions, 4 management treatments, and 30 weather years produced 12,960 individual simulations.

2.5 Findings

Table 2 shows the percent of the total variance (i.e. proportion of the sums of squares) explained by each factor in the simulation experiment. In all sites,

Table 2 Share of the total variance in simulated NO_3 loss attributed to simulation factors

Simulation factor	Site				
	Ames (%)	Gilmore (%)	Nashua (%)	Crawfordsville (%)	All (%)
Corn after corn					
Initial conditions	17	21	30	17	19
Weather-year	69	63	56	71	67
Management treatments	4	3	2	2	3
Interactions	11	13	13	10	11
Corn after soybean					
Initial conditions	6	12	20	6	8
Weather-year	85	75	65	86	83
Management treatments	1	1	1	1	1
Interactions	8	12	15	7	8
Soybean after corn					
Initial conditions	24	23	29	24	24
Weather-year	53	58	54	57	56
Management treatments	7	5	3	4	5
Interactions	16	15	14	15	15

weather-year was the factor that had the most impact on annual NO_3 loads (53–86% of the variation) followed by the soil initial conditions (6–30%), and then by management factors (2–7%). All interactive effects explained 7–16% of the total variation in simulated annual NO_3 loads.

Across all sites, annual NO_3 loads responded positively to the amount of spring rainfall and legacy soil NO_3, although the response differed substantially among crop phases (Fig. 1). Very few differences were noticeable between N fertilizer timing treatments (corn phases) or planting date (soybean phase) treatments, whereas these were greater between cover crop treatments. The corn-after-corn and the soybean-after-corn phases of the rotation had

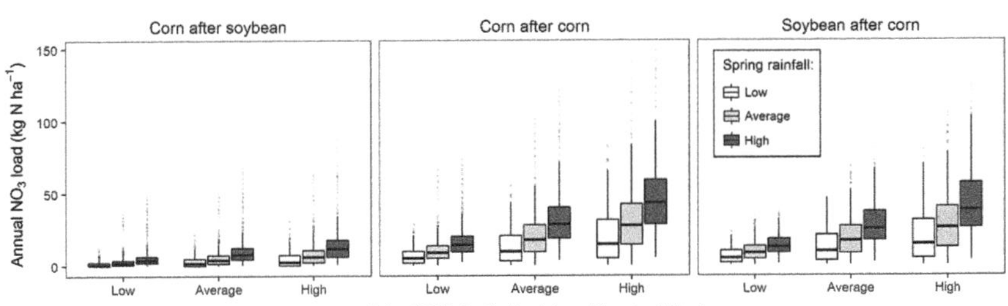

Figure 1 Simulated response of annual NO_3 loads to increasing spring rainfall and initial soil NO_3 content across all the sites and management treatments.

generally higher NO_3 losses than the corn-after-soybean phase (Fig. 1). This suggests that prior corn crop fertilization (e.g. legacy soil N from the previous crop) is an important contributor to the NO_3 losses, and that corn-soybean crop rotations have overall lower potential of annual NO_3 losses than continuous corn monocrops. This is consistent with experimental evidence (Christianson and Harmel, 2015) and other simulation studies (Martinez-Feria et al., 2018).

The scenarios with no cover crop, N fertilizer applied to corn at planting, and soybean with late planting (late May) averaged 7.2, 27, and 24 kg N ha^{-1} yr^{-1} annual NO_3 loads, in the corn after soybean, corn after corn, and soybean after corn phases, respectively. Using these as management baselines (Table 3), we see that improved practices reduced annual NO_3 loads more in wet springs. Likewise, reductions were greater in the soybean phase of the rotation. Few differences in the relative reduction were detected across levels of initial soil NO_3 content.

Even though management overall explained a small percent of the variation (Table 2), combining the establishment of a cover crop with other improved practices can still significantly reduce NO_3 losses when compared to a baseline management scenario, especially in wet years (Table 3).

Interestingly, the NO_3 loss mitigation potential of the cover crop was greater after corn than after soybean, perhaps due to the greater amount of residual NO_3 in the soil following corn. Based on these findings, we conclude that a corn-soybean rotation with cover crops, paired with early planting in soybeans and in-season N applications in corn is the best performing cropping system among the options examined in this case study.

Table 3 Reduction in annual NO_3 loads relative to baseline management

	Spring rainfall			Fall soil NO_3		
Treatment	Low (%)	Average (%)	High (%)	Low (%)	Average (%)	High (%)
Corn after corn						
Baseline (BL) + Split N	0	1	3	2	1	1
BL + Cover crop	21	26	30	25	27	26
BL + Split N + Cover crop	21	28	34	28	29	27
Corn after soybean						
BL + Split N	1	2	4	3	2	2
BL + Cover crop	9	7	21	16	12	12
BL + Split N + Cover crop	9	11	27	20	15	15
Soybean after corn						
BL + Early planting	8	4	2	4	5	5
BL + Cover crop	25	32	38	31	34	36
BL + Early planting + Cover crop	26	35	40	31	33	32

3 Sequential crop models: the example of adapting crop rotations to sustain yields and soil health under climate change

Changing climatic conditions will likely reshape the nature of future food production. In many regions of the world, the negative impacts of climate change on crop yields are already visible (Lobell et al., 2011; Porter et al., 2014). Yields are projected to decrease by the second half of the twenty-first century in many temperate and tropical areas (Challinor et al., 2014; Porter et al., 2014; UNFAO, 2016). Increasing microbial activity due to higher soil temperatures and decreasing crop residues due to low yields will also affect the ability of agricultural soils to store carbon (Powlson et al., 2014), and impoverish their fertility (Jarecki et al., 2018). While management practices can help farmers adapt, the interactions between management, soils, and climate remain poorly understood (Basso et al., 2018). The objective of this case study was to evaluate the performance of corn-based cropping systems under projected climate changes across a range of Midwestern soils. We aim to identify a system that could be best adapted to maintain crop yields and soil health in future climates.

3.1 Sites and soils

We used data from experimental sites located in eight states across the Midwest US: Iowa, Illinois, Indiana, Michigan, Minnesota, Missouri, Ohio, and Wisconsin (Table 4). Each site has dedicated field trials with site-specific treatments to study corn-based cropping systems in terms of productivity and environmental impacts. Soils in these sites range from low (MI and IN) to very high (MN) organic carbon content (Table 4). Measured soil texture, bulk density, and organic carbon content were available for the top 0–60 cm at most sites, and information below 60 cm was obtained from the SSURGO database. These data were used to calculate soil hydraulic properties with pedotransfer functions (Ritchie et al., 1999; Suleiman and Ritchie, 2001). Further details about the sites and experimental treatments are available in Necpalova et al. (2014).

3.2 Historic and future weather

Historical weather data (1979–2013) downscaled to the location of the experimental sites were obtained from the National Centers for Environmental Prediction North American Regional Reanalysis (NCEP-NARR) model (Mesinger et al., 2006). We generated time-series climate change weather for the 2070–2100 time frame under the representative concentration pathway (RCP) 2.6 and 6.0 emissions scenarios by simple transformations of historical NARR dataset (i.e., delta method; Table 5). The RCP 2.6 represents a 'best-case' climate

Table 4 Summary of location and soil characteristics at the sentinel sites

State	Site	Location		Soil type	Soil organic carbon
		Lat.	Long.		(Mg C ha^{-1}; 0-60 cm)
IA	ISUAG	42.00	−93.78	Loam and clay loam	137
IL	NWREC	40.93	−90.72	Silt loam and silty clay loam	153
IN	SEPAC	39.02	−85.54	Silt loam	47
MI	KBS	42.41	−85.37	Loam and sandy loam	48
MN	SWROC.B	44.35	−95.53	Silty clay loam	215
MO	Bradford	38.90	−92.20	Silty clay loam	76
OH	Hoytville	41.21	−83.76	Clay loam	99
WI	Marshfield	44.76	−90.09	Silt loam	173

change scenario, where global annual GHG emissions peak between 2010 and 2020 and decline thereafter, whereas the RCP 6.0 is a scenario where emissions peak around 2080, then stabilize (Meinshausen et al., 2011). The magnitude of the changes, at the annual level or as a seasonal pattern, was derived from the IPPC 5th assessment report (IPCC, 2014) and its actual translation at the US level, as reported in the National Climate Assessment of the US Global Change Research Program (Melillo et al., 2014). The three main changes applied to the historical records are shown in Table 5.

3.3 The system approach to land use sustainability (SALUS) model

The system approach to land use sustainability (SALUS) model is a process-based modeling system that simulates plant growth and development responses to environmental conditions (soil and weather), genetics, and management strategies in a sequential mode (Basso and Ritchie, 2015). The model uses daily values of incoming solar radiation (MJ m^{-2}), maximum and minimum air temperature (°C), and rainfall (mm), as well as soil and management information to simulate crop yields, and water, N and C cycling. More information about SALUS and its performance against measured data is available in the following studies: Albarenque et al., 2016; Basso et al., 2016, 2018; Basso and Ritchie, 2015.

3.4 Crop rotation scenarios

Four corn-based cropping systems were designed for evaluation (Table 6): (i) continuous corn (SC1-2); (ii) corn-soybean rotation with rye cover crop (Corn-cc-SB-cc; SC3); and (iii) corn-soybean-wheat rotation with rye cover crop

Table 5 Operations used to generate future weather (2070–2100) from historic weather (1979–2013)

Scenario	Variable	Period				Operation
		DJF	MAM	JJA	SON	
Baseline (BL)	-	NCEP-NARR (1979–2013)				-
RCP 2.6	Precipitation	1.1	1.1	0.95	1	Multiplying BL by coef.
	Temperature	-----------+3°C-----------				Adding fixed value to BL
	$[CO_2]$	-----------400 ppm-----------				Replacing BL value
RCP 6.0	Precipitation	1.2	1.2	0.9	1	Multiplying BL by coef.
	Temperature	-----------+6°C-----------				Adding fixed value
	$[CO_2]$	-----------540 ppm-----------				Replacing value

Table 6 Synthesis of corn-based system evaluated with the SALUS model

Scenario	Rotation[a]	Manure	Timing of N fertilization to corn[b]	Tillage
SC1	Corn	Fall manure	100% at planting	Conv. Till.
SC2	Corn	No manure	25% at planting and 75% at V6	No Till.
SC3	Corn-cc-SB-cc	Fall manure	25% at planting and 75% at V6	Minimum Till
SC4	Corn-cc-SB-WW-cc	Fall manure	25% at planting and 75% at V6	Minimum Till

[a] SB = Soybean, WW = winter wheat, cc = rye cover crop.
[b] Corn received 200 kg N ha^{-1} in all scenarios.

following corn and wheat (Corn-cc-SB-WW-cc; SC4). In addition, we evaluated two N management strategies common in the Midwest: (i) 200 kg N ha^{-1} applied to corn at planting and (ii) 50 kg N ha^{-1} at corn planting and 150 kg N ha^{-1} applied at corn V6 growing stage. In addition, scenarios 1, 3, and 4 also included fall manure applications prior to corn (e.g. 500 kg C ha^{-1} and 50 kg N ha^{-1}). Three tillage regimes were also evaluated: (i) conventional tillage in the 0–30 cm soil profile (SC1); (ii) no tillage in which fresh organic carbon and residues remained on the soil surface; and (iii) minimum tillage regime with shallow tillage (22 cm) once every rotation cycle to incorporate crop residues and manure when applied to corn.

3.5 Findings

Simulated grain yields at most of the sites were impacted negatively by climate change. For instance, yields under SC1 crop rotation over the 35-year period decreased on average by 5% and 20% under the RCP 2.6 and 6.0 climate scenarios, respectively. Yield losses due to climate change were more extreme in Ohio, Illinois, and Iowa, ranging from 22% to 46%. Only two sites, Michigan and Wisconsin, saw slight yield gains under the climate change scenarios, probably due to increased precipitation and the low soil water holding capacity at those sites (Table 4), where water and temperature stress limitations were not enough to overcome the positive effects of increased CO_2 (Table 5). However, these gains were partially or totally offset under RPC 6.0.

On average, corn yield losses under climate change were the greatest with rotation SC4 (13% and 32% under RPC 2.6 and 6.0, respectively). We should note that SC4, the most complex rotation that included cover crops and winter crops (Table 4), also was the one with the greatest yield under baseline conditions in most states. On the other hand, SC2 seems to be the least impacted (2% and 10% yield loss under RPC 2.6 and 6.0, respectively), although this cropping system had generally low yields even with the historical weather baseline (7% lower than SC1; Table 7).

As expected, climatic change had a negative impact on the amount of SOC storage. Table 8 shows the change in SOC over the 35 years of the simulation for four of the simulated crop rotations. On average, soils lost 20 and 26 Mg C ha^{-1} under RPC 2.6 and 6.0, respectively, in the SC1 scenario. This is more than double of that lost with the historical baseline climate. This trend was similar across all the crop rotation scenarios. The no-till scenario (SC2) was the only condition that resulted in SOC gains under the baseline, although these gains were partially or totally offset by climate change. Where no-tillage techniques were applied, and cover crops were part of more complex rotations, the SOC losses remained limited (Table 8).

All of this suggests that maximizing the amount of residues returned to the soil and minimizing tillage are broadly the most effective strategies for

Table 7 Comparison of corn grain yields of simulated crop rotations under different climate scenarios

Scenario		Mean change in corn yield relative to SC1 over 35-yr (%)		
Climate	Rotation	Min	Mean	Max
BL	SC2	−17.5(MI) and −14.2(IA)	−7.10	+4.9(MO) and +3.2(OH)
	SC3	−20.8(IA) and −8.5(MI)	−5.00	+2.1(OH) and +1.9(MO)
	SC4	−38.8(IL) and −5.7(IA)	−4.70	+6.24(MI) and +4.9(OH)
RCP 2.6	SC2	−16.6(MI) and −12.5(IA)	−1.10	+18.6(WI) and +8.0(IL)
	SC3	−20.7(MO) and −15.1(IA)	−1.50	+14.1(OH,WI) and +9.8(IL)
	SC4	−40.0(IL) and −32.5(IA)	−12.7	+10.0(OH) and +5.7(WI)
RCP 6.0	SC2	−12.1(MI) and −8.8(IA)	+3.5	+37.3(IL) and +13.1(WI)
	SC3	−43.4(IL) and −25.0(MO)	−11.6	+11.1(WI) and −0.4(OH)
	SC4	−41.7(IA) and −28.1(IL)	−17.0	+0.1(OH) and −1.7(IN)

Table 8 Effect of rotation and climate scenarios on the change in soil organic carbon (SOC) as simulated by SALUS

Scenario		35-yr change in SOC (Mg C ha^{-1})		
Rotation	Climate	Min	Mean	Max
SC1	BL	−44.9(IL) and −35.9(IA)	−10.7	+15.9(MN) and +12.7(IN)
	RCP 2.6	−54.4(IL) and −47.6(IA)	−20.4	+8.3(MN) and +4.7(IN)
	RCP 6.0	−60.4(IL) and −55.4(IA)	−26.6	+3.9(MN) and −0.3(IN)
SC2	BL	−4.2(IL) and +1.9(IA)	8.7	+18.3(MN) and +15.7(IN)
	RCP 2.6	−13.6(IL) and −8.0(IA)	4.2	+19.3(MN) and +16.3(IN)
	RCP 6.0	−19.7(IL) and −14.8(IA)	−0.7	+15.9(MN) and +11.6(IN)
SC3	BL	−30.4(IL, OH) and −28.9(IA)	−7.2	+19.6(IN) and +13.8(MN)
	RCP 2.6	−40.6(OH) and −38.8(IL, IA)	−14.8	+13.2(IN) and +7.6(MN)
	RCP 6.0	−47.4(OH) and −46.4(IL)	−19.9	+8.7(IN) and +5.4(MN)
SC4	BL	−39.2(IA) and −34.7(IL)	−11.8	+14.8(IN) and +8.8(MN)
	RCP 2.6	−40.7(OH) and −40.0(IA, IL)	−16.1	+10.4(IN) and +5.4(MN)
	RCP 6.0	−45.3(IA, OH) and −43 (IL)	−20.2	+4.9(IN) and +2.2(MN)

mitigating SOC changes under climate change, which agrees with literature (Basso et al., 2018; Poffenbarger et al., 2017). The 2- and 3-year rotations (SC3 and SC4) included low-residue crops, and it appears that manure applications and minimum tillage would not be able to offset the SOC losses. The decline of SOC over the long term introduced negative feedbacks (e.g. more evaporation, less soil N mineralization), which contributed to the lower corn yields under the SC4. Based on these findings, we conclude that among the evaluated systems, the no-till continuous corn system (SC2) had the highest potential to mitigate the negative yield and SOC impacts of climate change.

4 Conclusion: improving crop rotations through modeling

Crop rotations have been identified in many studies as a means to improve the agronomic and societal outcomes from crop production (Bennett et al., 2012; Davis et al., 2012; Karlen et al., 1994; Zentner et al., 2001) and as possible adaptation strategies to changing climates (Farina et al., 2018; Nendel et al., 2014; Reidsma et al., 2010; Teixeira et al., 2018). The case studies presented here examined simulations from various hypothetical crop rotations to explore pathways to increase the future sustainability and resilience of Midwestern corn production.

Here we show that crop rotation coupled with improved management is an effective way to mitigate NO_3 losses under wet springs (Table 3), which are expected to be more common (Bowles et al., 2018). Over the long term, increasing temperatures will more likely be detrimental for both corn yields and

SOC storage (Basso et al., 2018), and only a few of the crop sequence scenarios examined here seem to have the potential to reverse this trend. Critically, while the extended crop rotation that included corn, soybean, winter grains, and cover crops seemed to have overall corn yield benefits (compared to continuous corn monoculture) under historical weather (Table 7), these benefits seem to be negated under climate change because of lower amounts of crop residue input to the soil (greater SOC depletion) in these systems, compared to continuous corn (Table 8). It should be noted that these analyses cover only biophysical and geochemical aspects of crop rotations and not biotic benefits such as reduced pest and diseases or increased habitat for beneficial organisms, which can account for a significant share of the rotation effect (Bennett et al., 2012).

These results highlight the importance of considering crop rotation legacies when assessing the impact of drivers such as weather and soils, and of the value of simulation approaches in this respect. Experimental evidence for the frequently observed productivity and environmental advantages of crop sequences is difficult to discern, because the legacy effects are often not the result of one single factor, but of many. Moreover, these legacies are often simultaneously at play; thus the ultimate result depends on the magnitude of each individual effect and its interaction with weather conditions (Martinez-Feria et al., 2016). Accounting for such legacy effects is not only needed for accurately predicting short-term outcomes (e.g. crop yield responses to N fertilizer; Puntel et al., 2018), but has also shown to be very important for estimating long-term impacts, such as nutrient balances (Martinez-Feria et al., 2018) or soil carbon storage (Basso et al., 2018). The case studies presented here focus on a few rotation effects important in rainfed temperate regions (i.e. NO_3 loss, yield, and SOC change). In semi-arid regions where soil profile moisture is typically depleted by the end of each season, examining how crop rotations affect the carry-over soil moisture can help assess the probability of obtaining a given yield the next growing season (Nielsen et al., 2009). It also provides insight into possible adjustments to management (e.g. sowing date, sowing density, N fertilization, choice of cultivar or crop) or to irrigation strategies where irrigation water is available (Araya et al., 2017).

While the legacy effects (e.g. soil N and water content, or quantity and quality of crop residues) are not inherently accounted for when models are run with reset mode, these may be approximated by altering the soil initial conditions as part of the scenario analysis. This decoupling of the weather effects from the legacy factors allows for easier interpretation of modeling results, especially when the examination of the effect of year-to-year differences in weather patterns are the focus of study. Consider the example of the year 2013 in our reset mode case study. Iowa experienced a drought in 2012, which resulted in reduced yields and large amounts of residual soil NO_3 at harvest. Because of this, NO_3 losses in 2013 are heavily influenced by the previous year when running the model in sequential mode (as it does in reality). Thus, separating the effect of 2013 weather and management from the 2012 drought is rather difficult. If our interest is to find best management options within a crop rotation for different weather patterns (e.g. Table 3), the reset mode allows us to examine this independently from important rotation legacy factors, such as soil NO_3 left over by the previous crop (i.e. the initial soil NO_3 in the simulation). In this case, initial conditions should be representative and preferably explore a range of potential values, such as we do in this case study.

Predefining initial conditions every season, however, runs the risk of introducing uncertainty and biases. These have been seen to affect the ability of models to accurately predict yield, especially under extremely dry weather conditions (Teixeira et al., 2015). If there is confidence in the representation of the processes being simulated, then legacies should be sufficiently well captured by the sequential simulation mode. A recent study showed that yield simulation error decreased as the number of successive crops modeled increased, which was interpreted as evidence that the inadequacy of initial conditions becomes less important with each passing crop as the system reaches equilibrium (Gaydon et al., 2017). However, another study was not able to establish a clear advantage of sequential simulations over seasonal reset mode (Yin et al., 2017), which was attributed to uncertainty regarding how well the process was simulated.

In the context of long-term impacts, biases in the simulation of crop yields within decadal time frames lead to long-term biases in soil residue inputs, obscuring the potential for management to mitigate the negative climate change impacts on SOC (Basso et al., 2015). This could have been the case in our seasonal reset mode study if, for example, we had only considered initializing the model with low soil NO_3. In the corn-following-soybean phase of the rotation, this led to a higher NO_3 loss reduction effectiveness with cover crops, when compared to average and high levels (Table 3). Here the initial soil conditions were varied within reasonable ranges (i.e. 50% more or less initial soil N; soil water saturation 12 cm above or below the subsurface drain), which reflect variations observed in field data (Hirsh and Weil, 2019). Still, varying these legacy conditions contributed up to 30% of the variation in NO_3 losses, a much greater influence than the management treatments simulated (Table 2). To mitigate the introduction of potential biases, the model can be run for a 'spin-up' period (typically 10–20 years) to derive robust initial conditions. This is a common technique often used to define stable SOC pool fractions, which has been seen to improve SOC simulation (O'Leary et al., 2016). A benchmark for defining reasonable initial conditions could be an average of the last 5–10 years of the spin-up or by other approaches (Teixeira et al., 2015).

Though some progress has been made in recent years (Donatelli et al., 2017), most crop-soil models currently do not account for other important benefits of crop rotations such as breaking weeds and pests lifecycles, or the positive impacts to the diversity and function of microbial and beneficial insect communities. Yet these results highlight the value of simulation approaches to assess the impact of drivers such as weather and soils, and their feedbacks on crops grown within a rotation. As evaluation of cropping patterns extends to include the spatial configuration of the crop in the landscape (e.g. economic and environmental benefits of allocating low-productivity areas to native vegetation; Basso et al., 2019), crop models linked with geospatial technologies will continue to play a critical role in capturing interactions between soil processes, crop growth, management, climate variability, and spatial and temporal variation under future climate projections.

5 References

Abendroth, L. J., Herzmann, D. E., Chighladze, G., Kladivko, E. J., Helmers, M. J., Bowling, L., Castellano, M., Cruse, R. M., Dick, W. A., Fausey, N. R., Frankenberger, J., Gassmann, A. J., Kravchenko, A., Lal, R., Lauer, J. G., Mueller, D. S., Nafziger, E. D., Nkongolo, N., O'Neal, M., Sawyer, J. E., Scharf, P., Strock, J. S. and Villamil, M. B. 2017. Sustainable Corn CAP research data (USDA-NIFA Award No. 2011-68002-30190). National Agricultural Library - ARS - USDA. doi:10.15482/USDA.ADC/1411953.

Albarenque, S. M., Basso, B., Caviglia, O. P. and Melchiori, R. J. M. 2016. Spatio-temporal nitrogen fertilizer response in maize: field study and modeling approach. *Agronomy Journal* 108(5), 2110–22. doi:10.2134/agronj2016.02.0081.

Alhameid, A., Tobin, C., Maiga, A., Kumar, S., Osborne, S. and Schumacher, T. 2017. Intensified agroecosystems and changes in soil carbon dynamics. In: Al-Kaisi, M. M. and Lowery, B. (Eds), *Soil Health and Intensification of Agroecosystems*. Academic Press, pp. 195–214.

Araya, A., Kisekka, I.,Gowda, P. H. and Prasad, P. V. V. 2017. Evaluation of water-limited cropping systems in a semi-arid climate using DSSAT-CSM. *Agricultural Systems* 150, 86–98. doi:10.1016/j.agsy.2016.10.007.

Archontoulis, S. V., Miguez, F. E. and Moore, K. J. 2014a. Evaluating APSIM maize, soil water, soil nitrogen, manure, and soil temperature modules in the Midwestern United States. *Agronomy Journal* 106(3), 1025–40. doi:10.2134/agronj2013.0421.

Archontoulis, S. V., Miguez, F. E. and Moore, K. J. 2014b. A methodology and an optimization tool to calibrate phenology of short-day species included in the APSIM PLANT model: application to soybean. *Environmental Modelling and Software* 62, 465–77. doi:10.1016/j.envsoft.2014.04.009.

Basso, B. and Ritchie, J. T. 2015. Simulating crop growth and biogeochemical fluxes in response to land management using the SALUS model. In: Hamilton, S. K., Doll, J. E. and Robertson, G. P. (Eds), *The Ecology of Agricultural Landscapes: Long-Term Research on the Path to Sustainability*. Oxford University Press, pp. 252–74.

Basso, B., Hyndman, D. W., Kendall, A. D., Robertson, G. and Grace, R. 2015. Can impacts of climate change and agricultural adaption strategies be accurately quantified if crop models are annually reinitialized? *PLoS ONE* 10, e0127333.

Basso, B., Giola, P., Dumont, B., Migliorati, M. D. A., Cammarano, D., Pruneddu, G. and Giunta, F. 2016. Tradeoffs between maize silage yield and nitrate leaching in a Mediterranean nitrate-vulnerable zone under current and projected climate scenarios. *PLoS ONE* 11(1), e0146360. doi:10.1371/journal.pone.0146360.

Basso, B., Dumont, B., Maestrini, B., Shcherbak, I., Robertson, G. P., Porter, J. R., Smith, P., Paustian, K., Grace, P. R., Asseng, S., Bassu, S., Biernath, C., Boote, K. J., Cammarano, D., De Sanctis, G., Durand, J. -L., Ewert, F., Gayler, S., Hyndman, D. W., Kent, J., Martre, P., Nendel, C., Priesack, E., Ripoche, D., Ruane, A. C., Sharp, J., Thorburn, P. J., Hatfield, J. L., Jones, J. W. and Rosenzweig, C. 2018. Soil organic carbon and nitrogen feedbacks on crop yields under climate change. *Agricultural and Environmental Letters* 3(1). doi:10.2134/ael2018.05.0026.

Basso, B., Shuai, G., Zhang, J. and Robertson, G. P. 2019. Yield stability analysis reveals sources of large-scale nitrogen loss from the US Midwest. *Scientific Reports* 9, 5774. doi:10.1038/s41598-019-42271-1.

Bennett, A. J., Bending, G. D., Chandler, D., Hilton, S. and Mills, P. 2012. Meeting the demand for crop production: the challenge of yield decline in crops grown in short rotations. *Biological Reviews of the Cambridge Philosophical Society* 87(1), 52–71. doi:10.1111/j.1469-185X.2011.00184.x.

Berntsen, J., Petersen, B. M. and Olesen, J. E.2007. Simulating trends in crop yield and soil carbon in a long-term experiment–effects of rising CO_2, N deposition and improved cultivation. In: Kersebaum, K. Ch. (Ed.), *Modelling Water and Nutrient Dynamics in Soil-Crop Systems*. Springer Netherlands, pp. 197–207.

Blanco-Canqui, H. and Lal, R. 2009. Crop residue removal impacts on soil productivity and environmental quality. *Critical Reviews in Plant Sciences* 28(3), 139–63. doi:10.1080/07352680902776507.

Bowles, T. M., Atallah, S. S., Campbell, E. E., Gaudin, A. C. M., Wieder, W. R. and Grandy, A. S. 2018. Addressing agricultural nitrogen losses in a changing climate. *Nature Sustainability* 1(8), 399–408. doi:10.1038/s41893-018-0106-0.

Bullock, D. G. 1992. Crop rotation. *Critical Reviews in Plant Science* 11, 309–26.

Cassman, K. G., Dobermann, A. and Walters, D. T. 2002. Agroecosystems, nitrogen-use efficiency and nitrogen management. *Ambio* 31(2), 132–40. doi:10.1579/0044-7447-31.2.132.

Challinor, A. J., Watson, J., Lobell, D. B., Howden, S. M., Smith, D. R. and Chhetri, N. 2014. A meta-analysis of crop yield under climate change and adaptation. *Nature Climate Change* 4(4), 287–91. doi:10.1038/nclimate2153.

Christianson, L. E. and Harmel, R. D. 2015. 4R water quality impacts: an assessment and synthesis of forty years of drainage nitrogen losses. *Journal of Environmental Quality* 44(6), 1852–60. doi:10.2134/jeq2015.03.0170.

Christianson, R., Christianson, L., Wong, C., Helmers, M., McIsaac, G., Mulla, D. and McDonald, M. 2018. Beyond the nutrient strategies: common ground to accelerate agricultural water quality improvement in the upper Midwest. *Journal of Environmental Management* 206, 1072–80. doi:10.1016/j.jenvman.2017.11.051.

Curl, E. A. 1963. Control of plant diseases by crop rotation. *The Botanical Review* 29(4), 413–79. doi:10.1007/BF02860813.

David, M. B., Drinkwater, L. E. and McIsaac, G. F. 2010. Sources of nitrate yields in the Mississippi River Basin. *Journal of Environmental Quality* 39(5), 1657-67. doi:10.2134/jeq2010.0115.

Davis, A. S., Hill, J. D., Chase, C. A., Johanns, A. M. and Liebman, M. 2012. Increasing cropping system diversity balances productivity, profitability and environmental health. *PLoS ONE* 7(10), e47149. doi:10.1371/journal.pone.0047149.

Dietzel, R., Liebman, M., Ewing, R., Helmers, M. J., Horton, R., Jarchow, M. E. and Archontoulis, S. V. 2016. How efficiently do corn- and soybean-based cropping systems use water? A systems modeling analysis. *Global Change Biology* 22(2), 666-81. doi:10.1111/gcb.13101.

Donatelli, M., Magarey, R. D., Bregaglio, S., Willocquet, L., Whish, J. P. M. and Savary, S. 2017. Modelling the impacts of pests and diseases on agricultural systems. *Agricultural Systems* 155, 213-24. doi:10.1016/j.agsy.2017.01.019.

Ewert, F., Rötter, R. P., Bindi, M., Webber, H., Trnka, M., Kersebaum, K. C., Olesen, J. E., van Ittersum, M. K., Janssen, S., Rivington, M., Semenov, M. A., Wallach, D., Porter, J. R., Stewart, D., Verhagen, J., Gaiser, T., Palosuo, T., Tao, F., Nendel, C., Roggero, P. P., Bartošová, L. and Asseng, S. 2015. Crop modelling for integrated assessment of risk to food production from climate change. *Environmental Modelling and Software* 72, 287-303. doi:10.1016/j.envsoft.2014.12.003.

Farina, R., Di Bene, C., Piccini, C., Marchetti, A., Troccoli, A. and Francaviglia, R. 2018. Do crop rotations improve the adaptation of agricultural systems to climate change? A modeling approach to predict the effect of durum wheat-based rotations on soil organic carbon and nitrogen. In: Muñoz, M. A. and Zornoza, R. (Eds), *Soil Management and Climate Change*. Academic press, pp. 221-36.

Fletcher, A. L., Brown, H. E., Johnstone, P. R., de Ruiter, J. M. and Zyskowski, R. F. 2011. Making sense of yield trade-offs in a crop sequence: a New Zealand case study. *Field Crops Research* 124(2), 149-56. doi:10.1016/j.fcr.2011.01.007.

Gaydon, D. S., Balwinder-Singh, Wang, E., Poulton, P. L., Ahmad, B., Ahmed, F., Akhter, S., Ali, I., Amarasingha, R., Chaki, A. K., Chen, C., Choudhury, B. U., Darai, R., Das, A., Hochman, Z., Horan, H., Hosang, E. Y., Kumar, P. V., Khan, A. S. M. M. R., Laing, A. M., Liu, L., Malaviachichi, M. A. P. W. K., Mohapatra, K. P., Muttaleb, M. A., Power, B., Radanielson, A. M., Rai, G. S., Rashid, M. H., Rathanayake, W. M. U. K., Sarker, M. M. R., Sena, D. R., Shamim, M., Subash, N., Suriadi, A., Suriyagoda, L. D. B., Wang, G., Wang, J., Yadav, R. K. and Roth, C. H. 2017. Evaluation of the APSIM model in cropping systems of Asia. *Field Crops Research* 204, 52-75. doi:10.1016/j.fcr.2016.12.015.

Gillette, K., Malone, R. W., Kaspar, T. C., Ma, L., Parkin, T. B., Jaynes, D. B., Fang, Q. X., Hatfield, J. L., Feyereisen, G. W. and Kersebaum, K. C. 2018. N loss to drain flow and N_2O emissions from a corn-soybean rotation with winter rye. *The Science of the Total Environment* 618, 982-97. doi:10.1016/j.scitotenv.2017.09.054.

Hirsh, S. M. and Weil, R. R. 2019. Deep soil cores reveal large end-of-season residual mineral nitrogen pool. *Agricultural and Environmental Letters* 4(1). doi:10.2134/ael2018.10.0055.

Hlavinka, P., Trnka, M., Kersebaum, K. C., Čermak, P., Pohankova, E., ORSÁG, M., Pokorny, E., Fischer, M., Brtbucky, M. and Žalud, Z. 2014. Modelling of yields and soil nitrogen dynamics for crop rotations by HERMES under different climate and soil conditions in the Czech Republic. *The Journal of Agricultural Science* 152(2), 188-204. doi:10.1017/S0021859612001001.

Holzworth, D. P., Huth, N. I., DeVoil, P. G., Zurcher, E. J., Herrmann, N. I., McLean, G., Chenu, K., van Oosterom, E. J., Snow, V., Murphy, C., Moore, A. D., Brown, H., Whish, J. P. M., Verrall, S., Fainges, J., Bell, L. W., Peake, A. S., Poulton, P. L., Hochman, Z., Thorburn, P. J., Gaydon, D. S., Dalgliesh, N. P., Rodriguez, D., Cox, H., Chapman, S., Doherty, A., Teixeira, E., Sharp, J., Cichota, R., Vogeler, I., Li, F. Y., Wang, E., Hammer, G. L., Robertson, M. J., Dimes, J. P., Whitbread, A. M., Hunt, J., van Rees, H., McClelland, T., Carberry, P. S., Hargreaves, J. N. G., MacLeod, N., McDonald, C., Harsdorf, J., Wedgwood, S. and Keating, B. A. 2014. APSIM - evolution towards a new generation of agricultural systems simulation. *Environmental Modelling and Software* 62, 327-50. doi:10.1016/j.envsoft.2014.07.009.

IPCC. 2014. *Climate Change 2014: Synthesis Report. Contribution of Working Groups I, II and III to the Fifth Assessment Report of the Intergovernmental Panel on Climate Change*. Geneva, Switzerland.

Jarecki, M., Grant, B., Smith, W., Deen, B., Drury, C., VanderZaag, A., Qian, B., Yang, J. and Wagner-Riddle, C. 2018. Long-term trends in corn yields and soil carbon under diversified crop rotations. *Journal of Environmental Quality* 47(4), 635–43. doi:10.2134/jeq2017.08.0317.

Karlen, D. L., Varvel, G. E., Bullock, D. G. and Cruse, R. M. 1994. Crop rotations for the 21st century. *Advances in Agronomy* 53, 1–45. doi:10.1016/S0065-2113(08)60611-2.

Kay, B. D. 1990. Rates of change of soil structure under different cropping systems. *Advances in Soil Science* 12, 1–52. doi:10.1007/978-1-4612-3316-9_1.

Kollas, C., Kersebaum, K. C., Nendel, C., Manevski, K., Müller, C., Palosuo, T., Armas-Herrera, C. M., Beaudoin, N., Bindi, M., Charfeddine, M., Conradt, T., Constantin, J., Eitzinger, J., Ewert, F., Ferrise, R., Gaiser, T., Cortazar-Atauri, I. G. de, Giglio, L., Hlavinka, P., Hoffmann, H., Hoffmann, M. P., Launay, M., Manderscheid, R., Mary, B., Mirschel, W., Moriondo, M., Olesen, J. E., Öztürk, I., Pacholski, A., Ripoche-Wachter, D., Roggero, P. P., Roncossek, S., Rötter, R. P., Ruget, F., Sharif, B., Trnka, M., Ventrella, D., Waha, K., Wegehenkel, M., Weigel, H.-J. and Wu, L. 2015. Crop rotation modelling–a European model intercomparison. *European Journal of Agronomy* 70, 98–111. doi:10.1016/j.eja.2015.06.007.

Kovács, G. J., Németh, T. and Ritchie, J. T. 1995. Testing simulation models for the assessment of crop production and nitrate leaching in Hungary. *Agricultural Systems* 49(4), 385-97. doi:10.1016/0308-521X(95)00032-Z.

Leighty, C. E. 1938. *Crop Rotation, in Soils and Men: Agriculture Yearbook 1938*. United States Department of Agriculture, pp. 406–30.

Liebman, M. and Dyck, E. 1993. Crop rotation and intercropping strategies for weed management. *Ecological Applications: a Publication of the Ecological Society of America* 3(1), 92–122. doi:10.2307/1941795.

Lobell, D. B., Schlenker, W. and Costa-Roberts, J. 2011. Climate trends and global crop production since 1980. *Science* 333(6042), 616–20. doi:10.1126/science.1204531.

Martinez-Feria, R. 2018. Drivers of yield and nitrogen-loss tradeoffs: cropping system evaluations with process-based modeling. Iowa State University Graduate Theses and Dissertations.

Martinez-Feria, R. A., Dietzel, R., Liebman, M., Helmers, M. J. and Archontoulis, S. V. 2016. Rye cover crop effects on maize: a system-level analysis. *Field Crops Research* 196, 145–59. doi:10.1016/j.fcr.2016.06.016.

Martinez-Feria, R. A., Castellano, M. J., Dietzel, R. N., Helmers, M. J., Liebman, M., Huber, I. and Archontoulis, S. V. 2018. Linking crop- and soil-based approaches to evaluate system nitrogen-use efficiency and tradeoffs. *Agriculture, Ecosystems and Environment* 256, 131–43. doi:10.1016/j.agee.2018.01.002.

Médiène, S., Valantin-Morison, M., Sarthou, J.-P., Tourdonnet, S., Gosme, M., Bertrand, M., Roger-Estrade, J., Aubertot, J.-N., Rusch, A., Motisi, N., Pelosi, C. and Doré, T. 2011. Agroecosystem management and biotic interactions: a review. *Agronomy for Sustainable Development* 31(3), 491–514. doi:10.1007/s13593-011-0009-1.

Meinshausen, M., Smith, S. J., Calvin, K., Daniel, J. S., Kainuma, M. L. T., Lamarque, J., Matsumoto, K., Montzka, S. A., Raper, S. C. B., Riahi, K., Thomson, A., Velders, G. J. M. and van Vuuren, D. P. P. 2011. The RCP greenhouse gas concentrations and their extensions from 1765 to 2300. *Climatic Change* 109(1–2), 213–41. doi:10.1007/s10584-011-0156-z.

Melillo, J. M., Richmond, T. C. and Yohe, G. 2014. Highlights of climate change impacts in the United States: the third national climate assessment. doi:10.7930/J01Z429C. On.

Mesinger, F., DiMego, G., Kalnay, E., Mitchell, K., Shafran, P. C., Ebisuzaki, W., Jović, D., Woollen, J., Rogers, E., Berbery, E. H., Ek, M. B., Fan, Y., Grumbine, R., Higgins, W., Li, H., Lin, Y., Manikin, G., Parrish, D. and Shi, W. 2006. North American regional reanalysis. *Bulletin of the American Meteorological Society* 87(3), 343–60. doi:10.1175/BAMS-87-3-343.

Miller, D. R., Chen, S. Y., Porter, P. M., Johnson, G. A., Wyse, D. L., Stetina, S. R., Klossner, L. D. and Nelson, G. A. 2006. Rotation crop evaluation for management of the soybean cyst nematode in Minnesota. *Agronomy Journal* 98(3), 569–78. doi:10.2134/agronj2005.0185.

Necpalova, M., Anex, R. P., Kravchenko, A. N., Abendroth, L. J., Del Grosso, S. J., Dick, W. A., Helmers, M. J., Herzmann, D., Lauer, J. G., Nafziger, E. D., Sawyer, J. E., Scharf, P. C., Strock, J. S. and Villamil, M. B. 2014. What does it take to detect a change in soil carbon stock? A regional comparison of minimum detectable difference and experiment duration in the north central United States. *Journal of Soil and Water Conservation* 69(6), 517–31. doi:10.2489/jswc.69.6.517.

Nendel, C., Kersebaum, K. C., Mirschel, W. and Wenkel, K. O. 2014. Testing farm management options as climate change adaptation strategies using the MONICA model. *European Journal of Agronomy* 52, 47–56. doi:10.1016/j.eja.2012.09.005.

Nielsen, D. C., Vigil, M. F. and Benjamin, J. G. 2009. The variable response of dryland corn yield to soil water content at planting. *Agricultural Water Management* 96(2), 330–6. doi:10.1016/j.agwat.2008.08.011.

O'Leary, G. J., Liu, D. L., Ma, Y., Li, F. Y., McCaskill, M., Conyers, M., Dalal, R., Reeves, S., Page, K., Dang, Y. P. and Robertson, F. 2016. Modelling soil organic carbon 1. Performance of APSIM crop and pasture modules against long-term experimental data. *Geoderma* 264, 227–37. doi:10.1016/j.geoderma.2015.11.004.

Poffenbarger, H. J., Barker, D. W., Helmers, M. J., Miguez, F. E., Olk, D. C., Sawyer, J. E., Six, J. and Castellano, M. J. 2017. Maximum soil organic carbon storage in Midwest U.S. cropping systems when crops are optimally nitrogen-fertilized. *PLoS ONE* 12(3), e0172293. doi:10.1371/journal.pone.0172293.

Porter, J. R., Xie, L., Challinor, A. J., Cochrane, K., Howden, S. M., Iqbal, M. M., Lobell, D. B. and Travasso, M. I. 2014. Food security and food production systems. In: Field, C. B., Barros, V. R., Dokken, D. J., Mach, K. J., Mastrandrea, M. D., Bilir, T. E., Chatterjee, M., Ebi, K. L., Estrada, Y. O., Genova, R. C., Girma, B., Kissel, E. S., Levy, A. N., MacCracken, S., Mastrandrea, P. R. and White, L. L. (Eds), *Climate Change 2014: Impacts,*

Adaptation, and Vulnerability. Part A: Global and Sectoral Aspects. Contribution of Working Group II to the Fourth Assessment Report of the Intergovernmental Panel on Climate Change. Cambridge University Press, pp. 485–533.

Post, J., Habeck, A., Hattermann, F., Krysanova, V., Wechsung, F. and Suckow, F. 2007. Evaluation of water and nutrient dynamics in soil-crop systems using the eco-hydrological catchment model SWIM. In: Kersebaum, K. C. (Ed.), *Modelling Water and Nutrient Dynamics in Soil-Crop Systems*, pp. 129–46.

Powlson, D. S., Stirling, C. M., Jat, M. L., Gerard, B. G., Palm, C. A., Sanchez, P. A. and Cassman, K. G. 2014. Limited potential of no-till agriculture for climate change mitigation. *Nature Climate Change* 4(8), 678–83. doi:10.1038/nclimate2292.

Puntel, L. A., Sawyer, J. E., Barker, D. W., Dietzel, R., Poffenbarger, H., Castellano, M. J., Moore, K. J., Thorburn, P. J. and Archontoulis, S. V. 2016. Modeling long-term corn yield response to nitrogen rate and crop rotation. *Frontiers in Plant Science* 7, 1630. doi:10.3389/fpls.2016.01630.

Puntel, L. A., Sawyer, J. E., Barker, D. W., Thorburn, P. J., Castellano, M. J., Moore, K. J., VanLoocke, A., Heaton, E. A. and Archontoulis, S. V. 2018. A systems modeling approach to forecast corn economic optimum nitrogen rate. *Frontiers in Plant Science* 9, 436. doi:10.3389/fpls.2018.00436.

Randall, G. W. and Mulla, D. J. 2001. Nitrate nitrogen in surface waters as influenced by climatic conditions and agricultural practices. *Journal of Environmental Quality* 30(2), 337–44. doi:10.2134/jeq2001.302337x.

Reeves, D. W. 1994. Cover crops and rotations. In: Hatfield, J. L. and Steward, B. A. (Eds), *Crop Residue Management*. Lewis Publications, CRC Press, Boca Raton, FL, pp. 125–75.

Reidsma, P., Ewert, F., Lansink, A. O. and Leemans, R. 2010. Adaptation to climate change and climate variability in European agriculture: the importance of farm level responses. *European Journal of Agronomy* 32(1), 91–102. doi:10.1016/j.eja.2009.06.003.

Ritchie, J. T., Gerakis, A. and Suleiman, A. 1999. Simple model to estimate field-measured soil water limits. *Transactions of ASAE* 42, 1609–14.

Robertson, G. P. and Vitousek, P. M. 2009. Nitrogen in agriculture: balancing the cost of an essential resource. *Annual Review of Environment and Resources* 34(1), 97–125. doi:10.1146/annurev.environ.032108.105046.

Ryan, J., Pala, M., Masri, S., Singh, M. and Harris, H. 2008. Rainfed wheat-based rotations under Mediterranean conditions: crop sequences, nitrogen fertilization, and stubble grazing in relation to grain and straw quality. *European Journal of Agronomy* 28(2), 112–8. doi:10.1016/j.eja.2007.05.008.

Ryan, J., Ibrikci, H., Sommer, R. and McNeill, A. 2009. Nitrogen in rainfed and irrigated cropping systems in the Mediterranean region. *Advances in Agronomy* 136, 53–136. doi:10.1016/S0065-2113(09)04002-4.

Salado-Navarro, L. R. and Sinclair, T. R. 2009. Crop rotations in Argentina: analysis of water balance and yield using crop models. *Agricultural Systems* 102(1–3), 11–6. doi:10.1016/j.agsy.2009.06.004.

Sawyer, J. E., Nafziger, E., Randall, G., Brundy, L., Rehm, G. and Joern, B. 2006. Concepts and rationale for regional nitrogen rate guidelines for corn (PM2015). Iowa State University Extension and Outreach.

Soil Survey Staff. *Soil Survey Geographic (SSURGO) Database*.

Stevenson, F. C. and van Kessel, C. 1996. The nitrogen and non-nitrogen rotation benefits of pea to succeeding crops. *Canadian Journal of Plant Science* 76(4), 735–45. doi:10.4141/cjps96-126.

Suleiman, A. A. and Ritchie, J. T. 2001. Estimating saturated hydraulic conductivity from soil porosity. *Transactions of ASAE* 44, 235–339.

Teixeira, E. I., Brown, H. E., Sharp, J., Meenken, E. D. and Ewert, F. 2015. Evaluating methods to simulate crop rotations for climate impact assessments – a case study on the Canterbury plains of New Zealand. *Environmental Modelling and Software* 72, 304–13. doi:10.1016/j.envsoft.2015.05.012.

Teixeira, E. I., Ruiter, J. De, Ausseil, A. G., Daigneault, A., Johnstone, P., Holmes, A., Tait, A. and Ewert, F. 2018. Adapting crop rotations to climate change in regional impact modelling assessments. *The Science of the Total Environment* 616–617, 785–95. doi:10.1016/j.scitotenv.2017.10.247.

Thornton, P. E., Thornton, M. M., Mayer, B. W., Wei, Y., Devarakonda, R., Vose, R. S. and Cook, R. B. 2018. *Daymet: Daily Surface Weather Data on a 1-km Grid for North America, Version 3.*

United Nations Food and Agriculture Organization. 2016. *The State of Food and Agriculture 2016 (SOFA): Climate Change, Agriculture and Food Security, Livestock in the Balance.* ISBN: 978-92-5-107671-2 I.

Wallach, D., Makowski, D., Jones, J. W. and Brun, F. 2014. *Working with Dynamic Crop Models: Methods, Tools and Examples for Agriculture and Environment.* Academic Press.

White, J. W., Hoogenboom, G., Kimball, B. A. and Wall, G. W. 2011. Methodologies for simulating impacts of climate change on crop production. *Field Crops Research* 124(3), 357–68. doi:10.1016/j.fcr.2011.07.001.

Yin, X., Kersebaum, K. C., Kollas, C., Manevski, K., Baby, S., Beaudoin, N., Öztürk, I., Gaiser, T., Wu, L., Hoffmann, M., Charfeddine, M., Conradt, T., Constantin, J., Ewert, F., de Cortazar-Atauri, I. G., Giglio, L., Hlavinka, P., Hoffmann, H., Launay, M., Louarn, G., Manderscheid, R., Mary, B., Mirschel, W., Nendel, C., Pacholski, A., Palosuo, T., Ripoche-Wachter, D., P. Rötter, R., Ruget, F., Sharif, B., Trnka, M., Ventrella, D., Weigel, H. and E. Olesen, J. 2017. Performance of process-based models for simulation of grain N in crop rotations across Europe. *Agricultural Systems* 154, 63–77. doi:10.1016/j.agsy.2017.03.005.

Zentner, R. P., Campbell, C. A., Biederbeck, V. O., Miller, P. R., Selles, F. and Fernandez, M. R. 2001. In search of a sustainable cropping system for the semiarid Canadian prairies. *Journal of Sustainable Agriculture* 18(2–3), 117–36. doi:10.1300/J064v18n02_10.

Zentner, R. P., Wall, D. D., Nagy, C. N., Smith, E. G., Young, D. L., Miller, P. R., Campbell, C. A., McConkey, B. G., Brandt, S. A., Lafond, G. P., Johnston, A. M. and Derksen, D. A. 2002. Economics of crop diversification and soil tillage opportunities in the Canadian prairies. *Agronomy Journal* 94(2), 216–30. doi:10.2134/agronj2002.0216.

Zhang, X., Davidson, E. A., Mauzerall, D. L., Searchinger, T. D., Dumas, P. and Shen, Y. 2015 Managing nitrogen for sustainable development. *Nature* 528(7580), 51-9. doi:10.1038/nature15743.

Chapter 9

Integrating livestock production into whole-farm system models of mixed crop-livestock systems

Katrien Descheemaeker, Wageningen University and Research, The Netherlands; and Lindsay Bell, CSIRO Agriculture Flagship, Australia

1 Introduction

The worldwide importance of mixed crop-livestock farming systems is well described (Herrero et al., 2010). Especially in the developing world, mixed systems are dominant in terms of areal extent, food production and livelihood provision (Tarawali et al., 2011). Yet, trends indicate that crop-livestock farms are disappearing where market prices, availability of cheap inputs and agricultural policies push farms to specialize. These trends have been described for Europe and South America (Peyraud et al., 2014), the United States (Sulc and Franzluebbers, 2014) and Australia (Bell and Moore, 2012). At the farm level, labour constraints and high skill requirements for the complex management of integrated crop-livestock systems challenge these systems' maintenance or re-introduction (Ryschawy et al., 2017). Specialization has gone hand in hand with the intensification of agriculture and livestock production, leading to negative environmental impacts, including the loss of biodiversity, loss of soil organic matter, eutrophication, emissions and groundwater contamination (Lemaire et al., 2014; Peyraud et al., 2014). Because of the interactions between crops, animals and pastures, mixed farming systems are believed to be more efficient than specialized systems at recycling nutrients, improving soil quality and supporting heterogeneous landscapes and species-rich land uses (Lemaire et al., 2014; Franzluebbers et al., 2014a). Hence, recent attention in

http://dx.doi.org/10.19103/AS.2019.0061.12

the scientific literature has gone to investigating the sustainability of crop-livestock integration (Franzluebbers et al., 2014a,b; Ryschawy et al., 2014) and to exploring new modes of integration at farm and larger landscape levels (Martin et al., 2016; Ryschawy et al., 2017).

The numerous interactions between the components of crop-livestock systems have led many to describe mixed systems as highly complex (Thornton and Herrero, 2001; Parsons et al., 2011; van Wijk et al., 2014). Modelling tools can contribute to understanding the functioning of such complex systems, by disentangling the component interactions and simulating the emergent behaviour of the integrated components (Vayssières et al., 2009b). Additionally, models can be used to explore the potential impacts of trends and interventions, and to design improved systems at different scales, thus underpinning the needed transformation to sustainable agriculture worldwide.

This chapter first presents a description of mixed crop-livestock systems, their components and interactions. We then focus on whole-farm dynamic simulation models, and their typical model setup consisting of integrated component models. After discussing the various component models, we pay attention to how integrated models deal with their interactions. As part of a book focussed on crop modelling, this chapter identifies aspects and outputs that crop (and pasture) models need to incorporate for functional integration with a livestock component model. Likewise, we point to livestock-related processes influencing soil conditions and plant growth that can be captured by crop (and pasture) models. We end this chapter by illustrating concrete crop-livestock model applications based on two case studies from contrasting environments.

While paying attention to the management decisions in mixed crop-livestock systems, this chapter's focus is on the bio-physical aspects of crop-livestock systems. Hence, a detailed analysis of how whole-farm models include socio-economic aspects (e.g. access to resources, farm assets, input availability, market access, prices) is out of the scope of this chapter. With its focus on bio-physical dynamic simulation models, this chapter only points to other types of models (e.g. optimization or agent-based models), where such references add value. Finally, we refer specifically to ruminant livestock systems where grazing, feed-related and animal performance issues are discussed.

2 Key features of crop-livestock systems

Mixed crop-livestock systems are composed of a bio-physical sub-system, a decision-making sub-system (the farm manager) and an operating sub-system making use of resources, such as land, agricultural inputs and labour (e.g. Vayssières et al., 2009a; Snow et al., 2014). Others have described mixed systems as socio-ecological systems (e.g. Moraine et al., 2014), where the ecological sub-system interacts with a social system composed of stakeholders including farmers, value chain actors, advisory services and policy makers. The bio-physical or ecological sub-system itself consists of various components and its functioning depends on environmental factors (e.g. climate) and interactions between the components. The major bio-physical components interacting with each other include crops and pastures with the soils they grown on, livestock and excreta (Fig. 1). Depending on the system, livestock consume fodder crops, grain and/or crop residues and graze pastures and/or rangelands. Livestock produce excreta (manure and urine), which may be deposited while grazing or, after storage, be applied as a fertilizer on cropland and grazing land. Livestock also influence crop and grazing land by soil treading and vegetation trampling, defoliation of the vegetation and, in some systems, by providing draught power for land cultivation (feedback not included in Fig. 1). In addition, there are various effects of crop management and livestock production on the environment (e.g. erosion, nutrient leaching, greenhouse gas emissions), which feed back to farm performance directly and indirectly (e.g. because of legislative restrictions).

Mixed systems harbour a huge diversity in the kinds of crops, pastures and animals, the management and input-use intensity, and also in terms of the degree and scale of integration between the system components. For example, the bio-physical farm components can be integrated at the farm level, but at a larger territory scale the interaction between specialized crop and livestock farms can also be considered as a mixed system. Interactions can be described in space as co-located versus segregated and in time as synchronized versus rotated (Bell and Moore, 2012). Examples of co-location include animals grazing grass or crop residues on a farm or on farms of a community (Rufino et al., 2011), as opposed to a segregated system in which forage crops are grown in a separate location from where animals are kept. Synchronized temporal interactions exist in case of, for example, dual-purpose cropping, such as sweet sorghum varieties that produce high-quality stover in addition to grain (Blümmel et al., 2003), and the grazing of crops during their vegetative period after which they recover to produce grain (Dove and Kirkegaard, 2014). Another common synchronized interaction is the intercropping of crop and forage plants, such as maize with cowpea fodder varieties (e.g. Falconnier et al., 2016) or mucuna (e.g. Ngwira et al., 2012). Interaction through the rotation of crops and pasture takes place because of effects on (soil) pathogens, soil conditions (SOC, soil structure, etc.), nutrient dynamics and weed pressure (Angus et al., 2001; Franzluebbers et al., 2014b; Lemaire et al., 2014). The interactions can be described through the flow of material or energy (Moraine et al., 2014). Flow of materials, usually biomass, is considered when describing the amount of fodder produced by crops or forages and consumed by animals. Nutrient flows need to be quantified when assessing the nutritive value of the fodder, the

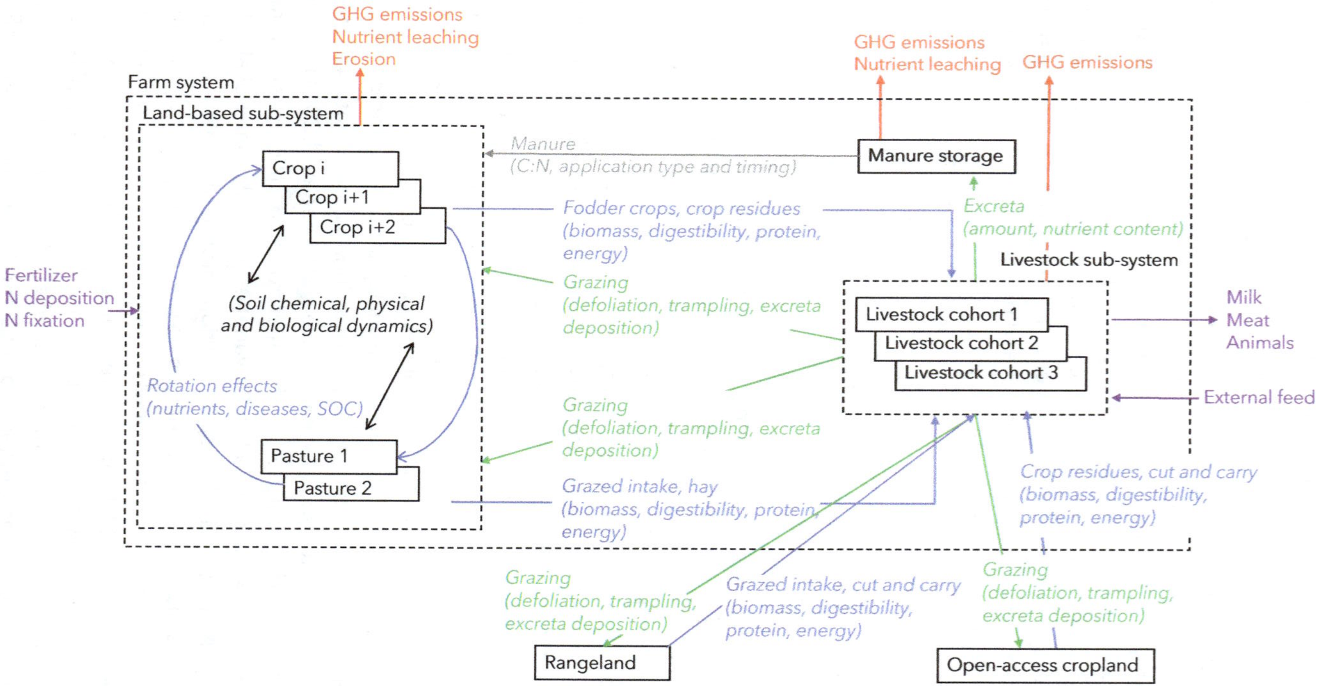

Figure 1 System diagram representing the key bio-physical components and interactions in a mixed crop–livestock farm. Bio-physical components in black text and boxes; sub-system and system boundaries in dashed black lines; losses in red; input and outputs in purple; interaction effects in green (originating from livestock), blue (originating from crops and pasture) and grey (originating from manure), with text in brackets indicating the key elements characterizing the interaction.

return of nutrients to crops and pasture in manure and urine, and the effects on soil fertility. Energy flows can be considered based on the energy content of the fodder and the energy delivered by draught animals when cultivating land.

3 Dynamic simulation models of whole-farm systems including livestock

The previous section illustrated how the bio-physical system components of mixed crop–livestock farming systems interact in many ways. In addition, the components are influenced by the environment and the interactions with the decision-making and operational sub-system of the farm. Hence, interventions (e.g. changes in management such as input use, timing of farm operations, allocation of farm resources) can have multiple effects on the system and its environment. Due to the complexity of interactions and feedbacks, the system's behaviour is emergent, which means that it cannot be deduced or simply scaled up from the component properties (Vayssières et al., 2009b; Kipling et al., 2016). Modelling tools help to better understand the functioning of such complex systems, and can be used to predict farm and component performance, support decisions and discussions among stakeholders, design improved systems and identify new research areas.

3.1 Component models

Similar to mixed farming systems integrating various components in reality, the models representing the real systems often integrate (existing) component models (Vayssières et al., 2009a; van Evert and Lamaker, 2007). The choice for the specific component models depends on the processes and their degree of detail and complexity that need to be represented. For example, depending on the required level of detail, animal feed intake from grazing can be approximated based on the stocking density and pasture biomass or, if data availability allows, it can be simulated more precisely by intake models (e.g. Cros et al., 2003). The capacity to simulate mixed farming systems also depends on the transfer of key information between the different model components. Hence, the component model choice depends on the models' inter-operability and flexibility for integration, which is facilitated if the model code is available. In the following sections, we briefly describe the main types of bio-physical component models and the important processes that they need to capture. We end with paying attention to the decision-making component.

3.1.1 Crops

With respect to crop growth models, we refer to the other chapters in this book for an overview of features, application domains and new developments. As an example, the Agricultural Production Systems Simulator model (APSIM) (Holzworth et al., 2014) is a widely used dynamic crop growth model that is integrated in various whole-farm simulation models such as the Integrated Analysis Tool (IAT) (Lisson et al., 2010), APSFarm (Rodriguez et al., 2011) and AUSFARM (Moore et al., 2007). These whole-farm models vary in how the feedbacks from livestock to crop production (e.g. return of nutrients through manure) are captured and in how farm management and decision-making are represented (e.g. based on intermediate model outputs or fixed). However, in all these cases, APSIM outputs on crop biomass are used in the livestock model component, as such influencing the simulated livestock production.

3.1.2 Pasture

Pasture models are numerous and diverse in terms of how they capture both the pasture-specific aspects such as species composition, reproduction and establishment, and the aspects related to pasture-animal interaction, such as defoliation and selective grazing effects on pasture quality and growth rate (see Section 3.2.1; and partial reviews by Snow et al., 2014; Wiegand et al., 2008). The diversity in pasture models can be characterized by differentiating between grassland-specific models (e.g. the GRAZPLAN pasture model (Moore et al., 1997), SPACSYS (Wu et al., 2007), PaSIM (Riedo et al., 1998)), crop models adapted to grasslands (e.g. STICS (Coucheney et al., 2015), APSIM (Holzworth et al., 2014)) and models based on plant functional types (see also below) (Kipling et al., 2016). Moulin et al. (2018) distinguish pasture models that describe monospecific swards (Duru et al., 2009; McCall and Bishop-Hurley, 2003) from others that simulate the dynamics of grass-legume mixtures (Lazzarotto et al., 2009), or multispecies communities (Jouven et al., 2006). Generally speaking, pasture models predict daily growth based on potential growth and environmental factors limiting growth (temperature, water, nutrients), with fewer models also taking growth-reducing factors (pests, weeds) into account. Although some pasture models merely predict biomass quantity, usually forage quality (e.g. digestibility, protein and fibre content) is also predicted as this is needed for a link with grazing and livestock performance models. By simulating plant structural components (e.g. leaves, stems, reproductive organs) separately, overall forage quality can be inferred from their proportions (Jouven et al., 2006) and the change in quality of these components, which can be predicted from plant phenological development (e.g. Bell et al., 2009). Whereas more complex models consider multiple species separately, thus enabling the prediction of seasonal changes in species composition (e.g. Moulin et al., 2018), other approaches simplify the computations by considering growth forms or assuming uniform, homogenous pastures. Rangelands, which are diverse in vegetation physiology and life forms and strongly influenced by plant-animal interactions, can be simulated with so-called plant functional type or biome models that also simulate biogeochemical processes. Examples include G-Range (Boone et al., 2018), SAVANNA (Coughenour, 1992) and Biome-BDGMuSo (Hidy et al., 2016).

3.1.3 Livestock

Models allowing the simulation of the animal component of mixed farm systems typically simulate animal performance in terms of the economic products of meat (and/or live animals), milk and fibre, as well as the by-product manure. Other important livestock services such as the provision of draught power, insurance functions or, in some societies, a variety of sociocultural functions (e.g. dowry, prestige) (Moll, 2005), are usually not part of model output, but may be inferred from it. Whereas the animal performance models come in various forms and levels of complexity, in general, they combine animal-related information with feed-related information, where the estimate of actual feed intake quantity and quality is a crucial element (Jones et al., 2017). Essential information about the animal includes the genetic potential in terms of for example morphometric characteristics and milk production potential. Combined with actual characteristics such as body weight, lactation and pregnancy status, these define the feed requirements. In terms of the feed, information about feed quantity needs to be combined with feed quality, for which digestibility, energy and crude protein content are essential. Information about the various feed constituents (fibre, mineral, amino acid, starch, fats, etc.) can also be used for a more complete estimate of nutritive value, but are not needed in all models. Commonly used feeding standards, such as the NRC system (NRC, 2001) of the United States and the AFRC system (AFRC, 1993) of the United Kingdom, are based on empirical functions estimating the animal feed requirements for different processes and activities (Tedeschi et al., 2014). These feeding standards served as the foundation of several decision support models, such as the Australian GRAZPLAN suite of models (Donnelly et al., 2002) and practical feed ration formulation. In some livestock models, the feeding standards are combined with process-based understanding of feed digestion and energy and protein needs of livestock (Jones et al., 2017). These models include knowledge on livestock nutrition to varying degrees of complexity, but what they have in common is that animal nutrient requirements are compared with nutrient supply. The latter depends on feed intake and energy and protein supply from digestion, which is determined by the passage through the gastrointestinal tract, the rumen function and the feed composition and digestibility. Nutrient excretion in the form of urine and manure typically forms part of the output of the mechanistic models, and routines for the calculation of greenhouse gas emissions from enteric fermentation are often included.

Besides the energy and protein supply from feed intake, animal performance is also influenced by environmental factors such as the climate, pathogens and diseases. With respect to the climate, empirical relations exist between temperature and humidity on the one hand and feed intake and utilization efficiency on the other hand (Gaughan et al., 2012). However, where available, mechanistic models of thermoregulation allow simulating the effect of climate on animal performance for a wide range of conditions, management and genotypes (van der Linden et al., 2019). With respect to diseases, animal performance is influenced by the spatial and temporal distribution and intensity of pathogens and diseases, which can be predicted with mechanistic models that capture the biology and ecology of the pathogens and their vectors and how these are influenced by environmental conditions (Kipling et al., 2016).

A final key aspect characterizing most livestock component models is the combination of animal performance prediction with herd demography. Whereas some models simulate growth and performance of each individual animal in the herd (e.g. LIVSIM (Rufino et al., 2009)), most models subdivide the herd in successive cohorts, which simplifies the computations especially for large herds. A cohort is a group of animals that can be thought of as identical in sex, age and/or weight, reproductive status, and genotype, and receive identical treatment in terms of feeding and other animal husbandry practices. Therefore, their simulated feed requirements, intake and resulting growth and reproductive performance are also identical. The herd dynamics are governed by reproductive (i.e. calving rate and age at first calving) and mortality parameters, as well as management characteristics related to mating, weaning, culling and replacement.

As mentioned before, a wide variety of livestock modelling approaches can be distinguished. Where some focus on the representation of animal growth based on feed availability and potential growth curves (e.g. GRASP (Rickert et al., 2000)), others quantify both energy and protein availability versus requirements for different animal functions, including maintenance and growth, wool and milk production and foetal growth (e.g. LIVSIM, GrazFeed (Freer et al., 1997)). There are also differences in the level of detail used to represent reproductive processes (e.g. a high level of detail in SITEL (Brun-Lafleur et al., 2013)), rumen processes (e.g. a high level of detail in models described by Baldwin (1995) and Hanigan et al. (2009)) and body tissue partitioning (e.g. a high level of detail in AgMOD (Johnson, 2016) and LiGAPS-Beef (van der Linden et al., 2019)). Furthermore, it is worth noting that some models explicitly include stochasticity (in terms of conception and mortality) in their simulations (e.g. LIVSIM, GRAZPLAN), and that there are differences with respect to the time step of the simulation, with typical ranges from a day (e.g. GRAZPLAN, AgMOD, LiGAPS-Beef) to a month (e.g. IFSM (Rotz et al., 2013), LIVSIM). Finally, livestock models can be categorized based on the livestock species and system orientation they focus on. Examples of beef models include LiGAPS-Beef (van der Linden et al., 2019), NODRIZA (Villalba et al., 2010) and the model described by Jouven et al. (2008). Dairy cattle models include GARUNS (Martin et al., 2019) and Karoline (Danfaer et al., 2006), with Tedeschi et al. (2014) providing a review. Pittroff and Kothmann (2001) provide a detailed comparison of feed intake models for sheep, for which GrazFeed (Freer et al, 1997) and the Texas A&M Sheep Production Systems Model (Blackburn and Cartwright, 1987) are examples of more widely used models. For goats, models have been developed and used for dairy-oriented systems (e.g. Guimarães et al., 2009) and meat-oriented systems (e.g. Amole et al., 2017).

3.1.4 Manure

As mentioned above, nutrient excretion in the form of manure and urine often forms part of the simulation output of animal performance models. In mixed crop-livestock systems, manure is a source of nutrients for crop production and organic carbon for soil fertility maintenance, but also a source of unwanted gaseous and nutrient losses to the environment (Sommer and Hutchings, 2001; Kebreab, 2013). To capture the nutrient-related interactions between the livestock, crop and soil components, whole-farm system models consider manure as a system component, of which the availability and quality are dynamically simulated over time. Decomposition and nutrient dynamics are the key processes here and depend on the specific characteristics of the handling process (the collection of manure from stables and kraals, its conditioning and treatment), the storage environment (i.e. aeration, temperature and moisture content influenced by storage management and the flooring, cover and roofing of the storage place), the application process (time, frequency and method of application) and the type of manure (liquid, raw solid, composted solid and directly deposited during grazing). Various approaches to simulating the dynamics in manure can be distinguished. Rufino et al. (2007) present a simple fuzzy logic system to capture the effects of different storage types on the evolution of manure mass and N content over time. Other approaches employ a mechanistic representation of organic matter decomposition governed by C:N ratio, moisture content and temperature (e.g. Parsons et al. (2011), relying on the Surface OM module of APSIM), or focus more on the representation of inflow and outflow quantities in a stock-flow model (e.g. MAGMA model by Guerrin (2001)). The inclusion of nutrient (mostly N and P in various forms) and CH_4 losses from manure in whole-farm models has become more common with the growing importance of predicting greenhouse gas emissions and other environmental hazards. The methods employed by these models vary from the implementation of IPCC factors to more mechanistic simulation modelling, and are reviewed for four whole-farm models (DairyWise, FarmGHG, FarmSim, $SIMS_{DAIRY}$) by Schils et al. (2007).

3.1.5 Farm management

Strategic, tactical and operational farm management have formed part of agricultural systems models since the early stages (Moore et al., 2014). In dynamic simulation models, farm management interventions can be simulated as discrete events that cause a change in the state of at least one of the farm components and, consequently, in the system as a whole. Farmers' practices can be captured by static input parameters or data tables (e.g. with information on feed rations, fertilizer application rates, land allocation, livestock numbers and stocking density). However, for a more realistic representation of the dynamic interactions between farmer actions and bio-physical processes, farm management can be dynamically simulated by a rule-based system that mimics decisions based on past and current environmental conditions and the state of other farm components (animals, plants, soils, manure). For example, the capacity to adjust livestock numbers according to the amount of feed available on the farm at a particular time. Moore et al. (2014) differentiate the latter approach from planning-based approaches in which simulations of future states are used to identify farm strategies that lead to desired consequences. One common challenge with simulating mixed farming enterprises is the high level of complexity required to adequately describe the various management decisions in every farm component. For this reason, often intermediate approaches which simplify (or set) some management decisions (e.g. crop management) while allowing others to alter (e.g. livestock stocking rate) are used.

Other types of models, such as linear programming models, are well suited to optimize strategic farm management decisions, and dynamic mathematical programming can be used to study the continuous adaptation of management over time (van Wijk et al., 2014; Janssen and van Ittersum, 2007). Some dynamic whole-farm models include a linear programming component, for example to optimize feed rations, as in IFSM (Rotz et al., 2013). Besides this, with their focus on capturing human decision-making, agent-based models are appropriate tools for analysing effects of decisions on and influences by the surrounding natural or social system (see e.g. An, 2012, for an overview).

3.2 Capturing the key interactions between the crop and livestock components of a mixed system

The previous section described the separate component models that usually form part of a mixed-farm model. The functioning and performance of the resulting whole-farm model depend not only on the component models but also on how the component models are linked, and which information from one component model is fed to and used in another component model. In this section, we discuss the critical interactions between the components of mixed farms with attention to the modelling challenges that arise when coupling the respective component models. Representing the key features governing component interactions may improve the realism of the mixed crop–livestock model predictions, but may come at the expense of increased complexity and reduced applicability. To a large extent, the question at hand and the intended use of the model determine the required degree of complexity and hence the balance between realism and pragmatism (Snow et al., 2014).

3.2.1 Grazing effects

Grazing is the main process governing feedbacks between livestock, plants and soils. Defoliation of vegetation directly impacts vegetation growth and composition and capturing the plant response and regrowth processes is required. This is particularly important for farming systems where crops are grown both for grazing and grain production. Long-season wheat varieties are commonly used as dual-purpose crops, for example in the US Southern Great Plains (Zhang et al., 2008) and southern Australia (Moore, 2009; Dove and Kirkegaard, 2014). Grazing wheat is usually done before the critical growth stage of the first hollow stem in order to avoid grain yield penalties. Nevertheless, the effects on plant growth and development are numerous and include a reduction in leaf biomass and area and delayed phenology. The first effect has consequences for photosynthesis and conversion to grain biomass, but also for evaporation and transpiration, with potential positive knock-on effects for moisture availability later in the season (Harrison et al., 2011). Delaying the phenology may have counteracting effects on final grain yield because of reducing the grain-filling period on the one hand and shifting the flowering stage outside the damaging frost period on the other hand. Cereal grazing impacts animal performance directly because the excellent nutritive value of the crop favours animal growth (Moore, 2009), and indirectly because of the deferment effect on pastures elsewhere from which the animal can benefit later in the season (Thomas et al., 2015). Modelling this plant-animal system has been done for different environments to unravel the complexities and trade-offs involved. With the exception of Moore et al. (2004), who adapted a pasture model to represent grain production, most studies have adapted existing crop growth models (e.g. CERES-Wheat (Rodriguez et al., 1990; Zhang et al., 2008) and APSIM (Moore, 2009; McCormick et al., 2015)) to accommodate the plant-animal interactions. The modifications are described in detail for CERES-Wheat, and include the reduction of (different pools of) plant biomass, plant N and leaf area based on animal intake, which is estimated from the available quantity and quality of the standing biomass. The standard model routines then take care of the simulation of knock-on effects on photosynthesis, evaporation and transpiration, which all influence grain yield in the end. In addition, an empirical approach mimics the effects of grazing after the first hollow stem stage on the loss of spikes (Zhang et al., 2008). A main difference with the APSIM approach is that the latter is modified to take effects on the phenological state into account (Moore, 2009), whereas this was deliberately not done in CERES-Wheat because of a reported lack of understanding of the process (Zhang et al., 2008). The modified crop models are linked to an existing livestock model (GRAZPLAN for the Australian cases) or a specifically developed model, which, in the case of Zhang et al. (2008), modifies animal intake for the time needed by animals to adjust to a new type of pasture.

The process and the intensity of defoliation are determined by elements related to the grazing animals, their management and the vegetation itself, with the mechanisms of vegetation response well understood (see reviews by Richards, 1993; Parsons and Chapman, 1999). Grazing animals prefer biomass with low fibre and high protein content that they can access at size-dependent preferred height. This selectivity is governed by animal-specific bite characteristics, such as mouth morphology, animal size and behaviour (e.g. Parsons et al., 2001; Snow et al., 2014). The available quantity and quality of the biomass determine the grazed intake. When pasture biomass grows beyond a certain point, its quality declines, which impairs animal intake and thus animal performance. Species composition also determines pasture growth and nutritive value, and consequently grazed intake. Finally, grazing intensity is an important management factor that affects biomass production and animal intake. It reduces pasture productivity at low-intensity levels due to canopy closure and competition for resources and at high-intensity levels because of less plant material available for photosynthesis (Bilotta et al., 2007). Increased grazing intensity also reduces the effects of animal selectivity. Models take different approaches to representing the above-mentioned processes of defoliation. Some models do not include selective grazing behaviour and remove pasture biomass homogenously based on estimated intake. Simplified predictions of intake can be derived as a function of forage quality and an animal-specific selectivity constant (see Bell et al., 2018, Eq. (1)). Other models mimic selectivity following a fill unit approach (Jarrige et al., 1986), based on the digestibility of different pasture patches (e.g. FASSET (Berntsen et al., 2003), GRAZPLAN), or the latter's fibre and protein content (e.g. IFSM). Incorporating sward heterogeneity in a pasture model is crucial for a dynamic link with a model of selective defoliation, and this can be done by differentiating plant structural compartments (e.g. green and dead leaves and sheaths, green and dead stems and reproductive organs) either for functional vegetation types (e.g. Jouven et al., 2006; Soussana et al., 2012) or for individual (groups of) species (Moulin et al., 2018; Johnson, 2016). Further complexity can be introduced by spatially explicit modelling of animal–plant interactions (e.g. Parsons et al., 2001) and by incorporating knowledge on bite-based interactions. The latter can be important when-detailed nutrient supply information is needed for mechanistic modelling of rumen processes, but it is less relevant in case of higher intensity and/or rotational grazing (Snow et al., 2014).

Finally, grazing animals also exert indirect effects through the excretion of dung and urine (see Section 3.2.2) and through treading of soil, trampling of vegetation and removal of vegetative cover. Treading leads to soil structural changes through compaction, pugging and poaching, which depend on animal species and weight, stocking density, vegetation cover and soil moisture (Bilotta et al., 2007). This can reduce soil porosity and increase soil strength and bulk density, decreasing infiltration, hydraulic conductivity and root growth rate. While these processes can be captured in most crop–soil models (e.g. Bell et al., 2011), they are rarely considered. Besides treading the soil, grazing animals also trample vegetation, a process taken into account by some pasture models by calculating the rate of fall of standing senesced material into the litter pool as a function of the stocking density (e.g. Moore et al., 1997). Increased erosion, partly induced by changed ground cover, is also an important consequence of livestock grazing on soil processes. The study by Thomas et al. (2018) is an example of a whole-farm modelling exercise where the crop, soil and surface residue modules of the APSIM crop growth model have been linked to the pasture and animal simulation modules of GRAZPLAN to investigate the trade-offs between production and soil erosion.

3.2.2 Representing manure dynamics

Manure is a key source of nutrients for plant growth and of organic carbon for soil fertility maintenance. Most crop models manage a fair representation of nutrient dynamics governed by mineral fertilizer application. With manure this is more complex, because it requires a mechanistic understanding of manure decomposition and release of nutrients during storage and after application (see Section 3.1.4). The animal feed composition and its digestion, the time and type of manure storage (e.g. covered or not; with or without adding of plant material), the type of manure application (e.g. broadcasted or incorporated), its timing and the applied manure type itself influence nutrient losses, manure quality and potential nutrient recovery by the plant (e.g. Chadwick et al., 2011; Rufino et al., 2007). To capture these nutrient-related processes, information from all system components (animal, manure, management, soil, crop) needs to be interchanged between the component models. Luckily, representing the actual interactions is relatively straightforward, apart from the grazing-mediated process of direct urine and dung deposition (see Section 3.2.3). Besides the recycling of nutrients, manure application also plays a role in soil organic carbon dynamics. As this is a slow process, capturing its effect on soil fertility and crop production requires long-term simulations of soil, crop and nutrient dynamics, without seasonal resets of soil conditions. Within a whole-farm context, manure application decisions may be influenced by labour availability, transport and storage capacities, and distance between the field and the storage place (Zingore et al., 2007; Guerrin, 2001), and can be captured through a link with the decision-making model component.

3.2.3 Animal mobility

Animals transfer nutrients across and between fields of farms and in landscapes due to their mobility. In addition, human-mediated nutrient transfers play a role in systems where animals are housed and excreta are collected, handled and then applied in specific fields or pastures or exported out of the farm (see Section 3.2.2). Random, small-scale patches of a high nutrient input are created by grazing animals mostly due to the deposition of urinary N (Snow et al., 2014). Furthermore, larger-scale systematic transfers of nutrients have been described in relation to preferred grazing and resting areas and structures or features (e.g. fences, drinking troughs) attracting animals. The latter are usually not represented in models, whereas attempts to incorporate the former are more common, especially where the aim is to capture environmental impacts. This is because, although the urinary patches are small, they affect leaching, (de)nitrification and volatilization, as well as pasture growth of considerable areas of a grazed pasture. Different pragmatic solutions to the modelling of the complex feature of urine patches are described by Snow et al. (2014) and rely on the separate modelling of areas affected and unaffected by excreta, which is governed by the stocking rate.

At a larger landscape scale, animal mobility is harnessed by pastoral herders to exploit the spatial and temporal variation in feed and water resources in rangeland ecosystems (Morton, 2007; Boone et al., 2011). However, with changes in land use and tenure, as well as settlement policies, herd mobility is increasingly being restricted (Vetter, 2005). Exploring the interactions between ecosystems services (e.g. the provision of feed and water; rangeland biodiversity; regulation of environmental flows) can be done with models that link the ecological with the animal-herder system. A complicating factor in this is the fact that rangelands are often common property resources (e.g. Gebremedhin et al., 2004), shared by several herds of different farms. Moreover, landscape and vegetation structure are heterogeneous, including multi-species trees and shrubs used for browsing or cut-and-carry of foliage. Hence, biomass availability and quality vary in space and time, making livestock herd mobility a key aspect to capture in a model.

Boone et al. (2011) report on such a modelling exercise where the SAVANNA ecosystem model is linked to a household model (DECUMA), where livestock performance and herd dynamics are simulated based on spatially explicit forage availability that can be accessed by mobile animals. A similar approach is implemented in the TERROIR model used to explore spatial and temporal agro-sylvo-pastoral system dynamics over time in the Senegal groundnut basin (Grillot et al., 2018). These examples illustrate that agent-based models provide a useful framework to link vegetation, animal and management component models (of varying degrees of complexity) and include livestock mobility.

3.2.4 Including feed quality in crop model outputs

Where crop products are used as animal feed, not only their quantity but also their nutritional quality determines animal performance. For forage crops and pastures, as their intended use is to feed animals, the nutritive value is usually dynamically simulated by the (pasture) model, and otherwise assumed based on well-known values. This may be different when information on crop residues or other crop outputs is taken from a crop growth model. Many crop growth models focus on biomass quantity of different crop parts, and in terms of quality, some include just N content.

Animals primarily need energy and proteins to function and be productive, and derive these preferably from easily digestible feed. Depending on the livestock sub-model, the feed-quality parameters needed to predict animal performance vary, but mostly include digestibility, crude protein content and metabolizable energy (ME) content. These are often not available from the existing crop models, so that values from the literature or databases (e.g. ILRI, n.d.) need to be used. For a dynamic coupling between crop and livestock component models, the output of the former should contain a minimum amount of information, such as the N content of the relevant plant parts, allowing the derivation or adjustment of feed-quality parameters. This is preferable over an assignment of fixed feed-quality parameters, because crop nutrient content depends on crop nutrition and other growth conditions, influenced by farm management (e.g. nutrient applications through manure) and environmental conditions. Ideally, crop and pasture models providing temporal predictions of digestibility, protein and energy content can be dynamically linked with ruminant livestock models. However, in most commonly used crop models, significant development is required to do this well.

3.2.5 Managing time scales and carry-over effects

Crop–livestock models combine the different time scales at which the key processes operate that drive farm component functioning. The most obvious example is the time frame within which the crop, pasture and animal components produce outputs and interact. Crop production is evaluated on a seasonal basis at harvest, while pastures are harvested or grazed multiple times and animal performance is expressed per lifetime, calving interval or lactation period. Further, discrete events interact with continuous processes over time, such as the use of seasonally harvested crop biomass to sustain animals' daily feed requirements. Another example is the regular addition of manure to a storage place, where gradual decomposition drives nutrient and carbon dynamics, and from where at a specific point in time, manure is applied to a field, where it gradually decomposes and influences soil-nutrient and carbon dynamics. Hence, allowing component models to operate and exchange information at a common time scale is critical. This often requires translation or compilation of outputs from one model (e.g. a crop model at a daily time step) to another (e.g. a livestock enterprise model requiring monthly forage input data).

Besides the differences in relevant time scales, crop models also differ fundamentally from pasture, livestock and whole-farm models in terms of the importance of start-up and carry-over effects. Starting conditions (e.g. soil fertility or moisture content, starting pasture biomass or species composition, herd composition) determine the farm system dynamics for a long period of several seasons, whereas this is usually limited to one season for annual crops, unless crop rotations or sequences are considered. This start-up issue is usually dealt with by starting with a stabilization period before the actual simulation run (e.g. Ghahramani and Moore, 2016). In line with this, so-called carry-over or cascading effects (e.g. Moore and Ghahramani, 2013) describe that what happens in one component model cascades to other components and, due to feedback mechanisms, potentially evolves to an unrealistic vicious or virtuous cycle. Likewise, errors may be propagated and reinforced. For example, an inaccurate estimate of forage quality from a crop model can have repercussions on livestock productivity and manure quality, with subsequent reduced soil-nutrient availability and crop productivity. Understanding model sensitivity helps to minimize such errors by paying attention to accurate estimation of key parameters and processes.

4 Crop-livestock model applications

Mixed crop-livestock models are used to answer different types of research questions, explore practical challenges and design improved systems. In general, simulation modelling enables understanding and quantifying the relationships and trade-offs between farm inputs, farm management, production and environmental effects. They are also powerful tools to expand the range of environmental, management and genotype conditions and combinations, which are limited in experimental set-ups.

The theoretical considerations of Section 2 and 3 clearly showed that mixed crop-livestock models can become very complex if all processes governing component dynamics and interactions are to be represented. In practice, modelling studies limit this complexity by focussing on representing the processes and interactions that are important for the question at hand and implementing pragmatic simplifications for other processes and interactions. This is justified if the conclusions drawn from the study take account of the implications of these simplifications, for which a sound understanding of the system behaviour is required.

The following two case studies from contrasting environments illustrate mixed crop-livestock modelling applications. Both link crop and pasture models with a livestock model and conduct multi-year simulations to assess seasonal variability and explore the effects of changes in farm management. Dynamic feedbacks from the livestock to the crop and pasture models are not included. The first case study adds to the limited body of literature on climate change effects on mixed crop-livestock systems, whereas the second case study looks in detail at the risk of feed gaps and how to mitigate them by diversifying the feed-base.

4.1 Case study 1: Simulating the effect of climate change and adaptation options on the livestock components in mixed farming systems of semi-arid Zimbabwe

Smallholder mixed crop-livestock systems predominate in many semi-arid and sub-humid regions of sub-Saharan African, contributing to rural livelihoods through the provision of meat, milk, crop-supporting services such as manure and traction and sociocultural services (Herrero et al., 2010; Tarawali et al., 2011). Whereas the vulnerability of these systems to climate change is widely acknowledged (Descheemaeker et al., 2016; Thornton and Herrero, 2014), uncertainties hinder the identification and implementation of promising adaptation measures. In particular, large knowledge gaps exist on the likely effects of climate change and adaptation options on livestock, with very few studies especially in the smallholder context. Climate change impacts the livestock component of mixed systems through temporal and spatial changes in feed quantity and quality, water resources and disease prevalence (Thornton et al., 2009; Godber and Wall, 2014), and these effects may play out differentially for the different livestock services and functions. Furthermore, in the diverse smallholder context, climate change effects are expected to be felt differently by different farm types.

We describe here a modelling framework for assessing impacts of climate change and improved management on the livestock component of mixed systems in the heterogeneous farm population of the Nkayi district in semi-arid

Zimbabwe. With poor soil fertility and low input use, current average maize yields are at a meagre 0.7 t ha^{-1} and other crops such as sorghum, millet and groundnut yield even less. About 60% of the households keep cattle and/or goats and donkeys, but animal productivity is poor and constrained by high mortality rates (Homann-Kee Tui et al., 2015). Low livestock productivity is partly explained by the poor feed-base, for which communal rangelands provide the major part, followed by low-quality crop residues (Homann-Kee Tui et al., 2013). With more than 76% of the rural population below the poverty line (ZimVAC, 2013) and food self-sufficiency achieved for only 3–10 months per year, rural households are extremely vulnerable to the adverse effects of climate change. This case study is representative for semi-arid southern Africa where the expected increase in temperature and rainfall variability threatens the predominantly rainfed farming systems (Masikati et al., 2015).

The modelling framework is embedded in the Agricultural Model Intercomparison and Improvement Project (AgMIP) Regional Integrated Assessment (RIA) approach, but focusses on the livestock component and its links with the crop and rangeland components of the farming system (Fig. 2). The framework integrates simulated crop and pasture production with farm-level information on cropland allocation, herd composition and soil, crop and livestock management practices, and feeds this into a livestock model. Through the effects on feed availability, the framework simulates the effects of climate

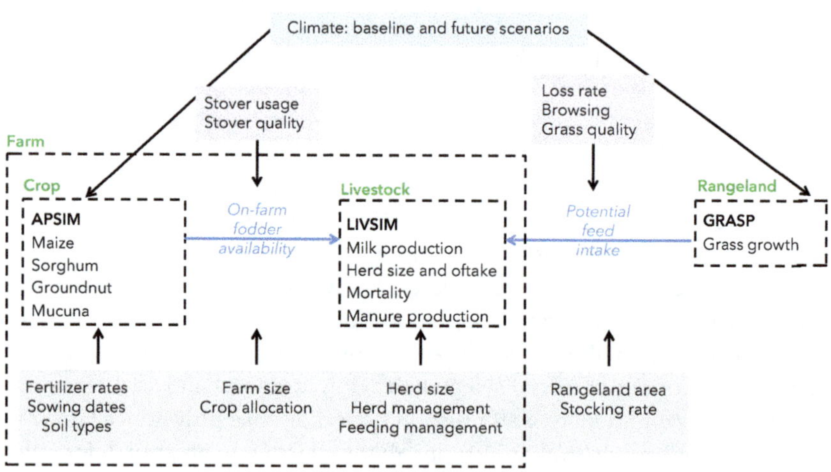

Figure 2 Modelling framework. System components are indicated with green font, with their boundaries as dashed lines. The component models are indicated in capital letters with the model outputs underneath. Parameters are indicated in purple boxes, farm and rangeland characteristics in the grey box, climate scenarios in the blue box. Black arrows indicate information flows between the various inputs and the component models. Interactions between the component models are indicated with blue arrows and italic font.

change and adaptation options on various livestock outputs for all cattle-keeping households in a farm population.

Here, we first give a short summary of the three-component models, followed by a description of the simulated scenarios. For crops, APSIM (Holzworth et al., 2014) was used to simulate grain and stover yields of maize, sorghum, groundnut and the forage crop mucuna (*Mucuna pruriens* (L.) DC.), after calibration and model evaluation (Masikati et al., 2013). Household-specific settings were used for fertilizer rate and sowing window, and three soil types were distinguished. For pasture, the GRASP module (McKeon et al., 2000; Rickert et al., 2000) was implemented in APSIM for the simulation of grass biomass production, and model calibration and evaluation was done based on reported biomass from literature (see Descheemaeker et al., 2018 for details). Feed intake from the rangelands was deduced as follows. Based on the simulated daily grass growth rate and an estimated stocking rate of 2 ha per Tropical Livestock Unit (TLU) (Masikati, 2011; Homann-Kee Tui et al., 2013), the monthly average available grass biomass per TLU was calculated. A 20% monthly loss rate for senescence, trampling and the non-utilized fraction was assumed. In case the grass biomass was insufficient, it was assumed that browsing would meet 20% of the dry matter (DM) requirements (Scoones, 1995; Illius et al., 2000). The latter was estimated by combining the daily ME requirements (roughly 45–65 MJ ME per animal) with grass ME content (ranging from 6.5 to 10.3 MJ kg−1 DM in the dry and the wet season, respectively). The livestock assessment focussed on cattle production, which was simulated with the LIVSIM model (LIVestock SIMulator, Rufino et al., 2009), previously tested for Zimbabwean conditions (Rufino et al., 2011). Based on genetic potential, feed quantity and quality and herd management rules, LIVSIM calculates body weight, milk and manure production for every animal, as well as herd dynamics with a monthly time step. Important model settings for feeding management included that rangeland grazing takes place throughout the year, that crop residues are fed from June to December and that 80% of the stover is available for animal feeding. Model sensibility and sensitivity tests are reported in Descheemaeker et al. (2018). The model framework was run for a baseline and six future climate scenarios, each consisting of 31 years, representing most detrimental ('hot-dry'), least detrimental ('cool-wet') and intermediate scenarios ('middle') for two representative concentration pathways (RCPs) by mid-century (Ruane and McDermid, 2017). Each climate scenario was run with the current farm configurations and with improved management. This consisted of a first 'Package a' of improved soil fertility management (mineral fertilizer at 20 kg N ha^{-1} plus manure at 1 t ha^{-1}) on the cereals, and a second 'Package b' which additionally altered the cropland allocation through a rotation with grain and fodder legumes. Mucuna biomass was partly fed to livestock (70%) and applied as mulch (30%) for soil fertility improvement. For each scenario, simulated values of milk, manure, herd size, mortality and offtake (i.e. sold and/or consumed animals) were assessed for all households, and integrated into a simple economic analysis. Herd size was used as an indication for the value derived from animal draft power.

Climate change affected both the rangeland and the on-farm fodder production, with decreasing average productivity and increasing year-to-year variability most strongly felt in the 'hot-dry' scenarios. Resulting from altered feed production, changes in animal fodder intake were simulated, ranging from a 9% decline compared to the baseline in the 'hot-dry' scenarios to a 5% increase in the 'cool-wet' scenario of RCP8.5. Effects on intake were more strongly felt on farms with larger stocking density. For all scenarios, large within-year fluctuations in body weight were simulated as cows lost weight in the dry season, especially due to a lack of protein. Livestock production was also variable between years, and depended on herd size and stocking density, which adversely affected individual animal productivity. The simulated annual revenue from the livestock component consisted mainly of income from milk, offtake and draft power, with manure contributing less. The aggregated livestock net revenue was most severely affected in the 'hot-dry' scenarios (Fig. 3), with declines ranging between 8% and 43% compared to the baseline. Even in the 'cool-wet' scenarios which resulted in overall positive effects, about 25–30% of the farm population faced (small) negative impacts (Fig. 3). Additionally, both the year-to-year variability and the mortality increased most strongly in the 'hot-dry' scenarios, highlighting the increased risk faced by farmers. Climate change was more strongly felt on farms with larger stocking density (Fig. 3), where animals faced more severe dry-season feed gaps. Also between the livestock functions, differences were discerned. For example, milk and offtake, which are influenced by animal productivity, were more strongly impacted by climate change than functions primarily determined by animal numbers, such as manure and revenue from draft power. This is explained by the fact that a decrease in fodder availability immediately affected the energy and protein available for producing milk, gestation and maintaining animal body condition, whereas the ability of animals to lose and re-gain weight meant that the herd size could be maintained in case of fluctuations in fodder availability.

The adaptation packages increased the on-farm fodder production and quality. In 'package a', annual feed intake increased by about 5%, but it was only with the inclusion of mucuna in 'package b' that the dry-season feed gaps were meaningfully alleviated and crude protein intake increased by 10–26% depending on the farm. These improvements were translated in reduced risk and better livestock revenues, ranging from a 12% to 20% increase compared to the current system with 'package a' and 'b', respectively. Livestock production was less sensitive to climate change under improved farm management, with declines in aggregated net revenue amounting to 28% compared to 43%

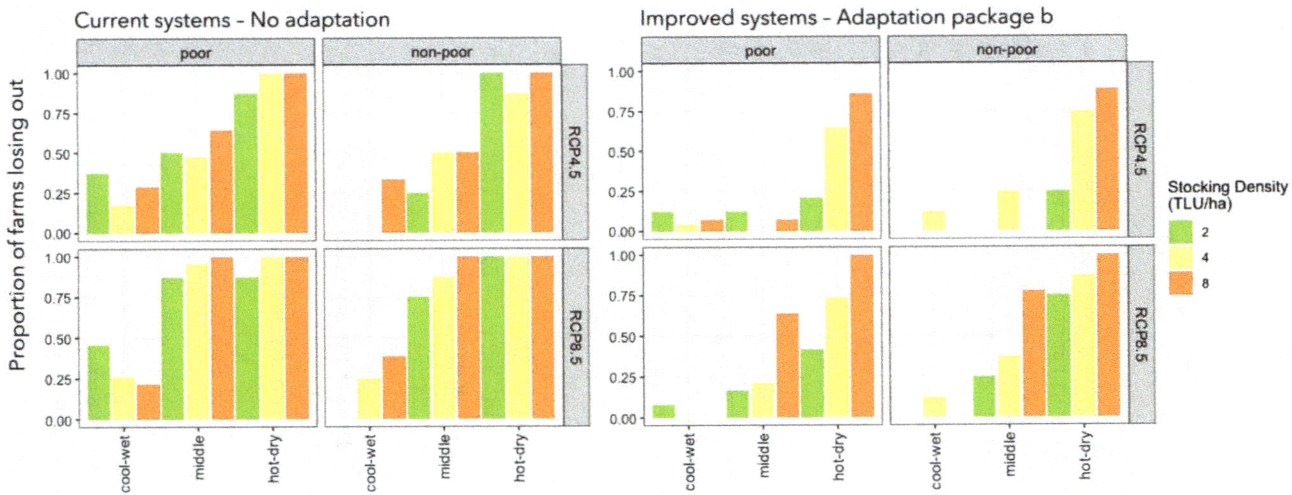

Figure 3 Proportion of farms losing livestock net revenue with different scenarios of climate change, representing relative 'cool-wet', intermediate ('middle') and 'hot-dry' conditions for two representative concentration pathways (RCP4.5 and RCP8.5) by mid-century, as compared to the current climate. Farms are categorized into two resource endowment groups (poor, with less than eight cattle head, and non-poor, with at least eight cattle head) and according to the livestock stocking density. The current system without adaptation (left) is contrasted with an improved system (right) that has adopted an adaptation package consisting of improved soil fertility management and the inclusion of a forage legume in the rotation.

under current management. Whereas most current farms in the population faced lower net revenues with climate change, adopting 'package b' would lower this proportion considerably, depending on the climate scenario and the stocking rate (Fig. 3).

By integrating crop, pasture and livestock models with farm management information, the modelling framework allowed the analysis of the effects of climate change and adaptation options on the livestock component in mixed farming systems. The framework considered the effects on livestock indirectly through the effects on feed composition and availability, thus ignoring direct effects of heat stress and other indirect effects through changes in water resources and pest and diseases. Besides a potential underestimation of climate change impact, this approach also hinges on the accurate representation of the feed-base. Not only crop model accuracy, but also a precise understanding of cropland allocation and feed management practices is crucial in that respect. An important source of uncertainty is the estimation of the feed intake from the rangelands. Here, we used a simple calculation scheme, largely driven by biomass growth rates from a grassland model and stocking rates. More detailed rangeland models could improve accuracy, if data scarcity, which currently precludes model parameterization and testing, would be addressed.

The simulations showed the sensitivity of the livestock component to changes in the climate and the propensity to increased risk, both contributing to the vulnerability of mixed crop-livestock systems. The modelling framework disentangled effects on different livestock performance indicators, which is essential as smallholder farmers keep livestock not only for meat and milk but for a variety of functions. In addition, by capturing the diverse effects of climate and adaptation options on heterogeneous farm populations, the framework can inform decision makers on which options to promote for different farm types to improve climate resilience.

4.2 Case study 2: Simulating diverse feed system options to reduce risks of feed gaps in Australian mixed crop-livestock systems

Most farms across Australia's broad-acre farming zone operate a mixture of grain crop and ruminant livestock enterprises, which provide a range of benefits to the resilience and sustainability of these production systems (Bell and Moore, 2012). These regions experience a highly variable climate both in terms of annual and seasonal rainfall, which presents a major management challenge (Nelson et al., 2010). The high climate variability induces high variability in forage growth and periods when animal demands are either over- or under-supplied and when either feed is wasted or livestock demand is not met. In particular, periods of under-supply of forage (i.e. a feed gap) are major problems requiring either costly supplementary feeding or a reduction in livestock numbers (Moore et al., 2009). These periods also impose high environmental risks of soil and pasture degradation (Thomas et al., 2018). Hence, farmers are often compelled to use conservative stocking rates to reduce their risk of feed deficits. Yet, a range of forage sources can be utilized on mixed crop-livestock farms that could be highly beneficial for managing periods of feed deficit. Evaluating the value of these forage sources is difficult and requires a systems approach capturing climate variability and its effects on forage supply in the farm feed-base, and how this matches against animal demand. Simple economic analyses, such as gross margins or partial budgets or static models, are inadequate to capture the interactions and dynamics of feed supply and demand over time and fail to capture climate-driven risks (Bell et al., 2008). Simulation models provide a useful tool to explore likely production outcomes over a range of climatic conditions and hence analyse risk and returns from alternative systems.

Here, we provide a case study of how various crop and pasture simulation models can be used to predict temporal changes in forage supply at the whole-farm level. By comparing supply with livestock feed demand, the performance and resilience of the different forage combinations were assessed. The balance of forage supply and demand was calculated on the basis of ME (protein was ignored here but could be equally incorporated) (Fig. 4). A database of long-term (1957–2010) simulated monthly production and quality of key forage sources were derived from forage and pasture simulation models; APSIM (oats, Lucerne, dual-purpose crops, crop residue and sown tropical pastures) (Holzworth et al., 2014), GrassGro (annual legume-based pastures, temperate perennial pastures) (Moore et al., 1997) and GRASP (tropical grass pastures) (McKeon et al., 2000). In all cases, simulations were designed to replicate common management practice for these various forage sources including sowing dates (for annual species), fertilizer inputs, grazing/cutting management and intensity (e.g. stocking rates). Locally relevant soils were used from the APSoil database (www.apsim.info/Products/APSoil.aspx) and historical climate information for simulated locations from the SILO database (Jeffrey et al., 2001). The provision of ME from each feed-base element was calculated from the net primary production for that forage and the forage quality in that month based on the average predicted annual cycle in forage quality (Fig. 4). The energy demand of the livestock enterprise was computed monthly based on expected annual cycles of growth, reproduction and lactation and following livestock feeding standards (Freer et al., 1997). For more detail on these calculations, see Bell et al. (2018).

The key feature of this modelling approach is that it computes the balance of energy supplied by the various forages used on the farm and the livestock enterprise on a monthly basis over a wide range of seasonal conditions (Fig. 5). Surplus forage was carried over to subsequent months, but to allow for losses

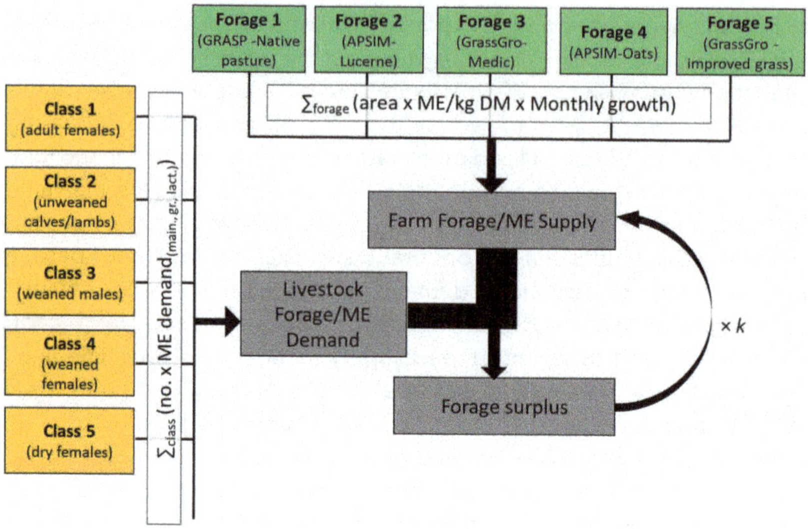

Figure 4 Computation approach for estimating monthly farm energy/forage balance from simulated growth and quality from various crop and pasture models compared to the demand for the whole livestock enterprise. The growth of various forages was simulated separately for a particular location (climate and soil) and the energy content and proportion of land allocated used to calculate the whole-farm energy supply. Similarly, the energy demand for all livestock classes was computed based on their energy requirements for maintenance, growth (based on expected trajectories) and lactation (based on annual reproductive cycle). After deduction of monthly demand, surplus forage was carried over into the subsequent month after correction for losses (the carry-over fraction k was assumed to be 66%).

and senescence only a portion (i.e. 66%) was carried forward (Fig. 4). Using this feed-balance calculation, three complementary statistics of feed gaps were predicted: the frequency of months when feed supply was less than feed demand, the frequency of periods when the farm feed balance was negative (i.e. including carry over) and the frequency of months when the stock of available forage on a farm fell below a minimum threshold of 500 kg DM/ha. This approach enables a prediction of the risk of feed gaps occurring under a range of combinations of livestock production intensity and mixtures of forages contributing to the farm feed-base in different production environments across Australia's mixed farming zone.

The analysis clearly showed that using a diversity of forages with complimentary growth patterns can significantly reduce the frequency and magnitude of feed gaps on mixed farms in Australia. Figure 6 shows predictions of the frequency that livestock demand exceeds monthly fresh forage supply (i.e. carry-over of forages is needed) or when edible forage availability falls below a threshold of 500 kg/ha (i.e. when livestock production

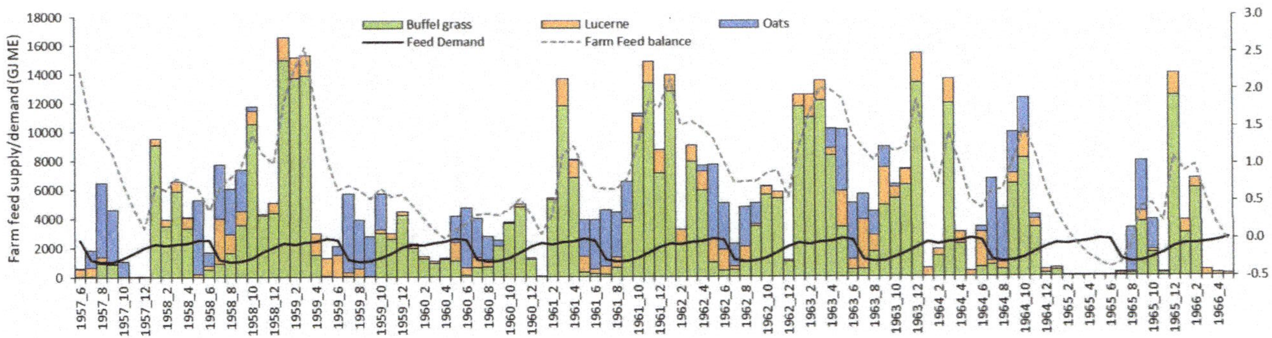

Figure 5 Illustration of predicted farm feed supply from various sources (bars), forage demand (solid black line) and farm feed balance (grey dashed line) in terms of metabolizable energy calculated on a monthly basis over a 10-year period showing periods of forage excess and deficit (when line exceeds bars). Predictions are for a self-replacing beef cattle breeding enterprise of 500 cows producing weaner calves for sale in June at Roma Queensland with a Buffel-grass dominated feed-base (75% area) complimented by lucerne and forage oats (12% area each) (as in Fig. 6c and d).

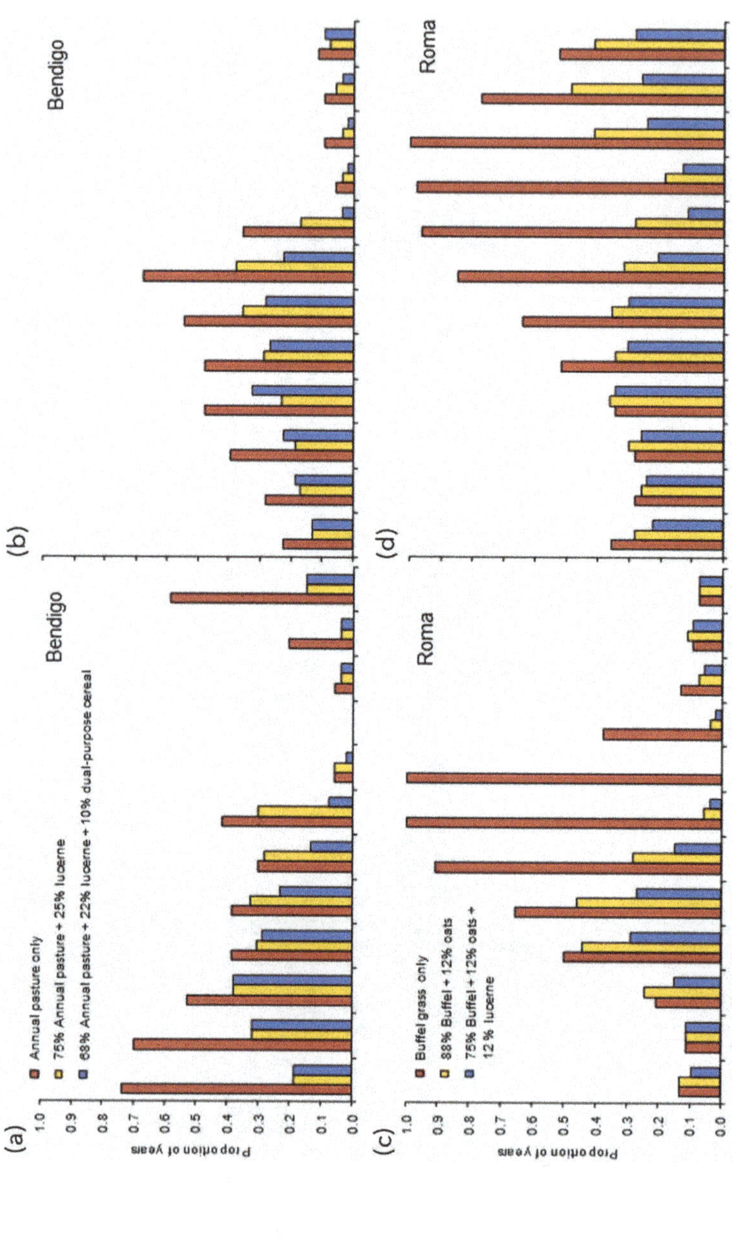

Figure 6 Frequency of monthly energy demand exceeding fresh supply (left) or when available forage falls below a threshold of 500 kg DM/ha (right) for common regional livestock enterprises (1957–2007) at two locations spanning Australia's crop-livestock zone. Red bars are for a feed-base consisting of only a base pasture at each location and the yellow and blue bars are when other complementary forage sources (Lucerne, oats, dual-purpose crops) are integrated as a portion of the feed-base.

would be reduced) for two contrasting locations and livestock-forage systems (Roma, Queensland and Bendigo, Victoria). In both situations, a system relying entirely on a single pasture type experiences periods of acute feed shortage. At Bendigo, annual pastures with a strong winter-dominated growth pattern have limited fresh growth from December to June (Fig. 6a), and this results in frequent farm feed deficits during late autumn (April–May) and early winter (June–July) (Fig. 6b). Replacing 25% of this annual pasture with a perennial pasture, such as lucerne, would provide significant growth over summer and thus strongly reduce the frequency of farm feed deficits during autumn. Further, the addition of a dual-purpose crop able to be grazed during early winter (May–July) would further reduce the frequency of feed gaps at this time. At Roma, the tropical grass (buffel) has limited growth during winter (May-September) leading to regular feed deficits during this period. The addition of a forage oat and/or Lucerne to provide more reliable forage supply during autumn and winter greatly reduces the frequency of feed deficits at this location (Fig. 6d).

This fairly simple modelling approach shows how using combinations of complementary forage sources can be used to strategically mitigate risks of feed gaps on mixed farms. Combining crop, pasture and a simple animal model allows one to demonstrate that forage supplied at different times of the year has different values, whereas it is often difficult to demonstrate or quantify this with field experiments. Critical to this is the capacity to capture the temporal variation in forage production and quality over time. Overall, such analysis shows that farm forage systems can be designed to maintain higher farm stocking rates without increasing risk of feed gaps.

5 Conclusion

Crop–livestock farming systems integrate various bio-physical components with farm decision-making. Whole-farm system models help to understand the complexity arising from this integration and are useful tools to explore the effects of interventions and to design improved systems. Usually mixed crop–livestock models are assembled from separate component models, which come in a huge diversity depending on the described processes and the level of detail with which they are captured. The interactions between farm components strongly determine whole-farm model functioning and performance, and often present modelling challenges to represent them. Key processes governing these interactions include grazing, manure dynamics and animal mobility, as well as the representation of feed quality in crop model outputs and the management of time scales and carry-over effects. Representing all farm components and their interactions at a high level of detail, may render crop–livestock models very complex, thus impairing the user's understanding. Yet, depending on the research or application question, pragmatic solutions exist to capture certain processes to the required degree of complexity. These can be implemented based on a sound knowledge of their implications on system feedbacks and behaviour.

6 References

AFRC. 1993. Energy and protein requirements of ruminants. An Advisory Manual Prepared by the AFRC Technical Committee on Response to Nutrients. CAB Publishing International, Wallingford, UK.

Amole, T. A., Zijlstra, M., Descheemaeker, K., Ayantunde, A. A. and Duncan, A. J. 2017. Assessment of lifetime performance of small ruminants under different feeding systems. *Animal* 11(5), 881–9. doi:10.1017/S1751731116002676.

An, L. 2012. Modeling human decisions in coupled human and natural systems: review of agent-based models. *Ecological Modelling* 229, 25–36. doi:10.1016/j.ecolmodel.2011.07.010.

Angus, J. F., Kirkegaard, J. A. and Peoples, M. B. 2001. Rotation, sequence and phase: research on crop and pasture sequences. *Proceedings 10th Australian Agronomy Conference*. Australian Society of Agronomy/The Regional Institute, Gosford, NSW. Available at: www.regional.org.au/au/asa/2001/plenary/4/angus.htm#TopOfPage.

Baldwin, R. L. 1995. *Modeling Ruminant Digestion and Metabolism*. Chapman & Hall, London.

Bell, L. W. and Moore, A. D. 2012. Integrated crop–livestock systems in Australian agriculture: trends, drivers and implications. *Agricultural Systems* 111, 1–12. doi:10.1016/j.agsy.2012.04.003.

Bell, L. W., Robertson, M. J., Revell, D. K., Lilley, J. M. and Moore, A. D. 2008. Approaches for assessing some attributes of feed-base systems in mixed farming enterprises. *Australian Journal of Experimental Agriculture* 48(7), 789–98. doi:10.1071/EA07421.

Bell, L. W., Hargreaves, J. N. G., Lawes, R. A. and Robertson, M. J. 2009. Sacrificial grazing of wheat crops - identifying tactics and opportunities in Western Australia's grain-belt using simulation approaches. *Animal Production Science* 49(10), 797–806. doi:10.1071/AN09014.

Bell, L. W., Kirkegaard, J. A., Swan, A., Hunt, J. R., Huth, N. I. and Fettell, N. A. 2011. Impacts of soil damage by grazing livestock on crop productivity. *Soil and Tillage Research* 113(1), 19–29. doi:10.1016/j.still.2011.02.003.

Bell, L. W., Moore, A. D. and Thomas, D. T. 2018. Integrating diverse forage sources reduces feed gaps on mixed crop-livestock farms. *Animal* 12(9), 1967–80. doi:10.1017/S1751731117003196.

Berntsen, J., Petersen, B. M., Jacobsen, B. H., Olesen, J. E. and Hutchings, N. J. 2003. Evaluating nitrogen taxation scenarios using the dynamic whole farm simulation model FASSET. *Agricultural Systems* 76(3), 817–39. doi:10.1016/S0308-521X(02)00111-7.

Bilotta, G. S., Brazier, R. E. and Haygarth, P. M. 2007. The impacts of grazing animals on the quality of soils, vegetation, and surface waters in intensively managed grasslands. *Advances in Agronomy* 94, 237–80. doi:10.1016/S0065-2113(06)94006-1.

Blackburn, H. D. and Cartwright, T. C. 1987. Description and validation of the Texas A&M sheep simulation model. *Journal of Animal Science* 65(2), 373-86. doi:10.2527/jas1987.652373x.

Blümmel, M., Zerbini, E., Reddy, B. V. S., Hash, C. T., Bidinger, F. and Khan, A. A. 2003. Improving the production and utilization of sorghum and pearl millet as livestock feed: progress towards dual-purpose genotypes. *Field Crops Research* 84(1-2), 143-58. doi:10.1016/S0378-4290(03)00146-1.

Boone, R. B., Galvin, K. A., BurnSilver, S. B., Thornton, P. K., Ojima, D. S. and Jawson, J. R. 2011. Using coupled simulation models to link pastoral decision making and ecosystem services. *Ecology and Society* 16(2), 6. doi:10.5751/ES-04035-160206.

Boone, R. B., Conant, R. T., Sircely, J., Thornton, P. K. and Herrero, M. 2018. Climate change impacts on selected global rangeland ecosystem services. *Global Change Biology* 24(3), 1382-93. doi:10.1111/gcb.13995.

Brun-Lafleur, L., Cutullic, E., Faverdin, P., Delaby, L. and Disenhaus, C. 2013. An individual reproduction model sensitive to milk yield and body condition in Holstein dairy cows. *Animal* 7(8), 1332-43. doi:10.1017/S1751731113000335.

Chadwick, D., Sommer, S., Thorman, R., Fangueiro, D., Cardenas, L., Amon, B. and Misselbrook, T. 2011. Manure management: implications for greenhouse gas emissions. *Animal Feed Science and Technology* 166-167, 514-31. doi:10.1016/j.anifeedsci.2011.04.036.

Coucheney, E., Buis, S., Launay, M., Constantin, J., Mary, B., García de Cortázar-Atauri, I., Ripoche, D., Beaudoin, N., Ruget, F., Andrianarisoa, K. S., Le Bas, C., Justes, E. and Léonard, J. 2015. Accuracy, robustness and behavior of the STICS soil-crop model for plant, water and nitrogen outputs: evaluation over a wide range of agro-environmental conditions in France. *Environmental Modelling and Software* 64, 177-90. doi:10.1016/j.envsoft.2014.11.024.

Coughenour, M. B. 1992. Spatial modeling and landscape characterization of an African pastoral ecosystem: a prototype model and its potential use for monitoring drought. In: McKenzie, D. H., Hyatt, D. E. and McDonald, V. J. (Eds). *Ecological Indicators*. Springer, New York, pp. 787-810.

Cros, M. J., Duru, M., Garcia, F. and Martin-Clouaire, R. 2003. A biophysical dairy farm model to evaluate rotational grazing management strategies. *Agronomie* 23(2), 105-22. doi:10.1051/agro:2002071.

Danfaer, A., Huhtanen, P., Udén, P., Sveinbjörnsson, J. and Volden, H. 2006. The Nordic dairy cow model, Karoline - Description. In: Kebreab, E., Dijkstra, J., Bannink, A., Gerrits, W. J. J. and France, J. (Eds), *Nutrient Digestion and Utilization in Farm Animals Modeling Approaches*. CABI Publishing, USA, pp. 383-406.

Descheemaeker, K., Oosting, S. J., Homann Kee-Tui, S., Masikati, P., Falconnier, G. N. and Giller, K. E. 2016. Climate change adaptation and mitigation in smallholder crop-livestock systems in sub-Saharan Africa: a call for integrated impact assessments. *Regional Environmental Change* 16(8), 2331-43. doi:10.1007/s10113-016-0957-8.

Descheemaeker, K., Zijlstra, M., Masikati, P., Crespo, O. and Homann-Kee Tui, S. 2018. Effects of climate change and adaptation on the livestock component of mixed farming systems: a modelling study from semi-arid Zimbabwe. *Agricultural Systems* 159, 282-95. doi:10.1016/j.agsy.2017.05.004.

Donnelly, J. R., Freer, M., Salmon, L., Moore, A. D., Simpson, R. J., Dove, H. and Bolger, T. P. 2002. Evolution of the GRAZPLAN decision support tools and adoption by the grazing industry in temperate Australia. *Agricultural Systems* 74(1), 115-39. doi:10.1016/S0308-521X(02)00024-0.

Dove, H. and Kirkegaard, J. 2014. Using dual-purpose crops in sheep-grazing systems. *Journal of the Science of Food and Agriculture* 94(7), 1276-83. doi:10.1002/jsfa.6527.

Duru, M., Adam, M., Cruz, P., Martin, G., Ansquer, P., Ducourtieux, C., Jouany, C., Theau, J. P. and Viegas, J. 2009. Modelling above-ground herbage mass for a wide range of grassland community types. *Ecological Modelling* 220(2), 209-25. doi:10.1016/j.ecolmodel.2008.09.015.

Falconnier, G. N., Descheemaeker, K., van Mourik, T. A. V. and Giller, K. E. 2016. Unravelling the causes of variability in crop yields and treatment responses for better tailoring of options for sustainable intensification in southern Mali. *Field Crops Research* 187, 113-26. doi:10.1016/j.fcr.2015.12.015.

Franzluebbers, A. J., Lemaire, G., de Faccio Carvalho, P. C., Sulc, R. M. and Dedieu, B. 2014a. Toward agricultural sustainability through integrated crop-livestock systems: environmental outcomes. *Agriculture, Ecosystems and Environment* 190, 1-3. doi:10.1016/j.agee.2014.04.028.

Franzluebbers, A. J., Sawchik, J. and Taboada, M. A. 2014b. Agronomic and environmental impacts of pasture-crop rotations in temperate North and South America. *Agriculture, Ecosystems and Environment* 190, 18-26. doi:10.1016/j.agee.2013.09.017.

Freer, M., Moore, A. D. and Donnelly, J. R. 1997. GRAZPLAN: decision support systems for Australian grazing enterprises – II. The animal biology model for feed intake, production and reproduction and the GrazFeed DSS. *Agricultural Systems* 54(1), 77-126. doi:10.1016/S0308-521X(96)00045-5.

Gaughan, J. B., Mader, T. L. and Gebremedhin, K. G. 2012. Rethinking heat index tools for livestock. In: Collier, R. J. and Collier, J. L. (Eds), *Environmental Physiology of Livestock*. Wiley- Blackwell, Chichester, pp. 243-65.

Gebremedhin, B., Pender, J. and Tesfay, G. 2004. Collective action for grazing land management in crop-livestock mixed systems in the highlands of northern Ethiopia. *Agricultural Systems* 82(3), 273-90. doi:10.1016/j.agsy.2004.07.004.

Ghahramani, A. and Moore, A. D. 2016. Impact of climate changes on existing crop-livestock farming systems. *Agricultural Systems* 146, 142-55. doi:10.1016/j.agsy.2016.05.011.

Godber, O. F. and Wall, R. 2014. Livestock and food security: vulnerability to population growth and climate change. *Global Change Biology* 20(10), 3092-102. doi:10.1111/gcb.12589.

Grillot, M., Vayssières, J. and Masse, D. 2018. Agent-based modelling as a time machine to assess nutrient cycling reorganization during past agrarian transitions in West Africa. *Agricultural Systems* 164, 133-51. doi:10.1016/j.agsy.2018.04.008.

Guerrin, F. 2001. MAGMA: a simulation model to help manage animal wastes at the farm level. *Computers and Electronics in Agriculture* 33(1), 35-54. doi:10.1016/S0168-1699(01)00171-5.

Guimarães, V. P., Tedeschi, L. O. and Rodrigues, M. T. 2009. Development of a mathematical model to study the impacts of production and management policies on the herd dynamics and profitability of dairy goats. *Agricultural Systems* 101(3), 186-96. doi:10.1016/j.agsy.2009.05.007.

Hanigan, M. D., Palliser, C. C. and Gregorini, P. 2009. Altering the representation of hormones and adding consideration of gestational metabolism in a metabolic cow model reduced prediction errors. *Journal of Dairy Science* 92(10), 5043–56. doi:10.3168/jds.2008-1922.

Harrison, M. T., Evans, J. R., Dove, H. and Moore, A. D. 2011. Recovery dynamics of rainfed winter wheat after livestock grazing 1. Growth rates, grain yields, soil water use and water-use efficiency. *Crop and Pasture Science* 62(11), 947–59. doi:10.1071/CP11234.

Herrero, M., Thornton, P. K., Notenbaert, A. M., Wood, S., Msangi, S., Freeman, H. A., Bossio, D., Dixon, J., Peters, M., van de Steeg, J., Lynam, J., Parthasarathy Rao, P., Macmillan, S., Gerard, B., McDermott, J., Seré, C. and Rosegrant, M. 2010. Smart investments in sustainable food production. Revisiting mixed crop-livestock systems. *Science* 327(5967), 822-5. doi:10.1126/science.1183725.

Hidy, D., Barcza, Z., Marjanović, H., Ostrogović Sever, M. Z., Dobor, L., Gelybó, G., Fodor, N., Pintér, K., Churkina, G., Running, S., Thornton, P., Bellocchi, G., Haszpra, L., Horváth, F., Suyker, A. and Nagy, Z. 2016. Terrestrial ecosystem process model Biome-BGCMuSo v4.0: summary of improvements and new modeling possibilities. *Geoscientific Model Development* 9(12), 4405–37. doi:10.5194/gmd-9-4405-2016.

Holzworth, D. P., Huth, N. I., deVoil, P. G., Zurcher, E. J., Herrmann, N. I., McLean, G., Chenu, K., van Oosterom, E. J., Snow, V., Murphy, C., Moore, A. D., Brown, H., Whish, J. P. M., Verrall, S., Fainges, J., Bell, L. W., Peake, A. S., Poulton, P. L., Hochman, Z., Thorburn, P. J., Gaydon, D. S., Dalgliesh, N. P., Rodriguez, D., Cox, H., Chapman, S., Doherty, A., Teixeira, E., Sharp, J., Cichota, R., Vogeler, I., Li, F. Y., Wang, E., Hammer, G. L., Robertson, M. J., Dimes, J. P., Whitbread, A. M., Hunt, J., van Rees, H., McClelland, T., Carberry, P. S., Hargreaves, J. N. G., MacLeod, N., McDonald, C., Harsdorf, J., Wedgwood, S. and Keating, B. A. 2014. APSIM – Evolution towards a new generation of agricultural systems simulation. *Environmental Modelling and Software* 62, 327–50. doi:10.1016/j.envsoft.2014.07.009.

Homann-Kee Tui, S., Blümmel, M., Valbuena, D., Chirima, A., Masikati, P., van Rooyen, A. F. and Kassie, G. T. 2013. Assessing the potential of dual-purpose maize in southern Africa: a multi-level approach. *Field Crops Research* 153, 37–51. doi:10.1016/j.fcr.2013.07.002.

Homann-Kee Tui, S., Valbuena, D., Masikati, P., Descheemaeker, K., Nyamangara, J., Claessens, L., Erenstein, O., van Rooyen, A. and Nkomboni, D. 2015. Economic trade-offs of biomass use in crop-livestock systems: exploring more sustainable options in semi-arid Zimbabwe. *Agricultural Systems* 134, 48–60. doi:10.1016/j.agsy.2014.06.009.

Illius, A. W., Magadzire, Z. and Mukungurutse, E. 2000. Browse production and utilisation. In: *Proceedings of the 3rd Livestock Production Programme Zimbabwe Projects Workshop*, Ingwe Lodge, Matobo, Zimbabwe, 26–28 September 2000.

ILRI. n.d. sub-Saharan Africa feed composition database. Available at: https://www.ilri.org/feedsdatabase/(accessed on 16 August 2019).

Janssen, S. and van Ittersum, M. K. 2007. Assessing farm innovations and responses to policies: a review of bio-economic farm models. *Agricultural Systems* 94(3), 622–36. doi:10.1016/j.agsy.2007.03.001.

Jarrige, R., Demarquilly, C., Dulphy, J. P., Holden, A., Robelin, J., Beranger, C., Geay, Y., Journet, M., Malterre, C., Micol, D. and Petit, M. 1986. The INRA "Fill Unit" system for predicting the voluntary intake of forage-based diets in ruminants: a review. *Journal of Animal Science* 63(6), 1737–58. doi:10.2527/jas1986.6361737x.

Jeffrey, S. J., Carter, J. O., Moodie, K. B. and Beswick, A. R. 2001. Using spatial interpolation to construct a comprehensive archive of Australian climate data. *Environmental Modelling and Software* 16(4), 309–30. doi:10.1016/S1364-8152(01)00008-1.

Johnson, I. R. 2016. DairyMod and the SGS Pasture Model: a mathematical description of the biophysical model structure. IMJ Consultants, Dorrigo, NSW, Australia. Available at: http://imj.com.au/sgs/(accessed on 15 August 2019).

Jones, J. W., Antle, J. M., Basso, B., Boote, K. J., Conant, R. T., Foster, I., Godfray, H. C. J., Herrero, M., Howitt, R. E., Janssen, S., Keating, B. A., Munoz-Carpena, R., Porter, C. H., Rosenzweig, C. and Wheeler, T. R. 2017. Toward a new generation of agricultural system data, models, and knowledge products: state of agricultural systems science. *Agricultural Systems* 155, 269–88. doi:10.1016/j.agsy.2016.09.021.

Jouven, M., Carrere, P. and Baumont, R. 2006. Model predicting dynamics of biomass, structure and digestibility of herbage in managed permanent pastures. 1. Model description. *Grass and Forage Science* 61(2), 112–24. doi:10.1111/j.1365-2494.2006.00515.x.

Jouven, M., Agabriel, J. and Baumont, R. 2008. A model predicting the seasonal dynamics of intake and production for suckler cows and their calves fed indoors or at pasture. *Animal Feed Science and Technology* 143(1–4), 256–79. doi:10.1016/j.anifeedsci.2007.05.014.

Kebreab, E. 2013. *Sustainable Animal Agriculture*. CAB International, Wallingford, p. 321.

Kipling, R. P., Bannink, A., Bellocchi, G., Dalgaard, T., Fox, N. J., Hutchings, N. J., Kjeldsen, C., Lacetera, N., Sinabell, F., Topp, C. F. E., van Oijen, M., Virkajärvi, P. and Scollan, N. D. 2016. Modeling European ruminant production systems: facing the challenges of climate change. *Agricultural Systems* 147, 24–37. doi:10.1016/j.agsy.2016.05.007.

Lazzarotto, P., Calanca, P. and Fuhrer, J. 2009. Dynamics of grass-clover mixtures-an analysis of the response to management with the PROductive GRASsland simulator (PROGRASS). *Ecological Modelling* 220(5), 703–24. doi:10.1016/j.ecolmodel.2008.11.023.

Lemaire, G., Franzluebbers, A., de Faccio Carvalho, P. C. and Dedieu, B. 2014. Integrated crop-livestock systems: strategies to achieve synergy between agricultural production and environmental quality. *Agriculture, Ecosystems and Environment* 190, 4–8. doi:10.1016/j.agee.2013.08.009.

Lisson, S., MacLeod, N., McDonald, C., Corfield, J., Pengelly, B., Wirajaswadi, L., Rahman, R., Bahar, S., Padjung, R., Razak, N., Puspadi, K., Dahlanuddin, Sutaryono, Y., Saenong, S., Panjaitan, T., Hadiawati, L., Ash, A. and Brennan, L. 2010. A participatory, farming systems approach to improving Bali cattle production in the smallholder crop-livestock systems of Eastern Indonesia. *Agricultural Systems* 103(7), 486-97. doi:10.1016/j.agsy.2010.05.002.

Martin, G., Moraine, M., Ryschawy, J., Magne, M.-A., Asai, M., Sarthou, J.-P., Duru, M. and Therond, O. 2016. Crop-livestock integration beyond the farm level: a review. *Agronomy for Sustainable Development* 36(3), 53. doi:10.1007/s13593-016-0390-x.

Martin, O., Blavy, P., Derks, M., Friggens, N. C. and Blanc, F. 2019. Coupling a reproductive function model to a productive function model to simulate lifetime performance in dairy cows. *Animal* 13(3), 570–9. doi:10.1017/S1751731118001830.

Masikati, P. 2011. Improving the water productivity of integrated crop-livestock systems in the semi-arid tropics of Zimbabwe: ex-ante analysis using simulation modeling. PhD thesis. ZEF, Bonn, p. 154.

Masikati, P., Manschadi, A., van Rooyen, A. and Hargreaves, J. 2013. Maize-mucuna rotation: a technology to improve water productivity in smallholder farming systems. *Agricultural Systems* 123, 62- 70.

Masikati, P., Homann-Kee Tui, S., Descheemaeker, K., Crespo, O., Walker, S., Lennard, C. J., Claessens, L., Gama, A. C., Famba, S., van Rooyen, A. F. and Valdivia, R. O. 2015. Crop-livestock intensification in the face of climate change: exploring opportunities to reduce risk and increase resilience in southern Africa by using an integrated multi-modeling approach. In: Rosenzweig, C. and Hillel, D. (Eds), *Handbook of Climate Change and Agroecosystems: the Agricultural Model Intercomparison and Improvement Project Integrated Crop and Economic Assessments, Part 2*. Imperial College Press, London, pp. 159-98.

McCall, D. G. and Bishop-Hurley, G. J. 2003. A pasture growth model for use in a whole-farm dairy production model. *Agricultural Systems* 76(3), 1183-205. doi:10.1016/S0308-521X(02)00104-X.

McCormick, J. I., Virgona, J. M., Lilley, J. M. and Kirkegaard, J. A. 2015. Evaluating the feasibility of dual-purpose canola in a medium-rainfall zone of south-eastern Australia: a simulation approach. *Crop and Pasture Science* 66(4), 318-15. doi:10.1071/CP13421.

McKeon, G. M., Ash, A. J., Hall, W. B. and Stafford Smith, M. 2000. Simulation of grazing strategies for beef production in north-east Queensland. In: Hammer, G., Nicholls, N. and Mitchell, C. (Eds), *Applications of Seasonal Climate Forecasting in Agricultural and Natural Ecosystems–the Australian Experience*. Kluwer Academic Press, Netherlands, pp. 227-52.

Moll, H. A. J. 2005. Costs and benefits of livestock systems and the role of market and nonmarket relationships. *Agricultural Economics* 32(2), 181-93. doi:10.1111/j.0169-5150.2005.00210.x.

Moore, A. D. 2009. Opportunities and trade-offs in dual-purpose cereals across the southern Australian mixed-farming zone: a modelling study. *Animal Production Science* 49(10), 759-68. doi:10.1071/AN09006.

Moore, A. D. and Ghahramani, A. 2013. Climate change and broadacre livestock production across southern Australia. 1. Impacts of climate change on pasture and livestock productivity, and on sustainable levels of profitability. *Global Change Biology* 19(5), 1440-55. doi:10.1111/gcb.12150.

Moore, A. D., Donnelly, J. R. and Freer, M. 1997. GRAZPLAN: decision support systems for Australian grazing enterprises. III. Pasture growth and soil moisture submodels, and the GrassGro DSS. *Agricultural Systems* 55(4), 535-82. doi:10.1016/S0308-521X(97)00023-1.

Moore, A. D., Salmon, L. and Dove, H. 2004. The whole-farm impact of including dual-purpose winter wheat and forage brassica crops in a grazing system: a simulation analysis. In: Fischer, R. A., Turner, N. C., Angus, J. F., McIntyre, C. L., Robertson, M. J., Borrell, A. K. and Lloyd, D. (Eds), *New Directions for a Diverse Planet. Proceedings 4th International Crop Science Congress*, Brisbane, Australia, 26 September - 1 October 2004. The Regional Institute, Gosford, NSW. Available at: http://www.cropscience.org.au/icsc2004/symposia/6/3/693_mooread.htm.

Moore, A. D., Holzworth, D. P., Herrmann, N. I., Huth, N. I. and Robertson, M. J. 2007. The Common Modelling Protocol: a hierarchical framework for simulation of agricultural and environmental systems. *Agricultural Systems* 95(1-3), 37-48. doi:10.1016/j.agsy.2007.03.006.

Moore, A. D., Bell, L. W. and Revell, D. K. 2009. Feed gaps in mixed farming systems: insights from the Grain & Graze program. *Animal Production Science* 49(10), 736-48. doi:10.1071/AN09010.

Moore, A. D., Holzworth, D. P., Herrmann, N. I., Brown, H. E., de Voil, P. G., Snow, V. O., Zurcher, E. J. and Huth, N. I. 2014. Modelling the manager: representing rule-based management in farming systems simulation models. *Environmental Modelling and Software* 62, 399-410. doi:10.1016/j.envsoft.2014.09.001.

Moraine, M., Duru, M., Nicholas, P., Leterme, P. and Therond, O. 2014. Farming system design for innovative crop-livestock integration in Europe. *Animal* 8(8), 1204-17. doi:10.1017/S1751731114001189.

Morton, J. F. 2007. The impact of climate change on smallholder and subsistence agriculture. *Proceedings of the National Academy of Sciences of the United States of America* 104(50), 19680-5. doi:10.1073/pnas.0701855104.

Moulin, T., Perasso, A. and Gillet, F. 2018. Modelling vegetation dynamics in managed grasslands: responses to drivers depend on species richness. *Ecological Modelling* 374, 22-36. doi:10.1016/j.ecolmodel.2018.02.013.

Nelson, R., Kokic, P., Crimp, S., Meinke, H. and Howden, S. M. 2010. The vulnerability of Australian rural communities to climate variability and change: Part I—conceptualising and measuring vulnerability. *Environmental Science and Policy* 13(1), 8-17. doi:10.1016/j.envsci.2009.09.006.

Ngwira, A. R., Aune, J. B. and Mkwinda, S. 2012. On-farm evaluation of yield and economic benefit of short term maize legume intercropping systems under conservation agriculture in Malawi. *Field Crops Research* 132, 149-57. doi:10.1016/j.fcr.2011.12.014.

NRC. 2001. *Nutrient Requirements of Dairy Cattle* (7th rev. edn.). National Research Council, Washington DC.

Parsons, A. J. and Chapman, D. F. 1999. The principles of pasture growth and utilization. In: Hopkins, A. (Ed.), *Grass: Its Production and Utilization* (3rd edn.). Blackwell Science for the British Grassland Society, Oxford, pp. 31-89.

Parsons, A. J., Schwinning, S. and Carrère, P. 2001. Plant growth functions and possible spatial and temporal scaling errors in models of herbivory. *Grass and Forage Science* 56(1), 21-34. doi:10.1046/j.1365-2494.2001.00243.x.

Parsons, D., Nicholson, C. F., Blake, R. W., Ketterings, Q. M., Ramírez-Aviles, L., Fox, D. G., Tedeschi, L. O. and Cherney, J. H. 2011. Development and evaluation of an integrated simulation model for assessing smallholder crop-livestock production in Yucatán, Mexico. *Agricultural Systems* 104(1), 1-12. doi:10.1016/j.agsy.2010.07.006.

Peyraud, J.-L., Taboada, M. and Delaby, L. 2014. Integrated crop and livestock systems in Western Europe and South America: a review. *European Journal of Agronomy* 57, 31-42. doi:10.1016/j.eja.2014.02.005.

Pittroff, W. and Kothmann, M. M. 2001. Quantitative prediction of feed intake in ruminants: I. Conceptual and mathematical analysis of models for sheep. *Livestock Production Science* 71(2-3), 131-50. doi:10.1016/S0301-6226(01)00216-0.

Richards, J. H. 1993. Physiology of plants recovering from defoliation. In: *Proceedings of the 17th International Grassland Congress 1993, Session: Plant Growth*. Palmerston North, New Zealand, vol. I. pp. 85-94.

Rickert, K. G., Stuth, J. W. and McKeon, G. M. 2000. Modelling pasture and animal production. In: 't Mannetje, L. and Jones, R. M. (Eds), *Field and Laboratory Methods for Grassland and Animal Production Research*. CABI Publishing, New York, pp. 29–66.

Riedo, M., Grub, A., Rosset, M. and Fuhrer, J. 1998. A pasture simulation model for dry matter production, and fluxes of carbon, nitrogen, water and energy. *Ecological Modelling* 105(2–3), 141–83. doi:10.1016/S0304-3800(97)00110-5.

Rodríguez, A., Trapp, J. N., Walker, O. L. and Bernardo, D. J. 1990. A wheat grazing systems model for the US Southern Plains: Part I–Model description and performance. *Agricultural Systems* 33(1), 41–59. doi:10.1016/0308-521X(90)90069-3.

Rodriguez, D., deVoil, P., Power, B., Cox, H., Crimp, S. and Meinke, H. 2011. The intrinsic plasticity of farm businesses and their resilience to change - an Australian example. *Field Crops Research* 124(2), 157–70. doi:10.1016/j.fcr.2011.02.012.

Rotz, C. A., Corson, M. S., Chianese, D. S., Montes, F., Hafner, S. D. and Coiner, C. U. 2013. *The Integrated Farm System Model Reference Manual, Version 4.0*. USDA-Agricultural Research Service, p. 191. Available at: https://www.ars.usda.gov/SP2UserFiles/Place/19020500/Reference%20Manual.pdf. (accessed on 16 August 2019).

Ruane, A. C. and McDermid, S. P. 2017. Selection of a representative subset of global climate models that captures the profile of regional changes for integrated climate impacts assessment. *Earth Perspectives* 4(1), 1. doi:10.1186/s40322-017-0036-4.

Rufino, M. C., Tittonell, P., van Wijk, M. T., Castellanos-Navarrete, A., Delve, R. J., de Ridder, N. and Giller, K. E. 2007. Manure as a key resource within smallholder farming systems: analysing farm-scale nutrient cycling efficiencies with the NUANCES framework. *Livestock Science* 112(3), 273–87. doi:10.1016/j.livsci.2007.09.011.

Rufino, M. C., Herrero, M., van Wijk, M. T., Hemerik, L., de Ridder, N. and Giller, K. E. 2009. Lifetime productivity of dairy cows in smallholder farming systems of the highlands of Central Kenya. *Animal* 3(7), 1044–56. doi:10.1017/S1751731109004248.

Rufino, M. C., Dury, J., Tittonell, P., van Wijk, M. T., Herrero, M., Zingore, S., Mapfumo, P. and Giller, K. E. 2011. Competing use of organic resources village-level interactions between farm types and climate variability in a communal area of NE Zimbabwe. *Agricultural Systems* 104(2), 175–90. doi:10.1016/j.agsy.2010.06.001.

Ryschawy, J., Joannon, A., Choisis, J. P., Gibon, A. and Le Gal, P. Y. 2014. Participative assessment of innovative technical scenarios for enhancing sustainability of French mixed crop-livestock farms. *Agricultural Systems* 129, 1-8. doi:10.1016/j.agsy.2014.05.004.

Ryschawy, J., Martin, G., Moraine, M., Duru, M. and Therond, O. 2017. Designing crop-livestock integration at different levels: Toward new agroecological models? *Nutrient Cycling in Agroecosystems* 108(1), 5-20. doi:10.1007/s10705-016-9815-9.

Schils, R. L. M., Olesen, J. E., del Prado, A. and Soussana, J. F. 2007. A review of farm level modelling approaches for mitigating greenhouse gas emissions from ruminant livestock systems. *Livestock Science* 112(3), 240–51. doi:10.1016/j.livsci.2007.09.005.

Scoones, I. 1995. Exploiting heterogeneity: habitat use by cattle in dryland Zimbabwe. *Journal of Arid Environments* 29(2), 221–37. doi:10.1016/S0140-1963(05)80092-8.

Snow, V. O., Rotz, C. A., Moore, A. D., Martin-Clouaire, R., Johnson, I. R., Hutchings, N. J. and Eckard, R. J. 2014. The challenges - and some solutions - to process-based modelling of grazed agricultural systems. *Environmental Modelling and Software* 62, 420–36. doi:10.1016/j.envsoft.2014.03.009.

Sommer, S. G. and Hutchings, N. J. 2001. Ammonia emission from field applied manure and its reduction–invited paper. *European Journal of Agronomy* 15(1), 1–15. doi:10.1016/S1161-0301(01)00112-5.

Soussana, J. F., Maire, V., Gross, N., Bachelet, B., Pagès, L., Martin, R., Hill, D. and Wirth, C. 2012. Gemini: a grassland model simulating the role of plant traits for community dynamics and ecosystem functioning. Parameterization and evaluation. *Ecological Modelling* 231, 134–45. doi:10.1016/j.ecolmodel.2012.02.002.

Sulc, R. M. and Franzluebbers, A. J. 2014. Exploring integrated crop-livestock systems in different ecoregions of the United States. *European Journal of Agronomy* 57, 21–30. doi:10.1016/j.eja.2013.10.007.

Tarawali, S., Herrero, M., Descheemaeker, K., Grings, E. and Blümmel, M. 2011. Pathways for sustainable development of mixed crop livestock systems: taking a livestock and pro-poor approach. *Livestock Science* 139(1–2), 11–21. doi:10.1016/j.livsci.2011.03.003.

Tedeschi, L. O., Cavalcanti, L. F. L., Fonseca, M. A., Herrero, M. and Thornton, P. K. 2014. The evolution and evaluation of dairy cattle models for predicting milk production: an agricultural model intercomparison and improvement project (AgMIP) for livestock. *Animal Production Science* 54(12), 2052–67. doi:10.1071/AN14620.

Thomas, D. T., Moore, A. D., Norman, H. C. and Revell, C. K. 2015. Small effects of deferment of annual pastures through grazing spring wheat crops in Western Australia can benefit livestock productivity. *Crop and Pasture Science* 66(4), 410–7. doi:10.1071/CP14090.

Thomas, D. T., Moore, A. D., Bell, L. W. and Webb, N. P. 2018. Ground cover, erosion risk and production implications of targeted management practices in Australian mixed farming systems: lessons from the Grain and Graze program. *Agricultural Systems* 162, 123–35. doi:10.1016/j.agsy.2018.02.001.

Thornton, P. K. and Herrero, M. 2001. Integrated crop-livestock simulation models for scenario analysis and impact assessment. *Agricultural Systems* 70(2–3), 581–602. doi:10.1016/S0308-521X(01)00060-9.

Thornton, P. K. and Herrero, M. 2014. Climate change adaptation in mixed crop-livestock systems in developing countries. *Global Food Security* 3(2), 99–107. doi:10.1016/j.gfs.2014.02.002.

Thornton, P. K., van de Steeg, J., Notenbaert, A. and Herrero, M. 2009. The impacts of climate change on livestock and livestock systems in developing countries: a review of what we know and what we need to know. *Agricultural Systems* 101(3), 113–27. doi:10.1016/j.agsy.2009.05.002.

van der Linden, A., van de Ven, G. W. J., Oosting, S. J., van Ittersum, M. K. and De Boer, I. J. M. 2019. LiGAPS-Beef, a mechanistic model to explore potential and feed-limited beef production 1: model description and illustration. *Animal* 13(4), 845–55. doi:10.1017/S1751731118001726.

van Evert, F. K. and Lamaker, E. J. J. 2007. The MODCOM framework for component-based simulation. In: *Farming Systems Design 2007, International Symposium on Methodologies for Integrated Analysis of Farm Production Systems*, 10–12 September, Catania, Italy, vol. 2. Field-farm Scale Design and Improvement, pp. 223–4.

van Wijk, M. T., Rufino, M. C., Enahoro, D., Parsons, D., Silvestri, S., Valdivia, R. O. and Herrero, M. 2014. Farm household models to analyse food security in a changing climate: a review. *Global Food Security* 3(2), 77–84. doi:10.1016/j.gfs.2014.05.001.

Vayssières, J., Guerrin, F., Paillat, J.-M. and Lecomte, P. 2009a. GAMEDE: a global activity model for evaluating the sustainability of dairy enterprises part I - whole-farm dynamic model. *Agricultural Systems* 101(3), 128–38. doi:10.1016/j.agsy.2009.05.001.

Vayssières, J., Bocquier, F. and Lecomte, P. 2009b. GAMEDE: a global activity model for evaluating the sustainability of dairy enterprises. Part II - Interactive simulation of various management strategies with diverse stakeholders. *Agricultural Systems* 101(3), 139–51. doi:10.1016/j.agsy.2009.05.006.

Vetter, S. 2005. Rangelands at equilibrium and non-equilibrium: recent developments in the debate. *Journal of Arid Environments* 62(2), 321–41. doi:10.1016/j.jaridenv.2004.11.015.

Villalba, D., Ripoll, G., Ruiz, R. and Bernués, A. 2010. Long-term stochastic simulation of mountain beef cattle herds under diverse management strategies. *Agricultural Systems* 103(4), 210–20. doi:10.1016/j.agsy.2010.01.003.

Wiegand, T., Wiegand, K. and Pütz, S. 2008. Grassland models. In: Jørgensen, S. E. and Fath, B. D. (Eds), *Encyclopedia of Ecology* (vol. 3). Elsevier, Oxford, pp. 1754–65.

Wu, L., McGechan, M. B., McRoberts, N., Baddeley, J. A. and Watson, C. A. 2007. SPACSYS: integration of a 3D root architecture component to carbon, nitrogen and water cycling- model description. *Ecological Modelling* 200(3-4), 343–59. doi:10.1016/j.ecolmodel.2006.08.010.

Zhang, X. -C., Phillips, W. A., Garbrecht, J. D., Steiner, J. L. and Hunt, L. A. 2008. A wheat grazing model for simulating grain and beef production: Part I–Model Development. *Agronomy Journal* 100(5), 1242–7. doi:10.2134/agronj2007.0372.

ZimVAC (Zimbabwe Vulnerability Assessment Committee). 2013. Rural livelihoods assessment, Harare. Available at: https://reliefweb.int/report/zimbabwe/zimbabwe-vulnerability-assessment-committee-zimvac-2013-rural-livelihoods-assessment (accessed on 17 October 2019).

Zingore, S., Murwira, H. K., Delve, R. J. and Giller, K. E. 2007. Influence of nutrient management strategies on variability of soil fertility, crop yields and nutrient balances on smallholder farms in Zimbabwe. *Agriculture, Ecosystems and Environment* 119(1-2), 112–26. doi:10.1016/j.agee.2006.06.019.

Chapter 10

Integrating economic simulation models with whole-farm system models for *ex ante* technology impact assessment

John M. Antle, Oregon State University, USA

1 Introduction

A major challenge facing the agricultural research community is to evaluate and improve the sustainability of agricultural systems. As the recent report by the National Academy of Sciences (2018) emphasizes, to meet this challenge we must understand and predict the behaviour of complex systems across diverse agro-ecological and socio-economic conditions that cannot be studied using experimental methods, but can be simulated using agricultural system models. Analysis of climate change impact and adaptation is the most obvious example of this type of research, but the same methodological challenges pervade all forward-looking efforts at technology design and evaluation. In this chapter, I refer to this type of analysis as *ex ante* technology impact assessment.

The key analytical challenge to this type of analysis is to predict the behaviour of agricultural systems in the multiple dimensions that are relevant

http://dx.doi.org/10.19103/AS.2019.0061.14

to their economic, environmental and social sustainability, under recent historically observed conditions and when they are modified to improve their performance, for example, by adopting improved crop varieties and associated management practices. Typically, it is difficult to improve performance in all relevant dimensions and thus changes in the management of agricultural systems involve *trade-offs* that must be evaluated and addressed in the context of agricultural, environmental and social policies. Antle et al. (2014) and Kanter et al. (2018) provide reviews of methods for analysis of agricultural system trade-offs at field, farm and regional scales.

Crop modellers approach modelling farm production systems as a biological process, and typically use experimental data to parameterize and calibrate their models. Economists approach modelling farm production systems from the perspective of the farm decision-maker, typically using observations of actual farm behaviour from farm survey or similar data. The farm decision-maker is assumed to make purposeful choices to manage the farm to achieve one or more well-defined objectives, subject to the constraints imposed by physical and biological processes and available physical, financial, technological and human resources. Bio-physical farm system models, such as the crop and livestock simulation models in APSIM, DSSAT and CROPSYST, can be viewed as a process-engineered representation of the production function. Rather than using process-based engineering functions, economists traditionally have used statistical methods to estimate parameters of production functions. Both approaches have strengths and weaknesses (Antle and Stöckle, 2017; Antle, 2019).

An advantage of process-based models is that they embody relationships that are invariant to a particular site or sample of data. This feature means that they can be used to solve one of the key challenges in *ex ante* technology impact assessment, that is to predict 'out of sample', or to predict outside the range of historically observed conditions. Of course, this will only be true to the degree that the models accurately represent fundamental processes, and it is clear from recent research that there is still a substantial gap between the capability of currently available models and reality. But even when they can reliably predict crop productivity, process-based models have other limitations, notably, they are complex and data-intensive, and as-yet available models do not embody some important elements of production systems such as human labour, the management goals and ability of the decision-maker and the effects of pests and diseases and pest management. Moreover, models are not available for some important crops and for complex rotations and inter-crops, and are not well integrated with livestock models. As a result, process-based models tend to produce biased predictions of actual crop production or productivity (yield), and users typically 'bias-adjust' the models to predict observed yields, but this limits their validity for out-of-sample prediction.

Strictly empirical, statistical production functions used in many economic studies have the advantage of being able to incorporate the full range of management inputs, and with sufficiently good data typically will do a better job of predicting within-sample for the simple reason that regression-based statistical models provide unbiased estimates of the sample mean. But these models have many limitations. For example, because they do not incorporate the processes that relate soil properties, weather and nutrients to crop growth, they may provide poor predictions of how yield responds to these factors. Additionally, effects of as-yet unobserved environmental changes, such as an increase in atmospheric CO_2 or ozone, cannot be reliably estimated using observational data. Likewise, they are unable to predict threshold effects outside the range of observed data, such as the effect of temperature extremes or droughts that are not represented in observed data. Available data rarely span sufficiently long time periods to allow reliable statistical estimates of the dynamic properties of production systems.

In this chapter, I discuss methods to combine process-based and economic production models to overcome some of the limitations of each approach with the goal of producing more useful models for technology impact assessment. I also discuss the data challenges that must be addressed to achieve more useful models for on-farm decision-making as well as for broader technology assessment and policy analysis. I define the scope of whole-farm production system modelling in the broader context of what economists call household production models. In Section 2, I introduce the economic impact evaluation paradigm, and then use that paradigm to discuss how the integration of process-based models with economic models can help address key methodological challenges. In Section 3, I discuss methods that have been developed to link crop simulation models to economic models in what I refer to as hybrid structural models. I also provide an example of how crop and economic simulation models can be linked to model new systems in new environments for *ex ante* technology impact assessment. I conclude with a discussion of the data challenge, and point to sources of further information on this topic.

2 Modelling whole-farm systems for sustainability research

Figure 1 presents the possible components and linkages in a household production model that includes an agricultural production subsystem (within the dashed oval on the left-hand side) and a household production subsystem (the dashed oval on the right-hand side). Economists describe this type of household production model as *non-separable* between production and other household decisions, meaning for example that outputs of the

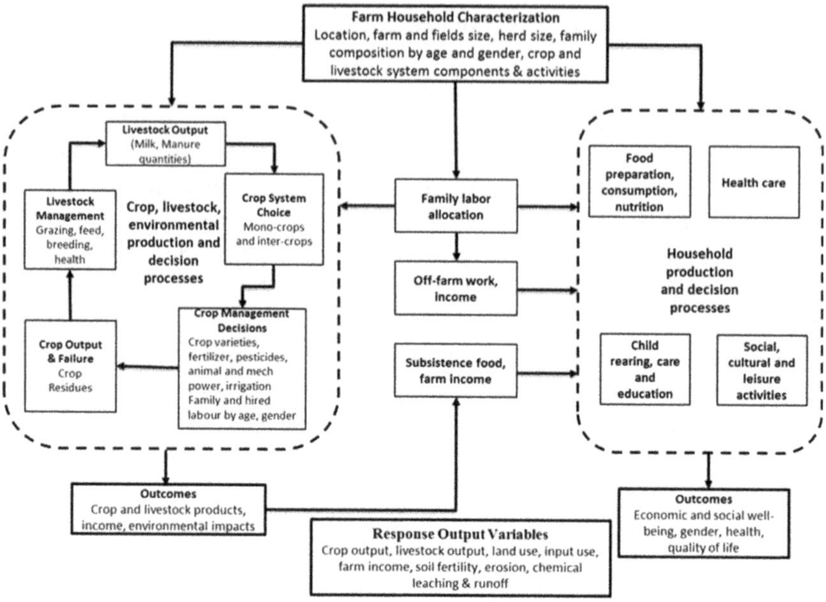

Figure 1 Components and linkages in household production model. Source: adapted from Jones et al. (2017).

agricultural production system enter into the household's other activities involving food consumption and nutrition, health, child-rearing and other activities relating to household members' well-being. Additionally, farm livelihood activities involving household labour and assets influence the provision of farm labour and availability of financial resources for input purchases, as well as saving and investment in land and other forms of capital, and other agricultural production activities. In actuality, these decisions occur over time and are interconnected in complex ways. Under these conditions, the decision-making for the agricultural production system cannot be separated from other household activities and thus should, in principle, be modelled jointly.

However, the data requirements and model complexity involved with non-separable farm household models often make this approach infeasible. Alternatively, the household can be viewed, at least as a first-order approximation, as making farm production decisions to generate income, and other household decisions can be viewed as being based only on household income. In this case, farm management decisions can be separated from other household decisions. In the remainder of this chapter, I will discuss modelling farm production systems with bio-physical and economic components that can be separated in this sense from other household decision-making.

An example of a production system modelling framework is presented in Fig. 2. This figure represents what Antle and Capalbo (2001) describe as an econometric process simulation model. As Fig. 2 shows, this framework links a crop simulation model (or more generally, a whole-farm bio-physical simulation

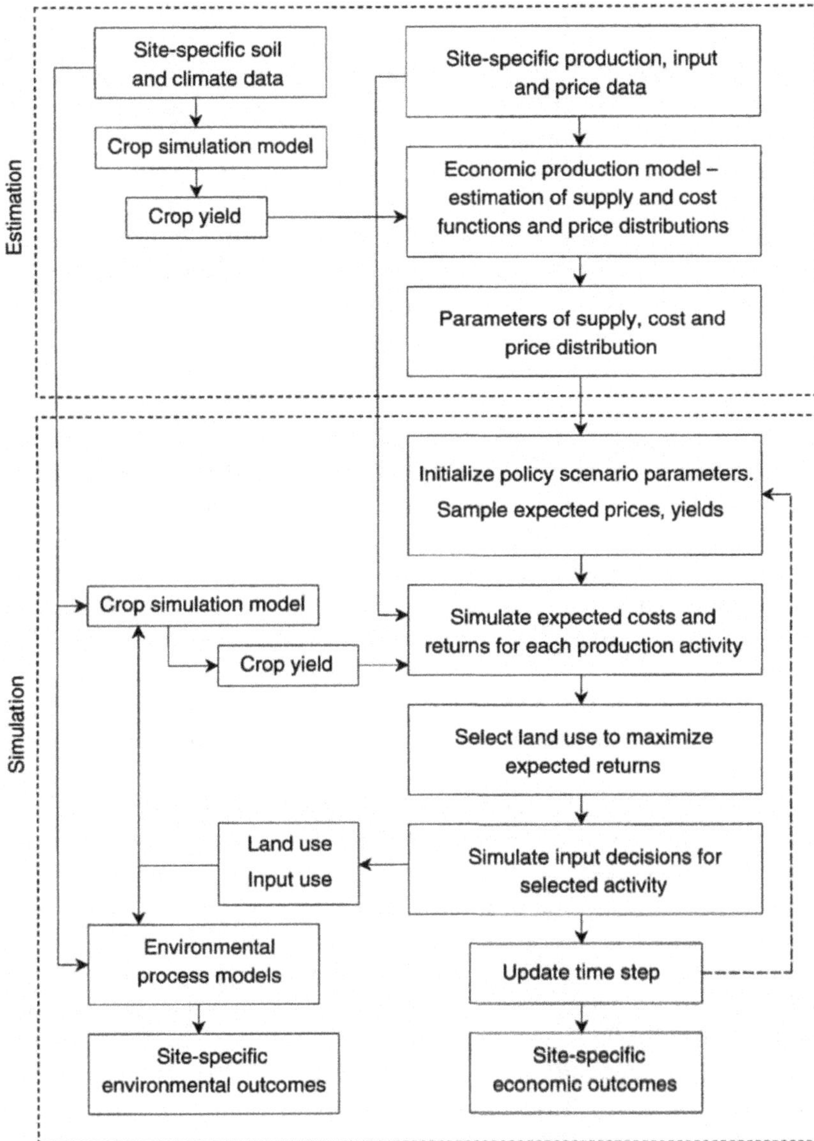

Figure 2 Crop model and economic model components of an econometric process simulation model. Source: adapted from Antle and Capalbo (2001).

model) to an economic simulation model. The economic model represents two types of production decisions: the choice of activity on each land unit (crop choice) and the management of that activity during the growing season (input use). The framework is dynamic and thus also suitable to model interactions across growing seasons, that is crop rotations. Many models in the literature, often referred to as a 'bio-economic' models, are formulated in this way (see van Wijk et al., 2014 for a survey of such models). Many of these models use a 'representative farm' model, and attempt to solve for 'optimal' management decisions. The approach portrayed in Fig. 2 instead uses an econometric model estimated with observations of actual behaviour, and then simulates the model to evaluate the effects of changes in the system, for example the introduction of a new crop variety.

The upper part of Fig. 2 shows that crop models can be used in two ways in this framework. First, the upper part of the figure shows that simulated crop yields can be incorporated in the statistical estimation of the econometric model to represent how soils and climate affect crop productivity and thus affect crop choice and input-use decisions. Second, the lower part of the figure illustrates how the simulation analysis is implemented. Site-specific simulated crop yields are input into the economic model, the economic model determines site-specific land-use and management decisions and these decisions can in turn feed forward into environmental models to estimate environmental impacts (e.g. models of soil erosion, chemical leaching or other ecosystem services). By simulating a representative set of sites across the landscape, this type of modelling can provide landscape-scale predictions of economic and environmental outcomes. If models are available for other household outcomes represented in Fig. 1 such as human health or nutrition, the analyst can also then simulate these population-level social impacts. Thus, this type of framework can be used to evaluate the sustainability of agricultural systems at the landscape scale.

The econometric process simulation approach in Fig. 2 represents a population of farms by simulating a sample of farms and aggregating results to the population level. An alternative modelling approach is to model the population of farm households directly by estimating population parameters rather than trying to model and simulate individual farms (Fig. 3). For example, in this approach, crop models can be used to simulate the spatial distributions of crop yields, and these yield distributions are combined with economic data from the farm population (e.g. farm survey data) to estimate the population mean and variance of economic returns to the actual production system. Scenario analysis is used to modify these distributions to represent how these distributions would change. For example, with the introduction of a new technology such as a new crop variety, a crop model can be used to simulate the yield distribution associated with the current variety (System

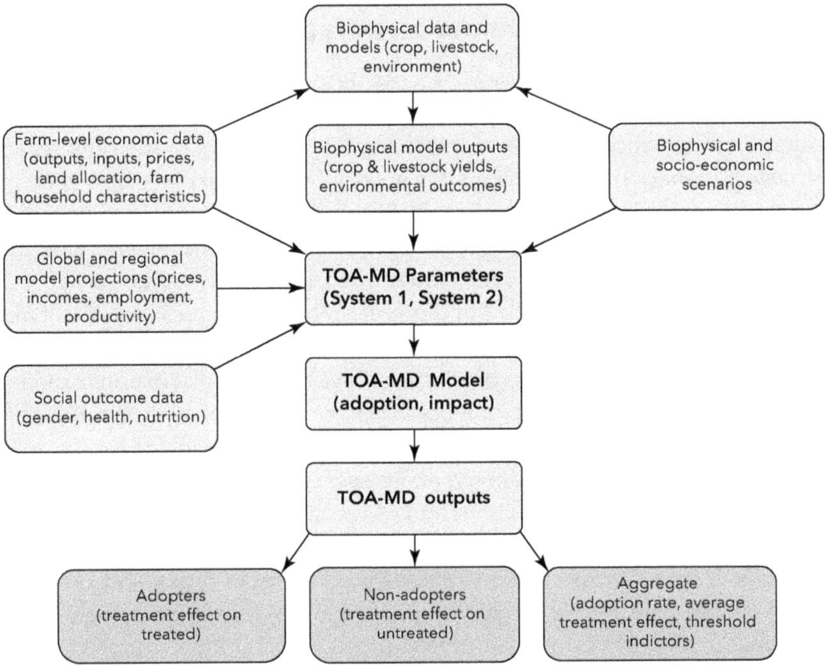

Figure 3 Implementation of the hybrid structural modelling approach using the trade-off analysis model for multidimensional impact assessment (TOA-MD). Source: adapted from Capalbo et al. (2017).

1 in the figure) as well as the new variety (System 2). As discussed below, these yield simulations can be used to construct relative yield distributions that can in turn be incorporated into the economic simulation of technology adoption (i.e. the choice between System 1 and System 2). In the economic analysis, the difference in economic performance between the actual system and the prototype or 'counterfactual' system is used to infer the potential adoption rate of the prototype system and the associated economic impacts. Scenarios can also be used to characterize other changes that would affect the adoption of a new technology, such as changes in prices and policy, or changes in the farm population such as an increase in average farm size. Also, environmental and social data can be incorporated in this approach to represent the changes in the distribution of environmental or social outcomes associated with a change in technology. This type of analysis is based on the statistical concepts that are the foundation of the econometric literature on policy impact evaluation (e.g. see Heckman, 2010; Antle, 2011; Antle et al., 2014). This population-based approach has been developed and implemented as the Tradeoff Analysis Model for Multi-dimensional Impact Assessment (TOA-MD).

3 The *ex ante* impact evaluation paradigm

Section 2 showed how bio-physical process-based models can be incorporated into an economic production system model and more general household models. In this section, I narrow the focus to the modelling of the outputs and associated management decisions for a given crop to explain how they are related to *ex ante* impact evaluation. I will use a stylized representation of an agricultural system to introduce the key methodological issues and how they can be applied to assess the sustainability of agricultural systems.

I define Y as the system output per unit area (kg/ha/time); here I will treat it as a scalar but more generally there can be multiple outputs in the system. Here I assume that outputs are produced individually on distinct units of land, with allocatable variable and fixed inputs (think of the typical farm field with a single crop), but more generally they may be jointly produced crops (i.e. intercrops), and there may be system dynamics associated with crop rotations. Just et al. (1983) provide a useful discussion of the standard economic framework for analysis of farm production with multiple distinct crop or livestock activities. I define X as a vector of choice variables such as input types and quantities (units/ha/time). The farmer's decision-making behaviour is described by the equation:

$$X = X\left[A, \alpha\right], \tag{1}$$

where A represents a set of covariates (such as input and output prices and farm characteristics) and α is the parameter vector that relates A to X. The production function is defined as:

$$Y = Y\left[X, B, \beta\right], \tag{2}$$

where B represents a set of covariates (e.g. soil characteristics, farmer characteristics such as education or experience) and stochastic factors (e.g. weather, pests), and β is the vector of parameters that relate X and B to Y. Combining Eqs. (1) and (2) gives the reduced form:

$$Y = S\left[A, B, \alpha, \beta\right]. \tag{3}$$

Note that X represents management decisions typically made before some of the realizations of B are observable by the farmers. For example, farmers make planting decisions before growing season weather occurs. The structure depends on assumptions made about the decision-making process of the farmers, such as how expectations are formed and the timing of decisions in relation to the realization of elements of B such as rainfall or pest infestations. Thus, in an explicitly dynamic model such as Antle and Hatchett (1986), the model has a recursive stochastic structure. Moreover, Eq. (2) represents output realizations; the farmer's decision-making typically depends on some function of Y, as in a model of expected profit or utility maximization.

We can also provide alternative interpretations of this general formulation. For example, in the conventional micro-economic model, we can interpret Y as profit rather than output, and the expectation of $Y[X, B, \beta]$ as the expected profit equation embedding the production function, output price and cost of production. X is a vector of variable inputs such as seed variety and fertilizer, B represents exogenous variables that can include prices, soil, weather and other biotic and abiotic factors affecting production.

Conventional representations of economic production functions express some elements of A and B as random variation independent of covariates, but some of the factors attributed to the error term are observable in principle. This gives rise to the problem of unobserved heterogeneity that has been intensively studied in the econometrics and statistics literatures. In contrast, weather is typically included in process-based models as an explicit model input. Likewise, with available weather data, weather distributions can be characterized for modelling *ex ante* decision-making, and can be observed and included in the model as an explanatory variable for modelling output realizations, as is typically done in process-based crop models. Similarly, the random variation in the behavioural Eq. (1) across farms or over time could be attributed to differences across individuals in psychological attitudes such as risk aversion, experience or knowledge, but in principle with appropriate data each farmers' risk attitudes, experience or knowledge could be quantified and included explicitly as an explanatory variable.

A key economic question is how the decision Eq. (1) is derived. The answer provided by economic theory derives from the assumption of maximizing behaviour by farm decision-makers. For example, a standard assumption is that X is chosen to maximize expected profit subject to the production function (Eq. (2)), and expected profit is defined as the mathematical expectation of profit. To implement this type of analysis, the elements of B that are unknown to the decision-maker at the time decisions are made, such as weather, are assigned a probability density $\phi(B|\omega)$ where ω represents the parameters of this density function (here treated as a scalar for simplicity). Also, define p as the price of output Y and define v as the price of input X (also treated as a scalar here for simplicity). Then the profit maximization problem is to choose X to maximize:

$$\int \{pY[X, B, \beta] - vX\}\phi(B|\omega)dB$$

Now interpreting Eq. (1) as the solution to this maximization problem, we can see that the parameter vector α is composed of the parameters in β and ω and the prices p and v. If we then substitute this solution for X into Eq. (2) we obtain Eq. (3) which economists call the farm's supply function. In this case, we can see that the supply function depends on output and input prices and other exogenous factors affecting productivity and input use (represented in this example by ω). For example, in addition to expected weather, factors such as the farmer's age, experience and education or factors affecting farmers' access to weather and price information could play a role. Other behavioural assumptions can be made that lead to models with other features. For example, farmers may be risk averse, in which case variables representing their risk attitudes could be included.

Various other extensions of this type of model are possible. An example is a semi-subsistence household production model where Y could be a vector of outcomes that includes farm output or income as well as nutritional outcomes of farm family members that depend on consumption of farm output. Environmental impacts also can be added to the analysis as a function of farm management decisions X and environmental factors captured in B, as portrayed in Figs. 2 and 3. Other generalizations are also possible, for example, to account for system dynamics (Antle and Stoorvogel, 2006).

4 Implications of the economic evaluation paradigm for agricultural system modelling

The *ex ante* technology impact assessment problem is to predict the behaviour of agricultural systems outside the range of historical observation. This problem is similar to the policy evaluation problem studied by economists and statisticians. Heckman (2010) argues that 'economic policy evaluation' can be defined in terms of three types of problems. The first two involve the evaluation of implemented interventions in the environment where they are observed (the problem of internal validity in *ex post* evaluation, P1), and the evaluation of interventions in a different but observable environment (the problem of external validity in *ex post* evaluation, P2). While these two problems are of some interest, the more relevant problem in most contexts is the evaluation of new interventions in environments *never historically observed* (the *ex ante* evaluation problem, P3).

Problem P3 is essentially the problem we face in *ex ante* technology impact assessment. Heckman (2010) argues that in order to solve P3, models are needed with sufficient structure to represent the effects of the intervention not only in the environment in which the parameters are estimated, but also in other environments and responses to new kinds of interventions. In other words, we need models based on parameters that are invariant to the changes being imposed on the system. This aspect is where process-based crop models can be useful because, at least in principle, their structure and parameters should be valid outside the historically observed values.

To illustrate, consider an experiment that randomizes a treatment $T = (0,1)$ among a set of individuals with observable characteristics Z, and let ε represent independent errors, and let γ be a parameter vector. The analyst estimates:

$$Y = R[T, Z, \gamma, \varepsilon]. \tag{4}$$

Interpreting the relevant measure of impact as the change in the dependent variable, the conventional statistical approach to impact evaluation addresses P1 by using randomized experimental data to estimate Eq. (4). This provides an estimate of the average treatment effect equal to the mean of Y for T = 1 minus the mean of Y for T = 0. However, this type of analysis cannot address P2 or P3, because the empirical parameters are specific to the particular experiment used to generate the data. Thus, because this procedure does not identify the underlying structural parameters α and β (see Eqs. (1) and (2)) that relate Y to X, B and A, it follows that γ embeds elements of these variables in a way that is specific to the experiment used to generate the data, and thus cannot be generalized to other environments where X, B and A are different. Additionally, since the behavioural Eq. (1) cannot be estimated with data from a randomized experiment (since T is randomly assigned, not chosen by the decision-maker), it is not possible to identify the relevant range of behaviour and estimate other relevant treatment effects such as the effect on those individuals who choose to adopt a new technology (treatment effect on the treated), or the effect over a policy-relevant range of a behaviour (what Heckman (2010) calls the policy-relevant treatment effect).

In contrast, suppose Eqs. (1) and (2) represent a structure representing the relevant behaviour. To evaluate P2, observations of A and B are needed in the environment that is distinct from the one used to estimate the parameters. For the estimates of α and β to be unbiased and stable across environments, the description of A and B must be correct and observations of them must be accurate. Solving P3 involves evaluation of as-yet unobserved interventions in novel environments. A novel environment can be the same location in the future, as in climate change assessments, or a different location with different conditions, or both. This type of analysis requires predicting A and B into the novel environment. If the new intervention involves a change in technology, the functions in Eqs. (1) and (2) must incorporate the technological change.

5 Evaluating novel and future systems using hybrid structural models

Problem P3 is the most challenging but also most important problem in agricultural systems analysis because it requires, first, parameterization of a model with sufficiently 'deep' structure to represent the changes in the technology (Eq. (2)) and associated behaviour (Eq. (1)); and second, it requires projection of exogenous variables *A* and *B* into the future, into novel environments. The solution to this second problem is the use of participatory scenario methods like the ones that are currently in use for climate impact assessment. This aspect is beyond the scope of this chapter. The reader may consult Valdivia et al. (2015) and Antle et al. (2017a) for applications of scenario methods in agricultural systems modelling used to address this part of *ex ante* assessment. Here, I focus on the development of what I will call *hybrid structural models* that are capable of projecting new agricultural systems into novel environments.

As noted above, Antle and Capalbo (2001) proposed econometric process simulation models as a way to combine mechanistic crop growth models with econometric production models to simulate agricultural systems out of sample, that is to address problems P2 and P3. The structure of an analysis using this type of model is illustrated in Fig. 2. However, to date none of the existing models has accomplished a full integration of bio-physical and economic-behavioural models. There are both conceptual and computational difficulties in implementing this kind of integration, for example crop simulation models typically work on a daily time step, whereas economic decision-making may be made on a longer time step according to operations such as land preparation, planting, irrigation, cultivation and harvest, as well as on longer time horizons involving inter-seasonal dynamics of crop rotations and long-term capital investment decisions. To effectively achieve this type of integration, it will most likely be necessary to create modular crop and economic models so that different components can be linked in a 'plug and play' system (Antle and Stoorvogel, 2006). Thus far, this modularization of both crop and economic models has not been achieved.

To illustrate methods for linking models without full integration, I consider the approach developed by Stoorvogel et al. (2004) and used technology impact and climate impact assessments such as Valdivia et al. (2012, 2017). Their goal was to replace the usual *ad hoc* inclusion of some bio-physical variables in econometric models with an approach that would incorporate the effects of soil, climate and the genetic characteristics of crops and livestock in a way that is consistent with the process-based knowledge in crop and livestock simulation models. An underlying assumption behind this approach is that crop simulation models represent the kind of bio-physical relationships that farmers know and incorporate into their decision-making.

We begin with an economic model as represented by Eqs. (1) and (2). Define Y_{it}^{s} as the yield for a crop produced by farm *i* in period *t* using technology *s*. Following the notation in Eq. (2), let $B_{it} = (B_{it}^{f}, e_{it})$, where B_{it}^{f} is the set of biotic (e.g. soil microorganisms) and abiotic (e.g. soil properties) factors and farm characteristics (e.g. fixed capital) that affect production, and e_{it} is the set of random factors that cannot be controlled by the farmer (pest infestations, weather). X_{it}^{s} is a vector of input and management decisions made by the farmer that are specific to this technology, and β is the parameter vector relating X_{it}^{s}, B_{it}^{f} and e_{it} to Y_{it}^{s}. Thus, the production function is:

$$Y_{it}^{s} = f^{s}\left[X_{it}^{s}, B_{it}^{f}, e_{it}, \beta\right]. \tag{5}$$

Following Eq. (1), define A_{it} as the farmer's information set used to determine X_{it}^{s} that includes B_{it}^{f} as well as output and input prices, and other relevant information such as past production experience, weather forecasts and so forth. Thus, management decisions are a function of this information, $X_{it}^{s}(A_{it}, \alpha)$.

Following Eq. (5), we can use a crop growth simulation to calculate a yield Y_{it}^{*} for a specified management X_{it}^{*} (e.g. average observed management) and stochastic variables e_{it}^{*} as $Y_{it}^{*} = f^{s}[X_{it}^{*}, B_{it}^{f}, e_{it}^{*}, \beta]$. Stoorvogel et al. (2004) refer to this yield estimate as the *inherent productivity* of the site, to distinguish it from an estimate of actual yield and interpret this quantity as representing what an informed farmer knows about the productivity of the site based on knowledge of its soils and climate. Now partition $B_{it}^{f} = (B_{it}^{a}, B_{it}^{b})$, where B_{it}^{b} is the subset of variables used in the crop growth model. Under the assumption that the farmer's expected output is a function of the site's inherent productivity, represented by Y_{it}^{*}, and the farmer's management decisions the statistical production function can be written:

$$Y_{it}^{s} = m^{s}\left[X_{it}^{s}\left(A_{it}\right), B_{it}^{a}, Y_{it}^{*}, \beta\right] + v_{it}^{s} = m^{s}\left[X_{it}^{s}\left(A_{it}\right), B_{it}^{a}, f^{s}\left[X_{it}^{*}, B_{it}^{b}, e_{it}^{*}, \beta\right], \beta^{m}\right] + v_{it}^{s}, \tag{6}$$

where β^{m} is the set of parameters relating X_{it}^{s}, B_{it}^{a} and Y_{it}^{*} to Y_{it}^{s}, and v_{it}^{s} is a random component with zero expectation. Thus, this procedure produces a special case of the statistical production function in which the bio-physical variables B_{it}^{b} and e_{it}^{*} are embodied in a sub-function that interacts with the management inputs chosen by the farm decision-maker. Economists say this type of function is 'weakly separable' in the bio-physical variables.

This form of the production function implies that the behavioural equations (output supply, input demand) depend on the bio-physical factors through inherent productivity. This structure induces an economic-adaptive response to the bio-physical variables B_{it}^{b} and e_{it}^{*} (e.g. changes in soils, climate etc.). However, note that the management variable X_{it}^{*} is held constant, so a feedback from economic behaviour to inherent productivity is not incorporated. A simpler approach that can be interpreted as a special case of the inherent productivity model makes the assumption of strong separability between crop model variables and management inputs, that is the production function takes the form:

$$Y_{it}^{s} = m^{s}\left[X_{it}^{s}\left(A_{it}\right), B_{it}^{a}, \beta^{m}\right] f^{s}\left[X_{it}^{*}, B_{it}^{b}, e_{it}^{*}, \beta\right] + v_{it}^{s}. \tag{7}$$

There are at least two ways to implement this type of model. Stoorvogel et al. (2004) describe incorporating simulated yields as an explanatory variable in an econometrically estimated production model (Eq. (6)) to account for the effects of the bio-physical factors represented by B_{it}^{b}. A disadvantage of this approach is that there must be sufficient variation in observed soil and weather data to reliably estimate the relationship between simulated and observed yields. In many cases, a lack of weather stations close to farms violates these conditions.

The most commonly used approach to link crop and economic models is based on the concept of relative yield. In this approach, we replace the function f^{s} in Eq. (7) with the relative yield:

$$R_{it} = f^{F}\left[X_{it}^{F}, B_{it}^{F}, e_{it}^{F}, \beta\right] / f^{H}\left[X_{it}^{H}, B_{it}^{H}, e_{it}^{H}, \beta\right], \tag{8}$$

where f^{H} is the simulated historical yield, f^{F} is the simulated future yield and we interpret $m^{H}[X_{it}^{H}(A_{it}^{H}), B_{it}^{H}, \beta^{X}]$ as the historically observed production function, so that future yield is predicted to be $Y_{it}^{F} = m^{H}[X_{it}^{H}(A_{it}^{H}), B_{it}^{H}, \beta^{X}]R_{it}$. This relative yield method has a number of advantages as well as some limitations for *ex ante* technology impact assessment (Antle et al., 2017c). On the positive side, this method can be implemented independently by crop modellers who simulate yields, and by economists who construct relative yield distributions to be combined with economic models. Also, if the crop model simulated yields are biased, the relative yield will be unbiased to the extent that the bias is proportional to the simulated yield. On the negative side, the relative yield method requires the assumption of strong separability between management decisions and bio-physical factors. It also requires observation of a yield for the 'base' system, for example for the historical yield, which is not the case when the alternative system involves new activities that have not yet been observed. One strategy for this situation is to use a 'simulation matching' method as discussed in Antle et al. (2017c), wherein an observable yield is used as a proxy for the unobservable yield.

6 Case study: a hybrid structural model of the wheat-fallow system in the US Pacific Northwest

Here I use an example from research on wheat-based systems in the US Pacific Northwest to illustrate the application of a hybrid model. This example uses data from the main winter wheat-producing region of the Pacific Northwest; see Antle et al. (2017c) for further details. In areas with relatively high precipitation, winter wheat is grown in an annual rotation with other crops without fallow (WWA). But continuous cropping using the WWA system in the areas with low precipitation depletes soil moisture and provides low yields. Thus, in drier areas a winter wheat-fallow (WWF) rotation is the predominant cropping system with wheat planted every other year in a given field. Even though only half the cropped land is planted each year under the WWF system, it is more profitable than the continuously cropped WWA system.

The simulation experiment presented here is designed to test the ability of the model to predict farmers' choices between the WWF and WWA systems. In this implementation of the hybrid structural model approach, the CropSyst model (Stöckle et al., 2003) is combined with the TOA-MD (Antle et al., 2014). As illustrated in Fig. 3, the CropSyst model is used to simulate the relative yield between the two systems, and the TOA-MD model simulates the adoption rate of the WWA system in the low-precipitation area where most farmers use WWF. Thus, in this experiment, the reference system is WWF used by farmers in the WWF region, and the 'treatment' is the WWA system. The goal is to test if the models can predict the observed proportion of farms using the WWA system in the WWF region. In the 2007 data used for this test, about 23% of the farms were using WWA in the WWF region, reflecting the lower productivity of the WWA system on average in the WWF region.

The CropSyst model was used to project yields for both WWF and WWA in the WWF region on a 4 km grid, and then these yields were used to calculate relative yields for WWA relative to WWF (Fig. 4). These relative yields were then combined with farm-level economic data from the 2007 agricultural census to parameterize the TOA-MD impact assessment model. The TOA-MD model uses the difference between expected economic returns to the two systems to simulate farmers' choices between systems. The TOA-MD model requires estimates of the parameters of the distribution of the difference in expected returns between the two systems (mean and variance, assuming normality), which can be calculated from the means, variances and covariance of the expected returns to each system. The relative yields are used to translate the observed wheat returns of farms using one system into the counterfactual returns of the other system, as discussed in the previous section.

The CropSyst simulations show, as expected, that the simulated WWF yield distribution in the WWF region has a much higher mean (about 50 bu/ac) than the simulated WWA yield distribution (about 31 bu/ac) in the WWF region. The resulting relative yield distribution indicates that the average relative yield (WWA/WWF) is about 0.7 and almost all relative yields are less than 1 (bottom panel of Fig. 4). When combined with economic data in the TOA-MD model, the economic simulation predicts an adoption rate of WWA of about 20%, which compares favourably to the observed adoption rate of 23%, confirming that the hybrid model predicts observed behaviour reasonably well.

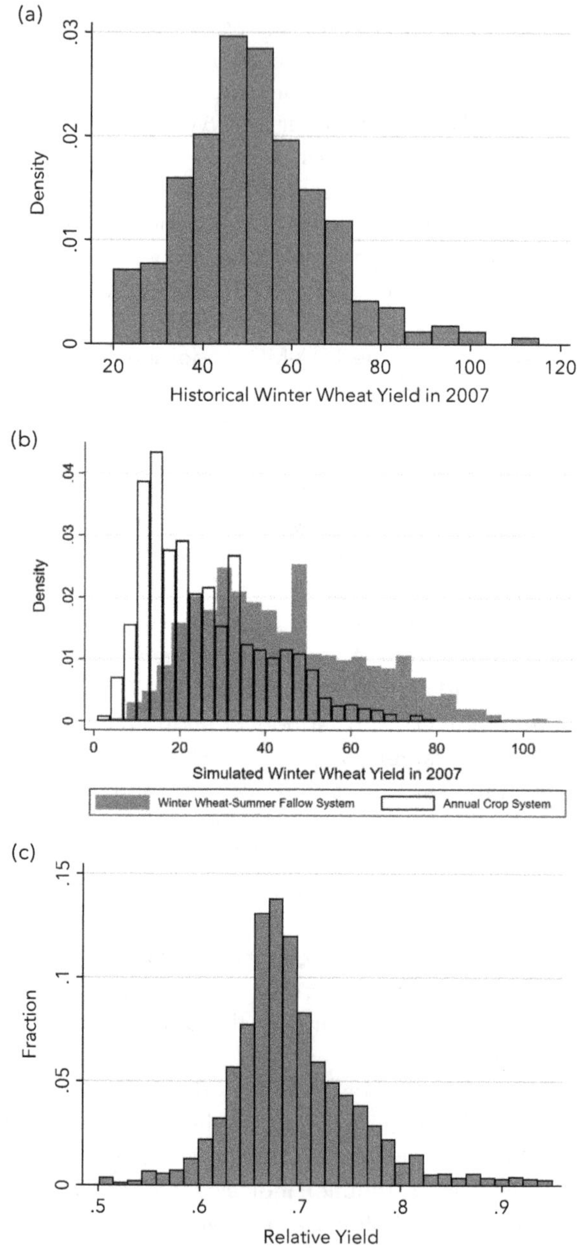

Figure 4 US Pacific Northwest Historical Winter Wheat Yield Distribution for 2007 (a); CropSyst Simulated Winter Wheat Yield Distributions for Winter Wheat Fallow and Annual Crop system in the Winter Wheat Fallow region (b); CropSyst simulated Relative Yield Distribution for Annual Crop system in the Winter Wheat Fallow region (c). Source: adapted from Antle (2019).

7 The data challenge for agricultural systems modelling

A key implication of the preceding discussion is that improved acquisition and use of data is a critical constraint on the use of agricultural systems models. This was also a main conclusion of a recent study of 'next-generation' agricultural systems (Antle, Jones and Rosenzweig, 2017). Better data are needed to further improve crop and livestock models in ways that are useful for both on-farm management decision-making and for use in research to develop and test new technologies, and to evaluate their productivity and sustainability.

My discussion above of the evaluation paradigm and its application to agricultural systems analysis demonstrated that better observational data, and the utilization of both mechanistic and empirical production models, could help address the key analytical challenges in assessing the productivity and sustainability of agricultural systems. Valid mechanistic models could be used for counterfactual identification; better observational data could address bias problems caused by unobserved heterogeneity and inaccurate and incomplete data; and the combination of mechanistic and empirical production system models could improve the analysis of new technologies.

Capalbo et al. (2017) discuss a prototype private-public data system that could meet this need. Both farm-level decision-making and broader, landscape-scale analysis depend on private data (site- and farm-specific characteristics of the land and the farm operation, and the site- and farm-specific management decisions that are made) as well as public data (weather, climate and other physical data describing a specific location, and prices and other publicly available economic data). They also provide an example of decision support tools that could be used to enable a data system at the farm level, and that could be linked to cloud-based data that could in turn be used for scientific research and policy decision-making. However, various issues would have to be addressed to implement this type of system. One issue is how to encourage growers to provide accurate data. This issue would presumably be resolved if the data are being entered into a management tool. To translate data into a generic format that would be FAIR (findable, accessible, interoperable and reusable), translators from commercial management tools into the generic format could be used. A key issue for most farmers is data confidentiality. This issue poses a critical challenge to be addressed if researchers are to be able to obtain the location-specific information needed to link economic and management data to bio-physical data such as weather and soils.

Today we are far from realizing an integrated private-public data infrastructure. Private data and related soft and hard infrastructure are being developed by a growing array of management advisory and technology companies. Data generated by individual producers or by private firms selling data or advisory services are not public and thus not findable or accessible, often even by farmers themselves. There are no established data standards being used, and thus data are not interoperable even when findable and accessible.

There are also many limitations of currently produced public data. Some of the data identified as public in Fig. 3, such as some weather, price and crop yield data, are open access or available for fees. However, many of the data related to agricultural production are collected for various government administrative purposes and are not intended to be used for research or for private decision-making. For example, much of the data collected by the National Agricultural Research Service are findable but are not easily accessible in a timely manner, and then only available in summary or aggregated form. Some of these data, such as Agricultural Census and Agricultural Research Management Survey, can now be accessed at the respondent level, but without location identifiers that are needed for many research purposes. Also, most data fail the interoperability standard, and thus must be processed by users to put in a form that can be usable by analytical software or models. A major shortcoming of the available production data is in terms of management data and cost of production data. For example, the census collects cost of production at the whole farm level, so it cannot be disaggregated to individual production activity (e.g. crop) level. Another major shortcoming is that none of the available data can be used to construct longitudinal data for multiple proximate growing seasons or years. The census data can be used to construct a panel, but is only carried out every 5 years.

Various efforts are underway to address these data challenges. For example, the international agricultural research centres' Big Data Initiative is working to develop data ontologies for agricultural systems data. The Global Open Data for Agriculture and Nutrition initiative is supported by a number of governmental organizations. In the United States, the National Institutes for Food and Agriculture is supporting the Food and Agriculture Cyberinformatics and Tools grant programme. Coble et al. (2018) provide an overview of big data in agriculture and the contributions that economists can make to its use in research and technology implementation.

It is now widely acknowledged that to advance agricultural systems science in order to achieve goals of agricultural system sustainability, there is a need for collaboration among disciplines to understand and analyse complex systems (National Academies of Science, 2018). The growing capabilities of mobile sensor technologies, unmanned aerial vehicles, cloud computing and other emerging technologies, show the need to include the data, engineering and computer sciences into the group of relevant disciplines for the further development of agricultural systems science.

A key question is how to organize and incentivize these new collaborations. One answer is for public funding of related research and development, as envisaged by NIFA's Food and Agriculture Cyberinformatics and Tools programme. Following ideas taken from the pharmaceutical industry, Antle et al. (2017b) propose the development of 'pre-competitive spaces' for public-good-related research and 'competitive spaces' for development of knowledge products in the private sector. To support the pre-competitive development of models and data, advances in collaborative research are also needed. For example, AgMIP has demonstrated the power and value of global teams developing data for model inter-comparisons that have led to substantial model improvement.

8 Future trends

There are a number of important directions for future research on integrated bio-physical and economic modelling of agricultural systems. These include:

- Closer coupling of bio-physical and economic models, possibly through the development of modular, open-source model components.
- Better integration of model components to represent system dynamics, within growing seasons (e.g. for irrigation or pest management decisions) and across growing seasons, for example to represent crop rotations and perennial crops.
- Use of crop simulation models to characterize production risk for economic decision models.
- Use of vocabularies and ontologies being developed for genetic and agronomic modelling in the development of ontologies for economic data.
- Developing methods for acquiring private data and maintaining confidentiality of data for use in research and policy analysis.

9 Conclusion

A major challenge facing the agricultural research community is to evaluate and improve the sustainability of agricultural systems. To meet this challenge, economic models are being combined with process-based agricultural system models for a variety of analytical purposes, ranging from farm-level decision support to global climate impact assessment. In this chapter, I discuss how process-based farm system models can be combined with economic models to assess and improve the sustainability of agricultural systems. This type of analysis is referred to as *ex ante* technology impact assessment. The key analytical challenge to this type of analysis is to assess the performance of modified or new agricultural systems in new environments.

In this chapter I argue that the best approach to dealing with these challenges is to combine process-based crop simulation models with empirical economic models. I propose a 'hybrid structural model' as a way to implement this idea that combines simulated yields from crop models with economic models of producer behaviour. I illustrate this type of model with an example that combines the CropSyst crop simulation model with the TOA-MD economic impact assessment model. I conclude with a discussion of the key role that new and better data also plays in making better modelling feasibility.

10 Where to look for further information

A number of useful survey papers provide a wealth of information about economic models and their use in analysis of agricultural systems. Just and Pope (2001) provide an overview of the production economics literature. Van Wijk et al. (2014) survey the 'bio-economic' modelling literature, in the context of climate impact and food security assessment. Kanter et al. (2018) provide an overview of farming system and economic models used to assess sustainability of agricultural systems and related trade-offs among production, economic and environmental outcomes. Antle and Stöckle (2017) survey the use of agronomic-economic models for climate impact assessment. Antle (2019) discusses the use of hybrid agronomic-economic models to overcome methodological and data challenges in impact assessment. Jones et al. (2017) and Antle et al. (2017b) provide surveys of agricultural systems science and opportunities to advance towards 'next-generation' data and models.

The reader can find a wide array of applications of crop and economic simulation studies using the hybrid structural approach implemented with the TOA-MD model. Most of these studies can be found in the applications library at: tradeoffs.oregonstate.edu. This type of analysis has been used widely for climate impact and adaptation analysis, including the suite of studies led by the Agricultural Model Inter-comparison and Improvement Project (AgMIP). The methods developed by AgMIP for regional integrated assessment of climate impacts using the hybrid structural modelling approach are described in detail at: http://www.agmip.org/regional-integrated-assessments-handbook/.

11 References

Antle, J. M. 2011. Parsimonious multi-dimensional impact assessment. *American Journal of Agricultural Economics* 93(5), 1292–311. doi:10.1093/ajae/aar052.

Antle, J. M. 2019. Data, economics and computational agricultural science. *American Journal of Agricultural Economics* 101(2), 365–82. doi:10.1093/ajae/aay103.

Antle, J. M. and Capalbo, S. M. 2001. Econometric-process models for integrated assessment of agricultural production systems. *American Journal of Agricultural Economics* 83(2), 389–401. doi:10.1111/0002-9092.00164.

Antle, J. M. and Hatchett, S. A. 1986. Dynamic input decisions in econometric production models. *American Journal of Agricultural Economics* 68(4), 939–49. doi:10.2307/1242140.

Antle, J. M. and Stöckle, C. O. 2017. Climate impacts on agriculture: insights from agronomic-economic analysis. *Review of Environmental Economics and Policy* 11(2), 299–318. doi:10.1093/reep/rex012.

Antle, J. M. and Stoorvogel, J. J. 2006. Incorporating systems dynamics and spatial heterogeneity in integrated assessment of agricultural production systems. *Environment and Development Economics* 11(1), 39-58. doi:10.1017/S1355770X05002639.

Antle, J., J. Jones and C. Rosenzweig. 2017. Next Generation Agricultural System Data, Models and Knowledge Products: Synthesis and Strategy. *Agricultural Systems* 155, 179–185.

Antle, J. M., Stoorvogel, J. J. and Valdivia, R. O. 2014. New parsimonious simulation methods and tools to assess future food and environmental security of farm populations. *Philosophical Transactions of the Royal Society of London. Series B, Biological Sciences* 369(1639), 20120280. doi:10.1098/rstb.2012.0280.

Antle, J. M., Mu, J. E., Zhang, H., Capalbo, S. M., Diebel, P. L., Eigenbrode, S. D., Kruger, C. E., Stöckle, C. O., Wulfhorst, J. D. and Abatzoglou, J. T. 2017a. Design and use of representative agricultural pathways in U.S. Pacific Northwest wheat-based systems. *Frontiers in Ecology and Evolution* 5. doi:10.3389/fevo.2017.00099.

Antle, J. M., Basso, B. O., Conant, R. T., Godfray, H. C. J., Jones, J. W., Herrero, M., Howitt, R. E., Keating, B. A., Munoz-Carpena, R., Rosenzweig, C. E., et al. 2017b. Towards a new generation of agricultural system data, models and knowledge products: design and improvement. *Agricultural Systems* 155, 255–68. doi:10.1016/j.agsy.2016.10.002.

Antle, J. M., Zhang, H., Mu, J. E., Abatzoglou, J. and Stöckle, C. O. 2017c. Methods to assess between-system adaptations to climate change: dryland wheat systems in the Pacific Northwest United States. *Agriculture, Ecosystems and Environment* 253, 195–207. doi:10.1016/j.agee.2017.03.017.

Capalbo, S. M., Antle, J. M. and Seavert, C. 2017. Next generation data systems and knowledge products to support agricultural producers and science-based policy decision making. *Agricultural Systems* 155, 191–9. doi:10.1016/j.agsy.2016.10.009.

Coble, K. H., A. K. Mishra, S. Ferrell and T. Griffin. 2018. Big Data in Agriculture: A Challenge for the Future. *Applied Economic Perspectives and Policy* 40, 79–96.

Heckman, J. J. 2010. Building bridges between structural and program evaluation approaches to evaluating policy. *Journal of Economic Literature* 48(2), 356–98. doi:10.1257/jel.48.2.356.

Jones, J. W., Antle, J. M., Basso, B. O., Boote, K., Conant, R. T., Foster, I., Godfray, H. C. J., Herrero, M., Howitt, R. E., Janssen, S., et al. 2017. Towards a new generation of agricultural system models, data, and knowledge products: state of agricultural systems science. *Agricultural Systems* 155, 269–88.

Just, R. E. and Pope, R. D. 2001. The agricultural producer: theory and statistical measurement. In: Gardner, B. L. and Rausser, G. C. (Eds), *Handbook of Agricultural Economics, Vol. 1A, Agricultural Production*. North-Holland, Amsterdam and London, pp. 631–741.

Just, R. E., Zilberman, D. and Hochman, E. 1983. Estimation of multicrop production functions. *American Journal of Agricultural Economics* 65(4), 770–80. doi:10.2307/1240465.

Kanter, D. R., Musumba, M., Wood, S. L. R., Palm, C., Antle, J., Balvanera, P., Dale, V. H., Havlik, P., Kline, K. L., Scholes, R. J., et al. 2018. Evaluating agricultural trade-offs in the age of sustainable development. *Agricultural Systems* 163, 73–88. doi:10.1016/j.agsy.2016.09.010.

National Academies of Sciences, Engineering, and Medicine. 2018. *Science Breakthroughs to Advance Food and Agricultural Research by 2030*. The National Academies Press, Washington DC. doi:10.17226/25059.

Stöckle, C. O., Donatelli, M. and Nelson, R. 2003. CropSyst, a cropping systems simulation model. *European Journal of Agronomy* 18(3–4), 289–307. doi:10.1016/S1161-0301(02)00109-0.

Stoorvogel, J. J., Antle, J. M., Crissman, C. C. and Bowen, W. 2004. The tradeoff analysis model: integrated bio-physical and economic modeling of agricultural production systems. *Agricultural Systems* 80(1), 43–66. doi:10.1016/j.agsy.2003.06.002.

Valdivia, R. O., Antle, J. M. and Stoorvogel, J. J. 2012. Coupling the tradeoff analysis model with a market equilibrium model to analyze economic and environmental outcomes of agricultural production systems. *Agricultural Systems* 110, 17–29. doi:10.1016/j.agsy.2012.03.003.

Valdivia, R. O., Antle, J. M., Rosenzweig, C., Ruane, A. C., Vervoort, J., Ashfaq, M., Hathie, I., Tui, S. H.-K., Mulwa, R., Nhemachena, C., et al. 2015. Representative agricultural pathways and scenarios for regional integrated assessment of climate change impact, vulnerability and adaptation. In: Rosenzweig, C. and Hillel, D. (Eds), *Handbook of Climate Change and Agroecosystems: The Agricultural Model Intercomparison and Improvement Project Integrated Crop and Economic Assessments, Part 1*. Imperial College Press, London.

Valdivia, R. O., Antle, J. M. and Stoorvogel, J. J. 2017. Designing and evaluating sustainable development pathways for semi-subsistence crop-livestock systems: lessons from Kenya. *Agricultural Economics* 48(S1), 11–26. doi:10.1111/agec.12383.

van Wijk, M. T., Rufino, M. C., Enahoro, D., Parsons, D., Silvestri, S., Valdivia, R. O. and Herrero, M. 2014. Farm household models to analyse food security in a changing climate: a review. *Global Food Security* 3(2), 77–84. doi:10.1016/j.gfs.2014.05.001.

Chapter 11

Developing climate-based decision support systems from agricultural systems models

Clyde W. Fraisse, University of Florida-Gainesville, USA; Norman E. Breuer, Catholic University Nuestra Señora de la Asunción, Paraguay; and Victor Cabrera, University of Wisconsin-Madison, USA

1 Introduction

Decision-making involves a number of steps that include gathering, processing, analyzing, and communicating information from different sources in time to act for a particular purpose. Decision support systems (DSSs) in production agriculture are computer software programs that make use of models and other information to make site-specific recommendations for farm management-related activities (Jones et al., 2017). DSSs are designed to help with the process of translating knowledge into actions and are formally defined as interactive computer-based systems that enable people to use IT communications, data, documents, knowledge, and models to solve problems and make decisions (Power, 2002).

Major advances have been made in developing DSSs for agriculture, particularly with respect to incorporating climate forecasts into farmers' decision-making process. Climate variability is a major source of production risk. The majority of crop failures in the United States and worldwide are associated with either a lack or excess of rainfall. Climate variability is also

http://dx.doi.org/10.19103/AS.2019.0061.15

associated with other sources of production risks such as pest and disease incidence. Unanticipated forces, such as inclement weather, drought conditions, crop failure or abundance, or pest or disease outbreak, can also lead to dramatic changes in crop and livestock prices and increase marketing risk. Nevertheless, climate forecasts alone can fall short in providing actionable information for improving farm-level decisions and policy-level interventions (Fraisse et al., 2006; Han et al., 2017). Climate forecasts must be linked to DSSs to help farmers improve strategic and tactical decisions to maximize benefits and minimize climate-related risks in the growing season.

Development of forecast information for farmers that can be used for decisions requires understanding of the climate impacts on plant growth and management operations. The response of crops to the different weather variables is complex and difficult to simulate in isolation. If one of the variables is limiting (e.g. temperatures that are too hot or too cold), then the effects of solar radiation or precipitation may not greatly affect the crop (Hollinger and Angel, 2009). However, in the majority of the situations, it is important to consider the interactions of environmental conditions with genetics and management practices to simulate potential crop outcomes.

Most climate-based DSSs integrate climate information such as historical weather data, short-term weather forecasts, and seasonal climate forecasts with agricultural system models to help farmers make better decisions and reduce risk. A few examples of climate-based DSSs include AgroClimate (http://www.agroclimate.org), the Useful to Usable (U2U) decision tools dashboard (https://mygeohub.org/groups/u2u/tools), the Yield Prophet platform based on the APSIM model (https://www.yieldprophet.com.au), and the PC-based Climate-Agriculture-Modeling and Decision Tool (CAMDT).

AgroClimate (Fraisse et al., 2006) was developed to provide extension agents, producers, and natural resource managers in the Southeastern United States with tools to aid their decision-making aimed at reducing risks associated with climate variability. AgroClimate provides users with climate forecasts and a variety of model-based tools from simple chill hours and growing degree-day accumulation models to disease risk alert systems and yield potential for alternative planting dates and climate scenarios based on the Decision Support System for Agrotechnology Transfer (DSSAT) suite of crop models. Many of the DSS tools available on AgroClimate take into consideration the effects of the El Niño Southern Oscillation (ENSO) on the climate of the Southeastern United States.

The U2U (Angel et al., 2017; Prokopy et al., 2017) decision tools dashboard focuses primarily on corn and was developed to provide both useful and usable climate information for the agricultural corn sector in the Midwestern United States. The project adopted an extensive coproduction of knowledge and decision-making approach that involved intense iteration with potential end users, including farmers and a variety of professional agricultural advisors, through focus groups and surveys, feedback at outreach events, and frequent informal interactions to develop both decision support tools and delivery mechanisms that met stakeholder needs.

Yield Prophet (Hochman et al., 2009) was developed in Australia and provides information on impacts of climate on crop growth and yield, disease occurrence, and recommended management practices based on several crop simulation models. The system offers three types of tools: (1) soil characterization tools to help farmers generate data on the physical and chemical characteristics of their farms' main soil types, and on the soil water and nutrient status at key decision points in the season; (2) systems simulation using the APSIM model combining seasonal climate outlooks with historical climate records to simulate probability outcomes for intended management actions relevant to an individual farmer's actual crop; and (3) seasonal climate forecasting tools that can provide the probabilities of rainfall and temperature for the upcoming 3-month period (e.g. the phase system of the Southern Oscillation Index (SOI)).

The CAMDT (Han et al., 2017) is PC-based, which limits its broader use by farmers and crop consultants. However, it provides an approach to integrate downscaled seasonal climate forecasts and crop simulation models. The current version is based on the DSSAT-CSM-Rice model, although there is potential of including other crop models in the future. CAMDT can display DSSAT outputs (e.g. yield and water stress), as well as translated outputs (e.g. risk of water stress and gross margins). According to Han et al. (2017), it is expected that the software can contribute to developing better-informed climate adaptation strategies by providing users the ability to easily assess scenarios of various agronomic practices, given an expected seasonal climate.

1.1 Strategic, tactical, and operational decisions

The importance of climate-based DSSs is greatly enhanced by the fact that the projected increase in climate variability due to climate change will require that farmers make significant changes to their farming systems. In order to implement effective adaptation strategies, farmers will have to maintain tactical and strategic flexibility in their farm systems (Cowan et al., 2013). However, maintaining flexibility requires tools to predict and analyze potential outcomes of alternative scenarios and making decisions that will affect agronomic and financial outputs in a farm. Decisions made in a farm can be categorized into three types of planning horizons: strategic, tactical, or operational. Strategies in farm businesses are decisions about what is going to be produced and how the business is going to be organized to produce it. *Strategic decisions* are major choices of actions and influence whole or a major part of a farm business enterprise. Once a farm has been organized to produce a set of outputs identified by strategy, tactics are used to produce these outputs. *Tactical decisions* relate to actions or methods used for the implementation of strategic decisions. Tactics are chosen from alternatives, all of which can serve the strategy under certain circumstances. *Operational decisions* relate to day-to-day operations of the farm enterprise. They have a short-term horizon and are based on facts regarding the events and do not require much business judgment (Cowan et al., 2013).

Operational decisions rely primarily on weather monitoring and forecast for the next hours or days. Operational decisions such as planting, spraying fungicide, turning on cold protection, or deciding about hay cutting can be improved using current weather monitoring and short-term weather forecast. While some operational decisions can be improved based only on weather forecast, other decisions such as spraying fungicides require present and expected weather conditions to be translated into disease risk levels to define an action. The best time to cut hay is normally after a few dry days with warm temperatures and when a few more dry days are in the forecast. However, determining other important factors in hay quality such as the state of maturity at harvest can be greatly improved with the help of a systems model. DSSs play an important role by effectively integrating environmental conditions, management practices, and genetics to provide science-based decisions with a higher probability for a successful outcome.

Strategic and tactical decisions are normally made before planting such as deciding what crop and cultivar to plant or the best time to plant. After planting, decisions tend to be more tactical or operational such as managing nutrients and pests, deciding how much and when to irrigate a crop, and harvesting (Table 1). The Midwest Corn Grower's Decision Calendar provides a good example of the nature and considerations required to make decisions over short- and long-term time frames (Takle et al., 2014).

Table 1 Typical strategic, tactical, and operational questions/decisions in a farm

Weather forecast Operational decisions	Seasonal forecast Tactical and strategic decisions
When should I plant my seeds?	What crop and variety to plant this season?
Should I apply N fertilizer to my fields?	How much should I invest in fertilizer? How to apply N?
Should I spray fungicide today?	Should I purchase or increase crop insurance coverage?
Will it be dry enough to harvest?	Marketing decisions
Should I cut hay today?	Should I invest in winter pasture or animal feed?
Should I turn on cold protection tonight?	

2 Importance of a participatory approach in DSS development

Toward the end of the twentieth century, it was assumed that the use of DSSs would become common practice in agricultural production. However, DSS use did not grow in a linear fashion with the spread of home computers. Many early DSSs had limited adoption (Parker, 1999; Hayman and Easdown, 2002; McCown et al., 2002; Cash and Buizer, 2005). The DSSAT (Jones et al., 2003) has been more widely adopted; however, its use remains mostly in the realm of researchers and consultants with high levels of education and training. Some reasons for non-adoption or short-term use cited were high levels of expectations on the part of farmers, lack of user-friendliness, the deterministic nature of outputs, and problems with upscaling (McCown et al., 2002). Substantial changes had to be incorporated into all phases of development if past deficiencies with regard to end user utilization were to be overcome. One of the most important aspects required may have been the intensive and effective participation and continuous feedback from all stakeholders involved. The Australian experience with FARMSCAPE (Carberry et al., 2002) represented one of the first successes of a system for farm-level developed using a participatory process, which had been posited by Cox (1996).

Extensive and comprehensive reviews have been conducted on participatory research on farmer responses to climate forecasts (Roncoli, 2006), stakeholder participation for environmental management (Reed, 2008), and more recently, tools and methods in participatory modeling (Voinov et al., 2018). Early research found that simply documenting the effects of climate variability and providing better climate forecasts to potential users was not sufficient for users to benefit fully from this information. Because of the complex interactions among biophysical, social, and institutional factors that affect agricultural systems, clients need decision aids and technical assistance to bridge the gap that still exists between available climate forecasts and their routine applications in agriculture (Hildebrand, 1999; Podestá et al., 1999; Jones et al., 2000; Hammer et al., 2001; Hansen, 2002; Hartmann et al., 2002). Such DSSs can help producers to better understand the possible responses to climate forecasts, and they must understand the risks associated with alternative responses in order to obtain benefits from a forecast (Jagtap et al., 2002; Letson et al., 2001, 2005). Initial scientific inquiry with regard to crop and farm management found that participatory approaches were not about science simply providing answers for management to practitioners, but rather employing cooperative learning to develop solutions (Ashby, 2002; Meinke et al., 2009).

Information alone, however, does not inspire change. At the policy level, stakeholders in governments, planners, communities, individuals, industry, and interest groups frequently disagree about the relevance and effectiveness of climate mitigation and adaptation strategies due to differences in culture and values. Thus, participatory methods that could incorporate a plurality of vulnerabilities and solutions were considered to be an effective approach for designing a DSS useful across several spatial and temporal scales and for many potential stakeholders (Rogers, 2003; Jacques et al., 2017). Direct participatory action research with farmers has helped establish credibility of models, simulation analyses, and the DSSs into which they are built. Researchers determined that software created for farmers, extension agents, and farm consultants should be user friendly, deliver what is asked for quickly and easily, be presented in an understandable format, address site-specific conditions, and be presented in real or near-real time (Archer et al., 2002; Cabrera et al., 2008).

The issue of scale, both temporal and spatial, is crucial to the development of useful and usable DSSs. Some systems, such as DSSAT, especially with its broadening into a global community through the AgMIP Project (Rosenzweig et al., 2014), are more amenable to global or regional food security applications and public policy formulation, under long-term climate change. Others, such as AgroClimate (Fraisse et al., 2006), are more suited to farm- or local-level decision-making for tactical or strategic planning for risk reduction on a short term or seasonal timescale.

DSS developers borrowed a key finding from systems analysts who had realized that a thorough understanding of complex problems is often gained *during* model building. The participation of the potential final users enables the researcher to enrich models, tools, and products. This is done by including subjective or experiential sources of knowledge and the objective knowledge derived from theories and empirical studies. A systems approach must include not only the biophysical component, but also the social, economic, and political environment of the farm together with a 'bottom-up' approach. Connectivity and dialogue among key players is essential for achieving relevant and significant intervention, because production decisions are not taken in a vacuum (Geurts and Joldersma, 2001; Marx et al., 2007; Breuer et al., 2008, 2009).

2.1 Defining participatory research and applying it to DSSs

Participatory research was first developed in the context of farming systems research for international rural development to increase technology adoption by involving farmers in developing and testing innovations (Hildebrand, 1981; Chambers,1994, 1995; Stringer et al., 2006). According to Marx et al. (2007), people rely heavily on experiential processing systems, and better understanding of experiential processing may lead to more comprehensible climate risk communication products. The rationale behind stakeholder participation is based upon two arguments: (a) that stakeholders have a fundamental right to be included in deliberations that will have an impact upon their lives, and (b) that listening to the voice of stakeholders and including them within a process of change can help make that change better (Morse, 2010).

Some of the main benefits posited for participatory research in climate services include: (1) it may be the best way to get a handle on stakeholder needs (rather than wants); (2) provides opportunities to network and discuss joint research activities in a (new) region; is useful for input, feedback, dissemination, and co-development of DSS and tools; provides a measure of accountability and transparency to all involved; may improve stakeholders' 'buy-in' and some measure of ownership of the project, thus potentially increasing adoption; may enhance the legitimacy of the investigation in the eyes of stakeholders finally; and may nurture equality by placing researchers and other stakeholders on a level playing field (Reed, 2008). An increasing awareness of the constraints and challenges that characterize farmers' decisions has prompted an interest in combining agronomic modeling with participatory approaches. Simulation models and decision support tools can also facilitate learning by farmers as well as scientists by enabling them to explore different options and the potential risks and gains associated with them (Roncoli, 2006; Crane et al., 2008, 2010; Bartels et al., 2013).

2.2 The SECC experience: participatory development of AgroClimate

The Southeast Climate Consortium (SECC) was a Regional Integrated Science and Assessment (RISA) Center funded by National Oceanic and Atmospheric Administration (NOAA). In 2004, the SECC released AgroClimate (Fraisse et al., 2006), a prototype DSS for climate risk reduction in agriculture. The process of participatory development of the website included regular surveys and workshops to assess producer and extension agent awareness of, attitudes toward, and interests (Breuer et al., 2008) in climate data, and to provide continuous feedback to the research and development team who were creating tools for the DSS. Based in part on concerns about the necessary differentiation between DSSs intended for use by scientists and those designed and aimed at directing behavioral change in farmers, policy makers, or extension agents, the SECC adopted participation as an appropriate focus (Breuer et al., 2008; Crane et al., 2010; Roncoli et al., 2012).

The explicit aim of participatory research in developing AgroClimate was to learn from and support agricultural producers' adaptation to seasonal climate variability. This was enhanced by translating raw climate forecasts into derived forecasts and finally to feasible management adjustment options, often suggested by producers based on their own experience and environments

(Roncoli, 2006; Breuer et al., 2008; Crane et al., 2008, 2010). These adjustments fine-tune or 'tweak' rather than change management. Derivative processes and products using the same methodology led to tools for use by wildfire managers (Roncoli et al., 2012) and water managers (Bolson et al., 2012).

Climate information delivery requires some understanding of its probabilistic nature. Probability distributions and probability of exceedance graphs were preferred by extension agents and farmers in Florida when compared to other graphic renditions of probabilistic forecasts, including terciles (Breuer et al., 2008). According to Hartmann et al. (2002), forecasters and their evaluations typically reflect forecaster, or scientist perspectives, rather than those of users. However, research shows that farmers' ability to interpret probability and integrate it into decisions can improve over time (Suarez and Patt, 2010; Hansen, 2002).

3 Examples of climate-based DSSs for crop and land management

Various DSSs have been developed for crop and land management, many of them related to the application of crop growth and development modeling under variable and changing climate conditions and to the application of precision agriculture technologies (Jones et al., 2003; Keating et al., 2003; Stöckle et al., 2003; Basso et al., 2013; Tackle et al., 2014; Lundström and Lindblom, 2018)

The evolution of agricultural systems modeling during the last 60 years is well described by Jones et al. (2017). The application of crop models for decision support is quite broad. It can be used to predict crop production on regional, national, and global scales to help governments define policies and trading companies strategize commercialization. Another application is the simulation of crop yield potential to help farmers select best planting dates under alternative climate scenarios. The majority of agricultural systems modeling programs such as DSSAT (Jones et al., 2003; Hoogenboom et al., 2012), APSIM (Keating et al., 2003), and CropSYS (Stöckle et al., 2003) include crop modules available for many grain and fiber crops grown in temperate and tropical areas.

An example of DSS tool for evaluating planting dates under alternative climate scenarios is the AgroClimate Planting Date Planner tool based on the DSSAT-CSM suite of crop models: http://agroclimate.org/tools/planting-date-planner/. This tool provides producers and extension agents with the ability to simulate 'what-if' scenarios that take into consideration the location, management practices, and current ENSO phase to define the planting dates with higher chances of crop success (Fig. 1). The tool also provides a table with the expected range of dates for crop flowering and maturity together with the probability of freezes occurring after planting for the various planting dates.

Figure 1 The planting date planner tool on AgroClimate is based on the DSSAT-CSM suite of models. This example shows that early planting (April 23) of rainfed cotton in Santa Rosa County, Florida (sandy loam soils) during La Niña years may increase the chances for above-average yields.

Another example of DSS tool for crop management is the crop season planning tool (http://cloud.agroclimate.org/tools/sowing-plan/v1/#/main) (Perondi et al., 2019). The tool aims at helping soybean, maize, and wheat growers make more informed sowing decisions, and lower the risk of production losses associated with the occurrence of extreme weather events at particular times during the growing season. This web-based tool simulates crop phenology and historical extreme weather events within the context of a crop(ping) season. The user defines a 'seasonal scenario' by selecting the field location, nearby weather station, sowing date, sowing depth, crop, and crop cultivar. The user can modify planting dates or varieties to minimize the risk of extreme events during critical crop development phases (Fig. 2).

4 Examples of climate-based DSSs for pest and disease management

Weather patterns, including high temperature and humidity, and the potential for daily rainfall, can create a near-perfect environment for the outbreak of fungal diseases. They can also impact the reproductive cycle of other pests and insects that function as disease vectors (Fraisse et al.,

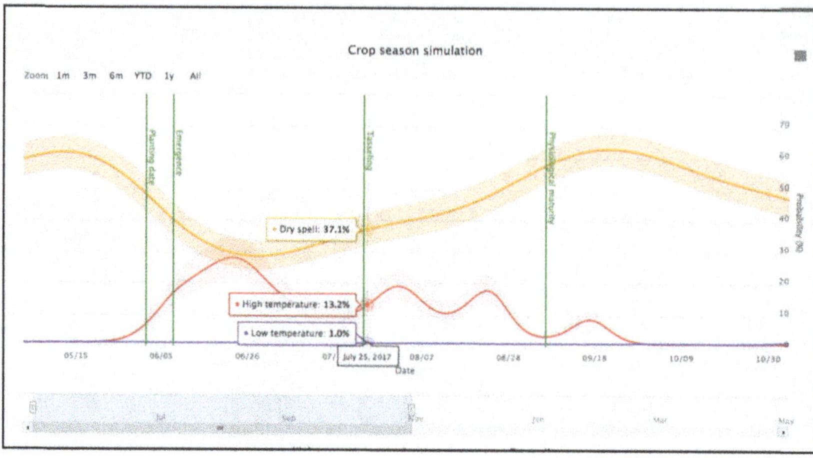

Figure 2 AgroClimate crop season planning tool showing the probability of extreme events during selected crop development phases.

2006). The prediction of plant diseases has emerged as a well-established component of epidemiology that is rapidly being incorporated into disease management (Beck et al., 1989; Pavan et al., 2011; Shtienberg, 2013; Small et al., 2015).

Most common pathogens in agricultural production systems, fungi and bacterium, require free water to grow and penetrate the plant tissue, starting the infection process, while temperature influences the speed of their metabolic reactions and development. The period with free water available on crop leaves is called leaf wetness duration (LWD) and is usually related to dew formation, rainfall, and irrigation. Favorable environmental conditions for disease development are specific for each host-pathogen interaction and are inferred by the combination of temperature and LWD. Traditionally, plant disease models have used LWD combined with temperature to predict infection and colonization, and then identify the risks of an epidemic. These types of models have been used with observed climate records to track the favorable periods, indicating tactics or strategies of control (Jabrzemski and Sutherland, 2006). Generally speaking, the closer the environment temperature is to the pathogen optimal development temperature, the shorter wetness duration necessary, and vice-versa.

Decisions related to disease management are generally related to the application or not of crop protection products. It is common for farmers to spray fungicide on a calendar basis, independent of conditions being favorable or not to the occurrence of infection. The BlightPro DSS developed by Small et al. (2015) for tomato and potato late blight management, a plant disease caused by the pathogen *Phytophthora infestans*, provides growers with location-specific management recommendations for fungicide application. The DSS uses observed and forecast weather data along with crop and management information to drive disease forecasting systems and a validated mechanistic model of the disease.

The Strawberry Advisory System (SAS) (Pavan et al., 2011) was developed to predict Anthracnose fruit rot (AFR) and Botrytis fruit rot (BFR) epidemics on strawberries and help strawberry producers in Florida avoid unnecessary applications of pesticide and reduce production costs. AFR, caused by *Colletotrichum acutatum* (Smith, 1998), and BFR, caused by *Botrytis cinerea* (Sutton, 1998), are the most threatening diseases for annual production of strawberries in Central Florida and worldwide. Growers can select the location closest to their plantings and SAS will provide a prediction of disease incidence and recommendations for fungicide applications. Users can also be provided warnings of the need to spray via e-mail or text messages. More recently, a mobile phone app of the original SAS system (Fig. 3) and an

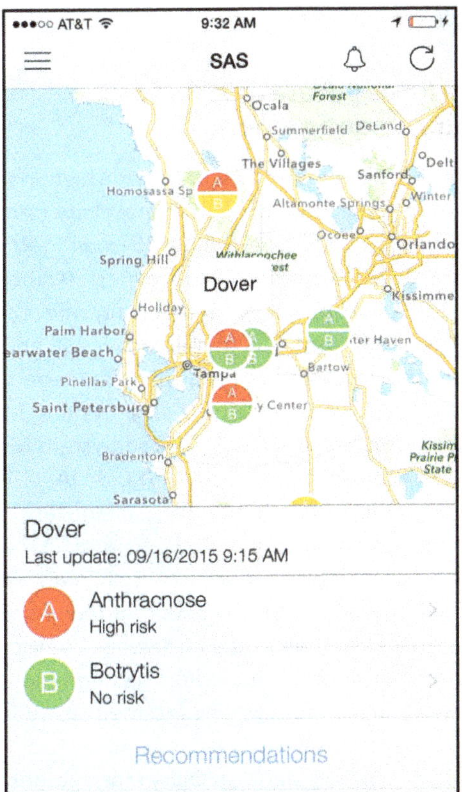

Figure 3 Mobile phone app of the Strawberry Advisory System (SAS) showing high risk for Anthracnose and low risk for Botrytis infection at the Dover station.

improved version were developed. The improved version (SAS Pro) integrates disease infection models, field application records, and fungicide resistance management guidelines to provide more specific product and spraying recommendations by automatically managing restrictions of fungicide use and chemical class.

The AgroClimate Citrus Advisory System (CAS) (http://agroclimate.org/tools/cas/) is another example of a DSS developed for management of plant diseases. Postbloom fruit drop (PFD), caused by the fungus *Colletotrichum acutatum*, affects all species and cultivars of citrus, but severity on a given cultivar varies according to the time of bloom in relation to rainfall (Peres and Dewdney, 2018). The system uses real-time weather data from Florida Automated Weather Network (FAWN) stations throughout the state to determine if risk conditions for PFD development are low (green), moderate (yellow), or high (red). Spray applications are recommended when sufficient bloom is present or developing to represent a significant portion of the total crop, no fungicide application has been made in the last 7–10 days, and environmental conditions are favorable for the development of the disease.

5 Examples of climate-based DSSs for livestock (dairy) management

Dairy farming is a highly dynamic and integrated production system that requires continuous and intense decision-making. Several dairy farm components that include (1) cattle, (2) crops, (3) soils, (4) climate, (5) management, (6) economics, and (7) environment are extremely interrelated (Cabrera, 2018). These components and their subcomponents dynamically affect and are affected by each other. Therefore, an efficient DSS framework within an integrated systems approach is critical for successful dairy farming management and decision-making (Cabrera, 2012). The Integrated Farm System Model (IFSM) (Rotz et al., 2018) is an expansion of the Dairy Forage System Model or DAFOSYM that started to be developed in the early 1980s. DAFOSYM included feed storage, animal performance, manure handling, tillage, and planting operation components in order to simulate a full dairy farm. It could also simulate grass, small grain, and soybean growth, harvest, and storage. Through a major revision, a beef animal component was added along with a crop farm option to form the IFSM. More recently, additional components were added to simulate environmental impacts, including gas emissions, nitrate leaching, and phosphorus runoff, and a life cycle assessment to determine the carbon footprint of production systems.

The University of Wisconsin-Madison Dairy Management program has also produced a suite of computerized decision support tools to assist dairy farm management that are openly and freely available at http://DairyMGT.info. Tools are customizable to be farm specific. Dairy farmers and dairy farm advisers are using these decision support tools to make better more informed decision-making. Better decision-making is translated into improved performance, reduced environmental impact, and enhanced profitability (Cabrera, 2018).

The dairy farm business faces important challenges regarding increased environmental scrutiny. An increasingly important dairy farm management task is to maintain a farm nutrient balance and therefore avoid overconcentration of nutrients in or around the farm. Opportunities exist to better utilize nutrients in dairy farming and not only improve the balance of nutrients coming in and going out of the farm, but also decrease fertilizer expenses and therefore environmental concerns (Cabrera, 2018). Depending on the farm herd and crop characteristics, additional expenses might be required to comply with environmental regulations. Economic assessment along with the environmental requirements promotes better decision-making. One of the UW-Dairy Management DSSs is the tool Dynamic North Florida Dairy Farm Mode or DyNoFlo Dairy (Fig. 4), which is an integrated, whole-farm, simulation and optimization model that maximizes the net economic return

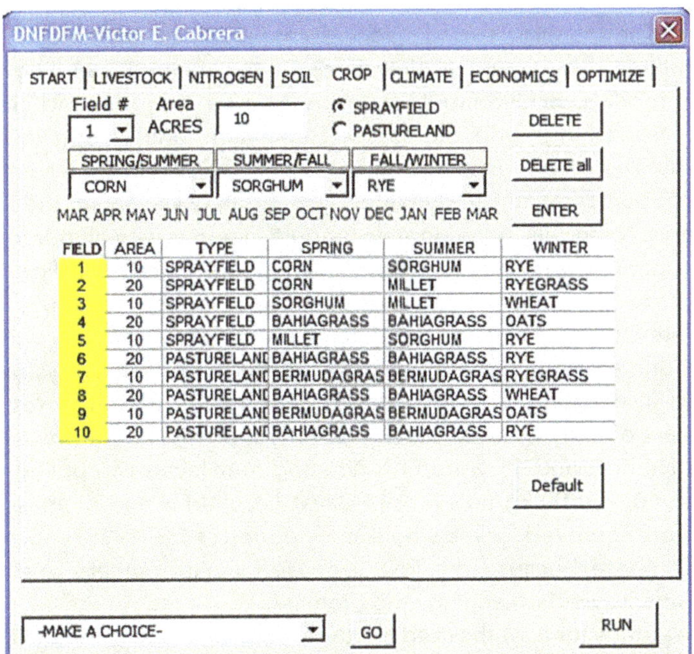

Figure 4 Screen display of the DyNoFlo Dairy model showing the crop forage systems control. Other controls (tabs) include the livestock module; the nitrogen waste management; the dairy farm location and soil type; the climatic component; the economic component; and the optimization algorithm. Source: adapted from Cabrera et al. (2005).

while minimizing nitrogen leaching to surface and groundwater sources. It helps dairy farmers project seasonally how much dairy cow manure (and consequently nutrients in the manure) will be produced and will be needed to be recycled.

The DyNoFlo Dairy integrates nutrient budgeting, crop, and optimization models created to assess N leaching from North Florida dairy farm systems and the economic impacts resulting from reducing it under different climatic conditions. The DSS, based on Excel® and Visual Basic® software, responds to dairy-specific environmental (climate and soils) and managerial characteristics (livestock management, waste management, crop systems management) and can be used to study the economic and ecologic sustainability of these systems. The DyNoFlo Dairy model is a dynamic adaptation of the framework 'balance' of nutrients in dairy farms, commonly used in Florida that is intended to be a tool for producers, regulatory agencies, and extension services (Cabrera et al., 2006b). The DyNoFlo Dairy model incorporates Markov chain probabilistic simulation of cow flows and crop simulation for historical climatic years ENSO, automated optimization of managerial options, participatory modeling, and user-friendliness. The livestock model simulates aging and distribution of cows, and manure N excretion. The manure N model receives inputs from the livestock model and simulates the manure N flow through the manure handling system. The crop models receive inputs from the manure N model and climate to simulate N leaching, crop biomass accumulation, and N uptake into the crop fields. The crop models are run on a daily basis and then summarized in monthly outcomes to run dynamically in monthly steps with the livestock and manure N models. An economic module interacts with all other models. Each component responds to a set of management practices. For the crop module, crop simulation models from the DSSAT (Jones et al., 2003) are used to translate climate, soil, and farm management practices into agricultural and environmental outcomes (Cabrera et al., 2006b). The crop systems were simulated in dairy sprayfields for ten types of soil found in North Florida, for 43 years of daily weather data (1956–1998, of which nine were La Niña, 23 were Neutral, and 11 were El Niño), and four levels of applied manure N (10, 20, 40, and 80 kg/ha every 15 days). Leached N and biomass (kg ha/month) were compiled for the whole study period (1956–1998) and classified by ENSO phase because the ENSO phase (climate) affects weather and weather affects yields, leaching, and biomass.

A case study for a synthesized North Florida dairy farm using the DyNoFlo Dairy model found substantial differences in the N leaching for different ENSO phases and other managerial factors; and the possibility of decreasing N leaching up to 25% while still maintaining profitability levels (Cabrera et al., 2005, 2006b).

6 Lessons learned on the development of climate-based DSSs

Forecast information can have value only if farmers can change their actions in beneficial ways based on the content of the information (Letson et al., 2005; Stern and Easterling, 1999). Surveys showed that DSSs can help guide decisions about what crops and varieties to plant, where on the farm and when a crop should be planted, fertilization rates, insurance purchases, input purchases, input applications, and marketing decisions. Climate factors of importance in relation to these decisions include temperature, rainfall, humidity, sunlight, chill unit accumulation, and growing degree days. Interviewees stressed the importance of extreme events such as storms, floods, droughts, frosts, freezes, and hail.

However, many farmers remain unsure about the validity and reliability of seasonal climate forecasts, especially regarding local variations. Diversity of agroecological conditions and farming systems directly influences the importance placed upon seasonal climate variability and decision-making by extension agents and farmers (Bartels et al., 2013). Producers insist that for a forecast to be used it must be at a level of sufficient resolution that it relates to their personal farm or county rather than for a region. Several producers identified within-season distribution of rainfall as more important than seasonal total, which is particularly important in areas of sandy soils having low water retention capacity. Typically, mixed crop and livestock farmers avoid risk by managing for adverse climate years as was noted by Thornton et al. (2004).

Decisions regarding management practices depend on many factors, not just climate. Many farmers are 'locked in' due to lack of flexibility in their farming systems. They express concern about the cost of changing management practices and insist that any recommendation should take the whole production system into account. If the potential benefits of responding to a climate forecast are relatively small per unit area, or if costs of implementing recommendations are high, then only farmers with plentiful resources on large farms might profitably use such recommendations. At the regional planning and policymaking levels, however, aggregate benefits are an important consideration in cost-benefit analysis of using or recommending climate-based decision support. Many growers state that influences on their decisions are highest from government regulations and market fluctuations. Other concerns include availability of labor and free trade agreements.

Small and organic farmers are on average, less interested in variety or crop recommendations than were larger farmers because they tend to mix crops and varieties in order to diversify and minimize risk, whereas larger farmers grew fewer crops and varieties in order to achieve economies of scale. Small and organic farmers reported an interest in getting recommendations for pest and disease management practices that depend on climate, and in getting detailed frost information (Furman et al., 2011, 2014).

DSS developers learned that internet use varies among different segments of the farmer population. Extension agents considered women and adolescents to be more likely to use the internet, therefore, they could be targeted as principal users of DSSs. In the Southeast United States, farmers' wives tend to handle many of the logistical details involved in the operation of the farm, a crucial gender observation. Older farmers with less computer experience may not be comfortable using web-based technology. The issue of access of and skillful use of the internet is even more crucial in developing country contexts.

7 Future trends and conclusion

Tailoring climate information to risk management problems is key methodological issue. Early and effective engagement with stakeholders is essential. Timely and effective implementation depends largely on institutional mapping (Hartmann, 2002; Breuer, 2008; Bartels et al., 2013). In the case of AgroClimate, using participatory research methodologies, the DSS was continually exposed to criticism and feedback to ensure the appropriateness of tools and qualitative information to a diverse community of farmers, policy makers, and extension agents. Foundational work (Mjelde et al., 1988; Stern and Easterling, 1999; Letson et al., 2005) had posited that forecasts are useful and have value only if they permit *ex ante* actions, such as altered choice of crop species and cultivars and shifting planting dates. Interviewees often agree that forecasts are helpful only if they arrive well in advance of when planting, input purchase, or stocking decisions are made. In the words of farmers interviewed 'timing of forecast availability may be more important than forecast accuracy.' Furthermore, site specificity or regional conditions are necessary and positively affect the usefulness of forecasts (Cabrera et al., 2006a; Breuer et al., 2010, 2011; Templeton et al., 2018) (Table 2).

Extension personnel and crop consultants (Prokopy, 2013) are 'users,' who, in turn, become 'producers' as they translate, repackage, or further analyze information for their clients. Members of boundary organizations are key to understanding farmer needs and adjustable management practices, and must be included in participatory DSS development together with producers and scientists. A focus on developing DSSs that promote user-driven risk management objectives where process and outcome are of equal weight can be achieved through participatory development and improvement. A collaborative, user-driven DSS might even better withstand an incorrect forecast (Cash and Buizer, 2005). It is likely that benefits such as higher adoption rates, more varied adaptation strategies, and user-trust in the provider institution are enhanced by the participatory, collaborative research, and development

Table 2 Summary of recommendations compiled from authors cited in this section for participatory DSS development[a]

- Focus on integrated climate-society system
- Decision-makers interested in information on a continuum of timescales
- Early and continuous partnership with users is essential
- Shared learning and joint problem-solving is key
- Outreach and dialogue programs should be priority activities
- Stress probabilistic nature of forecasts
- Building trust and credibility is a long-term endeavor
- Understanding the decision calendar and decision-making process is vital
- Continuous, interactive dialogue is needed
- Coproduction of knowledge improves adoption
- Experiences should be documented and shared
- A problem-focused approach engages stakeholders better
- Understanding place, context, history, and decision-making process are keys to success
- Useful and usable information responsive to different user needs
- Continuum of timescales, from weather through variability to change
- Climate information system is more useful than simple event forecasting

[a] Source: adapted from Reed (2008) and Breuer et al. (2008, 2009).

process. DSS use adds to systems' resilience and is a crucial component to adaptation to climate variability and change in agricultural production.

8 Where to look for further information

Although significant advances in the development of climate-based DSS have been made in recent years, it is clear that beyond effective engagement of stakeholders, future DSS development will need to integrate weather and climate information from different sources. Data generated from local farmer-owned weather stations to satellites will need to be integrated with forecasts and adequate models designed to help farmers make better strategic, tactical, and operational decisions. The future of agriculture will require the ability to gather and analyze big data generated by sources with distinct spatial and temporal resolutions. It is also clear that research efforts on the integration of these new data sources with models must be carried on in parallel with educational efforts designed to help farmers, agricultural managers, extension faculty, and consultants obtaining the technological education needed to make effective use of climate-based DSSs.

9 References

Angel, J. R., Widhalm, M., Todey, D., Massey, R. and Biehl, L. 2017. The U2U Corn Growing Degree Day tool: tracking corn growth across the US Corn Belt. *Clim. Risk Manag.* 15, 73–81. doi:10.1016/j.crm.2016.10.002.

Ashby, J. A. 2002. Integrating research on food and the environment: an exit strategy from the rational fool syndrome in agricultural science. *Ecol. Soc.* 5(2), 20. doi:10.5751/ES-00296-050220.

Bartels, W., Furman, C. A., Diehl, D. C., Royce, F. S., Dourte, D. R., Ortiz, B. V., Zierden, D. F., Irani, T. A., Fraisse, C. W. and Jones, J. W. 2013. Warming up to climate change: a participatory approach to engaging with agricultural stakeholders in the Southeast US. *Reg. Environ. Change* 13 (Suppl. 1), 45–55. doi:10.1007/s10113-012-0371-9.

Basso, B., Cammarano, D., Fiorentino, C. and Ritchie, J. T. 2013. Wheat yield response to spatially variable nitrogen fertilizer in Mediterranean environment. *Eur. J. Agron.* 51, 65–70. doi:10.1016/j.eja.2013.06.007.

Beck, H. W., Jones, P. and Jones, J. W. 1989. SOYBUG: an expert system for soybean insect pest management. *Agric. Syst.* 30(3), 269–86. doi:10.1016/0308-521X(89)90091-7.

Bolson, J., Martinez, C., Breuer, N., Srivastava, P. and Knox, P. 2012. Climate information use among Southeast U.S.A. water managers: beyond barrier identification and toward learning from regional water managers. *Reg. Env. Change* 1, 141–51.

Breuer, N. E., Cabrera, V. E., Ingram, K. T., Broad, K. and Hildebrand, P. E. 2008. AgClimate: a case study in participatory decision support system development. *Clim. Change* 87(3–4), 385–403. doi:10.1007/s10584-007-9323-7.

Breuer, N. E., Cabrera, V. E. and Hildebrand, P. E. 2009. Molding the pipeline into a loop: the participatory process of developing AgroClimate, a decision support system for climate risk reduction in agriculture. *J. Serv. Climatol.* 3, 1–12.

Breuer, N. E., Fraisse, C. W. and Cabrera, V. E. 2010. The cooperative extension service as a boundary organization for diffusion of climate forecasts: a 5-year study. *J. Ext.* 48, 4RIB7. Available at: http://www.joe.org/joe/2010august/rb7.php.

Breuer, N. E., Dinon, H., Boyles, R. and Wilkerson, G. 2011. North Carolina extension agent awareness of and interest in climate information. *J. Serv. Climatol.* 5, 1–20.

Cabrera, V. E. 2012. DairyMGT: a suite of decision support systems in dairy farm management. In: Jao, C. (Ed.), *Decision Support Systems*. Intech, Rijeka, Croatia.

Cabrera, V. E. 2018. Helping dairy farmers to improve economic performance utilizing data-driving decision support tools. *Animal* 12(1), 134–44. doi:10.1017/S1751731117001665.

Cabrera, V. E., Breuer, N. E., Hildebrand, P. E. and Letson, D. 2005. The dynamic north-Florida dairy farm model: a user-friendly computerized tool for increasing profits while minimizing environmental impacts. *Comput. Electron. Agric.* 49(2), 286–308. doi:10.1016/j.compag.2005.07.001.

Cabrera, V. E., Breuer, N. E., Bellow, J. and Fraisse, C. W. 2006a. *Extension Agents' Knowledge and Perceptions of Seasonal Climate Forecasts in Florida*. Southeast Climate Consortium Technical Report Series: SECC Technical Report 06-001, Gainesville, FL, 15pp.

Cabrera, V. E., Hildebrand, P. E., Jones, J. W., Letson, D. and de Vries, A. 2006b. An integrated North Florida dairy farm model to reduce environmental impacts under seasonal climate variability. *Agric. Ecosyst. Environ.* 113(1–4), 82–97. doi:10.1016/j.agee.2005.08.039.

Cabrera, V. E., Breuer, N. E. and Hildebrand, P. E. 2008. Participatory modeling of North Florida dairy farm systems: a method for building climate variability into farm models. *Clim. Change* 89(3–4), 395–409. doi:10.1007/s10584-007-9371-z.

Carberry, P. S. Hochman, Z., McCown, R. L., Dalgliesh, N. P., Foale, M. A., Poulton, P. L., Hargreaves, J. N. G., Hargreaves, D. M. G., Cawthray, S., Hillcoat, N. and Robertson,

M. J. 2002. The FARMSCAPE approach to decision support: Farmers', advisers', researchers' monitoring, simulation, communication and performance evaluation. *Agric. Syst.* 74, 141–77. doi:10.1016/S0308-521X(02)00025-2.

Cash, D. W. and Buizer, J. 2005. *Knowledge-Action Systems for Seasonal to Interannual Climate Forecasting: Summary of a Workshop*. National Academy Press, Washington DC.

Chambers, R. 1994. The origins and practice of participatory rural appraisal. *World Dev.* 22(7), 953–69. doi:10.1016/0305-750X(94)90141-4.

Chambers, R. 1995. *Whose Reality Counts?* Intermediate Technology Publications, Sussex, UK and Research/Extension Symposium, Manhattan, KS.

Cowan, L., Kaine, G. and Wright, V. 2013. The role of strategic and tactical flexibility in managing input variability on farms. *Syst. Res.* 30(4), 470–94. doi:10.1002/sres.2137.

Cox, P. G. 1996. Some issues in the design of agricultural decision support systems. *Agric. Syst.* 52(2–3), 355–81. doi:10.1016/0308-521X(96)00063-7.

Crane, T., Roncoli, C., Breuer, N., Broad, K., Paz, J., Fraisse, C., Ingram, K., Zierden, D., Hoogenboom, G. and O'Brien, J. 2008. Collaborative approaches to the development of climate-based decision support systems. In: *Proceedings of the 88th AMS Annual Meeting*, New Orleans, LA, 20–24 January 2008.

Crane, T., Roncoli, C., Paz, J., Breuer, N. E., Broad, K., Fraisse, C., Ingram, K., Zierden, D. and Hoogenboom 2010. Forecast skill and farmers' skills: seasonal climate forecasts and agricultural risk management in the Southeastern United States. *Wea. Clim. Soc.* 2, 44–59.

Fraisse, C. W., Breuer, N. E., Zierden, D., Bellow, J. G., Paz, J., Cabrera, V. E., Garcia y Garcia, A., Ingram, K. T., Hatch, U., Hoogenboom, G., et al. 2006. AgClimate: a climate forecast information system for agricultural risk management in the Southeastern USA. *Comput. Electron. Agric.* 53(1), 13–27. doi:10.1016/j.compag.2006.03.002.

Furman, C., Roncoli, C., Crane, T. and Hoogenboom, G. 2011. Beyond the 'fit': introducing climate forecasts among organic farmers in Georgia (United States). *Clim. Change* 109(3–4), 791–9. doi:10.1007/s10584-011-0238-y.

Furman, C., Roncoli, C., Nelson, D. R. and Hoogenboom, G. 2014. Growing food, growing a movement: climate adaptation and civic agriculture in the southeastern United States. *Agric. Hum. Values* 31(1), 69–82. doi:10.1007/s10460-013-9458-2.

Geurts, J. L. A. and Joldersma, C. 2001. Methodology for participatory policy analysis. *Eur. J. Oper. Res.* 128(2), 300–10. doi:10.1016/S0377-2217(00)00073-4.

Hammer, G. L., Hansen, J. W., Phillips, J. G., Mjelde, J. W., Hill, H., Love, A. and Potgieter, A. 2001. Advances in application of climate prediction in agriculture. *Agric. Sys.* 70(2–3), 515–53. doi:10.1016/S0308-521X(01)00058-0.

Han, E., Ines, A. V. M. and Baethgen, W. E. 2017. Climate-Agriculture-Modeling and Decision Tool (CAMDT): a software framework for climate risk management in agriculture. *Env. Modell. Softw.* 95, 102–14. doi:10.1016/j.envsoft.2017.06.024.

Hansen, J. W. 2002. Realizing the potential benefits of climate prediction in agriculture. *Agric. Syst.* 74(3), 309–30. doi:10.1016/S0308-521X(02)00043-4.

Hartmann, H. C., Pagano, T. C., Sorooshian, S. and Bales, R. 2002. Confidence builders: evaluating seasonal climate forecasts from user perspectives. *Bull. Amer. Meteor. Soc.* 83(5), 683–98. doi:10.1175/1520-0477(2002)083<0683:CBESCF>2.3.CO;2.

Hayman, P. T. and Easdown, W. J. 2002. An ecology of a DSS: reflections on managing wheat crops in the northeastern Australian grains region with WHEATMAN. *Agric. Syst.* 74(1), 57–77. doi:10.1016/S0308-521X(02)00018-5.

Hildebrand, P. E. 1981. Combining disciplines in rapid appraisal: the Sondeo approach. *Agric. Admin.* 8(6), 423–32. doi:10.1016/0309-586X(81)90037-6.

Hildebrand, P., Caudle, A., Cabrera, V., Downs, M., Langholtz, M., Mugisha, A., Sandals, R., Shriar, A. and Beach, K. 1999. *Potential Use of Long-Range Climate Forecasts by Agricultural Extension Advisors in Florida: A Sondeo Report (Staff Paper SP 99-9).* Food and Resource Economics Department, University of Florida, Gainesville, FL.

Hochman, Z., Rees, H. V., Carberry, P. S., Hunt, J. R., McCown, R. L., Gartmann, A., Holzworth, D., Rees, S. V., Dalgliesh, N. P., Long, W., et al. 2009. Re-inventing model-based decision support with Australian dryland farmers. 4. Yield Prophet® helps farmers monitor and manage crops in a variable climate. *Crop and Past Science* 60(11), 1057–70.

Hollinger, S. E. and Angel, J. R. 2009. Weather and crops. In: *Illinois Agronomy Handbook, Department of Crop Sciences.* University of Illinois at Urbana-Champaign, College of Agriculture, Cooperative Extension Service, pp. 1–12.

Hoogenboom, G., Jones, J. W., Wilkens, P. W., Porter, C. H., Boote, K. J., Hunt, L. A., Singh, U., Lizaso, J. L., White, J. W., Uryasev, O., et al. 2012. *Decision Support System for Agrotechnology Transfer (DSSAT)* (Version 4.5). CD-ROM. University of Hawaii, Honolulu, HI.

Jabrzemski, R. and Sutherland, A. 2006. An innovative approach to weather-based decision-support for agricultural models. In: *22nd International Conference on Interactive Information Processing Systems for Meteorology, Oceanography, and Hydrology.* American Meteorological Society, Washington DC.

Jacques, P. J., Broad, K., Butler, W., Emrich, C., Galindo, S., Knox, C., Rizzardi, K. W. and Ziewitz, K. 2017. Human dimensions and communication of Florida's climate. In: Chassignet, E. P., Jones, J. W., Misra, V. and Obeysekera, J. (Eds), *Florida's Climate: Changes, Variations, & Impacts.* Florida Climate Institute, Gainesville, FL. doi:10.17125/fci2017.

Jagtap, S. S., Jones, J. W., Hildebrand, P., Letson, D., O'Brien, J. J., Podestá, G., Zierden, D. and Zazueta, F. 2002. Responding to stakeholder's demands for climate information: from research to applications in Florida. *Agric. Syst.* 74(3), 415–30. doi:10.1016/S0308-521X(02)00048-3.

Jones, J. W., Hansen, J. W., Royce, F. S. and Messina, C. D. 2000. Potential benefits of climate forecasting to agriculture. *Agric. Ecosyst. Environ.* 82(1–3), 169–84. doi:10.1016/S0167-8809(00)00225-5.

Jones, J. W., Hoogenboom, G., Porter, C. H., Boote, K. J., Batchelor, W. D., Hunt, L. A., Wilkens, P. W., Singh, U., Gijsman, A. J. and Ritchie, J. T. 2003. The DSSAT cropping system model. *Eur. J. Agron.* 18(3–4), 235–65. doi:10.1016/S1161-0301(02)00107-7.

Jones, J. W., Antle, J. M., Basso, B., Boote, K. J., Conant, R. T., Foster, I., Godfray, H. C. J., Herrero, M., Howitt, R. E., Jansseni, S., et al. 2017. Brief history of agricultural systems modeling. *Agric. Syst.* 155, 240–54. doi:10.1016/j.agsy.2016.05.014.

Keating, B. A., Carberry, P. S., Hammer, G. L., Probert, M. E., Robertson, M. J., Holzworth, D., Huth, N. I., Hargreaves, J. N. G., Meinke, H., Hochman, Z., et al. 2003. An overview of APSIM, a model designed for farming systems simulation. *Eur. J. Agron.* 18(3–4), 267–88. doi:10.1016/S1161-0301(02)00108-9.

Letson, D., Hansen, J., Hildebrand, P., Jones, J. W., O'Brien, J. J., Podestá, G., Royce, F. and Zierden, D. 2001. *Florida's Agriculture and Climatic Variability: Reducing Vulnerability.* Florida Consortium Technical Report FCUM-2001-001.

Letson, D., Podestá, G. P., Messina, C. D. and Ferreyra, R. A. 2005. The uncertain value of perfect ENSO phase forecasts: stochastic agricultural prices and intra-phase climatic variations. *Clim. Change* 69(2–3), 163–96. doi:10.1007/s10584-005-1814-9.

Lundström, C. and Lindblom, J. 2018. Considering farmers' situated knowledge of using agricultural decision support systems (AgriDSS) to Foster farming practices: the case of CropSAT. *Agric. Syst.* 159, 9–20. doi:10.1016/j.agsy.2017.10.004.

Marx, S. M., Weber, E. U., Orlove, B. S., Leiserowitza, A., Krantza, D. H., Roncolia, C. and Phillips, J. 2007. Communication and mental processes: experiential and analytic processing of uncertain climate information. *Global Environ. Change* 17(1), 47–58. doi:10.1016/j.gloenvcha.2006.10.004.

McCown, R. L., Hochman, Z. and Carberry, P. S. 2002. Probing the enigma of the decision support system for farmers: learning from experience and from theory. *Agric. Syst.* 74(1), 1–10. doi:10.1016/S0308-521X(02)00021-5.

Meinke, H., Howden, S. M., Struik, P. C., Nelson, R., Rodriguez, D. and Chapman, S. C. 2009. Adaptation science for agriculture and natural resource management–urgency and theoretical basis. *Curr. Opin. Environ. Sustain.* 1(1), 69–76. doi:10.1016/j.cosust.2009.07.007.

Mjelde, J. W., Sonka, S. T., Dixon, B. L. and Lamb, P. J. 1988. Valuing forecast characteristics in a dynamic agricultural production system. *Am. J. Agric. Econ.* 70(3), 674–84. doi:10.2307/1241506.

Morse, R. S. 2010. Integrative public leadership: catalyzing collaboration to create public value. *Leadersh. Q.* 21(2), 231–45. doi:10.1016/j.leaqua.2010.01.004.

Parker, C. 1999. Decision support systems: lessons from past failures. *Farm Manag.* 10, 273–89.

Pavan, W., Fraisse, C. W. and Peres, N. A. 2011. Development of a web-based disease forecasting system for strawberries. *Comput. Electron. Agric.* 75(1), 169–75. doi:10.1016/j.compag.2010.10.013.

Peres, N. A. and Dewdney, M. M. 2018. 2018–2019 Florida citrus production guide: postbloom fruit drop. University of Florida Institute of Food and Agricultural Sciences EDIS PP-45. Available at: https://edis.ifas.ufl.edu/pdffiles/CG/CG00700.pdf.

Perondi, D., Fraisse, C. W., Staub, C. G., Cerbaro, V. A., Barreto, D. D., Pequeno, D. N. L., Mulvaney, M. J., Troy, P. and Pavan, W. 2019. Crop season planning tool: adjusting sowing decisions to reduce the risk of extreme weather events. *Comput. Electron. Agric.* 156, 62–70. doi:10.1016/j.compag.2018.11.013.

Podestá, G. P., Messina, C. D., Grondona, M. O. and Magrin, G. O. 1999. Associations between grain crop yields in centraleastern Argentina and El Niño-Southern Oscillation. *J. Appl. Meteorol.* 38(10), 1488–98. doi:10.1175/1520-0450(1999)038<1488:ABGCYI>2.0.CO;2.

Power, D. J. 2002. *Decision Support Systems: Concepts and Resources for Managers*. Greenwood Publishing Group Inc., 251pp.

Prokopy, L. S., Haigh, T., Mase, A. S., Angel, J., Hart, C., Knutson, C., Lemos, M. C., Lo, Y., McGuire, J., Morton, L. W., et al. 2013. Agricultural advisors: a receptive audience for weather and climate information? *Wea. Climate Soc.* 5(2), 162–7. doi:10.1175/WCAS-D-12-00036.1.

Prokopy, L. S., Carlton, J. S., Haigh, T., Lemos, M. C., Mase, A. S. and Widhalm, M. 2017. Useful to usable: developing usable climate science for agriculture. *Clim. Risk Manag.* 15, 1–7. doi:10.1016/j.crm.2016.10.004.

Reed, M. S. 2008. Stakeholder participation for environmental management: a literature review. *Biol. Conserv.* 141(10), 2417–31. doi:10.1016/j.biocon.2008.07.014.

Rogers, E. M. 2003. *Diffusion of Innovations*. Free Press, New York.

Roncoli, C. 2006. Ethnographic and participatory approaches to research on farmers' responses to climate predictions. *Clim. Res.* 33, 81–99. doi:10.3354/cr033081.

Roncoli, C., Breuer, N. E., Zierden, D., Broad, K., Fraisse, C. W. and Hoogenboom, G. 2012. The art of the science: climate forecasts and fire management in the Southeastern United States, climatic change. *Letters* (3–4), 1113–21.

Rosenzweig, C., Elliott, J., Deryng, D., Ruane, A. C., Müller, C., Arneth, A., Boote, K. J., Folberth, C., Glotter, M., Khabarov, N., et al. 2014. Assessing agricultural risks of climate change in the 21st century in a global gridded crop model intercomparison. *Proc. Natl. Acad. Sci. U. S. A.* 111(9), 3268–73. doi:10.1073/pnas.1222463110.

Rotz, C. A., Corson, M. S., Chianese, D. S., Montes, F., Hafner, S. D., Bonifacio, H. F. and Coiner, C. U. 2018. *The Integrated Farm System Model Reference Manual v. 4.4*. USDA Agricultural Research Service. Available at: https://www.ars.usda.gov/ARSUserFiles/80700500/Reference%20Manual.pdf.

Shtienberg, D. 2013. Will decision-support systems be widely used for the management of plant diseases? *Annu. Rev. Phytopathol.* 51, 1–16. doi:10.1146/annurev-phyto-082712-102244.

Small, I. M., Joseph, L. and Fry, W. E. 2015. Development and implementation of the BlightPro decision support system for potato and tomato late blight management. *Comput. Electron. Agric.* 115, 57–65. doi:10.1016/j.compag.2015.05.010.

Smith, B. 1998. Anthracnose fruit rot (black spot). In: Mass, J. L. (Ed.), *Compendium of Strawberry Diseases* (2nd edn.). APS Press, St. Paul, MN, pp. 31–3.

Stern, P. C. and Easterling, W. E. (Eds). 1999. *Making Climate Forecasts Matter*. National Academy Press, Washington DC.

Stöckle, C. O., Donatelli, M. and Nelson, R. 2003. CropSyst, a cropping systems simulation model. *Eur. J. Agron.* 18(3–4), 289–307. doi:10.1016/S1161-0301(02)00109-0.

Stringer, L. C., Dougill, A. J., Fraser, E., Hubacek, K., Prell, C. and Reed, M. S. 2006. Unpacking 'participation' in the adaptive management of social-ecological systems: a critical review. *Ecol. Soc.* 11(2), 39. Available at: http://www.ecologyandsociety.org/vol11/iss2/art39/.

Suarez, P. and Patt, A. G. 2010. Cognition, caution, and credibility: the risks of climate forecast application. *Risk, Decision, and Policy* 1, 75–89.

Sutton, J. 1998. Botrytis fruit rot (gray mold) and blossom blight. In: Mass, J. L. (Ed.), *Compendium of Strawberry Diseases* (2nd edn.). APS Press, St. Paul, MN, pp. 28–31.

Takle, E. S., Anderson, C. J., Andresen, J., Angel, J., Elmore, R. W., Gramig, B. M., Guinan, P., Hilberg, S., Kluck, D., Massey, R., et al. 2014. Climate forecasts for corn producer decision making. *Earth Interact.* 18(5), 1–8: doi:10.1175/2013EI000541.1.

Templeton, S., Hooper, A., Aldrich, H. and Breuer, N. 2018. Farmer interests in and uses of climate forecasts for Florida and the carolinas: conditional perspectives of extension personnel. *Wea. Clim. Soc.* 10(1), 103–20. doi:10.1175/WCAS-D-16-0057.1.

Thornton, P. K., Fawcett, R. H., Galvin, K. A., Boone, R. B., Hudson, J. W. and Vogel, C. H. 2004. Evaluating management options that use forecasts: modeling livestock production systems in the semi-arid zone of South Africa. *Clim. Res.* 26, 33–42.

Voinov, A., Jenni, K., Gray, S., Kolagani, N., Glynn, P. D., Bommel, P., Prell, C., Zellner, M., Paolisso, M., Jordan, R., et al. 2018. Tools and methods in participatory modeling: selecting the right tool for the job. *Environ. Modell. Softw.* 109, 232–55. doi:10.1016/j.envsoft.2018.08.028.

Chapter 12

Landscape models to support sustainable intensification of agroecological systems

C. Nendel and P. Zander, Leibniz Centre for Agricultural Landscape Research (ZALF), Germany

1 Introduction

The inclusion of a landscape perspective into agroecological systems modelling is the consequence of two different challenges that agriculture is facing. On one hand, further intensification of agriculture is needed to meet the growing demand of food and renewable resources. On the other hand, after years of intensification and optimisation of agricultural production systems, the current land use has led to a considerable decrease in species diversity and abundance with, for example, a dramatic decrease of insects (Hallmann et al., 2017; Vogel, 2017) and birds (DEFRA, 2018; Hallmann et al., 2014) in agricultural landscapes, to a considerable level of pollution in water bodies (EEA, 2018; USEPA, 2016) and to an alarming decline in fresh water resources around the globe (Boyer et al., 2006; Jury and Vaux, 2007; Mekonnen and Hoekstra, 2016; Vörösmarty et al., 2010). To protect our abiotic and genetic resources, the natural functioning of ecosystems and the diversity of landscapes, users of agricultural land must reduce resource use and impact on biodiversity and ecosystems. Addressing both challenges at the same time requires a perspective that takes a more detailed look into the heterogeneity of climate and soils and their role in agroecosystem functioning, the interactions between single plants and the

http://dx.doi.org/10.19103/AS.2019.0061.17

surrounding flora and fauna, the interdependencies between agricultural land and the surrounding natural and semi-natural areas and the role of humans and how they organise themselves in this system. This integrated view is essential for the development of concepts for a sustainable use of land.

Landscape changes are driven by a combination of institutional, cultural, economic, technological and natural factors (Plieninger et al., 2016) in which an adaptive management, different spatial scales, multi-functionality and numerous objectives interact (Sayer et al., 2013). Decisions towards adaptation to climate change or mitigation of climate impacts through agriculture are primarily made at the farm level. These farm-level decisions are responsible for the impacts of agriculture on natural resources and biodiversity and they are influenced by the social, economic and geo-ecological setting of the farm, its fields and its people. The system boundaries to describe the decision-making and the impact of the decisions go far beyond current mechanistic and one-dimensional crop and agroecosystem models, and often also beyond whole-farm models. For such purpose, integrated and scalable modelling approaches are in demand that address relevant landscape processes at the interface of different sub-systems. Beyond the interdisciplinary nature of modelling agricultural systems (Papajorgji and Panos, 2009), the integration of the landscape perspective into models requires the consideration of additional societal aspects present in *'an area perceived by people, whose character is the result of the action and interaction of natural and/or human factors'* (Council of Europe, 2000). In this context, farmers are actors among many others, and their role as landscape 'stewards' (Bieling and Plieninger, 2017) involves being producers, landowners and citizens at the same time (Farina, 2000).

In this chapter, we reflect on the different social, economic and natural features of agricultural landscapes that govern important processes in agroecosystems and discuss current modelling approaches that go beyond the possibilities which current agroecosystem models provide. We begin with the (i) motivation of modelling ecosystem services using a landscape approach (ii) present features of the current (iii) socio-economic and (iv) geo-ecological setting in landscapes and how they are considered in modelling approaches, then (v) address the modelling of biotic interactions and (vi) finally present a case study that demonstrates the state-of-the-art in landscape modelling.

2 Society and ecosystem services

The interactions between the different components of the ecosystem and society are currently captured in the concept of ecosystem services (ESS, Fisher et al., 2009) – the benefits humans freely gain from their natural environment. The concept distinguishes four broad categories: *provisioning*, such as the production of food and water; *regulating*, such as the control of climate and disease; *supporting*, such as nutrient cycles; and *cultural*, such as recreational benefits (Millennium Ecosystem Assessment, 2005). The presence of multiple actors leads to conflicting objectives, which are inherent to the concept of ESS, as not all services can be simultaneously attained at their maximum level. Maximisation of a single function may influence negatively other ecosystem functions and also human well-being (Seppelt et al., 2013). This often leads to a conflict between private and public interests that can be solved through regulations or compensation payments. Within the EU the Common Agricultural Policy (CAP) and the respective national embodiment of their laws, regulations and payments tries to govern land use in a way that supports a balanced provision of ESS. These regulations and especially regulations to improve biodiversity are under constant development.

Centuries ago, farmers were well aware of landscape features that impacted on the productivity of their fields (e.g. von der Marwitz, 1852). This included terrain sloping, terrain-induced soil type associations, wind breaking effects of hedges, shading and cold air emission by forest patches, just to name a few. Until recently, observed land use patterns and crop rotations represented much of this knowledge, including the choice of the best sites for the most productive land use (Nuppenau, 2018). Also, the beneficial effects of biodiversity, from pollination services to pest and disease control, have always been seen in the context of landscape-scale land use diversity. In conclusion there may be benefits of a small-scale and more precise management of crops for farmers and for society. To develop new management techniques and policies that enable and acknowledge the functioning of ecosystem services to support a more sustainable agriculture we have to understand in detail the interactions of ecosystems and management activities and how drivers and feedbacks influence actors decisions.

Economic transactions and matter flows within and between farms build the core dynamics of land use decisions. The landscape level perspective addresses the complex interactions between different spatial scales and human institutions (Sayer et al., 2013) developing land use management solutions that fit as much as possible to the needs of actors and communities (Seppelt et al., 2013). Therefore, modelling at landscape level, considering continuous learning and multiple perspectives (Sayer et al., 2013) is seen as a promising framework to analyse sustainability (Bieling and Plieninger, 2017) and to evaluate agro-environmental policies (Schönhart et al., 2011). As a consequence, any analysis of long-term agroecosystem dynamics requires a framing by biophysical (e.g. climate, terrain, soils) but also socio-economic conditions (e.g. agricultural markets, technologies, socio-economic narratives). Recently, integrated modelling of land use and management of ecosystems services has been used to depict impact chains, trade-offs and synergies in the spatial context, by linking data and models of multiple disciplines especially on the categories of *provisioning* and *regulating* and in lesser extent on *cultural* services of landscapes (Kirchner et al., 2015). Ideally, such analysis needs to explore (i) the response of plant growth and development to biophysical conditions, (ii) the interaction of plants among each other, (iii) the interaction of plants with the surrounding fauna, (iv) the impact of human action on plants, soils, terrain as well as energy and matter dynamics, (v) the impact of human action on land use and landscape structure and (vi) the interaction between different actors and (vii) the economic and social system setting which influence the actors. A bold vision - and for many questions of practical relevance a full representation as described may not be needed - but progress in landscape modelling is noted along this avenue, such as presented in the case studies at the end of this chapter.

3 Bio-economic and social interaction models

Agricultural land use is driven by policy and market conditions for farmers, access to markets and the agro-environmental conditions. Farmers continuously adapt to these conditions, they make investment decisions, plan crop rotations and crop management based on experience and expectations, taking the complex farm internal and external relations, matter and information flows, into account. Farmers' decision making based on their objectives and beliefs and these framing conditions largely determine land use change. In this process, farmers sometimes have to comply with a number of regulations regarding, for example, the use of chemical and genetic sources, the level of emissions allowed but also incentives to support specific ecosystem services that are not profitable but societally demanded. Farm models and especially bio-economic farm models are potentially relevant tools to assess ex ante the impact of different policies on farmers and thereby on ecosystem services (Reidsma et al., 2018; Uthes et al., 2008). Reidsma et al. (2018) found 202 articles published between 2007 and 2015 on applications of farm models for policy assessment, covering all relevant agricultural sectors from crop production over livestock to agroforestry-based systems.

While many studies state the objective 'to better understand farmer's decision making' (Reidsma et al., 2018), bio-economic models can also support the design of new farming and production systems (Groot et al., 2012). Most of the bio-economic farm models employ mathematical programming (MP), a technique that breaks the decision-making process of farmers down into mathematical equations that describe all production related activities and their interdependencies. It allows to account simultaneously for these interdependencies and to optimise the whole system for a given goal - in most applications the economic performance in terms of the total gross margin of the farm. Thereby, the more complex goal setting of individual farmers (e.g. have more spare time, or avoid heavy work load) is narrowed down to the economic performance. Individual farmers, however, adapt to changing conditions in different ways according to their resources and preferences. In many approaches this has been addressed by multi-objective programming, by using more detailed farmer-choices based restrictions and by introducing risk components etc. (Janssen and van Ittersum, 2007; Reidsma et al., 2018).

But farmers do not act alone, they are part of a community that, for example, share machinery, compete for land and may cooperate for a better marketing of their production. Agent-based models - as an extension of bio-economic models - often try to capture one or several of these interactions between farmers (e.g. Hampf et al., 2018). The agent-based model AGRIPOLIS is only one example that was originally developed to analyse structural change in response to different policy scenarios by modelling the exchange of land between farms through leasing or purchasing activities as the central interaction between farmers (Happe et al., 2006). At EU level the impact of externally defined structural change has been modelled by Espinosa et al (Espinosa et al., 2016). In more recent applications this approach was combined with behavioural experiments to analyse individual decision behaviour that substantially differed between individuals. With regards to the involvement of actors, the modelling approach can help to understand the complexity of stakeholder interaction (Le Pira et al., 2016) and support participatory decision-making processes among stakeholders. Challenges for the modelling of social interaction mechanisms at landscape level include both conceptual and methodological constraints. Firstly, the concepts and frameworks about landscape boundaries are still fuzzy and disperse (Bürgi et al., 2017) which makes the quantification of changes necessary (Plieninger et al., 2016). Secondly, the interaction between actors has traditionally been modelled as aggregated response patterns of groups of consumers and suppliers in trade models or land use change models (van Asselen and Verburg, 2013), as social networks (Le Pira et al., 2016) or using agent-based modelling (Le Pira et al., 2016; Zagaria et al., 2017).

Typically, farm-level agent-based and bio-economic modelling approaches try to include bio-physical impacts and interactions into the decision making by including more or less detailed bio-physical models or just their results (Carauta et al., 2018; Nuppenau, 2018; Peter et al., 2017). The way bio-physical assessments and interactions between different elements of the farm and of the landscape are modelled and included in economic approaches, differs largely. Brown (2000) describes the scope as ranging from simple inclusion of bio-physical performance indicators over bio-physical process models closely linked via defined interfaces to economic models to fully integrated models. The latter have the highest demand in programming and integrating expertise of different disciplines, while the first will not be able to integrate feedback loops in the modelling approach. Nevertheless, most farm-level modelling approaches follow a loose linking of models and often integrate only bio-physical indicators in a mathematical programming approach. Some examples exist that employ full agroecosystem simulation models to retrieve information about productivity and the use of fertiliser and irrigation water (e.g. Carauta et al., 2018). However, this does not necessarily mean that the linked agroecosystem model represents the same spatial scale as the farm-level model. To achieve this, agroecosystem models may need to consider a range of landscape-scale interactions.

4 Agroecological models in the landscape context

Agroecological research has developed a suite of different models to investigate systems behaviour under different conditions and to understand the role of specific processes within such systems (Ewert et al., 2015). At the biophysical side of the system, the majority of these models address the interplay of plant, soil and environment, with the latter including also human efforts to manage plants and soils for the purpose of producing food, fuels and fibres. The development of such models has been driven by alternating research questions, starting from general concerns about surface and groundwater quality in the 1960s (Biggar and Nielsen, 1962; Dutt and Tanji, 1962) to the proper use of nutrients (mostly nitrogen) in plant production (Dutt et al., 1972; Hagin and Amberger, 1974; Kaltofen, 1971; Seligman and van Keulen, 1981) and finally (among others) securing and enhancing yields under the threat of a changing climate (e.g. Bacsi et al., 1991; Favismortlock et al., 1991). Depending on their legacy, these models still have different foci and address primarily crop physiological processes (crop models) or soil-plant-atmosphere interaction, often also addressed as the interaction of genotype × environment × management (agroecosystem models). In simulating agricultural yields, the consumption of water, the fate of nutrients in the ecosystem, greenhouse gas emissions, and often also other related processes, these models support a vast range of applied questions in agricultural and environmental management, addressing aspects of climate change, food security and environmental pollution, among others (Ewert et al., 2015; Hatfield et al., 2011; Jones et al., 2017). However, from their mathematical formulation, most of these models are very similar, being mechanistic, dynamic and one-dimensional, i.e. representing the soil-plant-atmosphere continuum only in the vertical dimension, but not in terms of spatial area. This feature necessitates that any scaling to spatial representations need to be done with great care (Ewert et al., 2011). In most cases, the first scaling step is done implicitly by employing data for model calibration and testing, to which a spatial component is inherent, for example, by being sampled over a range of spatially distributed points in a field. Against this background, modellers often claim that their models represent a not sharply defined space, ranging from a

m^2 to a hectare, or even larger, and large-area simulations are often produced by simply applying a one-dimensional model in a grid of variable size across the desired space, which may cover a landscape, region, continent or even the whole globe (e.g. Bondeau et al., 2007; Elliott et al., 2015).

Large-scale gridded crop model applications have become a popular approach to assess continental and global climate change impacts on food security (Rosenzweig et al., 2014), especially in combination with global economic models (Nelson et al., 2014; Ruane et al., 2018). At this scale, the gap between the original domain of process representation of the biophysical part and the actual scale of application is largest, which requires a range of rough assumptions with respect to crop genotype properties and agricultural management (Ruane et al., 2017) and many of the detailed, field-level responses potentially available in the crop model structure vanish under the aggregation to larger scales (Porwollik et al., 2017; Schewe et al., 2019). Applying a sampling approach by simulating representative points distributed across larger areas provides an alternative if agricultural management is in focus (e.g. Asseng et al., 2015; Ruane et al., 2014). However, the combination of crop models with macro-economic models at large scales plays out an important methodological advantage in capturing global market mechanisms, which at some point outweighs the coarse representation of agriculture in these approaches (McDermid et al., 2017).

Current agroecosystem modellers have always argued that landscape-scale effects do not remain visible when simulating large areas at coarse resolution, which is why there have been published numerous studies that roughly estimate agricultural production, services and disservices across the globe or its continents using models with very rough assumptions on their input (e.g. Parry et al., 1999; Siebert and Döll, 2010; van Ittersum et al., 2013). However, when it comes to decision-making and the understanding of how management influences the agroecosystem, the relevant scale is at the landscape (Fig. 1).

Ecosystem services are influenced by environmental and management factors that change over time, but also over space. Often, these factors exhibit a spatial heterogeneity that explains significant differences in agroecosystem responses already at the m^2-scale (Rochette et al., 1991). Precision agriculture is the modern answer to such small-scale heterogeneity, aiming at managing the crop and the soil according to its current state at the smallest scale possible, while not losing overall operability.

4.1 Terrain

The most obvious feature of a landscape is terrain (Dietrich and Perron, 2006). It ranges from absolutely flat to highly mountainous, and already by its slope allows or disallows for certain land uses. While mechanised agriculture is

Figure 1 Landscape processes add to the scope of one-dimensional agroecosystem models: erosion, lateral water fluxes, shading of trees and hedges, interaction with living organisms, socio-economic processes and more.

already ruled out at slopes of approximately >20%, manually managed forms of smallholder agriculture (terracing) as well as orchards and viticulture can be found at even larger slopes. Steepest slopes, if not completely bare, are often used as grassland or forest. Terrain features are commonly applied as a factor in suitability assessments or as a rule in land use models (e.g. Rodriguez-Gallego et al., 2012). Land use models simulate the dynamics of land use under changing environmental, economic or societal conditions. The option to change an existing land use into a different, more promising one in the model is then often limited already by slope, if a threshold is defined above which certain land uses are disallowed. With the slope comes erosion, a great threat to sustainable land use, as erosion causes enormous losses of fertile top soils. Sartori et al. (2019) estimate that erosion reduces global agri-food production by 33.7 million tonnes with accompanying rises in agri-food world prices by 0.4–3.5%. The risk of erosion, and the necessity to reduce this risk, may be again a decisive factor in land use models, for example to keep forest on erosion-prone slopes to safeguard downhill human settlements (Baude et al., 2019). The same applies for the risk of landslides and avalanches. A selection of models is designed to predict erosion from a soil-loss perspective, and also these models use slope as an important factor (e.g. slope factor in the Universal Soil Loss Equation). Erosion models operate at different scales, which can be very small in the spatial, but also in the temporal dimension (Fiener et al., 2008; Schindewolf and Schmidt, 2012). Land use models, however, often operate at a coarse spatial scale, where landscape features, such as terrain, are only considered in a highly aggregated manner (Gollnow et al., 2018; Schaldach et al., 2011; Schulp et al., 2019).

At the landscape scale, terrain also influences the surface and subsurface flow of water. This finds consideration in landscape-scale hydrological models, in which slope defines the direction of water flow and the organisation of the hydrological network (Eddy et al., 2017; Jaber and Shukla, 2012; Trung et al., 2018). Hydrological models for small catchments (e.g. planning tools for water resources, hydropower and flood prevention) often operate at a scale where slope is an essential information (and needs to be provided through for example a digital elevation model).

Agroecosystem models use slope either because they are conceptually linked to an erosion model (e.g. EPIC, Williams, 1990), or it is considered as a factor to control infiltration of water into soil. Slope (and surface roughness) would here control the fraction of surface run-off after precipitation events, with the remaining fraction being available for infiltration, supplementing the soil water reservoir from which the crops benefit (e.g. MONICA; Nendel et al., 2011a). However, point-scale models cannot represent landscape-scale hydrological regimes as they lack lateral hydrological processes. Evidence suggests that lateral fluxes significantly affect the water balance and the spatial distribution of water availability to the vegetation at the landscape scale (Govind et al., 2011). A true land-scape scale, that is three-dimensional, application of agroecosystem models requires a conceptual link with a hydrological model that distributes the water across the surface, considering in- and outflow at each spatial unit, and infiltration losses. While a large number of hydrological and biogeochemical model couplings have been undertaken for forested watersheds (Govind et al., 2009; Tague and Band, 2004), a few examples of such combinations exist also for agroecosystems: the most prominent include the SWAT model for catchment-scale simulations (Arnold and Fohrer, 2005; Arnold et al., 1998; Parn et al., 2018). This model simulates vertical, but also lateral above- and below-ground water and nutrient fluxes in watersheds (basins and channels of different hierarchy levels) that are parcelled into a large number of polygons. It includes a fairly simple agroecosystem model to compute the nutrient input into the aquifers. But also the combination of STICS and CAWAQS (Flipo et al., 2007) has shown how such a link could operate. Even rarer are examples in which different ecosystems are included in the modelling approach. Govind et al. (2015) presented a case for a patchy landscape with forest stands and crop fields, aiming towards a representation of human-shaped heterogeneous landscapes.

Often, catchment models focus on the nutrient transport within the catchment aquifers to predict nutrient loads in surface waters and assume relatively coarse input information to do so (Arheimer and Brandt, 1998; Behrendt et al., 2002; Schoumans and Groenendijk, 2000). The vast majority of these applications deal with nitrogen leaching (Ackermann et al., 2016; Durand et al., 2015; Jordan and Smith, 2005), but also the effect of agricultural practices on the water cycle itself has been studied at this scale (Souchere et al., 1998). At the landscape scale, which is often smaller as compared to a river catchment, 3D hydrological simulations would facilitate simulations of water flows along hillslopes, above and below the surface, and help to explain the often lesser soil moisture availability at hilltops and the soil water accumulations in depressions, and also the respective consequences for plant growth: in dry years, plants in the depression benefit from the larger soil moisture availability, while plants at the hilltops experience drought (Fig. 2), and in wet years hilltop plants prosper, while plants in depressions are reduced in growth as their roots suffer from oxygen deficiency (Fig. 3). For the long-term, agroecosystem models would also simulate the different decomposition environments that result from the differences in soil moisture at contrasting terrain positions, and enable simulations of long-term soil organic matter turn-over regimes, with respective feedback to nutrient cycles and crop growth (Specka et al., 2016) As a simpler alternative, static proxies that derive from terrain features were suggested to provide landscape-scale soil moisture information (Beaudette et al., 2013).

4.2 Soil heterogeneity and its management

Terrain is a major control factor in long-term soil development. Erosion-induced pedogenesis can already be found in gently undulating landscapes

Figure 2 Severe drought has hit a rye crop at the hilltop (in the fore). Hillfoot plants (in the back) are less affected due to downhill water movement during early spring rainfall.

Figure 3 Crop failure in a previously flooded depression.

with slopes well below 5%, with eroded hilltops and accumulations of humus-rich former topsoils at the hillfoot depressions (Doetterl et al., 2012). This constant redistribution of organic matter produces a soil type pattern that has consequences for water holding capacity, nutrient availability and, finally, plant growth and yield. A similar effect is caused by shallow groundwater levels or underlying substrate horizons that originate from geological processes and that are different to the topsoil substrate (e.g. gravel or clay/loam horizons). The loamy ground moraine substrate in Northern European landscapes is such an example, which provides a larger soil water reservoir to the crop than the often very sandy topsoil. Especially under dry spells, the different depth of this reservoir below the soil surface shows through in the vegetation response, when crops more distant to the soil water reservoir begin to wilt, while neighbouring plants still prosper with their lowest roots well placed in the moist horizons.

This distribution of plant growth potentials and decomposition environments, paired with erosion, leads to a distinct landscape-scale pattern of soil organic matter and with this, soil carbon. As soil carbon has a significant share in the dynamic global carbon cycle (Crowther et al., 2016), its potential to transform into atmospheric CO_2 and to contribute to climate change is of interest to research (Minasny et al., 2017; van den Bygaart, 2018; van Oost et al., 2007). Especially the question whether soil carbon is a net sink or source of atmospheric CO_2 is in focus (Bloom et al., 2016; Ciais et al., 2010; Premke et al., 2016). Research suggests that a number of soil and terrain properties determine the sign of this carbon flux and that this sign may change under varying seasonal weather conditions (Specka et al., 2016; van den Bygaart et al., 2015), and a few models have been developed to investigate the cause and effect relationships that lead to either emission or sequestration of carbon in soils (e.g. Bell et al., 2012). As the scale of these processes is rather small, this domain is often approached with soil physical simulation models, which address the storage capacity of the soils, for example in relation to the soil's pore size distribution (Kuka et al., 2007) or using the soil pH value as a proxy for mineralogical conditions (Robertson et al., 2019).

With respect to agricultural systems, the soil organic matter balance has moved into focus, as agricultural management is an accepted lever of controlling soil organic carbon dynamics and sustaining agricultural production through fertile soils. However, the current debate sees confronting positions on whether or not adjusted tillage and residue management can extract significant amounts of carbon from the atmosphere and store it stably in soils (Chen et al., 2019; Minasny et al., 2017; van den Bygaart, 2018). Here, agroecosystem models with a strong soil organic matter routine can help to elucidate the long-term fate of soil organic carbon under different assumed futures and management strategies (e.g. Lemke et al., 2010; Stella et al., 2019; Zhao et al., 2013) and trade-offs with other environmental aims, such as the reduction of N_2O emissions (Chatskikh et al., 2008; Li et al., 2005). Soil heterogeneity, however, requires that these models are able to well reflect the consequences of small-scale changes in site properties for plant growth, soil organic matter decomposition and other soil processes, also including the management of the soils by the farmers (Wallor et al., 2018).

4.3 Transition zones and landscape elements

Transition zones denote all areas that are situated between two different, clearly defined land uses and still exhibit features of both, in forms of gradients. Schmidt et al. (2017) define transition zones as spatio-temporal variable entities with functional and structural gradients in between adjacent core matrices of ecosystems and find that transition zones can reach far into land use units, in a way that a major percentage of the core matrix features do not apply for the whole area that is defined for this land use. In terms of productivity, agricultural land units in fact produce less at their boundaries, if neighbouring forest patches (Fig. 4). However, Schmidt et al. (2019) also showed that under certain conditions, for example drought, agricultural land is more productive in its transition zones than in the core matrix. Transition zones have received little attention in modelling approaches, even though rough estimations show that in patchy landscapes transition zones make up a considerable percentage of the land area. Most important for agricultural land are the transition zones that span into adjacent forest patches, small water bodies and peri-urban areas. Very

Figure 4 The transition zone: maize loses out in competition for light and water against neighbouring trees at the field edge.

few studies address water, nutrient and micro-climatic gradients in transition zones, while a larger number of investigations have looked into gradients of biodiversity and related species dynamics in landscapes (e.g. da Silva et al., 2019; Hansen et al., 2019; Herse et al., 2018; Patten and Smith-Patten, 2012; Sample et al., 2018). Also with respect to the function of landscape elements, species population dynamics dominate the modelling literature (e.g. Heim et al., 2018; Ryberg and Fitzgerald, 2016; Zufiaurre et al., 2016), while the biophysical effects of landscape elements receive less attention and combined assessments are still rare (Meyer et al., 2012; Shaffer et al., 2019).

5 Biodiversity

Modelling flora and fauna in the landscape context adds another dimension into the already complex biophysical approaches to model agroecological systems. Animals and insects, but also wild plant species play important roles in the system functioning, by providing or contributing to relevant ecosystem services. Insects most prominently act as pollinators, ensuring yields for many agricultural plants (e.g. orchards, oil-seed rape and many legumes), or as pests, destroying considerable shares of agricultural produce, if not all (as often reported for armyworm *Spodoptera spec.* or locust invasions). Some insects have a role as disease vectors, spreading fungal diseases over crop stands, which then develop and damage the crop. On the other hand, insects also act as predators to pests, contributing actively to their control. The interplay between crops and insects involves a large range of species, which makes any attempt to model such system a great task. Larger animals appear in this context as herbivores, tramplers or diggers, damaging crops physically while browsing or just passing through. Smaller animals (e.g. rodents) are often perceived as root and leaf gnawer, leaving considerable crop damage when appearing in large numbers (Fischer et al., 2018). Wild plants, or 'weeds' as seen from the farmer's perspective, have the potential to interfere with the intentionally grown plants and reduce their growth through competition for light, water and nutrients or impede the harvesting process and are therefore subject to chemical or mechanical control prior to or during the growing period. Also here, a vast variety of different species is involved and a substantial number of different pathways to trouble the farmer and to impact the ecosystem.

Damages to the crop and competition for resources lead to reductions in crop yield and are therefore targeted by modelling approaches to estimate yield losses (Savary et al., 2006; Vyska et al., 2016). In areas where crop protection methods have not yet substantially developed or prove inefficient under rapid change of crop genotypes or cropping conditions, the share of pest, diseases and weeds in the responsibility for crop losses can be very large. For mechanistic crop models, impact mechanisms have been structured (Boote et al., 1983; Rabbinge and Rijsdijk, 1981; Savary et al., 2018) and exemplarily implemented (Luo et al., 1997; Pinnschmidt et al., 1995), but also more compact solutions have been presented (Savary et al., 2015; Willocquet et al., 2000).

The nature of pest and disease epidemics is not only polyetic, as inoculum builds up over many seasons, but has also a very strong landscape dimension, since landscape structure and heterogeneity strongly influence the ability of insects and pathogens to spread, and their ability and the conditions under which they survive (Karp et al., 2018; Meentemeyer et al., 2012; Plantegenest et al., 2007; Yuen and Mila, 2015). Control of pest and diseases at the landscape scale necessarily involves several farmers in an area, and therefore leads to collective choices and actions, that is spatial strategies (Savary and Willocquet, 2014). Similar aspects apply to the potential spread of genetically modified crops through the dispersal of their pollen (Angevin et al., 2008; Coleno et al., 2009; Leclerc et al., 2018). Very little quantitative modelling has been attempted at these scales that include the essential elements of crop health management and risk exposure. From the pest perspective, the probably most advanced individual-to-landscape approach to date has been presented by Bourhis et al. (2017) who developed a reaction-advection-diffusion population dynamics model accounting for the foraging process as a driver of an insect pest redistribution in space which is sensitive to landscape modifications. From the perspective of the crop to be protected, Rimbaud et al. (2018) developed a spatially explicit stochastic model able to assess the epidemiological and evolutionary outcomes of qualitative resistance and quantitative resistance traits against several components of pathogen aggressiveness. For pollinators, landscape-scale species distribution models exists (Groff et al., 2016; Polce et al., 2013), but with no direct link to fruit setting of crops.

Also, at the landscape scale, and important in agricultural landscapes, we find potential threats of modern agriculture to species. Exposure to agrochemicals has led to the development of spatial risk-exposure models based on population dynamics (e.g. Engelman et al., 2012; Liu et al., 2013) (Fig. 4).

6 Case study

Two recent software solutions for modelling agricultural systems at the landscape scale are presented in this section.

The probably most comprehensive and state-of-the-art case study of how landscape-scale modelling has helped to address existing problems in agriculture has recently been presented by Murgue et al. (2016). They presented a methodology to build a fine-resolution, dynamic model of a large agricultural landscape by mixing generic databases and local information sources. The methodology uses cognitive-task-analysis, heterogeneous participatory modelling, brainstorming and participatory mapping and the MAELIA simulation platform (Therond et al., 2014) to evaluate the model. This example was selected as it demonstrates a close-to-ideal methodological pathway that begins with the problem definition in a stakeholder process, then utilises advanced data sources (e.g. the Land Parcel Identification System that feeds into the European Integrated Administration and Control System IACS) to represent land use in a high conceptual resolution (i.e. crop rotations), integrates iterations with experts and local actors to further improve the data sources and to create rules for information that is not readily available and finally employs a simulation model to derive solutions for the stated problems.

A significant advancement is noted at two frontiers: First, the representation of an agricultural area through observed crop rotations at the field scale lifts large-area agroecosystem modelling to a new level. Departing from one-crop-grows-everywhere, as assumed in numerous agricultural modelling studies up until today (e.g. Webber et al., 2018), long-term simulations of single fields with observed crop sequences provide a much more realistic picture of for example soil carbon dynamics or water use. Through the application of remote sensing, satellite-derived crop maps will soon provide such information also for regions in which IACS data is not available or accessible (Griffiths et al., 2019). Agroecosystem models have already demonstrated that they largely capture the heterogeneity that results from the distributions of soil, climate and agricultural management, both at the regional (Maharjan et al., 2019; Zhao et al., 2016) and at the sub-field level (Wallor et al., 2018). Second, the integration of modelling approaches for different sub-systems, for which the state-of-the-art has previously been described by Ewert et al. (2009), advances here beyond the whole-farm modelling approach, integrating also catchment hydrology and water resource management.

Following a concept described by Duru et al. (2015), also a co-creation approach was included. This approach invited experts to add information to retrieved data (e.g. the observed and simplified crop rotations, local knowledge extracted from farmer interviews) and reformulate it. Also soil maps and farm categorisation was evaluated through experts before being fed into the platform. Crop management strategies were derived from farm surveys, including decision rules and determinants. All this adjusted data and formulated knowledge was then fed into the MAELIA simulation platform. MAELIA is a multi-agent modelling and simulation platform for assessing environmental, economic and social impacts of changes of for example water management, agricultural management, local policies, and global framework conditions, in any combination. MAELIA represents interactions between cropping and crop management strategies, the hydrological system and water resources management and allows to handle questions that arise in this context at the landscape level. The final output of the MAELIA simulation platform was then again reviewed by experts.

An earlier example of a landscape modelling approach was presented by Wenkel et al. (2013) for regions in Germany. Here, stakeholder communication was used to design the graphical user interface of the computer-aided decision support system, but was else not further integrated in the concept. The Landcare DSS builds on interactive simulation experience, in a way that users can define their own simulation tasks, execute them, analyse the results and then, based on this first round, change the simulation set-up, run again and compare the outputs of the two (or more) simulations. The zooming user interface (ZUI), which became popular by Google Maps®, allows the user to intuitively define the area of interest through zooming in or out of the map, and the DSS then offers simulation models for different tasks that match with the selected scale (Nendel et al., 2011b). If the user defined a larger area, for example at the level of a federal state, so-called REMICs are on display. These Regional Models of Intermediate Complexity are mostly statistical models, static by nature, which compute single indicators on the basis of regional geo-information and climate very quickly for a large region. This allows a swift system response, with almost no waiting time for the user in front of the screen. Simulations for irrigation water demand, yield expectations or grassland quality are produced within ten seconds, facilitating the gameplay feeling that emerges when using the models with different scenarios and options and comparing the respective results. At a more detailed spatial scale, more sophisticated models become available. The user may now select the agroecosystem model MONICA (Nendel et al., 2011a) to play in the same manner. MONICA is a speed-optimised software that is able to compute the highly interwoven system representation also within seconds, including multiple-decades simulations to produce a valid long-term statistics from which statements on the uncertainty of the different model outputs can be derived. Coupled to a farm economy coefficient generator, MONICA produces much more complex and integrated outputs as compared to the REMICs and allows the user to interactively play with crop sequences in rotation, irrigation water and nitrogen fertiliser input and other direct variable costs (Münch et al., 2014; Nendel et al., 2014). At the stage when this DSS was published, the approach did not include any landscape-scale interaction of processes then, such as mesoclimatic effects, lateral subcutaneous water flows, shading, erosion and matter transport, or any of this kind. However, the authors pointed at a possible avenue to integrate more models in a plug-and-play style that then also interact with each other.

In comparison with the MAELIA platform, the Landcare DSS is more focused on the technical solution of the 'look and feel', the handling of the software and the retrieval of information by the user, and of the seamless integration of additional models, while MAELIA puts more weight on the representation of the patterns and processes in the agricultural system. Both modelling solutions still have a way to go towards becoming a fully equipped and functional landscape model, but from both ends promising progress is noted.

7 Conclusion

This chapter has put agricultural modelling in the context of the landscape, in which many of the processes proceed that contribute to the provision of agricultural ecosystem services. The landscape scale is where approaches to simulate farm economics, crop-soil-atmosphere interaction, species population dynamics and many more share space. We highlighted that many of the ecosystem services that are delivered by agricultural systems have a landscape component and in attempts to represent such systems, these components need to find consideration. This being said, it is often not obvious how the consideration of the landscape scale may add value to simulations that also produce meaningful results in a narrower system representation and in many cases the landscape-scale influence is either not significant or the modelling approach required would add so much complexity to the simulation and the data requirements that the effort does not match the aim of the model. However, there are examples of modelling exercises that produce results from simulating small system sections, for which we could expect different outcomes if the landscape-scale processes had been considered. The simulation of pest and diseases impacts on crop yields at the plant scale, for example, is not very meaningful if the likelihood of pest or vector appearance in the crop stand is highly dependent on the habitat structure and on distances to the host plant.

Many solutions for challenges that agriculture currently faces have a landscape component, involving actors and process domains that are distributed in the landscape and that have different roles in the system. To understand these, and the pathways to put solutions into effect, agricultural landscape models may contribute by widening the view on the system and its feedback regulations. Sustainable crop production involves much more than just single farmers and the way they manage their fields. But only if we understand how and at which scale the different actors and processes play together, we can decide on the adequate level to implement solutions.

8 Future trends

We currently observe a number of trends along research avenues towards landscape modelling. With increasing computer power and big data analysis capabilities, the data required to assess agricultural systems at the landscape scale become now available. This includes monitoring through on-board sensors of agricultural machinery, but also long-term satellite missions with environmental sensors. The recent Sentinel satellite missions already have fuelled the identification of crops (e.g. Griffiths et al., 2019; Immitzer et al., 2016) and will soon also deliver other input information for simulation models (Demarez et al., 2019; Guzinski and Nieto, 2019). The scaling of 1D agroecosystem models has been systematically investigated (Hoffmann et al., 2015, 2016; Maharjan et al., 2019; Zhao et al., 2015, 2016) which provides more confidence in the use of such models in large-area applications. The advancement of social media data analysis delivers more and more information on people behaviour (Sherren et al., 2017; Wartmann and Purves, 2018), providing additional sources for social modelling. Finally, the awareness of feedback regulations between systems and the resulting need for interdisciplinary modelling efforts is increasing, and efforts in this direction, such as the Agricultural Model Intercomparison and Improvement Project (www.agmip.org, Rosenzweig et al., 2014) or the EU Joint Programming Initiative FACCE with its Modelling European Agriculture with Climate Change for Food Security knowledge hub (www.macsur.eu, Ewert et al., 2015; Kipling et al., 2016), foster this understanding and move modellers from different disciplines closer together. From this common understanding, new models will emerge that integrate more and more insights from neighbouring disciplines (Antle et al., 2017) and most likely identify the landscape as the arena of agricultural impacts and solutions.

9 Where to look for further information

Research institutions that focus on agricultural or environmental research in the context of landscapes are found in Switzerland (the Swiss Federal Institute for Forest, Snow and Landscape Research WS, www.wsl.ch) and Germany (the Leibniz Centre for Agricultural Landscape Research ZALF, www.zalf.de). Also the Centre for Ecology & Hydrology in the United Kingdom and different INRA units in France (e.g. https://www6.rennes.inra.fr/bagap_eng/), among others, partly cover landscape-related topics. Also, different aspects of landscapes, including modelling approaches, are addressed by a wide range of university faculties, that start from the geography, agroecology or landscape planning perspective. The Landscape Research Group is a non-profit organisation that works to advance interdisciplinary landscape research through dissemination and communication (https://landscaperesearch.org).

For modelling activities in the landscape context, please also observe INRA's openFLUID platform (https://www.openfluid-project.org) and the online compilation of Computational Modelling Software Frameworks (https://www.comses.net/resources/modeling-frameworks/).

Two books are currently in press that promise further up-to-date insights into the topic: 'Innovations in Landscape Research' by L. Mueller and 'Landscape Modelling and Decision Support' by W. Mirschel and colleagues (see Mueller (2019) and Mirschel et al. (2019) in the References section for full details).

10 References

Ackermann, A., Mahnkopf, J., Heidecke, C. and Venohr, M. 2016. Reducing agricultural nitrogen inputs in the German Baltic Sea catchment - trends and policy options. *Water Science and Technology: a Journal of the International Association on Water Pollution Research* 74(5), 1060-8. doi:10.2166/wst.2016.267.

Angevin, F., Klein, E. K., Choimet, C., Gauffreteau, A., Lavigne, C., Messéan, A. and Meynard, J. M. 2008. Modelling impacts of cropping systems and climate on maize cross-pollination in agricultural landscapes: the MAPOD model. *European Journal of Agronomy* 28(3), 471-84. doi:10.1016/j.eja.2007.11.010.

Antle, J. M., Jones, J. W. and Rosenzweig, C. 2017. Next generation agricultural system models and knowledge products: synthesis and strategy. *Agricultural Systems* 155, 179-85. doi:10.1016/j.agsy.2017.05.006.

Arheimer, B. and Brandt, M. 1998. Modelling nitrogen transport and retention in the catchments of southern Sweden. *Ambio - A Journal of the Human Environment* 27(6), 471-80.

Arnold, J. G. and Fohrer, N. 2005. SWAT2000: current capabilities and research opportunities in applied watershed modelling. *Hydrological Processes* 19(3), 563-72. doi:10.1002/hyp.5611.

Arnold, J. G., Srinivasan, R., Muttiah, R. S. and Williams, J. R. 1998. Large area hydrologic modeling and assessment - part 1: model development. *Journal of the American*

Water Resources Association 34(1), 73–89. doi:10.1111/j.1752-1688.1998.tb05961.x.

Asseng, S., Ewert, F., Martre, P., Rötter, R. P., Lobell, D. B., Cammarano, D., Kimball, B. A., Ottman, M. J., Wall, G. W., White, J. W., Reynolds, M. P., Alderman, P. D., Prasad, P. V. V., Aggarwal, P. K., Anothai, J., Basso, B., Biernath, C., Challinor, A. J., De Sanctis, G., Doltra, J., Fereres, E., Garcia-Vila, M., Gayler, S., Hoogenboom, G., Hunt, L. A., Izaurralde, R. C., Jabloun, M., Jones, C. D., Kersebaum, K. C., Koehler, A., Müller, C., Naresh Kumar, S., Nendel, C., O'Leary, G., Olesen, J. E., Palosuo, T., Priesack, E., Eyshi Rezaei, E., Ruane, A. C., Semenov, M. A., Shcherbak, I., Stöckle, C., Stratonovitch, P., Streck, T., Supit, I., Tao, F., Thorburn, P. J., Waha, K., Wang, E., Wallach, D., Wolf, J., Zhao, Z. and Zhu, Y. 2015. Rising temperatures reduce global wheat production. *Nature Climate Change* 5(2), 143–7. doi:10.1038/nclimate2470.

Bacsi, Z., Thornton, P. K. and Dent, J. B. 1991. Impacts of future climate change on Hungarian crop production – an application of crop growth simulation-models. *Agricultural Systems* 37(4), 435–50. doi:10.1016/0308-521X(91)90063-G.

Baude, M., Meyer, B. C. and Schindewolf, M. 2019. Land use change in an agricultural landscape causing degradation of soil based ecosystem services. *The Science of the Total Environment* 659, 1526–36. doi:10.1016/j.scitotenv.2018.12.455.

Beaudette, D. E., Dahlgren, R. A. and O'Geen, A. T. 2013. Terrain-shape indices for modeling soil moisture dynamics. *Soil Science Society of America Journal* 77(5), 1696–710. doi:10.2136/sssaj2013.02.0048.

Behrendt, H., Kornmilch, M., Opitz, D., Schmoll, O. and Scholz, G. 2002. Estimation of the nutrient inputs into river systems - experiences from German rivers. *Regional Environmental Change* 3(1–3), 107–17. doi:10.1007/s10113-002-0042-3.

Bell, M. J., Jones, E., Smith, J., Smith, P., Yeluripati, J., Augustin, J., Juszczak, R., Olejnik, J. and Sommer, M. 2012. Simulation of soil nitrogen, nitrous oxide emissions and mitigation scenarios at 3 European cropland sites using the ECOSSE model. *Nutrient Cycling in Agroecosystems* 92(2), 161–81. doi:10.1007/s10705-011-9479-4.

Bieling, C. and Plieninger, T. 2017. Leveraging landscape stewardship principles and ways forward. In: Bieling, C. and Plieninger, T. (Eds), *The Science and Practice of Landscape Stewardship*. Cambridge University Press, pp. 370–82.

Biggar, J. W. and Nielsen, D. R. 1962. Miscible displacement. II. Behavior of tracers. *Soil Science Society of America Journal* 26(2), 125–8. doi:10.2136/sssaj1962.03615995002600020010x.

Bloom, A. A., Exbrayat, J. F., van der Velde, I. R., Feng, L. and Williams, M. 2016. The decadal state of the terrestrial carbon cycle: global retrievals of terrestrial carbon allocation, pools, and residence times. *Proceedings of the National Academy of Sciences of the United States of America* 113(5), 1285–90. doi:10.1073/pnas.1515160113.

Bondeau, A., Smith, P. C., Zaehle, S., Schaphoff, S., Lucht, W., Cramer, W., Gerten, D., Lotze-campen, H., Müller, C., Reichstein, M. and Smith, B. 2007. Modelling the role of agriculture for the 20th century global terrestrial carbon balance. *Global Change Biology* 13(3), 679–706. doi:10.1111/j.1365-2486.2006.01305.x.

Boote, K. J., Jones, J. W., Mishoe, J. W. and Berger, R. D. 1983. Coupling pests to crop growth simulators to predict yield reductions. *Phytopathology* 73(11), 1581–7. doi:10.1094/Phyto-73-1581.

Bourhis, Y., Poggi, S., Mammeri, Y., Le Cointe, R., Cortesero, A. and Parisey, N. 2017. Foraging as the landscape grip for population dynamics-A mechanistic model applied to crop protection. *Ecological Modelling* 354, 26-36. doi:10.1016/j.ecolmodel.2017.03.005.

Boyer, E. W., Howarth, R. W., Galloway, J. N., Dentener, F. J., Green, P. A. and Vörösmarty, C. J. 2006. Riverine nitrogen export from the continents to the coasts. *Global Biogeochemical Cycles* 20(1), Gb1s91. doi:10.1029/2005GB002537.

Brown, D. R. 2000. *A Review of Bio-Economic Models*. Paper prepared for the Cornell African Food Security and Natural Resource Management (CAFSNRM) Program.

Bürgi, M., Ali, P., Chowdhury, A., Heinimann, A., Hett, C., Kienast, F., Mondal, M. K., Upreti, B. R. and Verburg, P. H. 2017. Integrated landscape approach: closing the gap between theory and application. *Sustainability* 9(8), 1371. doi:10.3390/su9081371.

Carauta, M., Latynskiy, E., Mössinger, J., Gil, J., Libera, A., Hampf, A., Monteiro, L., Siebold, M. and Berger, T. 2018. Can preferential credit programs speed up the adoption of low-carbon agricultural systems in Mato Grosso, Brazil? Results from bioeconomic microsimulation. *Regional Environmental Change* 18(1), 117-28. doi:10.1007/s10113-017-1104-x.

Chatskikh, D., Olesen, J. E., Hansen, E. M., Elsgaard, L. and Petersen, B. M. 2008. Effects of reduced tillage on net greenhouse gas fluxes from loamy sand soil under winter crops in Denmark. *Agriculture, Ecosystems and Environment* 128(1-2), 117-26. doi:10.1016/j.agee.2008.05.010.

Chen, J. H., Gong, Y., Wang, S., Guan, B., Balkovic, J. and Kraxner, F. 2019. To burn or retain crop residues on croplands? An integrated analysis of crop residue management in China. *The Science of the Total Environment* 662, 141-50. doi:10.1016/j.scitotenv.2019.01.150.

Ciais, P., Wattenbach, M., Vuichard, N., Smith, P., Piao, S. L., Don, A., Luyssaert, S., Janssens, I. A., Bondeau, A., Dechow, R., Leip, A., Smith, P., Beer, C., Van Der Werf, G. R., Gervois, S., Van Oost, K., Tomelleri, E., Freibauer, A. and Schulze, E. D. 2010. The European carbon balance. part 2: Croplands. *Global Change Biology* 16(5), 1409-28. doi:10.1111/j.1365-2486.2009.02055.x.

Coleno, F. C., Angevin, F. and Lecroart, B. 2009. A model to evaluate the consequences of GM and non-GM segregation scenarios on GM crop placement in the landscape and cross-pollination risk management. *Agricultural Systems* 101(1-2), 49-56. doi:10.1016/j.agsy.2009.03.002.

Council of Europe. 2000. *European Landscape Convention*. European Treaty Series No. 176, Florence.

Crowther, T. W., Todd-Brown, K. E. O., Rowe, C. W., Wieder, W. R., Carey, J. C., Machmuller, M. B., Snoek, B. L., Fang, S., Zhou, G., Allison, S. D., Blair, J. M., Bridgham, S. D., Burton, A. J., Carrillo, Y., Reich, P. B., Clark, J. S., Classen, A. T., Dijkstra, F. A., Elberling, B., Emmett, B. A., Estiarte, M., Frey, S. D., Guo, J., Harte, J., Jiang, L., Johnson, B. R., Kröel-Dulay, G., Larsen, K. S., Laudon, H., Lavallee, J. M., Luo, Y., Lupascu, M., Ma, L. N., Marhan, Michelsen, A., Mohan, J., Niu, S., Pendall, E., Peñuelas, J., Pfeifer-Meister, L., Poll, C., Reinsch, S., Reynolds, L. L., Schmidt, I. K., Sistla, S., Sokol, N. W., Templer, P. H., Treseder, K. K., Welker, J. M. and Bradford, M. A. 2016. Quantifying global soil carbon losses in response to warming. *Nature* 540(7631), 104-8. doi:10.1038/nature20150.

da Silva, P. G., Nunes, C. A., Ferreira, L. F., Braga, R. F., Beiroz, W., Perillo, L. N., Solar, R. R. C. and de Siqueira Neves, F. 2019. Patch and landscape effects on forest-dependent dung beetles are masked by matrix-tolerant dung beetles in a mountaintop rainforest archipelago. *The Science of the Total Environment* 651(1), 1321-31. doi:10.1016/j.scitotenv.2018.09.195.

DEFRA. 2018. *Wild Bird Populations in the UK, 1970–2016*. UK Department for Environment, Food and Rural Affairs, Biodiversity Statistics Team, York, UK.

Demarez, V., Helen, F., Marais-Sicre, C. and Baup, F. 2019. In-season mapping of irrigated crops using landsat 8 and Sentinel-1 time series. *Remote Sensing* 11(2), 118. doi:10.3390/rs11020118.

Dietrich, W. E. and Perron, J. T. 2006. The search for a topographic signature of life. *Nature* 439(7075), 411–8. doi:10.1038/nature04452.

Doetterl, S., Six, J., Van Wesemael, B. and Van Oost, K. 2012. Carbon cycling in eroding landscapes: geomorphic controls on soil organic C pool composition and C stabilization. *Global Change Biology* 18(7), 2218–32. doi:10.1111/j.1365-2486.2012.02680.x.

Durand, P., Moreau, P., Salmon-Monviola, J., Ruiz, L., Vertes, F. and Gascuel-Odoux, C. 2015. Modelling the interplay between nitrogen cycling processes and mitigation options in farming catchments. *The Journal of Agricultural Science* 153(6), 959–74. doi:10.1017/S0021859615000258.

Duru, M., Therond, O. and Fares, M. 2015. Designing agroecological transitions: a review. *Agronomy for Sustainable Development* 35(4), 1237–57. doi:10.1007/s13593-015-0318-x.

Dutt, G. R. and Tanji, K. K. 1962. Predicting concentrations of solutes in water percolated through a column of soil. *Journal of Geophysical Research* 67(9), 3437.

Dutt, G. R., Shaffer, M. J. and Moore, W. J. 1972. *Computer Simulation Model of Dynamic Bio-Physicochemical Processes in Soils*. Agricultural Experimental Station, University of Arizona.

Eddy, M. C., Moreda, F. G., Dykes, R. M., Bergenroth, B., Parks, A. and Rineer, J. 2017. The Watershed Flow and Allocation Model: An NHDPlus-Based Watershed Modeling Approach for Multiple Scales and Conditions. *JAWRA Journal of the American Water Resources Association* 53(1), 6–29. doi:10.1111/1752-1688.12496.

EEA. 2018. *European Waters - Assessment of Status and Pressures*. European Environment Agency, Luxembourg.

Elliott, J., Müller, C., Deryng, D., Chryssanthacopoulos, J., Boote, K. J., Büchner, M., Foster, I., Glotter, M., Heinke, J., Iizumi, T., Izaurralde, R. C., Mueller, N. D., Ray, D. K., Rosenzweig, C., Ruane, A. C. and Sheffield, J.. 2015. The Global Gridded Crop Model Intercomparison: data and modeling protocols for Phase 1 (v1.0). *Geoscientific Model Development* 8(2), 261–77. doi:10.5194/gmd-8-261-2015.

Engelman, C. A., Grant, W. E., Mora, M. A. and Woodin, M. 2012. Modelling effects of chemical exposure on birds wintering in agricultural landscapes: the western burrowing owl (*Athene cunicularia hypugaea*) as a case study. *Ecological Modelling* 224(1), 90–102. doi:10.1016/j.ecolmodel.2011.10.017.

Espinosa, M., Gocht, A., Heckelei, T. and Paloma, S. Gy 2016. Incorporating farm structural change in models assessing the Common Agricultural Policy: an application in the CAPRI farm type model. *Journal of Policy Modeling* 38(6), 1040–59. doi:10.1016/j.jpolmod.2016.03.005.

Ewert, F., van Ittersum, M. K., Bezlepkina, I., Therond, O., Andersen, E., Belhouchette, H., Bockstaller, C., Brouwer, F., Heckelei, T., Janssen, S., Knapen, R., Kuiper, M., Louhichi, K., Olsson, J. A., Turpin, N., Wery, J., Wien, J. E. and Wolf, J. 2009. A methodology for enhanced flexibility of integrated assessment in agriculture. *Environmental Science and Policy* 12(5), 546–61. doi:10.1016/j.envsci.2009.02.005.

Ewert, F., van Ittersum, M. K., Heckelei, T., Therond, O., Bezlepkina, I. and Andersen, E. 2011. Scale changes and model linking methods for integrated assessment of agri-environmental systems. *Agriculture, Ecosystems and Environment* 142(1-2), 6-17. doi:10.1016/j.agee.2011.05.016.

Ewert, F., Rötter, R. P., Bindi, M., Webber, H., Trnka, M., Kersebaum, K. C., Olesen, J. E., van Ittersum, M. K., Janssen, S., Rivington, M., Semenov, M. A., Wallach, D., Porter, J. R., Stewart, D., Verhagen, J., Gaiser, T., Palosuo, T., Tao, F., Nendel, C., Roggero, P. P., Bartošová, L. and Asseng, S. 2015. Crop modelling for integrated assessment of risk to food production from climate change. *Environmental Modelling and Software* 72, 287-303. doi:10.1016/j.envsoft.2014.12.003.

Farina, A. 2000. The cultural landscape as a model for the integration of ecology and economics. *BioScience* 50(4), 313-20. doi:10.1641/0006-3568(2000)050[0313:TCLAAM]2.3.CO;2.

Favismortlock, D. T., Evans, R., Boardman, J. and Harris, T. M. 1991. Climate change, winter-wheat yield and soil-erosion on the English south downs. *Agricultural Systems* 37(4), 415-33. doi:10.1016/0308-521X(91)90062-F.

Fiener, P., Govers, G. and Van Oost, K. V. 2008. Evaluation of a dynamic multi-class sediment transport model in a catchment under soil-conservation agriculture. *Earth Surface Processes and Landforms* 33(11), 1639-60. doi:10.1002/esp.1634.

Fischer, C., Gayer, C., Kurucz, K., Riesch, F., Tscharntke, T. and Batáry, P. 2018. Ecosystem services and disservices provided by small rodents in arable fields: effects of local and landscape management. *Journal of Applied Ecology* 55(2), 548-58. doi:10.1111/1365-2664.13016.

Fisher, B., Turner, R. K. and Morling, P. 2009. Defining and classifying ecosystem services for decision making. *Ecological Economics* 68(3), 643-53. doi:10.1016/j.ecolecon.2008.09.014.

Flipo, N., Even, S., Poulin, M., Thery, S. and Ledoux, E. 2007. Modeling nitrate fluxes at the catchment scale using the integrated tool CAWAQS. *The Science of the Total Environment* 375(1-3), 69-79. doi:10.1016/j.scitotenv.2006.12.016.

Gollnow, F., Gopel, J., Hissa, L. D. V., Schaldach, R. and Lakes, T. 2018. Scenarios of land-use change in a deforestation corridor in the Brazilian Amazon: combining two scales of analysis. *Regional Environmental Change* 18(1), 143-59. doi:10.1007/s10113-017-1129-1.

Govind, A., Chen, J. M., Margolis, H., Ju, W., Sonnentag, O. and Giasson, M. 2009. A spatially explicit hydro-ecological modeling framework (BEPS-TerrainLab V2.0): model description and test in a boreal ecosystem in Eastern North America. *Journal of Hydrology* 367(3-4), 200-16. doi:10.1016/j.jhydrol.2009.01.006.

Govind, A., Chen, J. M., McDonnell, J., Kumari, J. and Sonnentag, O. 2011. Effects of lateral hydrological processes on photosynthesis and evapotranspiration in a boreal ecosystem. *Ecohydrology* 4(3), 394-410. doi:10.1002/eco.141.

Govind, A., Cowling, S., Kumari, J., Rajan, N. and Al-Yaari, A. 2015. Distributed modeling of ecohydrological processes at high spatial resolution over a landscape having patches of managed forest stands and crop fields in SW Europe. *Ecological Modelling* 297, 126-40. doi:10.1016/j.ecolmodel.2014.10.019.

Griffiths, P., Nendel, C. and Hostert, P. 2019. Intra-annual reflectance composites from Sentinel-2 and Landsat for national-scale crop and land cover mapping. *Remote Sensing of Environment* 220, 135-51. doi:10.1016/j.rse.2018.10.031.

Groff, S. C., Loftin, C. S., Drummond, F., Bushmann, S. and McGill, B. 2016. Parameterization of the InVEST Crop Pollination Model to spatially predict abundance of wild blueberry (*Vaccinium angustifolium* Aiton) native bee pollinators in Maine, USA. *Environmental Modelling and Software* 79, 1-9. doi:10.1016/j.envsoft.2016.01.003.

Groot, J. C. J., Oomen, G. J. M. and Rossing, W. A. H. 2012. Multi-objective optimization and design of farming systems. *Agricultural Systems* 110, 63-77. doi:10.1016/j.agsy.2012.03.012.

Guzinski, R. and Nieto, H. 2019. Evaluating the feasibility of using Sentinel-2 and Sentinel-3 satellites for high-resolution evapotranspiration estimations. *Remote Sensing of Environment* 221, 157-72. doi:10.1016/j.rse.2018.11.019.

Hagin, J. and Amberger, A. 1974. *Contribution of Fertilizers and Manures to the N and P Load of Waters. A Computer Simulation*. Technische Universität München, 123p.

Hallmann, C. A., Foppen, R. P. B., van Turnhout, C. A. M., de Kroon, H. and Jongejans, E. 2014. Declines in insectivorous birds are associated with high neonicotinoid concentrations. *Nature* 511(7509), 341-3. doi:10.1038/nature13531.

Hallmann, C. A., Sorg, M., Jongejans, E., Siepel, H., Hofland, N., Schwan, H., Stenmans, W., Müller, A., Sumser, H., Hörren, T., Goulson, D. and de Kroon, H. 2017. More than 75 percent decline over 27 years in total flying insect biomass in protected areas. *PLoS ONE* 12(10), e0185809. doi:10.1371/journal.pone.0185809.

Hampf, A. C., Carauta, M., Latynskiy, E., Libera, A. A. D., Monteiro, L., Sentelhas, P., Troost, C., Berger, T. and Nendel, C. 2018. The biophysical and socio-economic dimension of yield gaps in the southern Amazon - a bio-economic modelling approach. *Agricultural Systems* 165, 1-13. doi:10.1016/j.agsy.2018.05.009.

Hansen, N. A., Sato, C. F., Michael, D. R., Lindenmayer, D. B. and Driscoll, D. A. 2019. Predation risk for reptiles is highest at remnant edges in agricultural landscapes. *Journal of Applied Ecology* 56(1), 31-43. doi:10.1111/1365-2664.13269.

Happe, K., Kellermann, K. and Balmann, A. 2006. Agent-based analysis of agricultural policies: an illustration of the agricultural policy simulator AgriPoliS, its adaptation and behavior. *Ecology and Society* 11(1), 49. doi:10.5751/ES-01741-110149.

Hatfield, J. L., Boote, K. J., Kimball, B. A., Ziska, L. H., Izaurralde, R. C., Ort, D., Thomson, A. M. and Wolfe, D. 2011. Climate impacts on agriculture: implications for crop production. *Agronomy Journal* 103(2), 351-70. doi:10.2134/agronj2010.0303.

Heim, O., Lenski, J., Schulze, J., Jung, K., Kramer-Schadt, S., Eccard, J. A. and Voigt, C. C. 2018. The relevance of vegetation structures and small water bodies for bats foraging above farmland. *Basic and Applied Ecology* 27, 9-19. doi:10.1016/j.baae.2017.12.001.

Herse, M. R., With, K. A. and Boyle, W. A. 2018. The importance of core habitat for a threatened species in changing landscapes. *Journal of Applied Ecology* 55(5), 2241-52. doi:10.1111/1365-2664.13234.

Hoffmann, H., Zhao, G., van Bussel, L., Enders, A., Specka, X., Sosa, C., Yeluripati, J., Tao, F., Constantin, J., Raynal, H., Teixeira, E., Grosz, B., Doro, L., Zhao, Z., Wang, E., Nendel, C., Kersebaum, K., Haas, E., Kiese, R., Klatt, S., Eckersten, H., Vanuytrecht, E., Kuhnert, M., Lewan, E., Rötter, R., Roggero, P., Wallach, D., Cammarano, D., Asseng, S., Krauss, G., Siebert, S., Gaiser, T. and Ewert, F. 2015. Variability of effects of spatial climate data aggregation on regional yield simulation by crop models. *Climate Research* 65, 53-69. doi:10.3354/cr01326.

Hoffmann, H., Zhao, G., Asseng, S., Bindi, M., Biernath, C., Constantin, J., Coucheney, E., Dechow, R., Doro, L., Eckersten, H., Gaiser, T., Grosz, B., Heinlein, F., Kassie, B. T., Kersebaum, K. C., Klein, C., Kuhnert, M., Lewan, E., Moriondo, M., Nendel, C., Priesack,

E., Raynal, H., Roggero, P. P., Rötter, R. P., Siebert, S., Specka, X., Tao, F., Teixeira, E., Trombi, G., Wallach, D., Weihermüller, L., Yeluripati, J. and Ewert, F. 2016. Impact of spatial soil and climate input data aggregation on regional yield simulations. *PLoS ONE* 11(4), e0151782. doi:10.1371/journal.pone.0151782.

Immitzer, M., Vuolo, F. and Atzberger, C. 2016. First experience with Sentinel-2 data for crop and tree species classifications in Central Europe. *Remote Sensing* 8(3). doi:10.3390/rs8030166.

Jaber, F. H. and Shukla, S. 2012. MIKE SHE: model use, calibration, and validation. *Transactions of the ASABE* 55(4), 1479–89. doi:10.13031/2013.42255.

Janssen, S. and van Ittersum, M. K. 2007. Assessing farm innovations and responses to policies: a review of bio-economic farm models. *Agricultural Systems* 94(3), 622–36. doi:10.1016/j.agsy.2007.03.001.

Jones, J. W., Antle, J. M., Basso, B., Boote, K. J., Conant, R. T., Foster, I., Godfray, H. C. J., Herrero, M., Howitt, R. E., Janssen, S., Keating, B. A., Munoz-Carpena, R., Porter, C. H., Rosenzweig, C. and Wheeler, T. R. 2017. Toward a new generation of agricultural system data, models, and knowledge products: state of agricultural systems science. *Agricultural Systems* 155, 269–88. doi:10.1016/j.agsy.2016.09.021.

Jordan, C. and Smith, R. V. 2005. Methods to predict the agricultural contribution to catchment nitrate loads: designation of nitrate vulnerable zones in Northern Ireland. *Journal of Hydrology* 304(1–4), 316–29. doi:10.1016/j.jhydrol.2004.07.037.

Jury, W. A. and Vaux, H. J. 2007. The emerging global water crisis: managing scarcity and conflict between water users. *Advances in Agronomy* 95, 1–76. doi:10.1016/S0065-2113(07)95001-4.

Kaltofen, H. 1971. Probleme der Systemanalyse und Modellierung auf dem Gebiet der Ertragsbildung. In: Kaltofen, H., Vetterlein, E., Unger, K. and Kundler, P. (Eds), *Problematik der Modellierung der Ertragsbildung*. Akademie-Verlag, Berlin, pp. 5–40.

Karp, D. S., Chaplin-Kramer, R., Meehan, T. D., Martin, E. A., DeClerck, F., Grab, H., Gratton, C., Hunt, L., Larsen, A. E., Martínez-Salinas, A., O'Rourke, M. E., Rusch, A., Poveda, K., Jonsson, M., Rosenheim, J. A., Schellhorn, N. A., Tscharntke, T., Wratten, S. D., Zhang, W., Iverson, A. L., Adler, L. S., Albrecht, M., Alignier, A., Angelella, G. M., Zubair Anjum, M., Avelino, J., Batáry, P., Baveco, J. M., Bianchi, F. J. J. A., Birkhofer, K., Bohnenblust, E. W., Bommarco, R., Brewer, M. J., Caballero-López, B., Carrière, Y., Carvalheiro, L. G., Cayuela, L., Centrella, M., Ćetković, A., Henri, D. C., Chabert, A., Costamagna, A. C., De la Mora, A., de Kraker, J., Desneux, N., Diehl, E., Diekötter, T., Dormann, C. F., Eckberg, J. O., Entling, M. H., Fiedler, D., Franck, P., Frank van Veen, F. J., Frank, T., Gagic, V., Garratt, M. P. D., Getachew, A., Gonthier, D. J., Goodell, P. B., Graziosi, I., Groves, R. L., Gurr, G. M., Hajian-Forooshani, Z., Heimpel, G. E., Herrmann, J. D., Huseth, A. S., Inclán, D. J., Ingrao, A. J., Iv, P., Jacot, K., Johnson, G. A., Jones, L., Kaiser, M., Kaser, J. M., Keasar, T., Kim, T. N., Kishinevsky, M., Landis, D. A., Lavandero, B., Lavigne, C., Le Ralec, A., Lemessa, D., Letourneau, D. K., Liere, H., Lu, Y., Lubin, Y., Luttermoser, T., Maas, B., Mace, K., Madeira, F., Mader, V., Cortesero, A. M., Marini, L., Martinez, E., Martinson, H. M., Menozzi, P., Mitchell, M. G. E., Miyashita, T., Molina, G. A. R., Molina-Montenegro, M. A., O'Neal, M. E., Opatovsky, I., Ortiz-Martinez, S., Nash, M., Östman, Ö, Ouin, A., Pak, D., Paredes, D., Parsa, S., Parry, H., Perez-Alvarez, R., Perović, D. J., Peterson, J. A., Petit, S., Philpott, S. M., Plantegenest, M., Plećaš, M., Pluess, T., Pons, X., Potts, S. G., Pywell, R. F., Ragsdale, D. W., Rand, T. A., Raymond, L., Ricci, B., Sargent, C.,

Sarthou, J. P., Saulais, J., Schäckermann, J., Schmidt, N. P., Schneider, G., Schüepp, C., Sivakoff, F. S., Smith, H. G., Stack Whitney, K., Stutz, S., Szendrei, Z., Takada, M. B., Taki, H., Tamburini, G., Thomson, L. J., Tricault, Y., Tsafack, N., Tschumi, M., Valantin-Morison, M., Van Trinh, M., van der Werf, W., Vierling, K. T., Werling, B. P., Wickens, J. B., Wickens, V. J., Woodcock, B. A., Wyckhuys, K., Xiao, H., Yasuda, M., Yoshioka, A. and Zou, Y. 2018. Crop pests and predators exhibit inconsistent responses to surrounding landscape composition. *Proceedings of the National Academy of Sciences of the United States of America* 115(33), E7863–70. doi:10.1073/pnas.1800042115.

Kipling, R. P., Virkajärvi, P., Breitsameter, L., Curnel, Y., De Swaef, T., Gustavsson, A. M., Hennart, S., Höglind, M., Järvenranta, K., Minet, J., Nendel, C., Persson, T., Picon-Cochard, C., Rolinski, S., Sandars, D. L., Scollan, N. D., Sebek, L., Seddaiu, G., Topp, C. F. E., Twardy, S., Van Middelkoop, J., Wu, L. and Bellocchi, G. 2016. Key challenges and priorities for modelling European grasslands under climate change. *The Science of the Total Environment* 566–567, 851–64. doi:10.1016/j.scitotenv.2016.05.144.

Kirchner, M., Schmidt, J., Kindermann, G., Kulmer, V., Mitter, H., Prettenthaler, F., Rüdisser, J., Schauppenlehner, T., Schönhart, M., Strauss, F., Tappeiner, U., Tasser, E. and Schmid, E. 2015. Ecosystem services and economic development in Austrian agricultural landscapes – the impact of policy and climate change scenarios on trade-offs and synergies. *Ecological Economics* 109, 161–74. doi:10.1016/j.ecolecon.2014.11.005.

Kuka, K., Franko, U. and Rühlmann, J. 2007. Modelling the impact of pore space distribution on carbon turnover. *Ecological Modelling* 208(2-4), 295–306. doi:10.1016/j.ecolmodel.2007.06.002.

Leclerc, M., Walkera, E., Messean, A. and Soubeyrand, S. 2018. Spatial exposure-hazard and landscape models for assessing the impact of GM crops on non-target organisms. *The Science of the Total Environment* 624, 470–9. doi:10.1016/j.scitotenv.2017.11.329.

Lemke, R. L., VandenBygaart, A. J., Campbell, C. A., Lafond, G. P. and Grant, B. 2010. Crop residue removal and fertilizer N: effects on soil organic carbon in a long-term crop rotation experiment on a Udic Boroll. *Agriculture, Ecosystems and Environment* 135(1–2), 42–51. doi:10.1016/j.agee.2009.08.010.

Le Pira, M., Ignaccolo, M., Inturri, G., Pluchino, A. and Rapisarda, A. 2016. Modelling stakeholder participation in transport planning. *Case Studies on Transport Policy* 4(3), 230–8.

Li, C. S., Frolking, S. and Butterbach-Bahl, K. 2005. Carbon sequestration in arable soils is likely to increase nitrous oxide emissions, offsetting reductions in climate radiative forcing. *Climatic Change* 72(3), 321–38. doi:10.1007/s10584-005-6791-5.

Liu, C., Sibly, R. M., Grimm, V. and Thorbek, P. 2013. Linking pesticide exposure and spatial dynamics: an individual-based model of wood mouse (*Apodemus sylvaticus*) populations in agricultural landscapes. *Ecological Modelling* 248, 92–102. doi:10.1016/j.ecolmodel.2012.09.016.

Luo, Y., Teng, P. S., Fabellar, N. G. and TeBeest, D. O. 1997. A rice-leaf blast combined model for simulation of epidemics and yield loss. *Agricultural Systems* 53(1), 27–39. doi:10.1016/S0308-521X(96)00038-8.

Maharjan, G. R., Hoffmann, H., Webber, H., Srivastava, A. K., Weihermüller, L., Villa, A., Coucheney, E., Lewan, E., Trombi, G., Moriondo, M., Bindi, M., Grosz, B., Dechow, R.,

Kuhnert, M., Doro, L., Kersebaum, K., Stella, T., Specka, X., Nendel, C., Constantin, J., Raynal, H., Ewert, F. and Gaiser, T. 2019. Effects of input data aggregation on simulated crop yields in temperate and Mediterranean climates. *European Journal of Agronomy* 103, 32–46. doi:10.1016/j.eja.2018.11.001.

McDermid, S. S., Mearns, L. O. and Ruane, A. C. 2017. Representing agriculture in Earth System Models: approaches and priorities for development. *Journal of Advances in Modeling Earth Systems* 9(5), 2230–65. doi:10.1002/2016MS000749.

Meentemeyer, R. K., Haas, S. E. and Vaclavik, T. 2012. Landscape epidemiology of emerging infectious diseases in natural and human-altered ecosystems. *Annual Review of Phytopathology* 50, 379–402. doi:10.1146/annurev-phyto-081211-172938.

Mekonnen, M. M. and Hoekstra, A. Y. 2016. Four billion people facing severe water scarcity. *Science Advances* 2(2), e1500323. doi:10.1126/sciadv.1500323.

Meyer, B. C., Wolf, T. and Grabaum, R. 2012. A multifunctional assessment method for compromise optimisation of linear landscape elements. *Ecological Indicators* 22, 53–63. doi:10.1016/j.ecolind.2011.07.020.

Millennium Ecosystem Assessment. 2005. *Ecosystems and Human Well-Being: Synthesis*. Island Press, Washington DC.

Minasny, B., Malone, B. P., McBratney, A. B., Angers, D. A., Arrouays, D., Chambers, A., Chaplot, V., Chen, Z., Cheng, K., Das, B. S., Field, D. J., Gimona, A., Hedley, C. B., Hong, S. Y., Mandal, B., Marchant, B. P., Martin, M., McConkey, B. G., Mulder, V. L., O'Rourke, S., Richer-de-Forges, A. C., Odeh, I., Padarian, J., Paustian, K., Pan, G., Poggio, L., Savin, I., Stolbovoy, V., Stockmann, U., Sulaeman, Y., Tsui, C., Vågen, T., van Wesemael, B. and Winowiecki, L. 2017. Soil carbon 4 per mille. *Geoderma* 292, 59–86. doi:10.1016/j.geoderma.2017.01.002.

Mirschel, W., Terleev, V. V. and Wenkel, K.-O. (Eds). 2019. *Landscape Modelling and Decision Support*. Innovations in Landscape Research, Springer Nature Switzerland.

Mueller, L. (Series Ed.). 2019. *Innovations in Landscape Research*. Springer (in press). ISSN: 2524-5155.

Münch, T., Berg, M., Mirschel, W., Wieland, R. and Nendel, C. 2014. Considering cost accountancy items in crop production simulations under climate change. *European Journal of Agronomy* 52, 57–68. doi:10.1016/j.eja.2013.01.005.

Murgue, C., Therond, O. and Leenhardt, D. 2016. Hybridizing local and generic information to model cropping system spatial distribution in an agricultural landscape. *Land Use Policy* 54, 339–54. doi:10.1016/j.landusepol.2016.02.020.

Nelson, G. C., Valin, H., Sands, R. D., Havlík, P., Ahammad, H., Deryng, D., Elliott, J., Fujimori, S., Hasegawa, T., Heyhoe, E., Kyle, P., Von Lampe, M., Lotze-Campen, H., Mason d'Croz, D., van Meijl, H., van der Mensbrugghe, D., Müller, C., Popp, A., Robertson, R., Robinson, S., Schmid, E., Schmitz, C., Tabeau, A. and Willenbockel, D. 2014. Climate change effects on agriculture: economic responses to biophysical shocks. *Proceedings of the National Academy of Sciences of the United States of America* 111(9), 3274–9. doi:10.1073/pnas.1222465110.

Nendel, C., Berg, M., Kersebaum, K. C., Mirschel, W., Specka, X., Wegehenkel, M., Wenkel, K. O. and Wieland, R. 2011a. The MONICA model: testing predictability for crop growth, soil moisture and nitrogen dynamics. *Ecological Modelling* 222(9), 1614–25. doi:10.1016/j.ecolmodel.2011.02.018.

Nendel, C., Berg, M., Kersebaum, K. C., Mirschel, W., Specka, X., Wenkel, K. O. and Wieland, R. 2011b. Landcare DSS – Decision support meets interactivity. *19th International Congress on Modelling and Simulation (Modsim2011)*, pp. 1244–50.

Nendel, C., Kersebaum, K. C., Mirschel, W. and Wenkel, K. O. 2014. Testing farm management options as climate change adaptation strategies using the MONICA model. *European Journal of Agronomy* 52, 47–56. doi:10.1016/j.eja.2012.09.005.

Nuppenau, E. A. 2018. Soil fertility management by transition matrices and crop rotation: on spatial and dynamic aspects in programming of ecosystem services. *Sustainability* 10(7), 2213. doi:10.3390/su10072213.

Papajorgji, P. J. and Panos, P. M. 2009. *Advances in Modeling Agricultural Systems*. Springer US, Boston, MA.

Parn, J., Henine, H., Kasak, K., Kauer, K., Sohar, K., Tournebize, J., Uuemaa, E., Välik, K. and Mander, Ü 2018. Nitrogen and phosphorus discharge from small agricultural catchments predicted from land use and hydroclimate. *Land Use Policy* 75, 260–8. doi:10.1016/j.landusepol.2018.03.048.

Parry, M., Rosenzweig, C., Iglesias, A., Fischer, G. and Livermore, M. 1999. Climate change and world food security: a new assessment. *Global Environmental Change-Human and Policy Dimensions* 9, S51–67. doi:10.1016/S0959-3780(99)00018-7.

Patten, M. A. and Smith-Patten, B. D. 2012. Testing the microclimate hypothesis: light environment and population trends of Neotropical birds. *Biological Conservation* 155, 85–93. doi:10.1016/j.biocon.2012.06.004.

Peter, C., Helming, K. and Nendel, C. 2017. Do greenhouse gas emission calculations from energy crop cultivation reflect actual agricultural management practices? – a review of carbon footprint calculators. *Renewable and Sustainable Energy Reviews* 67, 461–76. doi:10.1016/j.rser.2016.09.059.

Pinnschmidt, H. O., Batchelor, W. D. and Teng, P. S. 1995. Simulation of multiple species pest damage in rice using CERES-RICE. *Agricultural Systems* 48(2), 193–222. doi:10.1016/0308-521X(94)00012-G.

Plantegenest, M., Le May, C. and Fabre, F. 2007. Landscape epidemiology of plant diseases. *Journal of the Royal Society Interface* 4(16), 963–72. doi:10.1098/rsif.2007.1114.

Plieninger, T., Draux, H., Fagerholm, N., Bieling, C., Bürgi, M., Kizos, T., Kuemmerle, T., Primdahl, J. and Verburg, P. H. 2016. The driving forces of landscape change in Europe: a systematic review of the evidence. *Land Use Policy* 57, 204–14. doi:10.1016/j.landusepol.2016.04.040.

Polce, C., Termansen, M., Aguirre-Gutiérrez, J., Boatman, N. D., Budge, G. E., Crowe, A., Garratt, M. P., Pietravalle, S., Potts, S. G., Ramirez, J. A., Somerwill, K. E. and Biesmeijer, J. C. 2013. Species distribution models for crop pollination: A modelling framework applied to Great Britain. *PLoS ONE* 8(10), e76308. doi:10.1371/journal.pone.0076308.

Porwollik, V., Müller, C., Elliott, J., Chryssanthacopoulos, J., Iizumi, T., Ray, D. K., Ruane, A. C., Arneth, A., Balkovič, J., Ciais, P., Deryng, D., Folberth, C., Izaurralde, R. C., Jones, C. D., Khabarov, N., Lawrence, P. J., Liu, W., Pugh, T. A. M., Reddy, A., Sakurai, G., Schmid, E., Wang, X., de Wit, A. and Wu, X. 2017. Spatial and temporal uncertainty of crop yield aggregations. *European Journal of Agronomy* 88, 10–21. doi:10.1016/j.eja.2016.08.006.

Premke, K., Attermeyer, K., Augustin, J., Cabezas, A., Casper, P., Deumlich, D., Gelbrecht, J., Gerke, H. H., Gessler, A., Grossart, H., Hilt, S., Hupfer, M., Kalettka, T., Kayler, Z., Lischeid, G., Sommer, M. and Zak, D. 2016. The importance of landscape diversity for carbon fluxes at the landscape level: small-scale heterogeneity matters. *Wiley Interdisciplinary Reviews: Water* 3(4), 601–17. doi:10.1002/wat2.1147.

Rabbinge, R. and Rijsdijk, F. H. 1981. Disease and crop physiology: a modeller's point of view. In: Ayres, P. G. (Ed.), *Effects of Disease on the Physiology of the Growing Plant*. Cambridge University Press, Cambridge, pp. 201–20.

Reidsma, P., Janssen, S., Jansen, J. and van Ittersum, M. K. 2018. On the development and use of farm models for policy impact assessment in the European Union – a review. *Agricultural Systems* 159, 111–25. doi:10.1016/j.agsy.2017.10.012.

Rimbaud, L., Papaix, J., Rey, J. F., Barrett, L. G. and Thrall, P. H. 2018. Assessing the durability and efficiency of landscape-based strategies to deploy plant resistance to pathogens. *PLoS Computational Biology* 14(4), e1006067. doi:10.1371/journal.pcbi.1006067.

Robertson, A. D., Paustian, K., Ogle, S., Wallenstein, M. D., Lugato, E. and Cotrufo, M. F. 2019. Unifying soil organic matter formation and persistence frameworks: the MEMS model. *Biogeosciences* 16, 1225–48, doi:10.5194/bg-16-1225-2019.

Rochette, P., Desjardins, R. L. and Pattey, E. 1991. Spatial and temporal variability of soil respiration in agricultural fields. *Canadian Journal of Soil Science* 71(2), 189–96. doi:10.4141/cjss91-018.

Rodriguez-Gallego, L., Achkar, M. and Conde, D. 2012. Land suitability assessment in the catchment area of four southwestern Atlantic coastal lagoons: multicriteria and optimization modeling. *Environmental Management* 50(1), 140–52. doi:10.1007/s00267-012-9843-4.

Rosenzweig, C., Elliott, J., Deryng, D., Ruane, A. C., Müller, C., Arneth, A., Boote, K. J., Folberth, C., Glotter, M., Khabarov, N., Neumann, K., Piontek, F., Pugh, T. A., Schmid, E., Stehfest, E., Yang, H. and Jones, J. W. 2014. Assessing agricultural risks of climate change in the 21st century in a global gridded crop model intercomparison. *Proceedings of the National Academy of Sciences of the United States of America* 111(9), 3268–73. doi:10.1073/pnas.1222463110.

Ruane, A. C., McDermid, S., Rosenzweig, C., Baigorria, G. A., Jones, J. W., Romero, C. C. and Dewayne Cecil, L. 2014. Carbon-temperature-water change analysis for peanut production under climate change: a prototype for the AgMIP Coordinated Climate-Crop Modeling Project (C3MP). *Global Change Biology* 20(2), 394–407. doi:10.1111/gcb.12412.

Ruane, A. C., Rosenzweig, C., Asseng, S., Boote, K. J., Elliott, J., Ewert, F., Jones, J. W., Martre, P., McDermid, S. P., Müller, C., Snyder, A. and Thorburn, P. J. 2017. An AgMIP framework for improved agricultural representation in integrated assessment models. *Environmental Research Letters* 12(12). doi:10.1088/1748-9326/aa8da6.

Ruane, A. C., Antle, J., Elliott, J., Folberth, C., Hoogenboom, G., Mason-D'Croz, D., Müller, C., Porter, C., Phillips, M., Raymundo, R., Sands, R., Valdivia, R., White, J., Wiebe, K. and Rosenzweig, C. 2018. Biophysical and economic implications for agriculture of +1.5 degrees and+2.0 degrees C global warming using AgMIP Coordinated Global and Regional Assessments. *Climate Research* 76(1), 17–39. doi:10.3354/cr01520.

Ryberg, W. A. and Fitzgerald, L. A. 2016. Landscape composition, not connectivity, determines metacommunity structure across multiple scales. *Ecography* 39(10), 932–41. doi:10.1111/ecog.01321.

Sample, C., Fryxell, J. M., Bieri, J. A., Federico, P., Earl, J. E., Wiederholt, R., Mattsson, B. J., Flockhart, D. T. T., Nicol, S., Diffendorfer, J. E., Thogmartin, W. E., Erickson, R. A. and Norris, D. R. 2018. A general modeling framework for describing spatially structured population dynamics. *Ecology and Evolution* 8(1), 493–508. doi:10.1002/ece3.3685.

Sartori, M., Philippidis, G., Ferrari, E., Borrelli, P., Lugato, E., Montanarella, L. and Panagos, P. 2019. A linkage between the biophysical and the economic: assessing the global market impacts of soil erosion. *Land Use Policy* 86, 299–312. doi:10.1016/j.landusepol.2019.05.014.

Savary, S. and Willocquet, L. 2014. Simulation modeling in botanical epidemiology and crop loss analysis. *The Plant Health Instructor*. doi:10.1094/PHI-A-2014-0314-01.

Savary, S., Teng, P. S., Willocquet, L. and Nutter, F. W. 2006. Quantification and modeling of crop losses: a review of purposes. *Annual Review of Phytopathology* 44, 89–112. doi:10.1146/annurev.phyto.44.070505.143342.

Savary, S., Stetkiewicz, S., Brun, F. and Willocquet, L. 2015. Modelling and mapping potential epidemics of wheat diseases-examples on leaf rust and Septoria tritici blotch using EPIWHEAT. *European Journal of Plant Pathology* 142(4), 771–90. doi:10.1007/s10658-015-0650-7.

Savary, S., Nelson, A. D., Djurle, A., Esker, P. D., Sparks, A., Amorim, L., Bergamin Filho, A., Caffi, T., Castilla, N., Garrett, K., McRoberts, N., Rossi, V., Yuen, J. and Willocquet, L. 2018. Concepts, approaches, and avenues for modelling crop health and crop losses. *European Journal of Agronomy* 100, 4–18. doi:10.1016/j.eja.2018.04.003.

Sayer, J., Sunderland, T., Ghazoul, J., Pfund, J. L., Sheil, D., Meijaard, E., Venter, M., Boedhihartono, A. K., Day, M., Garcia, C., van Oosten, C. and Buck, L. E. 2013. Ten principles for a landscape approach to reconciling agriculture, conservation, and other competing land uses. *Proceedings of the National Academy of Sciences of the United States of America* 110(21), 8349–56. doi:10.1073/pnas.1210595110.

Schaldach, R., Alcamo, J., Koch, J., Kölking, C., Lapola, D. M., Schüngel, J. and Priess, J. A. 2011. An integrated approach to modelling land-use change on continental and global scales. *Environmental Modelling and Software* 26(8), 1041–51. doi:10.1016/j.envsoft.2011.02.013.

Schewe, J., Gosling, S. N., Reyer, C., Zhao, F., Ciais, P., Elliott, J., Francois, L., Huber, V., Lotze, H. K., Seneviratne, S. I., van Vliet, M. T. H., Vautard, R., Wada, Y., Breuer, L., Büchner, M., Carozza, D. A., Chang, J., Coll, M., Deryng, D., de Wit, A., Eddy, T. D., Folberth, C., Frieler, K., Friend, A. D., Gerten, D., Gudmundsson, L., Hanasaki, N., Ito, A., Khabarov, N., Kim, H., Lawrence, P., Morfopoulos, C., Müller, C., Müller Schmied, H., Orth, R., Ostberg, S., Pokhrel, Y., Pugh, T. A. M., Sakurai, G., Satoh, Y., Schmid, E., Stacke, T., Steenbeek, J., Steinkamp, J., Tang, Q., Tian, H., Tittensor, D. P., Volkholz, J., Wang, X. and Warszawski, L. 2019. State-of-the-art global models underestimate impacts from climate extremes. *Nature Communications* 10(1), 1005. doi:10.1038/s41467-019-08745-6.

Schindewolf, M. and Schmidt, J. 2012. Parameterization of the EROSION 2D/3D soil erosion model using a small-scale rainfall simulator and upstream runoff simulation. *CATENA* 91, 47–55. doi:10.1016/j.catena.2011.01.007.

Schmidt, M., Jochheim, H., Kersebaum, K. C., Lischeid, G. and Nendel, C. 2017. Gradients of microclimate, carbon and nitrogen in transition zones of fragmented landscapes – a review. *Agricultural and Forest Meteorology* 232, 659–71. doi:10.1016/j.agrformet.2016.10.022.

Schmidt, M., Nendel, C., Funk, R., Mitchell, M. G. E. and Lischeid, G. 2019. Modelling crop yield response to shading in field-to-forest transition zones in heterogeneous landscapes. *Agriculture* 9(1), 6. doi:10.3390/agriculture9010006.

Schönhart, M., Schauppenlehner, T., Schmid, E. and Muhar, A. 2011. Integration of bio-physical and economic models to analyze management intensity and landscape structure effects at farm and landscape level. *Agricultural Systems* 104(2), 122–34. doi:10.1016/j.agsy.2010.03.014.

Schoumans, O. F. and Groenendijk, P. 2000. Modeling soil phosphorus levels and phosphorus leaching from agricultural land in the Netherlands. *Journal of Environmental Quality* 29(1), 111–6. doi:10.2134/jeq2000.00472425002900010014x.

Schulp, C. J. E., Levers, C., Kuemmerle, T., Tieskens, K. F. and Verburg, P. H. 2019. Mapping and modelling past and future land use change in Europe's cultural landscapes. *Land Use Policy* 80, 332–44. doi:10.1016/j.landusepol.2018.04.030.

Seligman, N. C. and van Keulen, H. 1981. PAPRAN: a simulation model of annual pasture production limited by rainfall and nitrogen. In: Frissel, M. J. and van Veen, J. A. (Eds), *Simulation of Nitrogen Behaviour of Soil-Plant Systems*. PUDOC, Wageningen, pp. 192–221.

Seppelt, R., Lautenbach, S. and Volk, M. 2013. Identifying trade-offs between ecosystem services, land use, and biodiversity: a plea for combining scenario analysis and optimization on different spatial scales. *Current Opinion in Environmental Sustainability* 5(5), 458–63. doi:10.1016/j.cosust.2013.05.002.

Shaffer, J. A., Roth, C. L. and Mushet, D. M. 2019. Modeling effects of crop production, energy development and conservation-grassland loss on avian habitat. *PLoS ONE* 14(1), e0198382. doi:10.1371/journal.pone.0198382.

Sherren, K., Parkins, J. R., Smit, M., Holmlund, M. and Chen, Y. 2017. Digital archives, big data and image-based culturomics for social impact assessment: opportunities and challenges. *Environmental Impact Assessment Review* 67, 23–30. doi:10.1016/j.eiar.2017.08.002.

Siebert, S. and Döll, P. 2010. Quantifying blue and green virtual water contents in global crop production as well as potential production losses without irrigation. *Journal of Hydrology* 384(3–4), 198–217. doi:10.1016/j.jhydrol.2009.07.031.

Souchere, V., King, D., Daroussin, J., Papy, F. and Capillon, A. 1998. Effects of tillage on runoff directions: consequences on runoff contributing area within agricultural catchments. *Journal of Hydrology* 206(3–4), 256–67. doi:10.1016/S0022-1694(98)00103-6.

Specka, X., Nendel, C., Hagemann, U., Pohl, M., Hoffmann, M., Barkusky, D., Augustin, J., Sommer, M. and van Oost, K. 2016. Reproducing CO_2 exchange rates of a crop rotation at contrasting terrain positions using two different modelling approaches. *Soil and Tillage Research* 156, 219–29. doi:10.1016/j.still.2015.05.007.

Stella, T., Mouratiadou, I., Gaiser, T., Berg-Mohnicke, M., Wallor, E., Ewert, F. and Nendel, C. 2019. Estimating the contribution of crop residue management to soil organic carbon conservation. *Environmental Research Letters*. doi:10.1088/1748-9326/ab395c.

Tague, C. L. and Band, L. E. 2004. RHESSys: regional hydro-ecologic simulation system-an object-oriented approach to spatially distributed modeling of carbon, water, and nutrient cycling. *Earth Interactions* 8(19), 1–42. doi:10.1175/1087-3562(2004)8<1:RRHSSO>2.0.CO;2.

Therond, O., Sibertin-Blanc, C., Lardy, R., Gaudou, B., Balestrat, M., Hong, Y., Louail, T., Nguyen, V. B., Panzoli, D., Sanchez-Perez, J. M., Sauvage, S., Taillandier, P., Vavasseur, M. and Mazzega, P. 2014. Integrated modelling of social-ecological systems: the MAELIA high-resolution multi-agent platform to deal with water scarcity problems. In: Ames, D. P., Quinn, N. W. T. and Rizzoli, A. E. (Eds), *7th International Congress on Environment Modelling and Software (iEMSs)*, San Diego, CA, USA, p. 8.

Trung, N. Q., Polcher, J., Ducharne, A., Arsouze, T., Zhou, X., Schneider, A. and Fita, L. 2018. ORCHIDEE-ROUTING: revising the river routing scheme using a high-resolution hydrological database. *Geoscientific Model Development* 11(12), 4965-85. doi:10.5194/gmd-11-4965-2018.

USEPA. 2016. *National Lakes Assessment 2012: A Collaborative Survey of Lakes in the United States*. United States Environmental Protection Agency, Washington DC.

Uthes, S., Sattler, C., Reinhardt, F.-J., Piorr, A., Zander, P., Happe, K., Damgaard, M. and Osuch, A. 2008. Ecological effects of payment decoupling in a case study region in Germany. *Journal of Farm Management* 13(3), 219–30.

van Asselen, S. and Verburg, P. H. 2013. Land cover change or land-use intensification: simulating land system change with a global-scale land change model. *Global Change Biology* 19(12), 3648–67. doi:10.1111/gcb.12331.

van den Bygaart, A. J. 2018. Comments on soil carbon 4 per mille by Minasny et al. 2017. *Geoderma* 309, 113–4.

van den Bygaart, A. J., Gregorich, E. G. and Helgason, B. L. 2015. Cropland C erosion and burial: is buried soil organic matter biodegradable? *Geoderma* 239, 240-9.

van Ittersum, M. K., Cassman, K. G., Grassini, P., Wolf, J., Tittonell, P. and Hochman, Z. 2013. Yield gap analysis with local to global relevance - a review. *Field Crops Research* 143, 4–17. doi:10.1016/j.fcr.2012.09.009.

van Oost, K., Quine, T. A., Govers, G., De Gryze, S., Six, J., Harden, J. W., Ritchie, J. C., McCarty, G. W., Heckrath, G., Kosmas, C., Giraldez, J. V., da Silva, J. R. and Merckx, R. 2007. The impact of agricultural soil erosion on the global carbon cycle. *Science* 318(5850), 626–9. doi:10.1126/science.1145724.

Vogel, G. 2017. Where have all the insects gone? *Science* 356(6338), 576–9. doi:10.1126/science.356.6338.576.

von der Marwitz, F. A. L. 1852. *Aus dem Nachlasse Friedrich August Ludwig's von der Marwitz aus Friedersdorf. Erster Band. Lebendbeschreibung*. E. S. Mittler & Sohn, Berlin.

Vörösmarty, C. J., McIntyre, P. B., Gessner, M. O., Dudgeon, D., Prusevich, A., Green, P., Glidden, S., Bunn, S. E., Sullivan, C. A., Liermann, C. R. and Davies, P. M. 2010. Global threats to human water security and river biodiversity. *Nature* 467(7315), 555-61. doi:10.1038/nature09440.

Vyska, M., Cunniffe, N. and Gilligan, C. 2016. Trade-off between disease resistance and crop yield: a landscape-scale mathematical modelling perspective. *Journal of the Royal Society Interface* 13(123), 20160451.

Wallor, E., Kersebaum, K., Ventrella, D., Bindi, M., Cammarano, D., Coucheney, E., Gaiser, T., Garofalo, P., Giglio, L., Giola, P., Hoffmann, M. P., Iocola, I., Lana, M., Lewan, E., Maharjan, G. R., Moriondo, M., Mula, L., Nendel, C., Pohankova, E., Roggero, P. P., Trnka, M. and Trombi, G. 2018. The response of process-based agro-ecosystem models to within-field variability in site conditions. *Field Crops Research* 228, 1-19. doi:10.1016/j.fcr.2018.08.021.

Wartmann, F. M. and Purves, R. S. 2018. Investigating sense of place as a cultural ecosystem service in different landscapes through the lens of language. *Landscape and Urban Planning* 175, 169–83. doi:10.1016/j.landurbplan.2018.03.021.

Webber, H., Ewert, F., Olesen, J. E., Müller, C., Fronzek, S., Ruane, A. C., Bourgault, M., Martre, P., Ababaei, B., Bindi, M., Ferrise, R., Finger, R., Fodor, N., Gabaldón-Leal, C., Gaiser, T., Jabloun, M., Kersebaum, K. C., Lizaso, J. I., Lorite, I. J., Manceau, L., Moriondo, M., Nendel, C., Rodríguez, A., Ruiz-Ramos, M., Semenov, M. A., Siebert, S.,

Stella, T., Stratonovitch, P., Trombi, G. and Wallach, D. 2018. Diverging importance of drought stress for maize and winter wheat in Europe. *Nature Communications* 9(1), 4249. doi:10.1038/s41467-018-06525-2.

Wenkel, K. O., Berg, M., Mirschel, W., Wieland, R., Nendel, C. and Köstner, B. 2013. LandCaRe DSS - an interactive decision support system for climate change impact assessment and the analysis of potential agricultural land use adaptation strategies. *Journal of Environmental Management* 127, S168-83. doi:10.1016/j.jenvman.2013.02.051.

Williams, J. R. 1990. The Erosion-Productivity Impact Calculator (EPIC) model - a case-history. *Philosophical Transactions of the Royal Society of London. Series B: Biological Sciences* 329(1255), 421-8. doi:10.1098/rstb.1990.0184.

Willocquet, L., Savary, S., Fernandez, L., Elazegui, F. and Teng, P. 2000. Development and evaluation of a multiple-pest, production situation specific model to simulate yield losses of rice in tropical Asia. *Ecological Modelling* 131(2-3), 133-59. doi:10.1016/S0304-3800(00)00271-4.

Yuen, J. and Mila, A. 2015. Landscape-scale disease risk quantification and prediction. *Annual Review of Phytopathology* 53, 471-84. doi:10.1146/annurev-phyto-080614-120406.

Zagaria, C., Schulp, C. J. E., Kizos, T., Gounaridis, D. and Verburg, P. H. 2017. Cultural landscapes and behavioral transformations: an agent-based model for the simulation and discussion of alternative landscape futures in East Lesvos, Greece. *Land Use Policy* 65, 26-44. doi:10.1016/j.landusepol.2017.03.022.

Zhao, G., Bryan, B. A., King, D., Luo, Z., Wang, E., Song, X. and Yu, Q. 2013. Impact of agricultural management practices on soil organic carbon: simulation of Australian wheat systems. *Global Change Biology* 19(5), 1585-97. doi:10.1111/gcb.12145.

Zhao, G., Hoffmann, H., van Bussel, L., Enders, A., Specka, X., Sosa, C., Yeluripati, J., Tao, F., Constantin, J., Raynal, H., Teixeira, E., Grosz, B., Doro, L., Zhao, Z., Nendel, C., Kiese, R., Eckersten, H., Haas, E., Vanuytrecht, E., Wang, E., Kuhnert, M., Trombi, G., Moriondo, M., Bindi, M., Lewan, E., Bach, M., Kersebaum, K., Rötter, R., Roggero, P., Wallach, D., Cammarano, D., Asseng, S., Krauss, G., Siebert, S., Gaiser, T. and Ewert, F. 2015. Effect of weather data aggregation on regional crop simulation for different crops, production conditions, and response variables. *Climate Research* 65, 141-57. doi:10.3354/cr01301.

Zhao, G., Hoffmann, H., Yeluripati, J., Xenia, S., Nendel, C., Coucheney, E., Kuhnert, M., Tao, F., Constantin, J., Raynal, H., Teixeira, E., Grosz, B., Doro, L., Kiese, R., Eckersten, H., Haas, E., Cammarano, D., Kassie, B., Moriondo, M., Trombi, G., Bindi, M., Biernath, C., Heinlein, F., Klein, C., Priesack, E., Lewan, E., Kersebaum, K., Rötter, R., Roggero, P. P., Wallach, D., Asseng, S., Siebert, S., Gaiser, T. and Ewert, F. 2016. Evaluating the precision of eight spatial sampling schemes in estimating regional means of simulated yield for two crops. *Environmental Modelling and Software* 80, 100-12. doi:10.1016/j.envsoft.2016.02.022.

Zufiaurre, E., Abba, A., Bilenca, D. and Codesido, M. 2016. Role of landscape elements on recent distributional expansion of European starlings (Sturnus vulgaris) in agroecosystems of the Pampas, Argentina. *The Wilson Journal of Ornithology* 128(2), 306-13. doi:10.1676/wils-128-02-306-313.1.

Chapter 13

Agroecosystem models for delivering ecosystem services

Jerry L. Hatfield, Christian Dold, Erica J. Kistner-Thomas and Kenneth M. Wacha, USDA-ARS, USA

1 Introduction

Agriculture exists within the context of an ecosystem. There are as many different agricultural systems as there are ecosystems because each combination represents unique combinations of agricultural crops grown within a specific area in its ecological context. Agroecology represents the complex interactions among all of the components of the agricultural system across temporal and spatial scales. These interactions are often easier to describe qualitatively than quantitatively because it is difficult to determine the temporal and spatial dynamics of these interactions. Most of our models are fairly robust when we focus on the role of agriculture in provisioning because we are interested in how much we produce from a specific land area or practice; however, the linkage of specific management practices to regulatory, cultural or supporting ecosystem functions needs substantial improvement. The complexity of these linkages in the context of ecosystem functions relative to agricultural practices can be seen in Fig. 1 and provides a framework for how we could potentially structure our models.

Agroecology can be envisioned as a spatial-temporal dynamic representing the interaction of hydrological, biological, chemical and physical processes occurring within agricultural systems (Fig. 2). It is easier to conceptualize

http://dx.doi.org/10.19103/AS.2019.0061.18
Published by Burleigh Dodds Science Publishing Limited, 2020.

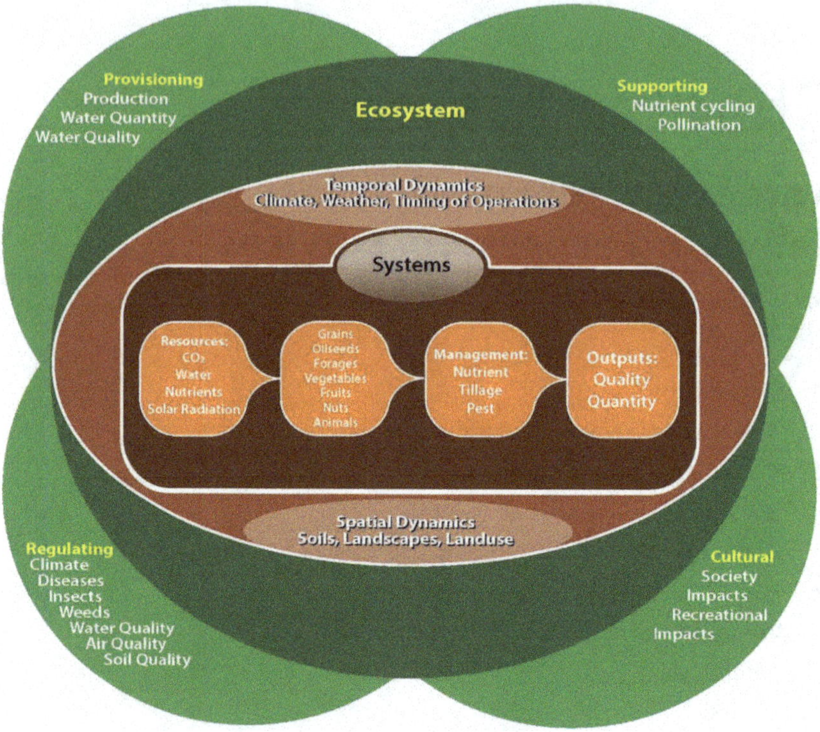

Figure 1 Depiction of agricultural systems with ecosystem services as the endpoints. Souce: Jerry L. Hatfield, USDA-ARS-NLAE, June 2018.

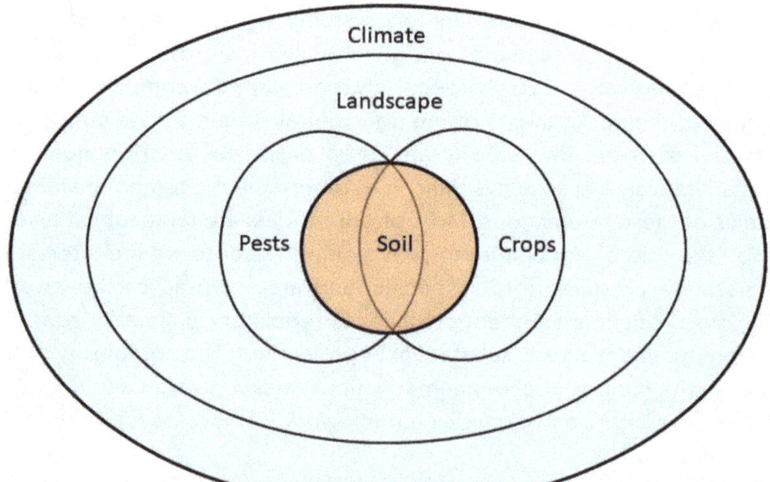

Figure 2 Simplified diagram of agricultural system components with crop and pest models in the context of landscape and climate scales.

these processes at the field scale but it is more difficult to envision a mosaic of these processes interacting in the landscape context. Throughout this chapter, our goal is to highlight the framework and the potential for developing agroecosystem models that can assess ecosystem services.

2 The need for an agroecosystem approach

Agroecosystems reflect the complexity of the natural world in which we live. In agriculture, we have relegated our approach to this complexity by considering an individual crop rather than an ecosystem. There are a range of crop simulation models used to evaluate the response of crop growth to the variety of factors, for example, management, soils and weather. CropSyst (www.ecologicalmodels.psu.edu/agroecology) is one example of an agroecological model that simulates the soil-water budget, nutrient (carbon, nitrogen) budgets, crop canopy development, root growth, dry matter production, yield, crop residue production and erosion. The question remains, why do we need agroecosystems in the context of ecosystem services? The answer is relatively simple; we need to evaluate agroecosystems in the context of the larger ecological context where we can begin to evaluate how agricultural systems impact the ecology and vice versa. Hatfield (2006) discussed the need to begin to approach agricultural systems from the view of the different functions agriculture provides to the ecosystem. If we consider how to quantify these interactions then the ecosystem functions depicted in Fig. 1 will become the endpoints for our different models.

The primary limitation to our understanding of the spatial and temporal dynamics impacting the soil, crop, pests and beneficial birds, animals and insects at the landscape scale is our inability to quantify the interactions among these components. At the field scale provisioning is often considered the dominant factor; however, once we place fields in the landscape context all factors begin to interact and become significant at varying times throughout the seasonal or rotational scale. Sekercioglu (2010) described ecosystem function and services as a unique set of interactions, when in balance create a series of positive impacts for all species, and when they are out of balance create problems of degradation in soils, lack of diversity of species, pollution, drought and disease and insect pressures. Seppelt (2006) presented the concept of agroecological models in the context of optimum control theory and used optimum fertilizer rates and crop rotations as an example of how this concept could be utilized in practice. This type of modelling approach follows the example of Costanza et al. (1993), where they used aggregated and complex models as descriptions of complexity, scale and hierarchy to quantify ecosystems. The major limitation in any modelling approach is being able to quantify the interactions among the components across a range of spatial and temporal scales. Wallor et al.

(2018) showed an example of using an agroecosystem model to evaluate crop management options and the impacts of climate change on agricultural production, food security and ecosystem services. They found the performance of models did not improve with an increased number of input parameters because of inconsistencies in empirically derived algorithms and model-specific sensitivity to spatial variation within fields. This one example shows the difficulty in being able to develop robust, universal models that are capable of simulating a wide range of soils, climates, crops and ecosystem variables. However, if we are to continue to progress in our ability to effectively manage natural resources in the context of ecological principles, we will have to continue to develop and improve simulation models capable of quantifying the interactions of all of the variables across the spatial and temporal domains that encompass agroecosystems. Throughout this chapter, we will highlight the current state of knowledge on the different components and detail the opportunities of utilizing agroecological models to enhance our understanding of agroecological systems.

3 Plants as part of the agroecosystem

Agroecosystems and agricultural landscapes are uniquely defined by their climate, soils and vegetation (Fig. 1), and there have been numerous crop models developed over the past decades to link climate data to crop growth and yield at different scales. Landscapes consist of a rich diversity of plant species that is determined by the agroclimatic properties of a given location. Hatfield and Prueger (2011) summarized how agroecological principles could be used to guide our understanding of how plants are coupled with the environment. Over time, crop models have provided the framework to quantify this coupling and how plants respond to the current environment and future climates. Di Paola et al. (2015) discussed 70 models simulating crop growth and yield, and Challinor et al. (2014) counted 1700 published crop simulations on the crop response to climate change between 2007 and 2012. Rauff and Bello (2015) explained the different types of crop models in relation to climate. In general, there are two main types of crop models: (1) mechanistic crop models, which attempt to explain the mechanisms between climate and dependent crop variables and (2) empirical models, which estimate the statistical and mathematical relationships between climate and crop parameters using empirical datasets. Lobell and Asseng (2017) stated that differences among models will eventually vanish as both types of models adapt approaches from each other. Another layer in climate models in agriculture is the spatial scale, which can be limited to field, landscape, regional and global scale, depending on the models. Thus, climate models coupled with crop models are theoretically capable of simulating the dynamics of agricultural landscapes. The vast abundance of crop models required the evaluation of uncertainties among modelling approaches. The Agricultural Model Implementation and Improvement Project (AgMIP) evaluates the effectiveness of crop models to predict the impact of climatic changes on crop growth and yield using a standardized protocol (Rosenzweig et al., 2013). Under the AgMIP framework, crop models have been compared for corn (*Zea mays*, Bassu et al., 2014), rice (*Oryza sativa*, Hasegawa et al., 2017), potato (*Solanum tuberosum*, Fleisher et al., 2017) and wheat (*Triticum aestivum*, Asseng et al., 2013). Martre et al. (2015) found a relative error of 24–38% among 27 models to simulate wheat grain yield. Lobell and Asseng (2017) identified the following future challenges/needs regarding the comparison of agroecological crop models: (1) more datasets and simulations are needed for temperature increases >2°C and precipitation effects; (2) comparison among models regarding the interaction effects of temperature increases and other variables is still lacking; and (3) include the impact of ambient CO_2 increase in empirical models. Rosenzweig et al. (2014) compared crop models such as the Environmental Policy Integrated Climate Model (EPIC), the Global Agro-Ecological Zone Model (GAEZ), the Crop Environment Resource Synthesis (CERES) and the Crop Template approach (CROPGRO) to evaluate the impact of climate change on global crop production. Coupling global gridded crop models should be done with caution because of the differences in model structure, processes, inputs, parameterizations, assumptions and outputs (Rosenzweig et al., 2014). They found that the use of ecosystem-based models in this analysis increased the uncertainty range when compared to previous studies using site-based models. They also concluded that improvements in the parameterization of the effects of increasing CO_2 on photosynthesis and transpiration will require additional research based on the observation that the global average response to climate was more similar when the CO_2 effects were removed from the models (Rosenzweig et al., 2014). Another limitation of crop models is the lack of data, especially in developing countries, and the simulation of complex cropping systems. The current state and history of crop models were summarized by Jones et al. (2017a,b). Below, a brief description, recent studies and challenges of some agroecological models are presented:

The Agricultural Production Systems sIMulator (APSIM) combines a framework of sub-models, such as crop growth and climate effect simulations (Holzworth et al., 2014). Jin et al. (2017) used APSIM to evaluate the impact of rising temperatures and CO_2 concentrations on rainfed corn and soybean (*Glycine max* (L.) Merr.) production in the Midwestern United States. Recently, Peng et al. (2018) combined the Community Land Model (CLM) with APSIM to predict heat stress impact on corn yields in the Midwestern United States. Gaydon et al. (2017) used APSIM to simulate diverse cropping systems over

12 Southeast Asian countries, and noted a wide range of model improvements needed, such as response to high temperatures and improved radiation use efficiency (RUE) values to simulate biomass accumulation, among others.

The GAEZ model categorizes areas suitable for crop production by climate and by specific growth limitations of crops, among others (Fischer et al., 2002; IIASA/FAO, 2012). One essential concept of GAEZ climate model is the length of the growing period, which is defined by temperature and water constraints. In addition, GAEZ estimates potential yield and applies constraint factors, such as heat or water stress, to calculate actual yield and yield gaps. Texeira et al. (2013) used GAEZ and estimated that 5 Mha of cropland suitable for corn production are at risk due to climate change induced heat stress at the flowering stage, and that yield declines are expected for different crops in the Northern hemisphere between 40°N and 60°N latitudes.

The EPIC model simulates crop growth and yield for over 80 crops in relation to climate, evapotranspiration (ET) and elevated CO_2 among others (Gassmann et al., 2004). Bhattarai et al. (2017) used EPIC to evaluate the impact on corn and soybean production on climate change. Lychuk et al. (2017a) simulated wheat, corn and soybean production under elevated CO_2. Lychuk et al. (2017b) also used EPIC to simulate yield for wheat, barley (*Hordeum vulgare* L.) and canola (*Brassica napus* L.), and found that EPIC was less accurate in simulating annual yield variation owing to weather extremes.

The Decision Support System for Agrotechnology Transfer (DSSAT) combines an ensemble of modules such as CERES and CROPGRO model families to simulate crop growth, and soil-plant-atmosphere and weather modules, which allows simulation of the impact of climate on plant production (Jones et al., 2003). In recent years, DSSAT had been used to evaluate impact of climate change on Durum wheat production in the Mediterranean region (Detorri et al., 2017), on corn in Northern China (Lin et al., 2015), on dry beans (*Phaseolus vulgaris* L.) in Central America (Eitzinger et al., 2017) and corn-cowpea (*Vigna unguiculata*) rotation systems in Malawi (Ngwira et al., 2014). Despite the ample use of DSSAT, improvements are still ongoing. Just recently, DeJonge and Thorp (2017) presented the implementation of a revised ET calculation in DSSAT.

The General Large-Area Model (GLAM) for annual crops' specific objective is to simulate the impact of weather on crop yield on large scales that represent regional or global simulations. GLAM's focus is only on weather impact and only planting date and soil type are considered (Challinor et al., 2004). The GLAM was used to investigate groundnut (*Arachis hypogaea* L.) cultivars in their resilience to climatic changes in India (Challinor et al., 2007) and to simulate wheat production under climate change scenarios in China (Challinor et al., 2010). The GLAM was also recently used to investigate the impact of climate change on corn production in West Africa (Parkes et al., 2018).

The demand on climate and crop models increases with the complexity of the agroecosystem. Agroforestry systems (AFS) represent complex production systems where several plant species are cultivated on the same parcel of land over the same, overlapping or different time periods. AFS combines one or several perennial plants such as trees or large forbs with crops or pastures, which further complicates the system by introducing various canopy stories. The planting design includes mixed stands, hedge rows or alleys and border trees such as in windbreaks or riparian buffers. Trees may be planted or selected, and canopies may be pruned in different seasons. Tall plant canopies will impact light distribution to the lower crop, and affect temperature, rainfall distribution, wind speed and relative humidity among others. In the past, researchers either used existing forestry or single-crop models (such as EPIC, CERES and APSIM) and adapted them to AFS or developed AFS-specific models for light distribution, CO_2 uptake, yield estimation and for other applications. Ellis et al. (2004) and Luedeling et al. (2016) gave an overview of available models for AFS, and Malézieux et al. (2009) reviewed models for both crop-crop and AFS multispecies systems. The most common models for tree-soil-crop interactions are WaNulCas (van Noordwijk and Lusiana, 1999; van Noordwijk et al., 2011), Yield-SAFE (van der Werf et al., 2007) and HyPar (Mobbs et al., 1997). The CO2FIX model is used for carbon sequestration calculations in AFS (Masera et al., 2003). Charbonnier et al. (2013) modelled light interception in coffee (*Coffea arabica*) AFS by adjusting the forestry-specific MAESTRA model (Wang and Jarvis, 1990). The challenges of modelling AFS are the complex spatial structure and the interannual and long-term changes in the system. Also, below-ground biomass, interactions among plant species and nutrient and water competition are challenging to model. As mentioned above, micrometeorological variables are affected by the upper plant story. For example, modelling incoming and absorbed solar radiation in AFS include a range from simple leaf area index approaches to complex 3D tree architecture approaches. The improvement of AFS models still requires more comprehensive long-term AFS datasets to validate and calibrate existing AFS models. It is important to further improve AFS models, as the introduction of perennial plants creates the demand for long-term prediction of crop performance, nutrient and water demands, light distribution and carbon sequestration.

4 Soils as part of the agroecosystem

Agricultural soils mediate the exchange of atmospheric, terrestrial, microbial and hydrologic fluxes (Wilson et al., 2018). Developing a more complete understanding of the spatial and temporal characteristics of these processes and their interconnection within agroecosystems remains a key scientific challenge. Many of the pioneering biogeochemical models (e.g. CENTURY,

RothC) developed for simulating soil-nutrient cycling originated from data collected in well-established, pseudo-equilibrated systems like grasslands (Parton, 1996; Smith et al., 1997). These modelling frameworks generally include a collection of sub-models used to simulate plant production, water balance, and carbon and nutrient cycling driven by climate, water and nutrient availability (Paris and Knapp, 1989). The spatial extent of these models is the soilscape, defined here as a cluster of soil polypedons, homogenous in terms of formative factors and composition (Hole, 1978; Schmidt et al., 2010). Many of these soilscape models do a reasonable job at simulating process occurring vertically within a soil column, specifically the production and decay of organic material (Li et al., 1997; Del Grosso et al., 2005), and can be applied at the sub-catchment scale or greater.

However, using a soilscape approach to simulate agroecosystem processes may be problematic because the mechanistic capabilities with which the models were built upon are not intended to capture the soil–water–nutrient interactions prevalent in a non-equilibrated, dynamic system. Grasslands typically retain a constant vegetative canopy cover that shields the soil surface from the effects of raindrop impact during rainfall events (Neave and Rayburg, 2007). Further, grassland soils comprised of dense rooting networks have been found to promote infiltration, reduce run-off and increase plant available water (Pan and Shangguan, 2006; Kato et al., 2009). On the other hand, row crop systems generally lack permanent surface cover, due in part to growing seasons and harvest operations, which leaves the soil surface periodically exposed and vulnerable to the impacts of raindrop energy and run-off processes to different degree based on management practices, storm intensity and gradient (Nearing et al., 2005; Stocking, 2017). In addition, management practices that utilize cultivation techniques apply mechanized forcing to the soil, which can break apart the soil structure and decrease the stability of soil aggregates (Six et al., 1999; Kasper et al., 2009). When weakened soil aggregates are subjected to hydrologic forcing, soil aggregates can collapse, clog pore spaces and restrict infiltration rates (Wacha et al., 2018).

Once rainfall rates begin to exceed the rate of infiltration, ponding and overland flow conditions can develop (Julien and Simons, 1985). As the depth of run-off increases along the downslope, so does the shear stress being applied to the soil surface and the carrying capacity of the flow (Papanicolaou et al., 2010; Wacha et al., 2018). During heavy rainfall events, soil particles and organic material dislodged from soil aggregates can be swept up in the flow and transported along the downslope (Fiener et al., 2015). These constituents can be redistributed (eroded, deposited) along various flowpaths of the landscape causing heterogeneity in soil texture, organic stocks and moisture conditions both spatially and temporally. The impact of redistribution can also negatively influence crop production as degraded, eroded soils have been found to be less productive and have smaller yields (Pimentel et al., 1995; Quinton et al., 2010). Soil texture variation has also been shown to influence soil microclimatic conditions (soil moisture and temperature), decomposition rates (Chappell et al., 2014), and the selectivity and enrichment of sediment size fractions being transported (Endale et al., 2017). Heterogeneity in surface moisture along the hillslope profile can also have implications on the partitioning of energy fluxes and water availability to plants (Famiglietti et al., 1998; Dold et al., 2017). Therefore, without considering landscape redistribution, estimates of soil, water and nutrient fluxes may be over or underestimated on the seasonal or crop rotational scale (Van Oost et al., 2007). In fact, applying a soilscape approach may lead to large error propagations or even camouflage the services/benefits that certain conservation practices may have, especially on highly erosive landscapes (Canadell et al., 2011).

Several well-documented watershed-scale hydrologic models have been developed to simulate the long-term effects of hydrological changes and management practices, including the Agricultural NonPoint Source Pollution model or AGNPS (Young et al., 1989) and the Soil Water Assessment Tool or SWAT (Arnold et al., 1998). Both models break apart the spatial extent of the watershed into homogenous land units (cells) based on similar soil, cover and management types, and use the SCS curve number (CN) approach to estimate run-off volume (Borah and Bera, 2003). The AnnAGNPS and SWAT models simulate erosion using modified forms of the Universal Soil Loss Equation (USLE), an empirical relation based on the product of erosivity, erodibility, topography and management factors. These models do a reasonable job at simulating long-term average conditions for agricultural watersheds, but lack the mechanistic capabilities to simulate event-based dynamics, deposition or the enrichment of redistributed material, due in part to a disconnection among the land unit cells, making it unable to account for landscape position (Wilson et al., 2016).

Physically based process models such as the Water Erosion Prediction Project or WEPP (Flanagan and Nearing, 1995) have been developed to better simulate rainfall/run-off erosion processes by accounting for spatial heterogeneity and temporal variability in soil properties (e.g. roughness, bulk density and hydraulic conductivity). The WEPP model breaks the hillslope profile into segments called overland flow elements (OFE), similar to SWAT, but provides connectivity between the flow elements to better reflect flow depth conditions and allow size fraction updates of material being mobilized, transported and deposited along the flowpath (Papanicolaou et al., 2018). Surface transport models like WEPP have been successfully coupled with biogeochemical models to improve nutrient cycling by providing updates to the soil from losses and gains from erosion/deposition processes (Yadav and Malanson, 2008; Papanicolaou et al., 2015), but are often limited in scale and lack mechanistic capability to simulate watershed level dynamics.

To build upon the abovementioned studies and move towards modelling frameworks that can accurately assess the services being supplied by an agroecosystem, we need to take our understanding of processes occurring at the hillslope/field scale and apply it to a larger, more meaningful extent. To do this we may consider the following. The underlying foundation of the framework should be built upon geospatial platforms similar to SWAT and GeoWEPP (Renschler, 2003) to allow the landscape to be discretized into a large number of interconnected cells (<soilscapes). To address continuous processes, each cell needs to possess the mechanistic capabilities to simulate vertical-based processes including crop production and the transfer and decay of organic material within the soil column. In addition, each cell should be able to simulate a wide range of crop rotations and climatic conditions, as well as soil–water interactions and influences on the soil microclimate.

To account for landscape redistribution, a surface hydrology model is needed that can accurately simulate overland flow by accounting for topographic curvature and roughness, size fraction mobilization, and the transport and deposition of soil and organic material across the landscape. Updating texture within the cells may improve soil–water sub-model routines that ultimately drive most crop production models. Because aggregate stability is a key parameter in regulating soil–water interactions, incorporating this may improve run-off estimations by accounting for pore-space clogging and restricted infiltration due to collapsing of the aggregate structure. Lastly, by improving the surface hydrology we may be able to capture moisture gradients across the landscape and better reflect microclimatic conditions, thereby improving estimates of soil-nutrient-water fluxes.

Determining and more importantly quantifying the services provided by an agroecosystem remains a giant research challenge. Approaching crop production, pest impact and hydropedology as separate entities will continue to misrepresent the true dynamic nature of an agroecosystem. Within agricultural landscapes there is a co-evolution of non-linear processes occurring at continuous (biologic) and discrete (episodic) time scales. It is apparent there are feedback between the crop-pest-soil mechanisms that need to be considered.

5 Pests, diseases, weeds and pollinators as part of the agroecosystem

Wide ranges of diseases, pests and weeds affect global crop production and cause considerable yield losses. Crop pollination is an important ecosystem service for the productivity of many agricultural crops depending on the presence of both wild and domestic pollinating insects. These organisms have long been recognized as integral components of the agroecosystem (Fig. 1). Extensive laboratory and field research has led to the development of numerous empirical models that often focus on a single biotic stressor or enhancer (pollination services by insects) in a specific cropping system. Over the last thirty years, these empirically derived datasets and models have given rise to several generic mechanistic modelling platforms that examine pollinator, crop pathogen, weed and pest dynamics over a wide range of spatial and temporal scales. Unfortunately, these models often focus mainly on the pest or pollinator component of the cropping system and lack detailed information on how the soil, weather and on-farm management practices are simultaneously impacting crop production. One viable solution would be the inclusion of biotic stressors and enhancers in more comprehensive agroecological modelling platforms such as APSIM. However, many crop simulation models still lack pollinator, pest or disease parameters (Donatelli et al., 2017). This section offers a brief overview of the current state of crop disease, pest, weed and pollinator models. For the sake of brevity, only insects will be discussed in terms of crop pests.

Various weed modelling approaches have been used to better understand weed ecology in agricultural systems and how crop-weed competition affects crop yield (Renton and Chauhan, 2017). Like the aforementioned crop models, many weed models take either an empirical or a mechanistic approach. Empirical approaches are data-driven and often employ statistical models (e.g. non-linear regressions) to describe and analyse crop-weed competition (Kropff and Lotz, 1992; Park et al., 2003; Gonzalez-Andujar et al., 2016). They also provide a simpler framework for practical support decisions regarding the timing of weed prevention practices compared to more complicated mechanistic models (Renton and Chauhan, 2017). Mechanistic weed-crop competition models such as INTERCOM (Kropff and van Laar, 1993) as well as modified versions of ALMANAC (Kiniry et al., 1992), CROPSIM (Chikoye et al., 1996) and APSIM (McCown et al., 1996; Probert et al., 1998) have proven to be relatively accurate in predicting crop-weed competition outcomes despite their use of relatively simple competition algorithms (Deen et al., 2003). These two plant species mechanistic models simulate growth of competing species (i.e. leaf area for each species) based on competition for light, water and nitrogen resources. Weed models have been used to study long-term weed population dynamics across numerous agricultural ecosystems. Holst et al. (2007) identified 134 original publications covering the weed population dynamics of 60 weed species across 40 cropping systems. However, many of these weed population dynamics models are lacking in biological and ecological complexity with most models being limited in application to strategic decision-making in targeted crop-weed systems. Regardless of limitations, weed population dynamics models are continuing to develop in terms of ecological and biological complexity such as the incorporation of evolutionary adaptation into weed population dynamics over generations. So far, this new genetic-based offshoot of previous weed population dynamics models has focussed on herbicide-resistant traits (Renton et al., 2015).

There have been numerous insect pest modelling approaches with phenology (Welch et al., 1978; Briere et al., 1999; Régnière et al., 2012), insect population dynamics (Yonow et al., 2004; Tonnang et al., 2009) as well as pest distribution and invasive pest risk mapping (Kriticos et al., 2015; Hong et al., 2015) being the most common approaches (Tonnang et al., 2017). Since temperature is one of the most important factors driving the developmental rate of arthropods, insect phenology and population models are often empirically derived from numerous experiments measuring insect development time (total and stage specific), mortality, fecundity and longevity under different temperatures (Briere et al., 1999; Régnière et al., 2012). Temperature-dependent phenology models can be quickly developed using a target insect's development threshold temperatures and growing degree day requirements for each life stage. Due to their simplicity, insect phenology models have long been used for decision support in pest management (Welch, 1978), leading to the development of numerous web-based decision support systems designed to help growers schedule scouting or pesticide applications (Damos, 2015). Insect phenology models also provide key input parameters for insect distribution models used in invasive pest risk mapping (Kriticos et al., 2015; Hong et al., 2015). Given the importance of the abiotic environment to insect ecology, daily weather data, particularly air temperature, precipitation and relative humidity are needed to predict insect pest densities over time (Magarey et al., 2001). User-friendly web-based platforms such as NAPPFAST (Magarey et al., 2015) and iPiPE (Isard, 2015) have successfully linked insect phenology models and infection models for simple pathogens to high-resolution gridded weather datasets enabling users to simulate the timing of insect pest and crop disease outbreaks. The models used in these systems are designed to simulate changes in pest occurrence over time for use in management operations such as scouting. In some instances it may be possible to estimate insect and disease-induced crop losses by converting these forecasts to projections of yield loss (Dillehay et al., 2005).

Modelling crop diseases in an agroecosystem context is especially challenging because pathogen interactions with the environment and their plant host(s) are extremely complex (Cunniffe et al., 2015). Like insects, pathogen life cycles are driven, in large part, by the abiotic environment with some epidemics only occurring under favourable weather conditions (van Mannen and Xu, 2003; Frantzen, 2007). Along with more standard temperature, precipitation and relative humidity inputs, many crop disease forecasting models also require information on leaf wetness. This free water on the host surface is necessary for many pathogens to successfully infect and reproduce in/upon the host plant. The availability of leaf wetness data remains a challenge for pest disease forecasting/modelling at more regional scales since this data is not traditionally collected at weather stations and is often limited to specific experimental trials (Lee et al., 2015). Other variables such as soil temperature, radiation, wind speed and direction are used in specialized models examining aerial transport of inoculum or soil pathogens. Many existing crop disease models are empirical and often require an extensive epidemiological dataset based upon laboratory, greenhouse or field experiments. Fortunately, laboratory experiments where host plants are inoculated under different temperature and wetness regimes can be used to parameterize generic infection models (Madden and Ellis, 1988; Magarey et al., 2005). Mechanistic-based infection models like the SEIR (Susceptible-Exposed-Infection-Removed) model represent a solid base for a generic baseline modelling effort (Zadoks, 1971). The SEIR model and related infections models (Magarey et al., 2005) are composed of multiple processes including infection, latency and sporulation in the case of aerially dispersed pathogens. These processes are influenced by the aforementioned abiotic factors which can be used as model climatic drivers. Individual processes are then combined to determine the epidemic process (i.e. disease dynamics in a population of host plants) (Frantzen, 2007). In addition, crop disease epidemics can be either monocyclic or polycyclic, depending on the number of infection cycles per crop cycle. Generic crop disease models like GENEPEST (Savary and Willocquet, 2014) and BioMA-Diseases (Donatelli, 2014; Bregaglio and Donatelli, 2015) can simulate epidemics by estimating, for example, the number of lesions per crop unit area. Accurately simulating crop loss from pathogens across a range of temporal and spatial scales remains challenging due to a lack of sufficiently detailed weather data. Furthermore, crop disease models are not always linked with agronomic data including key land management practices that influence pathogen prevalence (Ghorbani et al., 2008).

Crop loss models examine how pests and diseases affect crop growth and performance. These models follow the basic assumption that pest induced injuries may result in a decrease in crop quality or quantity (i.e. crop damage) (Savary et al., 2006; Esker et al., 2012). Crop damage may ultimately lead to a loss in economic returns from the damage itself or costs from management tactics aimed at reducing the damage (Zadoks, 1987). A damage function is needed to translate injury from a given pest into crop loss. Translating the damage function into a theoretical economic loss function is dependent upon the production system and market variations. Statistical models are most commonly used to define damage functions and economic thresholds (Esker et al., 2012). A major challenge in modelling crop loss from insects as well as diseases is the lack of standardized measure methods for assessing crop damage and a lack of model validation (Donatelli et al., 2017).

Given the limitations of current insect pest and disease models to accurately represent the full complexities of the agroecosystem, there have been multiple attempts to couple existing pest and crop disease models to more advanced mechanistic agroecological modelling platforms. The DSSAT crop models have generic coupling points whereby scouted pest damage can be entered into the model (Batchelor et al., 1993; Jones et al., 2003). Whish et al. (2015) have successfully linked the DYMEX population-modelling platform (Sutherst and Maywald, 1998; Maywald et al., 2014) with APSIM (Holzworth et al., 2014) to model reductions in wheat crop leaf area from rust (*Puccinia striiformis*). The modelling framework diseases in BioMA, a public domain software framework based on biophysical models in the domains of agriculture and environment (Donatelli et al., 2014b), includes modules simulating generic fungal disease epidemics, disease-induced crop damage and the impact of diseases control via agricultural management (Bregaglio and Donatelli, 2015).

Crop pollination is an essential but often poorly understood ecosystem service to agriculture that supports yield in 75% of globally important crops (Klein et al., 2007; Garibaldi et al., 2013). Despite the number of empirical studies examining how land use affects insect pollinator abundance, diversity and pollination services (Ricketts et al., 2008), there have been few attempts to model pollinator impacts in the agricultural landscape. Lonsdorf et al. (2009) developed the first quantitative model that predicts pollinator abundance at the landscape scale. The model uses land cover data, including nesting habitat, floral resource distribution and foraging distances, to predict relative abundances of wild pollinators on the farms requiring pollination services. Lonsdorf et al.'s (2009) model has been adapted to estimate wild bee pollinator abundance within key crop production regions across the United States (Koh et al., 2016) and Europe (Zulian et al., 2013) using a modified version of InVEST (Nelson et al., 2009). Two major limitations of the existing pollinator abundance models are (1) the lack of temporal inputs (i.e. weather, pesticide applications) which are known to influence pollinator population dynamics and (2) the inability of these models to link pollinator abundance to crop yield. Translating pollinator abundance to pollinator influence on crop yield is necessary for determining the economic value of pollination at the farm level. Unfortunately, our ability to predict the effect of pollination services on crop yield is severely limited by the lack of understanding of pollinator-yield effects and more research is greatly needed in this area.

6 Future trends and conclusion

Simulating agricultural production in the context of agroecosystems still remains challenging, as many interacting factors can influence crop production and ecosystem services. Further, we still lack sufficient long-term datasets to accurately predict impacts of crop production on larger spatial and temporal scales. Thereby, the simulation of certain agroecological factors are more developed than others. While crop-soil-weather models are able to simulate growth and yield of important annual crops with little uncertainty, other areas such as pests, diseases and pollinators are less developed. A major challenge in modelling the impacts of diseases, weeds, pests and pollinators on cropping systems is the huge diversity of organisms that affect cultivated crops including arthropods, plants, fungi, bacteria and viruses. Also, specialty crops, multispecies and agroforestry systems remain challenging. Researchers often try to adapt existing models to new management systems to overcome these problems.

A few model ensembles exist which are capable of simulating agroecosystems on the landscape level, including the soilscape, crop growth, climate, pest and diseases. One example of the complexity of this problem is provided in the recent paper by Basso et al. (2018) where they demonstrated that the interaction between climate change scenarios, soil fertility and crops had to be considered in order to quantify the impact of management and climate changes on the soil organic matter. Understanding the linkages among the components of the agroecosystem and ecosystem services will require understanding how components of the system interact with each other over time and space (Fig. 1). We can depict these interactions as shown in Fig. 3 where we can begin to understand that agroecosystem models are complex because the system we are trying to simulate is also complex and often too variable to comprehend. Therefore, we postulate that a more cohesive interdisciplinary research community including but not limited to plant pathologists, climatologists, soil scientists, botanists, entomologists, agronomists, social scientists and economists develop more holistic agricultural modelling approaches (Table 1).

The development and testing of agroecological models will require that we begin to place increased emphasis on quantifying the interactions among processes at a range of spatial and temporal scales.

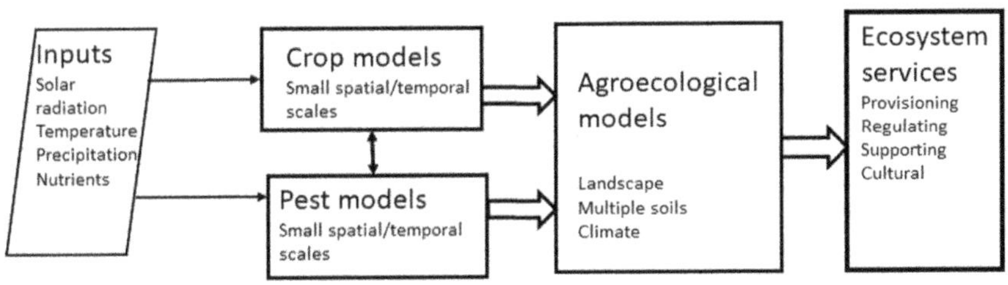

Figure 3 Flow diagram of the linkage between crop and pest models in agroecosystem models with the outcomes ecosystem services.

Table 1 List of agroecosystem models

Name	Description	References
AgMIP	The Agricultural Model Implementation and Improvement Project evaluates crop models to predict the impact of climatic changes on crop growth and yield using a standardized protocol.	Rosenzweig et al. (2013)
Agro-IBIS	Agro-Integrated Biosphere Simulator model simulates the energy, water, carbon and momentum balance of the soil-plant-atmosphere system.	Kucharik (2003)
ALMANAC	A process-oriented simulator of a cropping systems modified to include crop-weed competition.	Kiniry et al. (1992)
APEX	Agricultural Policy Environmental eXtender model for routing water, sediment, nutrients and pesticides across landscape.	Wang et al. (2012)
APSIM	Agricultural Production Systems Simulator framework utilizes sub-models to simulate crop growth, climate, soil effects and crop-weed completion, among others.	Probert et al. (1998), Holzworth et al. (2014)
BioMA-Diseases	A modelling framework that simulates fungal plant epidemic and subsequent impact of disease on crop production.	Donatelli (2014), Bregaglio and Donatelli (2015)
CO2FIX	Agroforestry systems (AFS) model for carbon sequestration.	Masera et al. (2003)
CROPSIM	The two-species version of CROPSIM simulates phenology and growth of competing species.	Chikoye et al. (1996)
CropSyst	Cropping Systems simulation model simulates soil–water budgets, phenology, biomass, soil erosion and crop growth.	Stockle et al. (1994)
DAISY	Mechanistic model that simulates physical and biological processes in agricultural fields.	Abrahamsen and Hansen (2000)
DayCent	Daily version of the CENTURY biogeochemical model.	Del Grosso et al. (2005)
DNDC	Denitrification-Decomposition model simulates carbon and nitrogen biochemistry in agroecosystems.	Li et al. (1997)
DSSAT	The Decision Support System for Agrotechnology Transfer is an ensemble of modules including CERES and CROPGRO model families to simulate crop growth. Additional weather, pest and soil modules among others.	Jones et al. (2003)
DYMEX	Process-based population modelling platform used to describe the life cycles of insects, weeds and diseases.	Maywald et al. (2014)
EPIC	Environmental Policy Integrated Climate model for crop growth and yield for over 80 crop types, includes sub-models for ET, ambient CO_2 and soil erosion, among others.	Wang et al. (2012)
GAEZ	Crop model based on the agroecological zone concept.	Fischer et al. (2002), IIASA/FAO (2012)
GENEPEST	A generic crop model that includes damage mechanisms for pests.	Savary and Willocquet (2014)
GLAM	Large-scale model specifically for impact of weather on crop yield.	Challinor et al. (2004)

Name	Description	References
HyPar	AFS model for tree-crop interactions.	Mobbs et al. (1997)
INTERCOM	Crop-weed competition based on leaf area production and distribution through the canopy in daily time steps.	Kropff and van Laar (1993)
InVEST - Pollinator Model	Integrated Valuation of Ecosystem Services and Trade-offs: Pollination module in the InVEST framework that estimates relative patterns of wild bee abundance and pollination contribution to crop yields.	Nelson et al. (2009), Lonsdorf et al. (2009)
iPiPE	The integrated Pest information Platform (iPiPE) project is an information technology platform that provides tools and models for managing and analysing data in order to generate products and commentary for integrated pest management (IPM). It includes a pest mapping and forecasting system known as ipmPIPE.	Isard et al. (2015)
MONICA	Model for Nitrogen and Carbon in Agro-ecosystems.	Nendel et al. (2011)
NAPPFAST	North Carolina State University/Animal and Plant Health Inspection Service Plant Pest Forecasting System was an internet system for pest and pathogen weather based forecasting and risk mapping. It was active from 2002 to 2013.	Magarey et al. (2015)
ROTHC	A model that accounts for the effects of soil type, temperature and plant canopy cover to simulate the turnover of organic carbon.	Coleman et al. (1997)
SALUS	System Approach to Land Use Sustainability model.	Basso et al. (2006)
SWAT	Soil Water Assessment Tool.	Douglas-Mankin et al. (2010)
WaNulCas	AFS model for tree-soil-crop interactions, computes diurnal light interception, water and nutrients under different management scenarios, complex and highly developed.	van Noordwijk et al. (2011)
Yield-SAFE	Agriculture, forestry and AFS model for tree-crop interactions, parsimonious model.	van der Werf et al. (2007)

We propose a series of potential areas where there would be fruitful progress towards constructing agroecological models applicable to a range of scales:

- Develop crop models capable of handling complex rotations with the goal of quantifying ecosystem services at the field and landscape scale.
- Develop models of ecosystem services (soil health, pollinator habitat, resilience to climatic variability and pest outbreaks, carbon sequestration) along with crop productivity in order to
    - Simulate trade-offs between ecosystem service and crop yield and
    - Provide short-term vs. long-term assessment of agroecosystems.
- Develop agroecological models capable of evaluating the impact of agricultural and landscape management options on all ecosystem services.

- Develop robust models of pest dynamics as part of the agricultural system in order to evaluate changes in pest populations as a result of changing management, soil degradation, climate change and shifts in cropping systems.
- Incorporate livestock into agricultural models in order to evaluate the potential impact of placing livestock systems on the landscape and effectively address the utilization of manure as a nutrient and soil quality resource.
- Develop models capable of addressing soil degradation and restoration as a function of soil management practices and link with climate scenarios to be able to address future impacts of climate change.
- Increase the capability of developing models that link fields to landscape processes in order to assess the role of interconnectivity for pollinator habitat and watershed health.
- Quantifying ecosystem services into a metric(s) for overall agroecological and landscape health.
- Create holistic models that can be used by policy makers and stakeholders to evaluate potential scenarios.

7 Where to look for further information

7.1 Further reading

- A good review of current trends in modelling insect pests and disease impacts on cropping systems is Donatelli et al. (2017).
- The current state and a history of climate and crop models are presented by Jones et al. (2017a,b).
- Ellis et al. (2004), Malézieux et al. (2009) and Luedeling et al. (2016) discussed modelling approaches in agroforestry systems and multi-species systems.
- Review of hydrological models (Devia et al., 2015).

7.2 Key journals/conferences

- *Agriculture, Ecosystems and Environment* is an international journal that features the latest scientific research examining the interface between agroecosystems and the natural environment.

7.3 Major international research projects

- Agricultural model intercomparison and improvement project is a major international effort linking the climate, crop and economic modelling communities in an effort to produce improved crop production models (AgMIP, www.agmip.org).

- Integrated Pest Information Platform for Extension and Education (iPiPE) is a large-scale US programme aiming to achieve more informed decision making in support of integrated pest management and food security (ipipe.org).

8 References

Abrahamsen, P. and Hansen, S. 2000. Daisy: an open soil-crop-atmosphere system model. *Environ. Model. Softw.* 15(3), 313–30.

Arnold, J. G., Srinivasan, R., Muttiah, R. S. and Williams, J. R. 1998. Large area hydrologic modeling and assessment part I: model development 1. *J. Am. Water Resources Assoc.* 34(1), 73–89. doi:10.1111/j.1752-1688.1998.tb05961.x.

Asseng, S., Ewert, F., Rosenzweig, C., Jones, J. W., Hatfield, J. L., Ruane, A. C., Boote, K. J., Thorburn, P. J., Rötter, R. P., Cammarano, D., et al. 2013. Uncertainty in simulating wheat yields under climate change. *Nat. Clim. Chang.* 3(9), 827–32. doi:10.1038/nclimate1916.

Basso, B., Ritchie, J. T., Grace, P. R. and Sartori, L. 2006. Simulation of tillage systems impact on soil biophysical properties using the SALUS model. *Ital. J. Agron.* 4, 677–88.

Basso, B., Dumont, B., Maestrini, B., Shcherbak, I., Robertson, G. P., Porter, J. R., Smith, P., Paustian, K., Grace, P. R., Asseng, S., et al. 2018. Soil organic carbon and nitrogen feedbacks on crop yields under climate change. *Agric. Environ. Lett.* 3(1). doi:10.2134/ael2018.05.0026.

Bassu, S., Brisson, N., Durand, J. L., Boote, K., Lizaso, J., Jones, J. W., Rosenzweig, C., Ruane, A. C., Adam, M., Baron, C., et al. 2014. How do various maize crop models vary in their responses to climate change factors? *Glob. Change Biol.* 20(7), 2301–20. doi:10.1111/gcb.12520.

Batchelor, W. D., Jones, J. W., Boote, K. J. and Pinnschmidt, H. O. 1993. Extending the use of crop models to study pest damage. *Trans. ASAE* 36, 551–8.

Bhattarai, M. D., Secchi, S. and Schoof, J. 2017. Projecting corn and soybeans yields under climate change in a corn belt watershed. *Agric. Sys.* 152, 90–9. doi:10.1016/j.agsy.2016.12.013.

Borah, D. K. and Bera, M. 2003. Watershed-scale hydrologic and nonpoint-source pollution models: review of mathematical bases. *Trans. ASAE* 46(6), 1553–66.

Bregaglio, S. and Donatelli, M. 2015. A set of software components for the simulation of plant airborne diseases. *Environ. Model. Softw.* 72, 426–44. doi:10.1016/j.envsoft.2015.05.011.

Briere, J.-F., Pracros, P., Le Roux, A.-Y. and Pierre, J.-S. 1999. A novel rate model of temperature-dependent development for arthropods. *Environ. Entomol.* 28(1), 22–9. doi:10.1093/ee/28.1.22.

Canadell, J. G., Ciais, P., Gurney, K., Le Quéré, C., Piao, S., Raupach, M. R. and Sabine, C. L. 2011. An international effort to quantify regional carbon fluxes. *Eos, Trans. Am. Geophys. Union* 92(10), 81–2. doi:10.1029/2011EO100001.

Challinor, A. J., Wheeler, T. R., Craufurd, P. Q., Slingo, J. M. and Grimes, D. I. F. 2004. Design and optimisation of a large-area process-based model for annual crops. *Agric. For. Meteorol.* 124(1), 99–120. doi:10.1016/j.agrformet.2004.01.002.

Challinor, A. J., Wheeler, T. R., Craufurd, P. Q., Ferro, C. A. T. and Stephenson, D. B. 2007. Adaptation of crops to climate change through genotypic responses to mean and extreme temperatures. *Agric. Ecosyst. Environ.* 119(1-2), 190-204. doi:10.1016/j.agee.2006.07.009.

Challinor, A. J., Simelton, E. S., Fraser, E. D. G., Hemming, D. and Collins, M. 2010. Increased crop failure due to climate change: assessing adaptation options using models and socio-economic data for wheat in China. *Environ. Res. Lett.* 5(3). doi:10.1088/1748-9326/5/3/034012.

Challinor, A. J., Watson, J., Lobell, D. B., Howden, S. M., Smith, D. R. and Chhetri, N. 2014. A meta-analysis of crop yield under climate change and adaptation. *Nat. Clim. Chang.* 4(4), 287-91. doi:10.1038/nclimate2153.

Chappell, A., Webb, N. P., Viscarra Rossel, R. A. and Bui, E. 2014. Australian net (1950s-1990) soil organic carbon erosion: implications for CO_2 emission and land-atmosphere modelling. *Biogeosciences* 11(18), 5235-44. doi:10.5194/bg-11-5235-2014.

Charbonnier, F., le Maire, G., Dreyer, E., Casanoves, F., Christina, M., Dauzat, J., Eitel, J. U. H., Vaast, P., Vierling, L. A. and Roupsard, O. 2013. Competition for light in heterogeneous canopies: application of MAESTRA to a coffee (*Coffea arabica* L.) agroforestry system. *Agric. For. Meteorol.* 181, 152-69. doi:10.1016/j.agrformet.2013.07.010.

Chikoye, D., Hunt, L. A. and Swanton, C. J. 1996. Simulation of competition for photosynthetically active radiation between common ragweed (*Ambrosia artemisiifolia*) and dry bean (*Phaseolus vulgaris*). *Weed Sci.* 44, 545-54.

Coleman, K., Jenkinson, D., Crocker, G. J., Grace, P. R., Klir, J., Körschens, M., Poulton, P. R. and Richter, D. D. 1997. Simulating trends in soil organic carbon in long-term experiments using RothC-26.3. *Geoderma* 81(1-2), 29-44.

Costanza, R., Wainger, L., Folke, C. and Mäler, K.-G. 1993. Modeling complex ecological economic systems. *BioScience* 43(8), 545-55. doi:10.2307/1311949.

Cunniffe, N. J., Koskella, B., Metcalf, C. J., Parnell, S., Gottwald, T. R. and Gilligan, C. A. 2015. Thirteen challenges in modelling plant diseases. *Epidemics* 10, 6-10. doi:10.1016/j.epidem.2014.06.002.

Damos, P. 2015. Modular structure of web-based decision support systems for integrated pest management. A review. *Agron. Sustain. Dev.* 35(4), 1347-72. doi:10.1007/s13593-015-0319-9.

Deen, W., Cousens, R., Warringa, J., Bastiaans, L., Carberry, P., Rebel, K., Riha, S., Murphy, C., Benjamin, L. R., Cloughley, C., et al. 2003. An evaluation of four crop:weed competition models using a common data set. *Weed Res.* 43(2), 116-29. doi:10.1046/j.1365-3180.2003.00323.x.

DeJonge, K. C. and Thorp, K. R. 2017. Implementing standardized reference evapotranspiration and dual crop coefficient approach in the DSSAT cropping system model. *Transactions of the ASABE* 60(6), 1965-81. doi:10.13031/trans.12321.

Del Grosso, S. J., Mosier, A. R., Parton, W. J. and Ojima, D. S. 2005. DAYCENT model analysis of past and contemporary soil N_2O and net greenhouse gas flux for major crops in the USA. *Soil Till. Res.* 83(1), 9-24. doi:10.1016/j.still.2005.02.007.

Dettori, M., Cesaraccio, C. and Duce, P. 2017. Simulation of climate change impacts on production and phenology of durum wheat in Mediterranean environments using CERES-Wheat model. *Field Crops Res.* 206, 43-53. doi:10.1016/j.fcr.2017.02.013.

Devia, G. K., Ganasri, B. P. and Dwarakish, G. S. 2015. A review on hydrological models. *Aquat. Procedia* 4, 1001-7. doi:10.1016/j.aqpro.2015.02.126.

Di Paola, A., Valentini, R. and Santini, M. 2015. An overview of available crop growth and yield models for studies and assessments in agriculture. *J. Sci. Food Agric.* 96(3), 709-14. doi:10.1002/jsfa.7359.

Dillehay, B. L., Calvin, D. D., Roth, G. W., Hyde, J. A., Kuldau, G. A., Kratochvil, R. J., Russo, J. M. and Voight, D. G. 2005. Verification of a European corn borer (Lepidoptera: Crambidae) loss equation in the major corn production region of the Northeastern United States. *J. Econ. Entomol.* 98(1), 103–12. doi:10.1093/jee/98.1.103.

Dold, C., Hatfield, J. L., Prueger, J., Sauer, T., Büyükcangaz, H. and Rondinelli, W. 2017. Long-term application of the crop water stress index in midwest agro-ecosystems. *Agron. J.* 109(5), 2172–81. doi:10.2134/agronj2016.09.0494.

Donatelli, M. 2014. BioMA - biophysical model application framework. Available at: https://en.wikipedia.org/wiki/BioMA.

Donatelli, M., Magarey, R. D., Bregaglio, S., Willocquet, L., Whish, J. P. M. and Savary, S. 2017. Modelling the impacts of pests and diseases on agricultural systems. *Agric. Syst.* 155, 213–24. doi:10.1016/j.agsy.2017.01.019.

Douglas-Mankin, K. R., Srinivasan, R. and Arnold, J. G. 2010. Soil and Water Assessment Tool (SWAT) model: current developments and applications. *Trans. ASABE* 53(5), 1423–31. doi:10.13031/2013.34915.

Eitzinger, A., Läderach, P., Rodriguez, B., Fisher, M., Beebe, S., Sonder, K. and Schmidt, A. 2017. Assessing high-impact spots of climate change: spatial yield simulations with Decision Support System for Agrotechnology Transfer (DSSAT) model. *Mitig. Adapt. Strateg. Glob. Chang.* 22(5), 743–60. doi:10.1007/s11027-015-9696-2.

Ellis, E. A., Bentrup, G. and Schoeneberger, M. M. 2004. Computer-based tools for decision support in agroforestry: current state and future needs. *Agroforest. Syst.* 61, 401–21. doi:10.1023/b:Agfo.0000029015.64463.65.

Endale, D. M., Potter, T. L., Strickland, T. C. and Bosch, D. D. 2017. Sediment-bound total organic carbon and total organic nitrogen losses from conventional and strip tillage cropping systems. *Soil Till. Res.* 171, 25–34. doi:10.1016/j.still.2017.04.004.

Esker, P. D., Savary, S. and McRoberts, N. 2012. Crop loss analysis and global food supply: focusing now on required harvests. *CAB Rev.* 7(052), 1–14.

Famiglietti, J. S., Rudnicki, J. W. and Rodell, M. 1998. Variability in surface moisture content along a hillslope transect: Rattlesnake Hill, Texas. *J. Hydrol.* 210(1-4), 259–81. doi:10.1016/S0022-1694(98)00187-5.

Fiener, P., Dlugoß, V. and Van Oost, K. 2015. Erosion-induced carbon redistribution, burial and mineralisation–is the episodic nature of erosion processes important? *Catena* 133, 282–92. doi:10.1016/j.catena.2015.05.027.

Fischer, G., Van Velthuizen, H., Shah, M. and Nachtergaele, F. O. 2002. Global agro-ecological assessment for agriculture in the 21st century: methodology and results. RR-02-02. IIASA, Laxenburg, Austria and FAO, Rome, Italy, 119pp.

Flanagan, D. C. and Nearing, M. A. 1995. USDA-water erosion prediction project: hillslope profile and watershed model documentation (vol. 10). NSERL report. USDA-ARS National Soil Erosion Research Laboratory, West Lafayette, IN.

Fleisher, D. H., Condori, B., Quiroz, R., Alva, A., Asseng, S., Barreda, C., Bindi, M., Boote, K. J., Ferrise, R., Franke, A. C., et al. 2017. A potato model inter-comparison across varying climates and productivity levels. *Glob. Change Biol.* 23(3), 1258–81. doi:10.1111/gcb.13411.

Frantzen, J. 2007. *Epidemiology and Plant Ecology: Principles and Applications.* World Scientific Publishing Company, Singapore.

Garibaldi, L. A., Steffan-Dewenter, I., Winfree, R., Aizen, M. A., Bommarco, R., Cunningham, S. A., Kremen, C., Carvalheiro, L. G., Harder, L. D., Afik, O., et al. 2013. Wild pollinators enhance fruit set of crops regardless of honey bee abundance. *Science* 339(6127), 1608–11. doi:10.1126/science.1230200.

Gassman, P. W., Williams, J. R., Benson, V. W., Cesar Izaurralde, R., Hauck, L. M., Jones, A. C., Atwood, J., Kiniry, J. R. and Flowers, J. D. 2004. Historical development and applications of the EPIC and APEX models. In: *ASAE/CSAE Meeting*. ASAE, St. Joseph, MI.

Gaydon, D. S., Balwinder, S., Wang, E., Poulton, P. L., Ahmad, B., Ahmed, F., Akhter, S., Ali, I., Amarasingha, R., Chaki, A. K., et al. 2017. Evaluation of the APSIM model in cropping systems of Asia. *Field Crops Res.* 204, 52–75. doi:10.1016/j.fcr.2016.12.015.

Ghorbani, R., Wilcockson, S., Koocheki, A. and Leifert, C. 2008. Soil management for sustainable crop disease control: a review. *Environ. Chem. Lett.* 6(3), 149–62. doi:10.1007/s10311-008-0147-0.

Gonzalez-Andujar, J. L., Chantre, G. R., Morvillo, C., Blanco, A. M. and Forcella, F. 2016. Predicting field weed emergence with empirical models and soft computing techniques. *Weed Res.* 56(6), 415–23. doi:10.1111/wre.12223.

Hasegawa, T., Li, T., Yin, X., Zhu, Y., Boote, K., Baker, J., Bregaglio, S., Buis, S., Confalonieri, R., Fugice, J., et al. 2017. Causes of variation among rice models in yield response to CO_2 examined with free-air CO_2 enrichment and growth chamber experiments. *Sci. Rep.* 7(1), 14858. doi:10.1038/s41598-017-13582-y.

Hatfield, J. L. 2006. Multifunctionality of agriculture and farming system design: perspectives from the United States. *Bibliotecha Fragm. Agron.* 11, 43–52.

Hatfield, J. L. and Prueger, J. H. 2011. Agroecology: implications for plant response to climate change. In: Yadav, S. S., Redden, R. J., Hatfield, J. L., Lotze-Campen, H. and Hall, A. E. (Eds), *Crop Adaptation to Climate Change*. John Wiley and Sons, Ltd, Chichester, UK, pp. 27–43.

Hole, F. D. 1978. An approach to landscape analysis with emphasis on soils. *Geoderma* 21(1), 1–23. doi:10.1016/0016-7061(78)90002-2.

Holst, N., Rasmussen, I. A. and Bastiaans, L. 2007. Field weed population dynamics: a review of model approaches and applications. *Weed Res.* 47(1), 1–14. doi:10.1111/j.1365-3180.2007.00534.x.

Holzworth, D. P., Huth, N. I., deVoil, P. G., Zurcher, E. J., Herrmann, N. I., McLean, G., Chenu, K., van Oosterom, E. J., Snow, V., Murphy, C., et al. 2014. APSIM – evolution towards a new generation of agricultural systems simulation. *Environ. Model. Softw.* 62, 327–50. doi:10.1016/j.envsoft.2014.07.009.

Hong, S. C., Magarey, R. D., Borchert, D. M., Vargas, R. I. and Souder, S. K. 2015. Site-specific temporal and spatial validation of a generic plant pest forecast system with observations of *Bactrocera dorsalis* (oriental fruit fly). *NeoBiota* 27, 37–67. doi:10.3897/neobiota.27.5177.

IIASA/FAO. 2012. *Global Agro-ecological Zones (GAEZ v3. 0)*. IIASA, Laxenburg, Austria and FAO, Rome, Italy.

Isard, S. A., Russo, J. M., Magarey, R. D., Golod, J. and VanKirk, J. R. 2015. Integrated Pest Information Platform for Extension and Education (iPiPE): progress through sharing. *J. Integr. Pest Manag.* 6(1), 15. doi:10.1093/jipm/pmv013.

Jin, Z., Zhuang, Q., Wang, J., Archontoulis, S. V., Zobel, Z. and Kotamarthi, V. R. 2017. The combined and separate impacts of climate extremes on the current and future US rainfed maize and soybean production under elevated CO_2. *Glob. Change Biol.* 23(7), 2687–704. doi:10.1111/gcb.13617.

Jones, J. W., Hoogenboom, G., Porter, C. H., Boote, K. J., Batchelor, W. D., Hunt, L. A., Wilkens, P. W., Singh, U., Gijsman, A. J. and Ritchie, J. T. 2003. The DSSAT cropping system model. *Eur. J. Agron.* 18(3-4), 235-65. doi:10.1016/S1161-0301(02)00107-7.

Jones, J. W., Antle, J. M., Basso, B., Boote, K. J., Conant, R. T., Foster, I., Godfray, H. C. J., Herrero, M., Howitt, R. E., Janssen, S., et al. 2017a. Brief history of agricultural systems modeling. *Agric. Sys.* 155, 240-54. doi:10.1016/j.agsy.2016.05.014.

Jones, J. W., Antle, J. M., Basso, B., Boote, K. J., Conant, R. T., Foster, I., Godfray, H. C. J., Herrero, M., Howitt, R. E., Janssen, S., et al. 2017b. Toward a new generation of agricultural system data, models, and knowledge products: state of agricultural systems science. *Agric. Sys.* 155, 269-88. doi:10.1016/j.agsy.2016.09.021.

Julien, P. Y. and Simons, D. B. 1985. Sediment transport capacity of overland flow. *Trans. ASAE* 28(3), 755-62.

Kasper, M., Buchan, G. D., Mentler, A. and Blum, W. E. H. 2009. Influence of soil tillage systems on aggregate stability and the distribution of C and N in different aggregate fractions. *Soil Till. Res.* 105(2), 192-9. doi:10.1016/j.still.2009.08.002.

Kato, H., Onda, Y., Tanaka, Y. and Asano, M. 2009. Field measurement of infiltration rate using an oscillating nozzle rainfall simulator in the cold, semiarid grassland of Mongolia. *CATENA* 76(3), 173-81. doi:10.1016/j.catena.2008.11.003.

Kiniry, J. R., Williams, J. R., Gassman, P. W. and Debaeke, P. 1992. A general, process oriented model for two competing plant species. *Trans. ASAE* 35, 801-10.

Klein, A. M., Vaissiere, B. E., Cane, J. H., Steffan-Dewenter, I., Cunningham, S. A., Kremen, C. and Tscharntke, T. 2007. Importance of pollinators inchanging landscapes for world crops. *Proc. R. Soc. B Biol. Sci.* 274(1608), 303-13. doi:10.1098/rspb.2006.3721.

Koh, I., Lonsdorf, E. V., Williams, N. M., Brittain, C., Isaacs, R., Gibbs, J. and Ricketts, T. H. 2016. Modeling the status, trends, and impacts of wild bee abundance in the United States. *Proc. Natl. Acad. Sci. U.S.A.* 113(1), 140-5. doi:10.1073/pnas.1517685113.

Kriticos, D. J., Maywald, G. F., Yonow, T., Zurcher, E. J., Herrmann, N. I. and Sutherst, R. W. 2015. *CLIMEX Version 4: Exploring the Effects of Climate on Plants, Animals and Diseases*. CSIRO, Canberra, Australia.

Kropff, M. J. and Lotz, L. A. P. 1992. Systems approaches to quantify crop-weed interactions and their application in weed management. *Agric. Sys.* 40(1-3), 265-82. doi:10.1016/0308-521X(92)90024-I.

Kropff, M. J. and van Laar, H. H. 1993. *Modelling Crop-Weed Interactions*. CAB International, Wallingford, UK.

Lee, K. J., Kang, J. Y., Lee, D. Y., Jang, S. W., Lee, M. S., Lee, B. W. and Kim, K. S. 2015. Use of an empirical model to estimate leaf wetness duration for operation of a disease warning system under a shade in a ginseng field. *Plant Dis.* 100, 25-31.

Li, C., Frolking, S., Crocker, G. J., Grace, P. R., Klír, J., Körchens, M. and Poulton, P. R. 1997. Simulating trends in soil organic carbon in long-term experiments using the DNDC model. *Geoderma* 81(1-2), 45-60. doi:10.1016/S0016-7061(97)00080-3.

Kucharik, C. J. (2003). Evaluation of a process-based agro-ecosystem model (Agro-IBIS) across the US Corn Belt: simulations of the interannual variability in maize yield. *Earth Interact.* 7(14), 1-33.

Lin, Y., Wu, W. and Ge, Q. 2015. CERES-maize model-based simulation of climate change impacts on maize yields and potential adaptive measures in Heilongjiang Province, China. *J. Sci. Food Agric.* 95(14), 2838-49. doi:10.1002/jsfa.7024.

Lobell, D. B. and Asseng, S. 2017. Comparing estimates of climate change impacts from process-based and statistical crop models. *Environ. Res. Lett.* 12(1). doi:10.1088/1748-9326/aa518a.

Lonsdorf, E., Kremen, C., Ricketts, T., Winfree, R., Williams, N. and Greenleaf, S. 2009. Modelling pollination services across agricultural landscapes. *Ann. Bot.* 103(9), 1589–600. doi:10.1093/aob/mcp069.

Luedeling, E., Smethurst, P. J., Baudron, F., Bayala, J., Huth, N. I., van Noordwijk, M., Ong, C. K., Mulia, R., Lusiana, B., Muthuri, C., et al. 2016. Field-scale modeling of tree-crop interactions: challenges and development needs. *Agric. Sys.* 142, 51–69. doi:10.1016/j.agsy.2015.11.005.

Lychuk, T. E., Hill, R. L., Izaurralde, R. C., Momen, B. and Thomson, A. M. 2017a. Evaluation of climate change impacts and effectiveness of adaptation options on crop yield in the Southeastern United States. *Field Crops Res.* 214, 228–38. doi:10.1016/j.fcr.2017.09.020.

Lychuk, T. E., Moulin, A. P., Johnson, E. N., Olfert, O. O., Brandt, S. A. and Izaurralde, R. C. 2017b. Evaluation of the Environmental Policy Integrated Climate Model on predicting crop yield in the Canadian Prairies: a case study. *Can. J. Soil Sci.* 97, 692–702. doi:10.1139/cjss-2017-0044.

Madden, L. V. and Ellis, M. A. 1988. How to develop plant disease forecasters. In: Kranz, J. and Rotem, J. (Eds), *Experimental Techniques in Plant Disease Epidemiology*. Springer-Verlag, New York, pp. 191–208.

Magarey, R. D., Seem, R. C., Russo, J. M., Zack, J. W., Waight, K. T., Travis, J. W. and Oudemans, P. V. 2001. Site-specific weather information without on-site sensors. *Plant Dis.* 85(12), 1216–26. doi:10.1094/PDIS.2001.85.12.1216.

Magarey, R. D., Sutton, T. B. and Thayer, C. L. 2005. A simple generic infection model for foliar fungal plant pathogens. *Phytopathology* 95(1), 92–100. doi:10.1094/PHYTO-95-0092.

Magarey, R. D., Borchert, D. M., Fowler, G. A. and Hong, S. C. 2015. The NCSU/APHIS plant pest forecasting system (NAPPFAST). In: Venette, R. (Ed.), *Pest Risk Modeling and Mapping for Invasive Alien Species*. CAB International, Wallingford, UK.

Malézieux, E., Crozat, Y., Dupraz, C., Laurans, M., Makowski, D., Ozier-Lafontaine, H., Rapidel, B., Tourdonnet, S. and Valantin-Morison, M. 2009. Mixing plant species in cropping systems: concepts, tools and models. A review. *Agron. Sustain. Dev.* 29(1), 43–62. doi:10.1051/agro:2007057.

Martre, P., Wallach, D., Asseng, S., Ewert, F., Jones, J. W., Rötter, R. P., Boote, K. J., Ruane, A. C., Thorburn, P. J., Cammarano, D., et al. 2015. Multimodel ensembles of wheat growth: many models are better than one. *Glob. Change Biol.* 21(2), 911–25. doi:10.1111/gcb.12768.

Masera, O. R., Garza-Caligaris, J. F., Kanninen, M., Karjalainen, T., Liski, J., Nabuurs, G. J., et al. 2003. Modeling carbon sequestration in afforestation, agroforestry and forest management projects: the CO2FIX V.2 approach. *Ecol. Modell.* 164, 177–99. doi:10.1016/S0304-3800(02)00419-2.

Maywald, G. F., Sutherst, R. W., Kriticos, D. J., Herrmann, N. I., Yonow, T. and Zalucki, M. P. 2014. *DYMEX Version 4: Model Simulator*. CSIRO, Canberra, Australia.

McCown, R. L., Hammer, G. L., Hargreaves, J. N. G., Holzworth, D. P. and Freebairn, D. M. 1996. APSIM: a novel software system for model development, model testing and simulation in agricultural systems research. *Agric. Syst.* 50(3), 255–71. doi:10.1016/0308-521X(94)00055-V.

Mobbs, D., Crout, N., Lawson, G. and Cannell, M. 1997. Structure and applications of the HyPAR model. *Agrofor. Forum (UK)* 8, 10-13.

Nearing, M. A., Jetten, V., Baffaut, C., Cerdan, O., Couturier, A., Hernandez, M., Le Bissonnais, Y., Nichols, M. H., Nunes, J. P., Renschler, C. S., et al. 2005. Modeling response of soil erosion and runoff to changes in precipitation and cover. *Catena* 61(2-3), 131-54.

Neave, M. and Rayburg, S. 2007. A field investigation into the effects of progressive rainfall-induced soil seal and crust development on runoff and erosion rates: the impact of surface cover. *Geomorphology* 87(4), 378-90. doi:10.1016/j.geomorph.2006.10.007.

Nelson, E., Mendoza, G., Regetz, J., Polasky, S., Tallis, H., Cameron, D., Chan, K. M., Daily, G. C., Goldstein, J., Kareiva, P. M., et al. 2009. Modeling multiple ecosystem services, biodiversity conservation, commodity production, and tradeoffs at landscape scales. *Front. Ecol. Environ.* 7(1), 4-11. doi:10.890/080023.

Nendel, C., Berg, M., Kersebaum, K. C., Mirschel, W., Specka, X., Wegehenkel, M., Wenkel, K. O. and Wieland, R. 2011. The MONICA model: testing predictability for crop growth, soil moisture and nitrogen dynamics. *Ecol. Model.* 222(9), 1614-25.

Ngwira, A. R., Aune, J. B. and Thierfelder, C. 2014. DSSAT modelling of conservation agriculture maize response to climate change in Malawi. *Soil Till. Res.* 143, 85-94. doi:10.1016/j.still.2014.05.003.

Pan, C. and Shangguan, Z. 2006. Runoff hydraulic characteristics and sediment generation in sloped grassplots under simulated rainfall conditions. *J. Hydrol.* 331(1-2), 178-85. doi:10.1016/j.jhydrol.2006.05.011.

Papanicolaou, A. N., Sanford, J. T., Dermisis, D. C. and Mancilla, G. A. 2010. A 1-D morphodynamic model for rill erosion. *Water Resour. Res.* 46(9). doi:10.1029/2009WR008486.

Papanicolaou, A. N. T., Wacha, K. M., Abban, B. K., Wilson, C. G., Hatfield, J. L., Stanier, C. O. and Filley, T. R. 2015. From soilscapes to landscapes: a landscape-oriented approach to simulate soil organic carbon dynamics in intensively managed landscapes. *J. Geophys. Res. Biogeosci.* 120(11), 2375-401. doi:10.1002/2015JG003078.

Papanicolaou, A. N., Abban, B. K. B., Dermisis, D. C., Giannopoulos, C. P., Flanagan, D. C., Frankenberger, J. R. and Wacha, K. M. 2018. Flow resistance interactions on hillslopes with heterogeneous attributes: effects on runoff hydrograph characteristics. *Water Resour. Res.* 54(1), 359-80. doi:10.1002/2017WR021109.

Paris, Q. and Knapp, K. 1989. Estimation of von Liebig response functions. *Am. J. Agric. Econ.* 71(1), 178-86. doi:10.2307/1241786.

Park, S. E., Benjamin, L. R. and Watkinson, A. R. 2003. The theory and application of plant competition models: an agronomic perspective. *Ann. Bot.* 92(6), 741-8. doi:10.1093/aob/mcg204.

Parkes, B., Sultan, B. and Ciais, P. 2018. The impact of future climate change and potential adaptation methods on Maize yields in West Africa. *Clim. Change* 151(2), 205-17. doi:10.1007/s10584-018-2290-3.

Parton, W. J. 1996. The CENTURY model. In: Powlson, D. S., Smith, P. and Smith, J. U. (Eds), *Evaluation of Soil Organic Matter Models*. Springer, Berlin and Heidelberg, Germany, pp. 283-91.

Peng, B., Guan, K., Chen, M., Lawrence, D. M., Pokhrel, Y., Suyker, A., Arkebauer, T. and Lu, Y. 2018. Improving maize growth processes in the community land model: implementation and evaluation. *Agric. For. Meteorol.* 250-251, 64-89. doi:10.1016/j.agrformet.2017.11.012.

Pimentel, D., Harvey, C., Resosudarmo, P., Sinclair, K., Kurz, D., McNair, M., Crist, S., Shpritz, L., Fitton, L., Saffouri, R., et al. 1995. Environmental and economic costs of soil erosion and conservation benefits. *Science* 267(5201), 1117–23. doi:10.1126/science.267.5201.1117.

Probert, M. E., Carberry, P. S., McCown, R. L. and Turpin, J. E. 1998. Simulation of legume cereal systems using APSIM. *Aust. J. Agric. Res.* 49(3), 317–28. doi:10.1071/A97070.

Quinton, J. N., Govers, G., Van Oost, K. and Bardgett, R. D. 2010. The impact of agricultural soil erosion on biogeochemical cycling. *Nat. Geosci.* 3(5), 311–4. doi:10.1038/ngeo838.

Rauff, K. O. and Bello, R. 2015. A review of crop growth simulation models as tools for agricultural meteorology. *Agric.Sci.* 6, 1098–105.

Régnière, J., Powell, J., Bentz, B. and Nealis, V. 2012. Effects of temperature on development, survival and reproduction of insects: experimental design, data analysis and modeling. *J. Insect Physiol.* 58(5), 634–47. doi:10.1016/j.jinsphys.2012.01.010.

Renschler, C. S. 2003. Designing geo-spatial interfaces to scale process models: the GeoWEPP approach. *Hydrol. Process.* 17(5), 1005–17. doi:10.1002/hyp.1177.

Renton, M. and Chauhan, B. S. 2017. Modelling crop-weed competition: why, what, how and what lies ahead? *Crop Prot.* 95, 101–8. doi:10.1016/j.cropro.2016.09.003.

Renton, M., Lawes, R., Metcalf, T. and Robertson, M. 2015. Considering long-term ecological effects on future land-use options when making tactical break-crop decisions in cropping systems. *Crop Pasture Sci.* 66(6), 610–21. doi:10.1071/CP14135.

Ricketts, T. H., Regetz, J., Steffan-Dewenter, I., Cunningham, S. A., Kremen, C., Bogdanski, A., Gemmill-Herren, B., Greenleaf, S. S., Klein, A. M., Mayfield, M. M., et al. 2008. Landscape effects on crop pollination services: are there general patterns? *Ecol. Lett.* 11(5), 499–515. doi:10.1111/j.1461-0248.2008.01157.x.

Rosenzweig, C., Jones, J. W., Hatfield, J. L., Ruane, A. C., Boote, K. J., Thorburn, P., Antle, J. M., Nelson, G. C., Porter, C., Janssen, S., et al. 2013. The Agricultural Model Intercomparison and Improvement Project (AgMIP): protocols and pilot studies. *Agric. For. Meteorol.* 170, 166–82. doi:10.1016/j.agrformet.2012.09.011.

Rosenzweig, C., Elliott, J., Deryng, D., Ruane, A. C., Müller, C., Arneth, A., Boote, K. J., Folberth, C., Glotter, M., Khabarov, N., et al. 2014. Assessing agricultural risks of climate change in the 21st century in a global gridded crop model intercomparison. *Proc. Natl. Acad. Sci. U.S.A.* 111(9), 3268–73. doi:10.1073/pnas.1222463110.

Savary, S. and Willocquet, L. 2014. Simulation modeling in botanical epidemiology and crop loss analysis. *The Plant Health Instructor*. APSnet Education Center. Available at: https://www.apsnet.org/edcenter/advanced/topics/BotanicalEpidemiology/Pages/ModelingYieldLossesDuetoPests.aspx.

Savary, S., Teng, P. S., Willocquet, L. and Nutter, F. W. 2006. Quantification and modeling of crop losses: a review of purposes. *Annu. Rev. Phytopathol.* 44(1), 89–112. doi:10.1146/annurev.phyto.44.070505.143342.

Schmidt, K., Behrens, T., Friedrich, K. and Scholten, T. 2010. A method to generate soilscapes from soil maps. *J. Plant Nutr. Soil Sci.* 173(2), 163–72. doi:10.1002/jpln.200800208.

Sekercioglu, C. H. 2010. Ecosystem functions and services. In: Sodhi, N. S. and Ehrlich, P. R. (Eds), *Conservation Biology for All*. Oxford Scholarship (Online). Chapter 3. doi:10.1093/acprof:oso/9780199554232.003.0004.

Seppelt, R. 2006. Agroecosystem management. In: Pintér, J. (Ed.), *Global Optimization: Scientific and Engineering Case Studies*. SpringerLink, Springer, Boston, MA, pp. 413–39. doi:10.1007/0-387-30927-6_17.

Six, J., Elliott, E. T. and Paustian, K. 1999. Aggregate and soil organic matter dynamics under conventional and no-tillage systems. *Soil Sci. Soc. Am. J.* 63(5), 1350–8. doi:10.2136/sssaj1999.6351350x.

Smith, P., Smith, J. U., Powlson, D. S., McGill, W. B., Arah, J. R. M., Chertov, O. G., Coleman, K., Franko, U., Frolking, S., Jenkinson, D. S., et al. 1997. A comparison of the performance of nine soil organic matter models using datasets from seven long-term experiments. *Geoderma* 81(1–2), 153–225. doi:10.1016/S0016-7061(97)00087-6.

Stocking, M. A. 2017. Assessing vegetative cover and management effects. In: Skidmore, E. L. (Ed.), *Soil Erosion Research Methods* (2nd edn.). Routledge, UK, pp. 211–34.

Stockle, C. O., Martin, S. A. and Campbell, G. S. 1994. CropSyst, a cropping systems simulation model: water/nitrogen budgets and crop yield. *Agric. Syst.* 46(3), 335–59.

Sutherst, R. W. and Maywald, G. F. 1998. DYMEX modelling workshops: a national, collaborative approach to pest risk analysis and IPM in Australia. *Pest Manag. Future Chall.* 2, 57–62.

Teixeira, E. I., Fischer, G., van Velthuizen, H., Walter, C. and Ewert, F. 2013. Global hot-spots of heat stress on agricultural crops due to climate change. *Agric. For. Meteorol.* 170, 206–15. doi:10.1016/j.agrformet.2011.09.002.

Tonnang, H. E. Z., Nedorezov, L. V., Ochanda, H., Owino, J. and Löhr, B. 2009. Assessing the impact of biological control of *Plutella xylostella* through the application of Lotka-Volterra model. *Ecol. Modell.* 220(1), 60–70. doi:10.1016/j.ecolmodel.2008.09.002.

Tonnang, H. E. Z., Hervé, B. D. B., Biber-Freudenberger, L., Salifu, D., Subramanian, S., Ngowi, V. B., Guimapi, R. Y. A., Anani, B., Kakmeni, F. M. M., Affognon, H., et al. 2017. Advances in crop insect modelling methods – towards a whole system approach. *Ecol. Modell.* 354, 88–103. doi:10.1016/j.ecolmodel.2017.03.015.

van der Werf, W., Keesman, K., Burgess, P., Graves, A., Pilbeam, D., Incoll, L. D., Metselaar, K., Mayus, M., Stappers, R., van Keulen, H., et al. 2007. Yield-SAFE: a parameter-sparse, process-based dynamic model for predicting resource capture, growth, and production in agroforestry systems. *Ecol. Eng.* 29(4), 419–33. doi:10.1016/j.ecoleng.2006.09.017.

van Maanen, A. and Xu, X.-M. 2003. Modelling plant disease epidemics. *Eur. J. Plant Pathol.* 109(7), 669–82. doi:10.1023/A:1026018005613.

van Noordwijk, M. and Lusiana, B. 1999. WaNuLCAS, a model of water, nutrient and light capture in agroforestry systems. In: Auclair, D. and Dupraz, C. (Eds), *Agroforestry for Sustainable Land-Use Fundamental Research and Modelling with Emphasis on Temperate and Mediterranean Applications: Selected Papers from a Workshop Held in Montpellier, France, 23–29 June 1997*. Springer, Netherlands, Dordrecht, pp. 217–42.

van Noordwijk, M., Lusiana, B., Khasanah, N. and Mulia, R. 2011. *WaNuLCAS Version 4.0, Background on a Model of Water, Nutrient and Light Capture in Agroforestry Systems*. World Agroforestry Center–ICRAF, SEA Regional Office, Bogor, Indonesia, p. 224. ISBN: 978-979-3198-59-0.

Van Oost, K., Quine, T. A., Govers, G., De Gryze, S., Six, J., Harden, J. W., Ritchie, J. C., McCarty, G. W., Heckrath, G., Kosmas, C., et al. 2007. The impact of agricultural soil erosion on the global carbon cycle. *Science* 318(5850), 626–9. doi:10.1126/science.1145724.

Wacha, K., Papanicolaou, A., Giannopoulos, C., Abban, B., Wilson, C., Zhou, S., Hatfield, J., Filley, T. and Hou, T. 2018. The role of hydraulic connectivity and management on soil aggregate size and stability in the Clear Creek Watershed, Iowa. *Geosciences* 8(12), 470. doi:10.3390/geosciences8120470.

Wallor, E., Kersebaum, K.-C., Ventrella, D., Bindi, M., Cammarano, D., Coucheney, E., Gaiser, T., Garofalo, P., Giglio, L., Giola, P., et al. 2018. The response of process-based agro-ecosystem models to within-field variability in site conditions. *Field Crops Res.* 228, 1–19. doi:10.1016/j.fcr.2018.08.021.

Wang, Y. P. and Jarvis, P. G. 1990. Description and validation of an array model – MAESTRO. *Agric. For. Meteorol.* 51(3–4), 257–80. doi:10.1016/0168-1923(90)90112-J.

Wang, X., Williams, J. R., Gassman, P. W., Baffaut, C., Izaurralde, R. C., Jeong, J. and Kiniry, J. R. 2012. EPIC and APEX: model use, calibration, and validation. *Transactions of the ASABE* 55(4), 1447–62.

Welch, S. M., Croft, B. A., Brunner, J. F. and Michels, M. F. 1978. Pete: an extension phenology modeling system for management of multi-species pest complex. *Environ. Entomol.* 7(4), 487–94. doi:10.1093/ee/7.4.487.

Whish, J. P. M., Herrmann, N. I., White, N. A., Moore, A. D. and Kriticos, D. J. 2015. Integrating pest population models with biophysical cropmodels to better represent the farming system. *Environ. Model. Softw.* 72, 418–25. doi:10.1016/j.envsoft.2014.10.010.

Wilson, C. G., Wacha, K. M., Papanicolaou, A. N. T., Sander, H. A., Freudenberg, V. B., Abban, B. K. B. and Zhao, C. 2016. Dynamic assessment of current management in an intensively managed agroecosystem. *J. Contemp. Water Res. Educ.* 158(1), 148–71. doi:10.1111/j.1936-704X.2016.03225.x.

Wilson, C. G., Abban, B., Keefer, L. L., Wacha, K., Dermisis, D., Giannopoulos, C., Zhou, S., Goodwell, A. E., Woo, D. K., Yan, Q., et al. 2018. The intensively managed landscape critical zone observatory: a scientific testbed for understanding critical zone processes in agroecosystems. *Vadose Zone J.* 17(1). doi:10.2136/vzj2018.04.0088.

Yadav, V. and Malanson, G. 2008. Spatially explicit historical land use land cover and soil organic carbon transformations in Southern Illinois. *Agric. Ecosyst. Environ.* 123(4), 280–92. doi:10.1016/j.agee.2007.07.010.

Yonow, T., Zalucki, M. P., Sutherst, R. W., Dominiak, B. C., Maywald, G. F., Maelzer, D. A. and Kriticos, D. J. 2004. Modelling the population dynamics of the Queensland fruit fly, *Bactrocera* (Dacus) *tryoni*: a cohort-based approach incorporating the effects of weather. *Ecol. Modell.* 173(1), 9-30. doi:10.1016/S0304-3800(03)00306-5.

Young, R. A., Onstad, C. A., Bosch, D. D. and Anderson, W. P. 1989. AGNPS: a nonpoint-source pollution model for evaluating agricultural watersheds. *J. Soil Water Conserv.* 44(2), 168–73.

Zadoks, J. C. 1971. Systems analysis and the dynamics of epidemics. *Phytopathology* 61, 600–10.

Zadoks, J. C. 1987. The concept of thresholds: warning, action, and damage thresholds. In: Teng, P. S. (Ed.), *Crop Loss Assessment and Pest Management*. APS Press, St. Paul, MN, pp. 168–75.

Zulian, G., Maes, J. and Paracchini, M. L. 2013. Linking land cover data and crop yields for mapping and assessment of pollination services in Europe. *Land* 2(3), 472-92. doi:10.3390/land2030472.

Part 3

Modelling issues

Chapter 14

Data for developing, testing, and applying crop and farm models

Frits K. van Evert, Wageningen University & Research, Agrosystems Research, The Netherlands

1 Introduction

In this chapter we consider the data needed for developing, testing, and applying crop and farm models. Crop models here include the large number of discrete-event models with a time step of one day that have been developed in the past 50 years. For example, 27 wheat models were listed (Asseng et al., 2013). For potato, a similar number of models can be counted (Fleisher et al., 2017; Raymundo et al., 2014).

The data used to develop and test crop models include observations on the crop (e.g. biomass and nutrient content of various plant parts, leaf area index, light interception by the canopy, height, root depth), soil (physical, chemical, and biological characteristics), and weather.

Farm models deal with a larger set of processes, such as livestock, weeds, crop rotations, multi-year dynamics of soil organic matter, economics, labor availability, and farm machinery availability (e.g. Rotz et al., 1999). While crop models are almost always formulated as daily-time step discrete event models, for farm models often linear programming is used.

To develop and test farm models the same data are needed as for crop modeling and in addition, data may be needed on productivity of animals (e.g. milk, meat, wool), presence and abundance of weeds, prices of inputs (labor, fertilizers, pesticides, boxes, and bags) and outputs, and the cost and capacity of farming equipment.

http://dx.doi.org/10.19103/AS.2019.0061.19

For scientists, the interest in working with crop and farm models derives first and foremost from the ability of using a model to concisely and quantitatively summarize knowledge about the system under study. Seen in this light, a model is the ideal hypothesis: it is precise, quantitative, can make predictions, and can be falsified. When a model is used in this way, the data needed for developing and testing must be accurate and precise in order to be able to distinguish between the effects of implementing alternative hypotheses. Data collected in small experiments where detailed measurements can be made are most useful.

Models can be used to support farm management decisions. Early examples include Epipre (Zadoks, 1981) and the Gossym/Comax system (McKinion et al., 1989). Today the term precision agriculture is often used for any farming that uses technology to apply the right amount of the right input at the right time and at the right place. Often some form of modeling is used in precision agriculture techniques (Kempenaar et al., 2017). Precision agriculture techniques can increase the profitability and sustainability of potato farming in The Netherlands by 20% (Van Evert et al., 2017b). Currently, simple models are used but work is ongoing to use detailed crop models.

Models can be used to support strategic decisions, for example, to evaluate scenarios with respect to fertilizer management. Models can also be used to support tactical and operational decisions, for example whether to apply water, fungicide, nitrogen, and how much. The data requirements to operate models for farm management decisions are similar to the data requirements for research purposes. When a model is used on-farm it has probably already been tested and calibrated, so there is no need for high-frequency, detailed measurements. However, some of the input data will need to be provided by the farmer: data about the soil, crop management data such as planting date and rate, and application of fertilizers, irrigation water and crop protection, and local weather data (especially rainfall gauge data). For tactical and operational uses, moreover, the data need to be available in near real-time.

The European MARS crop yield forecasting system (Boogaard et al., 2002; Genovese, 1998; Van der Velde and Nisini, 2019; Van Diepen et al., 1998) is an example of using crop growth models to generate regional forecasts to support policy. For this kind of modeling, regional soils and weather data are needed. Also regional yields are needed in order to calibrate the model.

Finally, much effort is under way to predict the effects of a changed climate on crop yields using crop models (e.g. Adams et al., 1990; Asseng et al., 2019; Raymundo et al., 2018). Global soils and weather data are needed for this kind of modeling. In addition, future climate change scenarios as predicted by the various Global Circulation Models are needed.

The remainder of this chapter is organized as follows. In Section 2 attention is given to measurements, the source of data. In Section 3 we describe technologies used for storing and accessing data. Section 4 is dedicated to combining data and models to perform precision agriculture. Conclusions are drawn and recommendations are given in Sections 5 and 6.

2 Measurements

The earliest crop models were developed using basic agronomic experiments. Typically, a crop was grown on one site, perhaps for two or more years, and some management factors were varied such as amount of fertilizer (usually N), irrigation, or planting date. Measurements typically included LAI, weight of various plant parts, final yield, soil physical and chemical characteristics, and soil water content.

Today, crop and farm modelers have access to much richer datasets. In experiments, new sensors and especially automated sensors are used to collect more detailed information and with greater temporal resolution. Also the advent of new sensors has made it possible to collect data on commercial farms.

Some of the new measurements, such as yield monitor data from harvesters, are directly usable. Measurements such as soil electrical conductivity give great insight into temporal and spatial variation but can only be used indirectly in crop modeling.

2.1 Crop

2.1.1 Canopy reflectance

The crop canopy is important in modeling because it is the driver for growth. Measurement via growth sampling is labor intensive. The crop canopy can be assessed non-destructively. The reflectance of incident light by a crop canopy is influenced by the amount of biomass, by the N content (kg m^{-2} ground area), chlorophyll concentration (μg cm^{-2} leaf area), and to a lesser extent by canopy structure. A reflectance measurement can thus be used to estimate one or more of the above traits. In potato, LAI and biomass are directly linked, thus canopy reflectance allows the measurement of LAI (Bouman et al., 1992). Compared to destructive sampling, measuring crop reflectance is quick, cheap, and can be used to measure a large area at high resolution.

Reflectance is commonly measured with cameras on board of satellites, airplanes, and drones. It is also measured with non-imaging devices that are handheld or mounted on tractors. Remote and proximal sensing give the same results.

There are two major ways of interpreting reflectance measurements, namely (1) using vegetation indices and (2) using inverse modeling:

1 A vegetation index (VI) is calculated with a formula which combines reflectances measured in two or more narrow spectral bands (Hatfield

et al., 2008). Commonly used VIs include normalized difference vegetation index (NDVI) (Deering, 1978), weighted difference vegetation index (WDVI) (Clevers, 1989), normalized difference red edge (NDRE) (Barnes et al., 2000), and chlorophyll index (CI) (Gitelson et al., 2003). VIs can be used with the relatively cheap (multi-spectral) instruments that measure reflectance in only a few wavebands.

2 Inverse modeling is a method in which a model of canopy reflectance is used to simulate reflectance. The parameters describing the canopy (biomass, LAI, chlorophyll content, and so on) are then systematically varied until the simulated reflectance spectrum matches the measured reflectance spectrum.

The accuracy and precision of determination of biomass or N uptake using a VI are often not as high as one might desire. For potato, N uptake using WDVI was measured to be +/−30 kg ha^{-1} (Van Evert et al., 2012). Also for potato, but using the CI and the Meris terrestrial chlorophyll index (MTCI), canopy nitrogen content was estimated with an RMSE of 16 kg ha^{-1} (Clevers and Gitelson, 2013). The same authors found similar accuracies for maize, soybean, and grassland.

LAI and ground cover of potato were estimated using WDVI with an absolute error of less than 6 percentage points (Bouman et al., 1992).

Reflectance sensors need to be properly calibrated in order to give good measurements. The calibration of satellite and airborne sensors is obviously beyond the reach of the farmer. For the calibration of drone-mounted sensors detailed procedures are given. For tractor-mounted sensors the issue is less clear. For the Greenseeker, no calibration procedure is available, and no method to check the calibration is given by the manufacturer, other than returning the sensor. When six Greenseekers were tested side-by-side, they showed great variability (Van der Schans et al., 2012). For the N-Sensor a calibration procedure is available that can be done on-farm. The manufacturer of the CropCircle recommends returning the instrument to the factory for recalibration.

The differences between the commercially available proximal sensors mean that some sensors are more suitable for a given task than others. For example, the Greenseeker measures NDVI. In potatoes, this VI is capable of detecting the differences in crop senescence that are important for VRA haulm killing. In the same crop, however, NDVI is not sensitive to the differences in N uptake that exist in June/July when N sidedress is applied.

Remote sensing of canopy reflectance, using air- or space-borne instruments, is available. For use in agriculture, the Sentinel-II satellites recently launched by the European Space Agency are particularly useful because of the inclusion of a red-edge band which is sensitive to crop nitrogen. For The Netherlands, Sentinel-II imagery is processed operationally by Wageningen University & Research under contract to the Ministry of Economic Affairs.

Processed images have been corrected for atmospheric conditions (including cloud cover). They are made available via at www.groenmonitor.nl as well as via a web service for programmatic access.

Several manufacturers offer lightweight multi-spectral cameras suitable for mounting on a small unmanned aerial vehicle (UAV, or 'drone'). These cameras allow common VIs such as NDVI, WDVI, NDRE, and CI to be calculated. Drone images have a higher resolution (0.05–0.5 m) than satellite imagery (10–100 m). Drone images are more expensive (per ha) than satellite images but have the advantage that they can be obtained when clouds prevent the acquisition of satellite images. In The Netherlands where the sky is overcast 50% of the time, the Bioscope initiative (http://www.projectbioscope.eu/) aims to provide attractively priced weekly images of crop growth by using Sentinel-II images when available and supplementing them with drone images when necessary.

2.1.2 Yield monitor

A grain yield monitor measures grain mass flow as the grain is being harvested. Often other characteristics such as moisture or protein content are also measured. Coupled with a global navigation satellite system (GNSS) such as the global positioning system (GPS) or Galileo, this allows for the creation of yield maps. The number of unreliable data points in yield monitor data can be large. The method of data cleaning may influence the final yield map (Kharel et al., 2019).

For wheat and maize, yield monitors work very well. For root and tuber crops such as potato and sugar beet this is much more difficult because the harvestable parts of the crop need to be lifted from the soil. The operator of a potato harvester will try to optimize the speed of harvesting and the result of sieving by keeping the load on the sieving belt as constant as possible. This means that the driving speed must constantly be varied which makes it difficult to match load cell readings with the position in the field from which the potatoes were removed. Beets or tubers need to be separated from clods and stones. On heavy soil, the harvested product may include a significant amount of tare. In poor harvesting conditions, not all potatoes are always removed from the soil.

2.1.3 Radar estimates of biomass

Satellite-based measurement of canopy reflectance in the optical part of the spectrum is easily disturbed by cloud cover. Low-frequency microwaves (1–10 GHz) penetrate cloud cover. In addition, they allow nighttime measurements. The possibility of measuring crop biomass and/or LAI using radar was mentioned decades ago (e.g. Luciani et al., 1994; Ulaby et al., 1984).

Backscatter of microwaves is influenced by characteristics of the canopy and of the soil. Canopy characteristics include amount of biomass, geometry of the canopy, LAI, and water content of the canopy (Steele-Dunne et al., 2017). Soil characteristics include soil surface roughness and soil water content (Steele-Dunne et al., 2017). This means that in order to interpret the radar data, it is necessary to have a model of the backscattering processes. Information about crop and soil is then obtained from the measurement by inverting the model. Uncertainties in the model and noise in the data lead to the result that some information about the crop can be extracted, but at present this information is not accurate enough to be of interest to farmers.

This situation was improved with the launch by the European Space Agency of Sentinel-1A in 2014 and Sentinel-1B in 2016. These satellites carry a radar instrument with high resolution. Using Sentinel-1 data and random forest regression, it was possible to estimate the LAI of wheat in India with an RMSE of 0.3 and dry biomass with an RMSE of approximately 1 t ha^{-1} (Kumar et al., 2018). Also for wheat in Denmark good results are reported but quantitative results for this work-in-progress are not yet available (Christiansen et al., 2018).

2.1.4 Canopy temperature via thermal images

The development of lightweight thermal cameras allows the use of drones for acquisition of thermal imagery (e.g. Zarco-Tejada et al., 2012). Several cameras are now available commercially. Spatial and temporal variation in canopy temperature is a useful indicator of suboptimal growth conditions but cannot be linked directly to crop model output. However, it can be used for inverse modeling (Böttcher and Kage, 2005, 2007).

2.1.5 Canopy wilting

Sugar beet exhibits diurnal wilting even under moderate water stress. It was investigated whether sugar beet can be used as an intermediate sensor for quantifying potential soil water availability within the root zone (Zhang et al., 2011). The authors obtained good results using satellite images and a classification technique that relies on differences in greenness and brightness to detect wilting in sugar beet in England.

2.1.6 Grassland

An overview of remote sensing of grassland is given by Wachendorf et al. (2018). In general, canopy reflectance is not as well correlated with harvestable N or biomass as in arable crops. Possibly this is because when a pasture is grazed or harvested, a significant amount of biomass remains on the field as stubble; moreover, the amount of biomass in the stubble increases during the growing season. Dairy farmers have long relied on grass height measurements using a rising plate as an indicator of standing biomass. This measurement has been automated (McSweeney et al., 2019). Another solution relies on ultrasonic distance measurement (www.pasturereader.com.au).

2.1.7 Field phenotyping

Plant phenotyping is a general term meant to catch a variety of modern technologies to measure plant characteristics, typically on individual plants in controlled environments. Field phenotyping is a recent development where phenotyping is performed in the field. The variety of the types of measurements, coupled with high spatial and temporal frequency, is a very promising development for crop modeling. A recent review is available (Roitsch et al., 2019).

Researchers have mounted sensors on a variety of vehicles, ranging from bicycles to small tractors and high-clearance tractors. Some of these are capable of autonomous movement (Bao et al., 2019; Madec et al., 2017).

A number of field phenotyping systems use a fixed installation that typically covers 1000–10 000 m^2. A gantry system is available that carries multiple sensors over a field (https://www.lemnatec.com/products/high-throughput-phenotyping-solutions/field-scanalyzer/). At ETH Zürich, a rope suspended carrier system was built which holds multiple sensors that can be positioned anywhere over a field of approximately one hectare. The system includes visible, near-infrared, and thermal cameras, multi-spectral point sensors and a laser scanner (http://www.kp.ethz.ch/infrastructure/FIP.html).

2.2 National, regional, and global crop yield statistics

Statistics on crop yield and crop acreages can be downloaded from the websites of various authorities such as the European Union, FAO, and national governments. Data about agriculture in the EU are available from Eurostat (https://ec.europa.eu/eurostat/web/agriculture/overview). Global data are available from FAOSTAT (https://ec.europa.eu/eurostat/web/agriculture/overview).

2.3 Soil

The soil is an important determinant of crop growth. Soils are complex and can be described using many properties. In many cases the relationship between soil property and crop performance (e.g. yield) cannot be simply quantified. Even so, maps that quantify spatial variation in soil properties may be helpful to crop modelers when they provide a clue to the processes that result in spatial variation in crop performance. Below we list some soil properties and measurement methods that are of interest to crop modelers.

2.3.1 Electrical conductivity

Some of the soil properties that are important in crop modeling influence the apparent electrical conductivity (EC_a) of soil. These properties include the amount of water in the soil, the chargeable soil particles such as clay and organic matter, the presence of salts or salinity, and temperature. Bulk density and pore size distribution influence soil water content.

Soil properties that can be measured (indirectly) via EC_a therefore include clay content, organic matter content, bulk density, and pore size distribution (Adamchuk et al., 2017; Knotters et al., 2017; Triantafilis and Lesch, 2005). Related soil properties can sometimes be approximated locally using that relationship (Heil and Schmidhalter, 2017). Since soil moisture, clay content, and organic matter content all cause a higher EC one can argue that EC_a itself is an indicative measure of soil water availability.

2.3.2 VIS/NIR spectroscopy

The reflectance of light incident on a soil is influenced by the constituents of the soil. Different chemical bonds or mineralogical properties absorb radiation at specific wavelengths (e.g. chlorophyll, cellulose, lignin, kaolinite, carbonates). If the spectrum is measured at those wavelengths, the amount present can be quantified after calibration with samples. Relevant wavelengths for soil are located in the near (NIR) and the mid-infrared (MIR). For field measurements at present only NIR tools are commercially available. However, the influence of varying soil moisture and granularity of the samples limits the accuracy of measurements in the field. Therefore the calibration of spectra with soil properties (joined in spectral libraries) is performed on relevant lab-measured samples and then applied to field measurements. Precision and uncertainty of predictions depend on the instrument, measurement conditions and protocol, quality and relevance of the spectral library used, and modeling algorithms.

Soil properties that can usually be measured reasonably well (R^2 of 0.7 or 0.9 if a suitable and good quality library is available) are soil organic carbon (SOC), organic matter, total carbon, clay content, carbonate content, pH, and CEC. Other properties such as clay type and nutrients can be predicted too, but tend to have a higher uncertainty (R^2 of 0.6 or lower in lab conditions).

The applications at present mainly focus on soil organic carbon monitoring for climate, updating soil maps, in field measurement of SOC for precision agriculture, and more extensive soil property measurement using NIR and MIR in the lab. Other properties are still often mapped by other instruments or lab analysis. NIR or MIR systems are typically part of a multi-sensor setup for e.g. fertilizer advice (Lab-in-a-box - SoilCares) (Ackerson et al., 2017; Lobsey and Viscarra Rossel, 2016; Roudier et al., 2015).

2.3.3 Gamma rays

Gamma-ray spectrometry or radiometry is based on the passive measurement of naturally occurring radioactivity in the Earth's surface with a scintillation crystal. Radiation is emitted by nuclides (^{40}K, ^{238}U, ^{232}Th) with long half-life times that are present in minerals in rocks and soils or emitted by nuclear testing or accidents such as Chernobyl (^{137}Cs).

Advantages of this technique are the quantitative, robust measurement of texture (especially clay and loam content), a depth range that equals the tillage layer, and a relative insensitivity to moisture.

2.3.4 Direct measurement of soil properties using ion-selective electrodes

The possibility of ion-selective electrodes for on-the-go measurement of soil properties has been described by many authors (e.g. Adamchuk et al., 2005). To the best of our knowledge, currently only for pH such a sensor is available.[1] Tests in the Netherlands have shown that accuracy is 0.1–0.2 pH units and that the sensor gives systematically (slightly) higher values than measured in the lab with pH-KCl (Schans and Berg, 2013). Direct measurement of soil nitrate concentration would be of much greater use to modelers but such a sensor is not available.

2.3.5 Landscape position

Landscape position is an important determinant of crop performance (Maestrini and Basso, 2018). A digital elevation model (DEM) from which landscape position for any point can be inferred can be constructed in a number of different ways, including by relying on the elevation data logged by the GNSS mounted on farm machinery (Tagarakis et al., 2019).

2.3.6 Interpretation of soil scans using inverse modeling

Crop model inversion has been proposed to determine soil hydrological properties (Guérif et al., 2006; Sreelash et al., 2017). Often satellite imagery is used, but soil scans can also be used (Van Evert et al., 2018a).

1 https://veristech.com/the-sensors/msp3

2.3.7 Automated soil moisture measurements

The automated measurement of soil water content is now a routine with sensors being offered by a number of companies. An experiment field can probably be equipped with a sufficient number of soil moisture sensors to capture the spatial variability in soil and rainfall pattern. The spatial variation in precipitation is large and it will not be possible to install a sufficient number of precipitation gauges to capture this variation on all but the smallest commercial farms. In The Netherlands, rainfall radar is available at a resolution of 500 m and this has been used to overcome the limited number of precipitation gauges (Van Evert et al., 2018a).

2.3.8 Soil maps

When detailed soil measurements are not available, the modeler has no other option than to use general soil maps. SoilGrids[2] provides a global soil map (250 m resolution). The soil survey geographic database SSURGO[3] has information on soil properties for the US. BOFEK[4] is a 1:50000 map for The Netherlands, which has soil properties specifically aimed at hydrologic modeling (Wösten, 2012).

2.4 Weather

Automated weather stations are affordable and ubiquitous. Consequently, for most field experiments, reliable weather data are recorded. For regional or global simulations, gridded weather data are offered by a variety of companies including MeteoGroup[5] and The Weather Company.[6]

2.5 Commercial farms as a source of data

The opportunity to collect data on commercial farms has increased in recent years. As a consequence, commercial farms are becoming an important source of data for modelers. An example is the Van den Borne potato farm in The Netherlands[7] whose owners started experimenting with precision agriculture in 2002. The detailed records have attracted the attention of researchers and have led to a large number of scientific papers. Also in the US, data from commercial growers have been used in research (Maestrini and Basso, 2018). Some issues have been reviewed (van Evert et al., 2017a).

2 https://soilgrids.org
3 https://www.nrcs.usda.gov/wps/portal/nrcs/detail/soils/survey/geo/?cid=nrcs142p2_053627
4 https://www.wur.nl/nl/show/Bodemfysische-Eenhedenkaart-BOFEK2012.htm
5 https://www.meteogroup.com
6 https://www.ibm.com/weather
7 https://www.vandenborneaardappelen.com/

In the European Union, farmers are required to record which crop is grown on each field, in order to allow the administration of various subsidies. This information is collected at the national level and made publicly available. This information can of course be used by modelers.

3 Principles and practices for storing, accessing, and exchanging data

3.1 Data storage

Crop modelers recognized early the importance of careful organization of experiment data (Wallach, 1984). In due time a template was developed to record the data collected in the kind of experiment most often done to support development and testing of models (Hunt and Boote, 1998; Hunt et al., 1994). This developed into the ICASA data standards (White et al., 2013).

Various initiatives have developed systems for reporting and storing data from field research (Bostick et al., 2004; Van Evert et al., 1999a,b), but no single system is widely used (White and Van Evert, 2008). Currently, agronomic data are maintained and distributed in a variety of formats, including text files, spreadsheets, and relational databases.

Some systems use a relational database (RDBMS), which is a versatile tool to store and manipulate data. The technology of relational databases has been well developed during the last 50 years. RDBMS is now fast, cheap, reliable, and scalable. Structured query language (SQL) is a powerful tool for the manipulation of data in RDBMS. SQL is far from perfect, and it suffers from the fact that every vendor offers a slightly different version, but it is a dependable and capable workhorse. Postgresql[8] is a full-featured open-source RDBMS.

Despite their versatility, some kinds of data are best not stored in a RDBMS. Large binary objects such as drone images (up to several GB) are best handled by storing the image as file and storing the relevant metadata (e.g. acquisition data, geographic extent, wavebands, and vegetation indices) in a database for convenient querying. Data can then be extracted from images as needed.

3.2 Data exchange

3.2.1 Files and file formats

The most basic method of exchanging data is in the form of files. Files can be transferred in many ways, including by putting them on a physically transportable medium such as an external hard drive or a USB stick (thumb

8 www.postgresql.org

drive), via email, by copying them onto a network drive or a cloud service, or via a cloud-based file exchange service.

Several file formats are used to exchange agricultural data. The most commonly used formats are discussed here. Plain text files are intended in principle to be read and understood by humans. Tabular data are often given in comma-separated (CSV) files. In such a file each line of text corresponds to a table row, and a special character is used to separate the column values in a line. The format is based on the RFC4180[9] open standard. A problem with comma-separated files is that several conventions exist for the character used to separate the whole and fractional parts of a number (a period in the US versus a comma in Europe), as well as to separate columns (a comma in the US versus a semicolon in Europe). Some software is able to guess which format is being used but clearly there is ample potential for misinterpretation.

JavaScript Object Notation (JSON)[10] is a human-readable, open-standard file format that does not suffer from the above-mentioned limitations. In addition, JSON supports the use of custom schemas to declare the structure of the data. Schemas can be used to perform certain kinds of validations programmatically and help minimize errors. JSON is widely supported and libraries for processing it are available for many computer languages.

In agriculture much data is geo-referenced. Examples of geo-referenced data are: which crop is grown on a particular field, crop yield as measured with a yield monitor on a combine harvester, or crop reflectance as measured with a satellite. With geo-referenced data it is important to be explicit about the coordinate reference system (CRS) that is used. In many cases the CRS used will be 1984 revision of the World Geodetic System (WGS84) but many other CRSs are in use. If coordinates measured in the WGS84 are accidentally interpreted as, for example, coordinates in the European Terrestrial Reference System 1989 (ETRS89) or in the North American Datum of 1983 (NAD83), an error will be made. Moreover, this error will depend on time because WGS 84 is fixed relative to the Earth as a whole, whereas ETRS89 is fixed relative to the Eurasian Plate and NAD83 is fixed relative to the North American Plate.

For information that is related to a field, we need to record the boundaries of the field, in other words, we need to store a polygon. The de facto standard for this kind of geo-referenced data is the ESRI Shapefile.[11]

A satellite image records information in a rectangular grid. The Shapefile format is capable of storing gridded information but in practice the GeoTIFF[12] format is often used for rasterized information. GeoTIFF is an extension of the

9 https://tools.ietf.org/html/rfc4180
10 https://en.wikipedia.org/wiki/JSON
11 https://en.wikipedia.org/wiki/Shapefile
12 https://en.wikipedia.org/wiki/GeoTIFF

general image format TIFF and it allows geo-referencing information to be embedded in a TIFF file.

GeoJSON[13] is an extension of the JSON format meant to accommodate geographic information.

Relational database management systems (RDBMS) typically use a proprietary format which is also dependent on the specific version of the software. Relational data often are heavily indexed. Indexing enables fast querying but increases the size of the database file(s) - it is not uncommon that the indexed database is several times larger than the data itself. Files from an RDMBS are therefore unsuitable for exchanging. If data held in an RDBMS are to be exchanged, the data can be exported into another format (e.g. CSV, JSON) but in this way the relations between tables in the RDBMS will be lost. Another option is to use an SQL dump of the database. An SQL dump contains all the statements that are needed to re-create the database and populate it with the data. SQL is a standard language but most RDBMSs implement only a subset of the full standard and at the same time add some specific features. It is thus unlikely that an SQL dump from the RDBMS of one vendor can be read by the RDBMS of another vendor without at least some modifications. It is not even certain that a dump can be read by a later version of the same RDBMS. However, in the case of an open-source RDMBS such as Postgresql[14] it is highly likely that the code to read old versions will remain available.

Hierarchical Data Format[15] (version 5, not compatible with earlier versions) is a common dataset format with the ability to store data in multidimensional arrays, grouped into collections and/or hierarchies. Relationships between data in the arrays can be saved, but the format does not allow for storing structured metadata. The format is open and can be read in a variety of applications, but it is very difficult to process without the use of HDF5 software. NetCDF[16] is an extension of HDF5 that adds the ability to add weather-related metadata. It can be processed using HDF5 tools. NetCDF is used widely by climate scientists.

3.2.2 Web services

A web service is a generic mechanism for machine-to-machine communication using http. In practice, a web service commonly provides a web-based interface to a database server. The web service can be reached via its http address, e.g. https://rest.soilgrids.org/query/describe. The response can be in any format but many web services, including the one in the previous sentence, return responses that are formatted as JSON. Typically, some information must be

13 https://en.wikipedia.org/wiki/GeoJSON
14 https://www.postgresql.org/
15 https://en.wikipedia.org/wiki/Hierarchical_Data_Format
16 https://en.wikipedia.org/wiki/NetCDF

passed to the web service in order to specify what information is requested. A web service may expect requests to be specified via query parameters. For example, URL https://rest.soilgrids.org/query?lon=5.39&lat=51.57 can be used to request soil information for a location in The Netherlands. A web service may also expect requests to be specified via the POST protocol, in which case the request info is written in a file which is then sent to the request.

Web services enable fast, continuous, automated flow of data. Of course web services can be stacked on top of each other. In this way processing pipelines can be built that consists of many separate pieces of software, which may run on Internet-connected computers all over the world.

Web services may require authentication before they return information, either in order to be able to charge for their service or to protect sensitive information. A discussion of authentication methods is considered outside the scope of this chapter.

3.2.3 Automatic upload of data

There are many automated measuring instruments. Examples are automated weather stations, soil moisture and temperature sensors, and groundwater level sensors. If a field-deployed sensor can be connected to the Internet, whether via wired LAN, WiFi, 3/4/5G, or LoRa, then it becomes possible to upload the data automatically. A web service may be programmed to accept data from a specific type of sensor or from many different types of sensors. Akkerweb for example is able to accept upload from a number of sensor types, including Yara N-Sensor and the Multispec-4C multi-spectral camera that is used on drones.

3.2.4 Exchange algorithms instead of data

We can take it as a given that for many calculations we need to combine data that are located on more than one computer. A need to exchange data arises if we want to perform all computation on one computer. There may exist several reasons not to want to move data, for example because there is concern whether the privacy and security of the data can be maintained during transport or when it is stored on the other computer. Or because the data to be transferred are very large. In either case, it has been suggested that these concerns can be alleviated by exchanging the algorithm rather than data.

The crop modeling community is familiar with a form of this approach. In The agricultural model intercomparison and improvement project (AgMIP) (Rosenzweig et al., 2013), model runs are performed by the owners of models and results are collected in a central point. This method could be extended to calibration of models. If a large amount of observational data were stored in several repositories, it would be possible to set up a web service that runs a model that is sent to it on the data that are locally available and reports the goodness-of-fit. A central location could collect the results from several locations and decide on the parameters to be used for the next iteration step.

3.2.5 Data exchange in the commercial domain

The farm management information system (FMIS) is a class of software that allows farmers to collect and store data about fields, crops, and farm management. Unfortunately there is usually no easy way to transfer data from one FMIS to another, or combine data from two or more FMIS, which can result in lock-in effects. Although some of the major manufacturers offer application programmer's interfaces (API) built around the data model of their system, this still leaves the difficult task of translating from the source data model to the target data model to the user.

A generic approach is taken by AgGateway, a non-profit consortium with more than 100 members, whose recent agricultural data application programming toolkit (ADAPT) proposes a unified data model which can serve to exchange data between any two FMIS. The Open Ag Data Alliance and DKE Agrirouter are similar efforts.

It is important to define what the meaning is of a data item that is exchanged. Several vocabularies and data standards have been available for many years. Each is supported by its own community, but none of them seems to have gained widespread attention. This includes Agro-XML, EDI-Teelt++, and the EPPO Global Database.[17]

3.3 Linked data

Whether data are stored in text files, in an RDMBS, or accessed via a web service, almost always implicit knowledge is needed to interpret the data. This makes it all but impossible to fully automate processing. Manual intervention will be needed, for example, to use such a data source as input to a model, or to compare model output with it.

Consider a typical table in an RDBMS (Fig. 1) which has columns with names that seem to refer to well-understood concepts, such as 'experiment,' 'plot,' 'date,' and 'lai.' Persons with a background in agronomy will probably assume that the column labelled 'lai' contains measurements of Leaf Area Index, but we cannot be sure. The column 'tubers_fresh_g_per_m2' seems to indicate that it contains measurement of the fresh weight of tubers, expressed in g m^{-2}. However, this is an assumption that depends on the knowledge of English. Finally, suppose we wanted to combine data from several databases,

17 https://gd.eppo.int

Table 'crop_droevendaal'				
experiment	plot	date	lai	tubers_fresh_g_per_m2
15	14	26 June 1998	2.87	1412
15	38	26 June 1998	3.13	1481
15	55	26 June 1998	3.33	1622

Figure 1 Fragment of one table from a published dataset on a series of experiments on nitrogen fertilization in potato. Source: adapted from Van Evert et al. (2011).

each of which contained data on the fresh weight of potato tubers, but with each database using a different column name (in addition to 'tubers_fresh_g_per_m2,' there might be 'tubers_gram_m2,' 'potato_g_per_m2,' and 'tubers_fresh_dt_per_ha'). Even if we could correctly interpret these column names, it would require a lot of work to combine the data.

This is a generic problem for which linked data have been proposed as a generic solution (Berners-Lee et al., 2001). In order to use linked data, first every concept is given a name. Such a name starts with 'http,' e.g. http://www.wurvoc.org/resource/om-2/kilogram, and when that name is looked up, you get information about the concept in a standard format.

Second, all information is expressed using statements in the form 'subject-predicate-object,' where the subject, predicate, and object have also been named using the scheme mentioned in the previous sentence. For example, based on Fig. 1, we could define http://example.org/my_data_point1, http://example.org/lai, and http://example.org/has_quantity. These definitions could then be used to make the statement 'my_data_point1'-'has quantity'-'lai.' With a further definition, we could also make the statement 'my_data_point1'-'has value'-'2.87.' Statements such as these are called 'triples.' The scheme itself is called resource description framework (RDF)[18]. Figure 2 gives an impression of this process. An engaging text book is available (Allemang and Hendler, 2011).

The scheme to use a fixed list of definitions and triples to describe data is only really useful if generic lists of definitions can be used and re-used by many researchers. Such a list of definitions is called a vocabulary. If the vocabulary additionally contains relations between definitions, it is referred to as an ontology (Gruber, 1993).

Ontologies are an actively developing research field. It is important to note that an ontology is in fact a model of a domain of interest. Any model is an abstraction (or 'simplification') of the real thing. There is no single abstraction that is the right one-but in a community of users a consensus could arise that one ontology is more useful than all others. However, at this point in time, even

18 https://en.wikipedia.org/wiki/Resource_Description_Framework

```
@prefix ex: <http://example.org/ >.

ex:my_data_point1    ex:has_quantity        ex:lai
ex:my_data_point1    ex:has_value           2.87
ex:my_data_point1    ex:measurement_date    26 June 1998
ex:my_data_point1    ex:measured_on         ex:plot14
ex:experiment        ex:has_plot            ex:plot14
```

Figure 2 Toy example of data published as RDF triples. Please note the use of @prefix to limit the verbosity of the text.

for something as basic as units and measures, competing ontologies are being used: QUDT[19] and OM[20] (Rijgersberg et al., 2013).

For arable agriculture, there are currently no accepted ontologies despite initiatives such as PlantOntology (Plant Ontology™ Consortium, 2002; Shrestha et al., 2010). But this does not need to stop anybody from publishing data using an ad-hoc vocabulary or ontology. Several existing vocabularies may be helpful. The ICASA Master Variable List[21] contains many useful concepts but it has the weakness that its definitions typically combine quantity and unit in one concept. For example, the list specifies that 'grain_dry_weight' is expressed in units kg ha^{-1}. Another useful vocabulary is the EPPO Global Database.[22]

Triples are often stored in specialized software called triplestores or RDF stores. Currently, notable offerings include rd4j[23] and Virtuoso[24] (but keep in mind that this is a rapidly evolving landscape). In all but the most trivial of cases, the performance of triplestores is orders of magnitude worse than RDBMSs.

The process of converting data stored in an RDBMS can in principle be automated but at present no mature software products are available. D2RQ[25] is experimental software that allows the user to specify a mapping from the schema of relational database to RDF triples. When D2RQ was used to access a relational database with agronomic data (Van Evert et al., 2011), no problems were encountered (unpublished work).

If the use of triples to share agricultural data becomes widespread, then at some point in the future there may be hundreds or thousands of triplestores available on the Internet for a modeler to use. There would be two major ways of using the data. First, a query could be set up that accesses the triplestores directly. This would be inefficient but might just work for small amounts of data.

19 http://qudt.org/vocab/unit#Percent
20 http://www.wurvoc.org/vocabularies/om-1.8/percent
21 http://research.agmip.org/display/dev/ICASA+Master+Variable+List
22 https://gd.eppo.int/
23 http://rdf4j.org
24 https://virtuoso.openlinksw.com/
25 http://d2rq.org/

Second, it would be possible to harvest selected triples from several (hundreds, thousands, or more) sources and store the combined data locally in an RDBMS for fast analysis.

In summary, linked data promise to solve many of the problems in describing data and working with it. Unfortunately, its practical use is currently limited because neither the ontologies nor the software for efficiently storing and manipulating triples is available.

3.4 Data processing

It is rare that data are ready to be used. In many cases, it needs to be retrieved, filtered, transformed, and reorganized. Spreadsheet software is widely used and it is surprising how inventive some people are in using them but in fact for most data manipulation tasks more appropriate software tools are available. The decision which tool to use will be based at least in part on personal preferences.

In many instances data can be obtained through a web service using programs such as wget[26] or curl[27]. For example, the command line code in Fig. 3 retrieves daily maximum and minimum temperature data for the month of January 2018, for weather station De Bilt in The Netherlands, and stores it in a comma-separated file.

When data are in text files, a tool such as gawk[28] makes it possible to handle simple data-reformatting jobs with just a few lines of code. Log files produced by the Yara N-Sensor may contain one or more sections. Each section starts with a few lines with information about the specific sensor used and the configuration settings used. This is followed by a section of hundreds or thousands of lines with individual measurements (Fig. 4). A few lines of awk code convert the data to a fully tabular format that can easily be imported into a database.

When data are in an RDBMS, then SQL can be used to manipulate the data. SQL suffers from many drawbacks, including awful syntax, the fact that each vendor uses a slightly different language, and it is a procedural language rather than declarative, but it is very powerful. Views can be used to store often-used queries. Procedures and functions can be used to store transformations that cannot be expressed using views.

When data are in an RDF store, queries are expressed in SPARQL Protocol and RDF Query Language (SPARQL). SPARQL is very powerful but the efficiency of current implementations is much lower than with SQL/RDBMS.

26 https://www.gnu.org/software/wget/
27 https://curl.haxx.se/
28 https://www.gnu.org/software/gawk/

```
wget -O myfile.csv --post-data=
"stns=235&vars=TN:TX&byear=2018&bmonth=1&bday=1&eyear=2018&emonth=1&eday=31"
http://projects.knmi.nl/klimatologie/daggegevens/getdata_dag.cgi
```

Figure 3 Using the wget command line utility to retrieve weather data from the Netherlands Meteorological Service KNMI.

```
(a) original file

Filename,SysTime,Number,FilePos,Date,GPSTime,Latitude,Longitude,Altitude,Satellites,Quality,Age,XPos,YPos,S
peed,S1,S2,SN,SW,GBI,ApplRateN,ThinCrop,ApplFactor,OpMode,ActiveMode,CalStrip,SNL,SNR,ApplRateNL,ApplRateNR
,S1L_calc,S1R_calc
General information
  Sensor type: ALS-USB (S/N 125)
  Controller: LH5000
  Application mode: 0
  Working width: 24.00 m
  Signal delay: 1 s
  Response time: 1 s
  Agent: N
  Agent content: 27.0 %
Operation mode specific information (mode = 1)
  Crop type: Potatoes at growth stage 40
  Minimum: 5 kg N/ha  Maximum: 50 kg N/ha  Standard: 30 kg N/ha
  Reference rate: 20 kg N/ha  SN reference: 6.9
  Cutoff: 10.0  Slope: -1.3
00125_a_160907_15,0,2,1124,7-9-2016 11:35:56,34572.00,52.22670330,4.62876660,-
6.7,9,1,0.00,316639.21,5802966.45,,9.25,0.00,11.3,0.0,13.9,8.97,0,100,0,1,0,0.0,0.0,14.24,14.24,,
00125_a_160907_15,0,3,1430,7-9-2016 11:35:57,34573.00,52.22670330,4.62876660,-
6.7,9,1,0.00,316639.21,5802966.45,0.00,9.38,0.00,11.6,0.0,14.1,11.08,0,100,0,1,0,0.0,0.0,13.92,13.92,,


(b) AWK program

BEGIN {
   FS=","
   sensor="Sensor"
   croptype="Croptype"
}
/Sensor type/ { sensor = $0 }
/Crop type/ { croptype = $0 }

{ if (NF>10) {
      print sensor "," croptype "," $0
   }
}


(c) output

Sensor,Croptype,Filename,SysTime,Number,FilePos,Date,GPSTime,Latitude,Longitude,Altitude,Satellites,Quality
,Age,XPos,YPos,Speed,S1,S2,SN,SW,GBI,ApplRateN,ThinCrop,ApplFactor,OpMode,ActiveMode,CalStrip,SNL,SNR,ApplR
ateNL,ApplRateNR,S1L_calc,S1R_calc
  Sensor type: ALS-USB (S/N 125),  Crop type: Potatoes at growth stage 40,00125_a_160907_15,0,2,1124,7-9-
2016 11:35:56,34572.00,52.22670330,4.62876660,-
6.7,9,1,0.00,316639.21,5802966.45,,9.25,0.00,11.3,0.0,13.9,8.97,0,100,0,1,0,0.0,0.0,14.24,14.24,,
  Sensor type: ALS-USB (S/N 125),  Crop type: Potatoes at growth stage 40,00125_a_160907_15,0,3,1430,7-9-
2016 11:35:57,34573.00,52.22670330,4.62876660,-
6.7,9,1,0.00,316639.21,5802966.45,0.00,9.38,0.00,11.6,0.0,14.1,11.08,0,100,0,1,0,0.0,0.0,13.92,13.92,,
```

Figure 4 Using the command line utility gawk to convert a Yara N-Sensor log file to tabular format.

3.5 Ethical and social considerations

There are several good reasons to make experiment data available. First and foremost, the core of the scientific method is that others are able to reproduce one's results and extend it. Only by providing full details about one's work, including the data collected, will others be able to verify, interpret, reproduce, and extend that work. Simply including a statement in a scientific paper that says 'data are available upon request' is not sufficient: it will cease to work once the scientist has moved to another job, has left the field, or has passed away. A good reason to make data available is that it tends to increase the number of citations (Gleditsch et al., 2003). Once data have been published, responding to a request for the data is as simple as sending a link.

There are also several good reasons not to make scientific data available. It requires additional effort to make the data available once the data have been analyzed and papers have been published. Typically, some documentation has to be added to make sure the data can be understood by others than the original team. Permanent storage space has to be organized and paid for. Scientific data are valuable. Some of that value is realized by making it available (see above), but unfortunately, some of the value may be lost to the original when it is made available. If the original team is planning to write more papers using the data, there may be an incentive not to publish the data just yet. The team may be planning to use the data as leverage to negotiate collaboration with another team or as an asset in writing grant proposals. Finally, the team may be of the opinion that others will not be able to interpret the data correctly without interaction with the original researchers.

The arguments for making experiment data available are more or less straightforward. Intensively monitored farms may be the long-term experiments of the future. When on-farm collected data become an important vehicle for scientific progress, some of the arguments that apply to sharing scientific data become applicable to sharing farm data (van Evert et al., 2017a). Ethical, legal, and social considerations are more complex for data from commercial farms. The ethical matrix method (Mepham, 2005) has been used to explore the sharing of data in the context of commercial farming (van Evert et al., 2017a).

Social and legal issues related to agricultural data, such as data ownership, exchange, control, security, etc. have been addressed in a recent European Parliament report (Kritikos, 2017). Most social and legal challenges stem from the wide range of actors involved in the farm data chain and the fragmented and uneven character of the data ecosystem.

While it is clear that farmers' personal data are protected by current personal data regulations the ownership of equipment-generated data raises concerns among farmers and other agricultural stakeholders. Most companies state that farmers own the data they produce but this does not give the whole picture. Once a farmer's data are aggregated with other farmers' data, companies consider the aggregated data their property, which is where problems may arise. For example, measuring yield with a drone or a combine harvester produces data that are not personal (and therefore not protected) but may nevertheless give an indication of farm income (which is protected, personal data). Also farmers may waive data ownership rights by signing service agreements which they have not read (Rasmussen, 2016) or which are imposed on them in an unfair manner because they are the weaker party.[29]

29 Bertolo, S. Building a European Data economy. EIP-AGRI Workshop. Bratislava, 4-5 April 2017

3.6 Agribusiness and data sharing

Much of the data in commercial farming that is of interest to modelers are recorded with equipment from the large equipment manufacturers. This includes prescription maps and as-applied maps, yield maps, drone images, soil scans, soil moisture and temperature data, and farmer-owned weather stations. Several manufacturers are building proprietary systems to collect, store, and use data. Smaller companies are forming consortia such as AgriRouter[30] in order to maintain some degree of independence. Small companies and farmers' organizations are trying to develop standards that will make it possible to exchange data between equipment and software from different vendors (AgDataAlliance, EdiTeelt, API-AGRO, AgroXML).

There is a large number of farm management information systems (FMISs) and decision software on the market, including offerings by major agricultural companies (FieldView, Encirca, SMS, AFS Connect, Trimble Ag Software, and John Deere Operations Center), companies with FMIS as the main product (365Farmnet, Isagri, FarmLogs, SST, Crop-R, and CropVision), and consortium of stakeholders (API-AGRO). Further, Akkerweb, FertiWeb, and sigAGROasesor are examples of decision-centric software offered by a consortium of stakeholders, Adapt-N is decision-centric software offered by a company, and AgroSens focuses on data-driven decision-making.

4 Using data and models in precision agriculture

Models are being used by farmers in several ways. All ways are based on the premise that a model captures the important processes with a sufficient degree of realism and that the model has been adequately parameterized. If this is the case, the behavior of the model is very similar to the behavior of the real system and the model can be used as a learning tool and to inform decision-making. Below we discuss how models can be used in learning, for strategic purposes, and for in-season real-time decision making. An overview of systems discussed is given in Table 1.

4.1 Learning

A model can be used as a teaching tool when it is used to explore the effect of changing a management-related input such as planting date or the amount of fertilizer applied. A problem with this approach is that the effect of the planting date or fertilizer application is influenced much by the weather. A crop model can therefore not be used in training in the same way as a flight simulator or a car

30 https://my-agrirouter.com/en/

Table 1 Selected crop model-based advisory systems for use in farming practice

What	Data requirements						In-season	Strategic	Learning	Reference
	Weather (T, S, P)	Weather (more)	Soil	Groundwater	Management	Other inputs				
Adapt-N crop growth model	x		x		x	Prices	x			Melkonian et al. (2008)
Akkerweb late blight app	x	Long-wave radiation			x		x			www.akkerweb.eu
Akkerweb N sidedress	x				x	Crop reflectance Expected yield	x			Booij et al. (2017)
Irrigation recommendation	x		x	x		Amount of feed available on the farm Prices	x			https://www.zlto.nl/beregeningssignaal
Tipstar crop growth model	x		x		x		x			Jansen (2008)
Strategic simulations	x	x			x			x		Basso et al. (2010)
EPIPRE									x	Zadoks (1981)

driving simulator is used. The effect of different realizations of the weather can be quantified by repeating the simulation many times, each time using weather data from a different year in the past. If we simulate the effect of planting date on yield in this way, the combined simulations result in a distribution of yields for any given planting date. This information could be summarized in a graph with curves for several quantiles of the distributions, for example such that bands can be drawn which contain 50% and 90% of the simulated yields. However, information presented in this way is often poorly understood. A thoughtful, in-depth discussion of how to use models in agriculture is available (McCown et al., 2006, 2012; McCown, 2001; McCown and Parton, 2006).

4.2 Strategic use

A model can be used to explore several farming strategies in a systematic way (including the effect of year-to-year variations in weather) in order to make a decision about which strategy to use. The strategic question to be answered can be, for example, which planting date gives the most attractive trade-off between risk and profit (Farré et al., 2002), which fertilizer application is fitting for a particular soil (Basso et al., 2010), or whether a proposed new fertilizer strategy will be of benefit (Tovihoudji et al., 2019).

4.3 In-season, real-time recommendations

For many years there has been an interest to use crop models during the growing season to monitor the status of crop and soil, to make forecasts, and to make recommendations. From the eighties until now a large number of advisory systems have been proposed, a selection of which is presented in Table 1. Early examples include Epipre (Zadoks, 1981) for disease management in wheat and GOSSYM/COMAX for cotton (McKinion et al., 1989). These systems were meant to be decision-supporting systems, but in practice they functioned as learning tools.

The Tipstar system (Jansen, 2008) consists of a potato model and an in-season optimization procedure to decide when and how much nitrogen and water should be applied to prevent limitation of crop growth. Tipstar recommendations were as good as the decisions by experienced farmers (Jansen et al., 2003). The system could not be used on large scale in farming because it was not feasible to collect in practice the detailed soil information and weather and crop management data that were needed as input.

The Akkerweb sidedress N app is a simple app that gives a recommendation for the amount of sidedress N in potatoes. A farmer, who is planning to use the app, will apply two-thirds of the recommended amount of N at planting. Then, around 1 July, the amount of N already taken up by the crop is determined via a measurement of canopy reflectance. The amount of N taken up by the crop is compared with the amount of N that would have been taken up by a crop growing without N limitation. The difference is applied as sidedress N. This system maintains yield and results in an average reduction in N use of 15% (Booij et al., 2017; Van Evert et al., 2012, 2016). The amount of N taken up by a crop growing at the potential rate is currently modeled using a logistic growth curve driven by thermal time. The maximum N uptake (the upper asymptote of the curve) is set by the farmer based on past performance. Work is under way to replace the logistic curve with the output of a crop model.

Potato late blight is the most serious disease of potato, globally responsible for an annual loss of around M€4800 (Haverkort et al., 2008). Traditionally, potato late blight is controlled using frequent fungicide applications (Cooke et al., 2011). Fungicide input may be reduced, and quality of control increased, by optimizing the timing of fungicide applications. The Akkerweb late blight app is a decision support system implementing a preventive late blight control strategy aiming to only apply fungicides just before predicted late blight infection events. The app uses local, forecast, and historical hourly weather data to identify infection events in the near future and near past. It recommends the use of a preventive fungicide just before an infection event is predicted. It recommends the use of a curative fungicide when calculations show that an infection took place in the last 24 hours. Finally, it recommends the use of eradicant fungicide combinations on older, untreated latent and active infections. The late blight app needs real-time hourly weather data as well as 10 days of forecast hourly data. Unusually for crop models, the app requires incoming long-wave radiation as one of the inputs to calculate dew formation on the leaves.

Potatoes are harvested before the natural maturity of the crop. Typically, the above-ground part of the crop is killed about three weeks before the harvest date to ensure that the skin of the tubers has hardened when the crop is harvested. The most common method to kill the potato is to use defoliant herbicides (Kempenaar and Struijk, 2008). Herbicide dose can be based on the above-ground amount and activity of the biomass. The more biomass there is, the higher the dose of the defoliant should be (Kempenaar et al., 2014). Above-ground biomass can be measured for example using crop reflectance sensors. In this way a reduction in herbicide use between 20 and 47% can be achieved (Kempenaar et al., 2017; Van Evert et al., 2017b). The potato haulm killing app allows farmers to generate a task map for variable rate herbicide application based on a satellite or a drone image.

Adapt-N is an N recommendation system for maize in the United States based on a crop model (Melkonian et al., 2008; Sela et al., 2018). Adapt-N[31] is

31 www.adapt-n.com

offered commercially. It uses high-resolution gridded soil data as well as high-resolution gridded weather data. It can provide recommendations for flat rate fertilization, for variable rate fertilization, and for sidedress application.

In the Netherlands, an online, farm-level irrigation recommendation system is available for dairy farmers[32]. The system uses a crop model to monitor soil water status for each grass and (silage) maize field and gives an alert when irrigation is necessary. Grassland is often located in areas with groundwater close to the surface, therefore an important input to the model is current groundwater level obtained from monitoring wells. The system takes into account the amount of feed available and needed, as well as prices of feed, electricity, and labor.

4.4 Delivering actionable information to farmers

Setting up a system that uses data and crop models to deliver actionable information to farmers entails requirements: evidence-based decision rules; technical infrastructure to gather, process, and deliver information; trust; perception of value delivered; satisfactory legal framework; and revenue. Many initiatives in the past have failed because one or more of these requirements were not satisfied or could not continue to be satisfied.

Akkerweb[33] is a recent development which makes a conscious effort to meet the requirements. It is developed out of a DSS for nematode control in potatoes (Been and Schomaker, 2004; Been et al., 2007). A brief overview is available (Been and Molendijk, 2017; Van Evert et al., 2018b). Akkerweb was developed by a consortium consisting of Agrifirm, a farmers' cooperative, and Wageningen UR, a research institute. Akkerweb offers 30+ evidence-based decision rules, the quality of which is guaranteed by WUR staff. Farmers' priorities and interests are safeguarded by Agrifirm. Development and operation of IT infrastructure is in the hands of a software company.

5 Future trends and conclusion

In order for data to be there, it has to be measured first. Thus this chapter started by discussing modern techniques to collect data. There is a glaring lack of information about what happens below-ground. In general we have little direct information about where roots grow. In soils with compacted layers that restrict root growth, or shallow soils, we typically do not know the spatial variability of these properties. They need to be inferred from indirect observations on leaf area development, wilting, and ultimately, yield. This situation is even more

32 www.beregeningssignaal.nl
33 www.akkerweb.eu

dire for root and tuber crops where it is generally not possible to follow yield formation as closely as we can in grain crops.

We still face serious challenges in storing, exchanging, and using data. A single company, university, or farm can probably meet all data needs using an RDBMS. However most of the value of data will only be realized if it is combined with other data. For that, our current tools are woefully inadequate. In this area, first of all, we need vocabularies and ontologies so that data can be fully described in a machine-readable manner. Second, we need conversion tools that are capable of exporting relational data as sets of triples; these triple sets can then be combined using current linked data technologies (see Section 3.3). The conversion tools must also be capable of importing triple sets into an RDBMS in order to allow efficient processing of datasets resulting from the combination of several sources. Third, we need a global, machine-readable register of datasets.

There are also some very encouraging developments. The availability of modern instruments and data recording technologies brings commercial farms to the foreground as the new experiment stations (van Evert et al., 2017a). Modelers have traditionally used designed experiments to develop and test models. It seems highly likely that in the near future, they will collaborate with farmers and use new data sources such as optical and radar satellites, rainfall radar, and automated phenotyping. This will ask for smart ways of using inverse modeling and other data assimilation techniques.

6 Where to look for further information

Shannon et al. (2018) has several chapters on measurements (yield, soil, pests, remote sensing) and on data management. More recently, Castrignano et al. (2020) covers some of the same topics and focuses specifically on the internet of things for agriculture.

Allemang and Hendler (2011) is a very readable introduction to ontologies. The PHIS software (http://www.phis.inra.fr) is an ontology-based data management system for plant phenomics. PHIS is being used by the EU-funded networking and infrastructure project EPPN2020 (https://eppn2020.plant-phenotyping.eu/) as well as by the Dutch-funded infrastructure project Netherlands Plant Ecophenotyping Centre (NPEC) (https://www.npec.nl/).

Up-to-date information on using crop and livestock modeling in precision agriculture can be found at the European Conference on Precision Agriculture (ECPA) (http://ecpa2019.agrotic.org/) and the International Conference on Precision Agriculture (ICPA) (https://www.ispag.org/) which take place in alternate years.

Up-to-date information on using modeling for predicting and understanding the effects of a changing climate can be obtained through the work of the US-based Agricultural Model Intercomparison and Improvement Project (AgMIP) (https://agmip.org/) and the Australia-based APSIM Initiative (https://www.apsim.info/).

7 References

Ackerson, J. P., Morgan, C. L. S. and Ge, Y. 2017. Penetrometer-mounted VisNIR spectroscopy: application of EPO-PLS to in situ VisNIR spectra. *Geoderma* 286, 131–8. doi:10.1016/j.geoderma.2016.10.018.

Adamchuk, V. I., Lund, E. D., Sethuramasamyraja, B., Morgan, M. T., Dobermann, A. and Marx, D. B. 2005. Direct measurement of soil chemical properties on-the-go using ion-selective electrodes. *Computers and Electronics in Agriculture* 48(3), 272–94. doi:10.1016/j.compag.2005.05.001.

Adamchuk, V. I., Allred, B., Doolittle, J., Grote, K. and Viscarra Rossel, R. A. 2017. Tools for proximal soil sensing. In: Ditzler, C., Scheffe, K. and Monger, H. C. (Eds), *Soil Survey Manual*. USDA Handbook 18. Government Printing Office, Washington DC, pp. 355–87. Chapter 6.

Adams, R. M., Rosenzweig, C., Peart, R. M., Ritchie, J. T., McCarl, B. A., Glyer, J. D., Curry, R. B., Jones, J. W., Boote, K. J. and Allen, L. H. 1990. Global climate change and US agriculture. *Nature* 345(6272), 219–24. doi:10.1038/345219a0.

Allemang, D. and Hendler, J. 2011. *Semantic Web for the Working Ontologist: Effective Modeling in RDFS and OWL*. Elsevier, Amsterdam [etc.].

Asseng, S., Ewert, F., Rosenzweig, C., Jones, J. W., Hatfield, J. L., Ruane, A. C., Boote, K. J., Thorburn, P. J., Rötter, R. P., Cammarano, D., Brisson, N., Basso, B., Martre, P., Aggarwal, P. K., Angulo, C., Bertuzzi, P., Biernath, C., Challinor, A. J., Doltra, J., Gayler, S., Goldberg, R., Grant, R., Heng, L., Hooker, J., Hunt, L. A., Ingwersen, J., Izaurralde, R. C., Kersebaum, K. C., Müller, C., Naresh Kumar, S., Nendel, C., O'Leary, G., Olesen, J. E., Osborne, T. M., Palosuo, T., Priesack, E., Ripoche, D., Semenov, M. A., Shcherbak, I., Steduto, P., Stöckle, C., Stratonovitch, P., Streck, T., Supit, I., Tao, F., Travasso, M., Waha, K., Wallach, D., White, J. W., Williams, J. R. and Wolf, J. 2013. Uncertainty in simulating wheat yields under climate change. *Nature Climate Change* 3(9), 827–32. doi:10.1038/nclimate1916.

Asseng, S., Martre, P., Maiorano, A., Rötter, R. P., O'Leary, G. J., Fitzgerald, G. J., Girousse, C., Motzo, R., Giunta, F., Babar, M. A., Reynolds, M. P., Kheir, A. M. S., Thorburn, P. J., Waha, K., Ruane, A. C., Aggarwal, P. K., Ahmed, M., Balkovič, J., Basso, B., Biernath, C., Bindi, M., Cammarano, D., Challinor, A. J., De Sanctis, G., Dumont, B., Eyshi Rezaei, E., Fereres, E., Ferrise, R., Garcia-Vila, M., Gayler, S., Gao, Y., Horan, H., Hoogenboom, G., Izaurralde, R. C., Jabloun, M., Jones, C. D., Kassie, B. T., Kersebaum, K. C., Klein, C., Koehler, A. K., Liu, B., Minoli, S., Montesino San Martin, M., Müller, C., Naresh Kumar, S., Nendel, C., Olesen, J. E., Palosuo, T., Porter, J. R., Priesack, E., Ripoche, D., Semenov, M. A., Stöckle, C., Stratonovitch, P., Streck, T., Supit, I., Tao, F., Van der Velde, M., Wallach, D., Wang, E., Webber, H., Wolf, J., Xiao, L., Zhang, Z., Zhao, Z., Zhu, Y. and Ewert, F. 2019. Climate change impact and adaptation for wheat protein. *Global Change Biology* 25(1), 155–73. doi:10.1111/gcb.14481.

Bao, Y., Tang, L., Breitzman, M. W., Salas Fernandez, M. G. and Schnable, P. S. 2019. Field-based robotic phenotyping of sorghum plant architecture using stereo vision. *Journal of Field Robotics* 36(2), 397–415. doi:10.1002/rob.21830.

Barnes, E., Clarke, T., Richards, S., Colaizzi, P., Haberland, J., Kostrzewski, M., Waller, P., Choi, C., Riley, E., Thompson, T., Lascano, R. J., Li, H. and Moran, M. S. 2000. Coincident detection of crop water stress, nitrogen status and canopy density using ground based multispectral data. *Proceedings of the Fifth International Conference on Precision Agriculture*, Bloomington, MN, USA.

Basso, B., Cammarano, D., Troccoli, A., Chen, D. and Ritchie, J. T. 2010. Long-term wheat response to nitrogen in a rainfed Mediterranean environment: field data and simulation analysis. *European Journal of Agronomy* 33(2), 132–8. doi:10.1016/j.eja.2010.04.004.

Been, T. H. and Molendijk, L. P. G. 2017. Akkerweb: a view of your plots. Wageningen Plant Research, Wageningen, the Netherlands. Available at: http://edepot.wur.nl/427908.

Been, T. H. and Schomaker, C. H. 2004. A geo-referenced Decision Support System for nematodes in potatoes. In: MacKerron, D. K. L. and Haverkort, A. J. (Eds), *Decision Support Systems in Potato Production : Bringing Models to Practice*. Wageningen Academic Publishers, Wageningen, the Netherlands, pp. 154–67.

Been, T. H., Schomaker, C. H. and Molendijk, L. P. G. 2007. NemaDecide, a decision support system for the management of potato cyst nematodes. *Phytopathology* 97, S152.

Berners-Lee, T., Hendler, J. and Lassila, O. 2001. The semantic web. *Scientific American* 284, 34–43. doi:10.1038/scientificamerican0501-34.

Boogaard, H., Eerens, H., Supit, I., Van Diepen, C., Piccard, I. and Kempeneers, P. 2002. Description of the MARS crop yield forecasting system (MCYFS). Study contract: 02-F01FED. Joint Research Centre.

Booij, J. A., Van Evert, F. K., van Geel, W. C. A., Kroonen-Backbier, B. M. A. and Kempenaar, C. 2017. Roll-out of online application for N sidedress recommendations in potato. EFITA, Montpellier. Available at: http://library.wur.nl/WebQuery/wurpubs/fulltext/445495.

Bostick, W. M., Koo, J., Walen, V. K., Jones, J. W. and Hoogenboom, G. 2004. A web-based data exchange system for crop model applications. *Agronomy Journal* 96(3), 853–6. doi:10.2134/agronj2004.0853.

Böttcher, U. and Kage, H. 2005. Using sensor information on drought stress for a site specific calibration of a wheat simulation model. *5th, European Conference on Precision Agriculture*, Uppsala, Sweden. Wageningen Academic Publishers, Wageningen, the Netherlands, 803–10.

Böttcher, U. and Kage, H. 2007. Deriving parameters for soil water holding capacity from canopy temperature measurements and model inversion. *ASA/CSSA/SSSA Annual Meeting Abstracts*. ASA, Madison, WI.

Bouman, B. A. M., Uenk, D. and Haverkort, A. J. 1992. The estimation of ground cover of potato by reflectance measurements. *Potato Research* 35(2), 111–25. doi:10.1007/BF02357604.

Castrignano, A., Buttafuoco, G., Khosla, R., Mouazen, A. M., Moshou, D. and Naud, O. 2020. *Agricultural Internet of Things and Decision Support for Smart Farming*. Elsevier.

Christiansen, M. P., Laursen, M. S., Mikkelsen, B. F., Teimouri, N., Jørgensen, R. N. and Sørensen, C. A. G. 2018. Current potentials and challenges using Sentinel-1 for broadacre field remote sensing. AGENG2018.

Clevers, J. G. P. W. 1989. The application of a weighted infrared-red vegetation index for estimating leaf-area index by correcting for soil-moisture. *Remote Sensing of Environment* 29(1), 25–37. doi:10.1016/0034-4257(89)90076-X.

Clevers, J. G. P. W. and Gitelson, A. A. 2013. Remote estimation of crop and grass chlorophyll and nitrogen content using red-edge bands on sentinel-2 and-3. *International Journal of Applied Earth Observation and Geoinformation* 23, 344–51. doi:10.1016/j.jag.2012.10.008.

Cooke, L. R., Schepers, H. T. A. M., Hermansen, A., Bain, R. A., Bradshaw, N. J., Ritchie, F., Shaw, D. S., Evenhuis, A., Kessel, G. J. T., Wander, J. G. N., Andersson, B., Hansen, J. G., Hannukkala, A., Nærstad, R. and Nielsen, B. J. 2011. Epidemiology and integrated control of potato late blight in Europe. *Potato Research* 54(2), 183–222. doi:10.1007/s11540-011-9187-0.

Deering, D. W. 1978. Rangeland reflectance characteristics measured by aircraft and spacecraft sensors. Ph.D., Texas A&M University, College Station, TX.

Farré, I., Robertson, M. J., Walton, G. H. and Asseng, S. 2002. Simulating phenology and yield response of canola to sowing date in Western Australia using the APSIM model. *Australian Journal of Agricultural Research* 53(10), 1155–64. doi:10.1071/AR02031.

Fleisher, D. H., Condori, B., Quiroz, R., Alva, A., Asseng, S., Barreda, C., Bindi, M., Boote, K. J., Ferrise, R., Franke, A. C., Govindakrishnan, P. M., Harahagazwe, D., Hoogenboom, G., Naresh Kumar, S., Merante, P., Nendel, C., Olesen, J. E., Parker, P. S., Raes, D., Raymundo, R., Ruane, A. C., Stockle, C., Supit, I., Vanuytrecht, E., Wolf, J. and Woli, P. 2017. A potato model intercomparison across varying climates and productivity levels. *Global Change Biology* 23(3), 1258–81. doi:10.1111/gcb.13411.

Genovese, G. 1998. The methodology, the results and the evaluation of the MARS crop yield forecasting system. *Agrometeorological Applications for Regional Crop Monitoring and Production Assessment*, pp. 67–119.

Gitelson, A. A., Gritz, Y. and Merzlyak, M. N. 2003. Relationships between leaf chlorophyll content and spectral reflectance and algorithms for non-destructive chlorophyll assessment in higher plant leaves. *Journal of Plant Physiology* 160(3), 271–82. doi:10.1078/0176-1617-00887.

Gleditsch, N. P., Metelits, C. and Strand, H. 2003. Posting your data: will you be scooped or will you be famous. *International Studies Perspectives* 4, 89–97.

Gruber, T. R. 1993. A translation approach to portable ontology specifications. *Knowledge Acquisition* 5(2), 199–220. doi:10.1006/knac.1993.1008.

Guérif, M., Houles, V., Makowski, D. and Lauvernet, C. 2006. Data assimilation and parameter estimation for precision agriculture using the crop model STICS. In: Wallach, D., Makowski, D. and Jones, J. W. (Eds), *Working with Dynamic Crop Models: Evaluating, Analyzing, Parameterizing and Using Them*. Elsevier, pp. 391–8. Chapter 17.

Hatfield, J. L., Gitelson, A. A., Schepers, J. S. and Walthall, C. L. 2008. Application of spectral remote sensing for agronomic decisions. *Agronomy Journal* 100(Suppl. 3), S117–31. doi:10.2134/agronj2006.0370c.

Haverkort, A. J., Boonekamp, P. M., Hutten, R., Jacobsen, E., Lotz, L. A. P., Kessel, G. J. T., Visser, R. G. F. and van der Vossen, E. A. G. 2008. Societal costs of late blight in potato and prospects of durable resistance through cisgenic modification. *Potato Research* 51(1), 47–57. doi:10.1007/s11540-008-9089-y.

Heil, K. and Schmidhalter, U. 2017. The application of EM38: determination of soil parameters, selection of soil sampling points and use in agriculture and archaeology. *Sensors (Basel)* 17(11). doi:10.3390/s17112540.

Hunt, L. A. and Boote, K. J. 1998. Data for model operation, calibration, and evaluation. In: Tsuji, G. Y., Hoogenboom, G. and Thornton, P. K. (Eds), *Understanding Options for*

Agricultural Production. Kluwer Academic Publishers, Dordrecht, the Netherlands, pp. 9–39.

Hunt, L. A., Jones, J. W., Hoogenboom, G., Godwin, D. C., Singh, U., Pickering, N. B., Thornton, P. K., Boote, K. J. and Ritchie, J. T. 1994. Input and output file structures for crop simulation models. In: Uhlir, P. F. and Carter, G. C. (Eds), *Crop Modeling and Related Environmental Data: A Focus on Applications for Arid and Semiarid Regions in Developing Countries*. CODATA Commission on Global Change, Paris, France, pp. 35–72.

Jansen, D. M. 2008. Beschrijving van TIPSTAR: hét simulatiemodel voor groei en productie van zetmeelaardappelen. Nota 547. Plant Research International, Wageningen, the Netherlands. Available at: http://edepot.wur.nl/27135.

Jansen, D. M., Davies, J. A. and Steenhuizen, J. W. 2003. Testen van Tipstar in de praktijk. Report 244. Plant Research International, Wageningen, the Netherlands.

Kempenaar, C. and Struijk, P. C. 2008. The canon of potato science: haulm killing. *Potato Research* 50, 341–5. doi:10.1007/s11540-008-9082-5.

Kempenaar, C., Van Evert, F. K. and Been, T. 2014. Use of vegetation indices in variable rate application of potato haulm killing herbicides. *12th International Conference on Precision Agriculture (ICPA)*, Sacramento, CA, USA.

Kempenaar, C., Been, T., Booij, J., van Evert, F., Michielsen, J. M. and Kocks, C. 2017. Advances in variable rate technology application in potato in the Netherlands. *Potato Research* 60(3), 295–305. doi:10.1007/s11540-018-9357-4.

Kharel, T. P., Swink, S. N., Maresma, A., Youngerman, C., Kharel, D., Czymmek, K. J. and Ketterings, Q. M. 2019. Yield monitor data cleaning is essential for accurate corn grain and silage yield determination. *Agronomy Journal* 111(2). doi:10.2134/agronj2018.05.0317.

Knotters, M., van Egmond, F., Bakker, G., Walvoort, D. and Brouwer, F. 2017. *A Selection of Sensing Techniques for Mapping Soil Hydraulic Properties*. Wageningen Environmental Research, Wageningen, the Netherlands.

Kritikos, M. 2017. Precision agriculture in Europe: legal, social and ethical considerations. European Parliamentary Research Service. doi:10.2861/278.

Kumar, P., Prasad, R., Gupta, D. K., Mishra, V. N., Vishwakarma, A. K., Yadav, V. P., Bala, R., Choudhary, A. and Avtar, R. 2018. Estimation of winter wheat crop growth parameters using time series Sentinel-1A SAR data. *Geocarto International* 33(9), 942–56. doi:10.1080/10106049.2017.1316781.

Lobsey, C. R. and Viscarra Rossel, R. A. 2016. Sensing of soil bulk density for more accurate carbon accounting. *European Journal of Soil Science* 67(4), 504–13. doi:10.1111/ejss.12355.

Luciani, S., Paloscia, S., Pampaloni, P., Schiavon, G., Sigismondi, S. and Solimini, D. 1994. Sensitivity of microwave backscattering to crop biomass. *Proceedings of IGARSS '94 - 1994 IEEE International Geoscience and Remote Sensing Symposium* (vol. 1843), pp. 1844–6.

Madec, S., Baret, F., de Solan, B., Thomas, S., Dutartre, D., Jezequel, S., Hemmerlé, M., Colombeau, G. and Comar, A. 2017. High-throughput phenotyping of plant height: comparing unmanned aerial vehicles and ground LiDAR estimates. *Frontiers in Plant Science* 8, 2002. doi:10.3389/fpls.2017.02002.

Maestrini, B. and Basso, B. 2018. Predicting spatial patterns of within-field crop yield variability. *Field Crops Research* 219, 106–12. doi:10.1016/j.fcr.2018.01.028.

McCown, R. L. 2001. Learning to bridge the gap between science-based decision support and the practice of farming: evolution in paradigms of model-based research and intervention from design to dialogue. *Australian Journal of Agricultural Research* 52(5), 549–71. doi:10.1071/AR00119.

McCown, R. L. and Parton, K. A. 2006. Learning from the historical failure of farm management models to aid management practice. part 2. Three systems approaches. *Australian Journal of Agricultural Research* 57(2), 157–72. doi:10.1071/AR05052.

McCown, R. L., Brennan, L. E. and Parton, K. A. 2006. Learning from the historical failure of farm management models to aid management practice. part 1. The rise and demise of theoretical models of farm economics. *Australian Journal of Agricultural Research* 57(2), 143–56. doi:10.1071/AR05051.

McCown, R. L., Carberry, P. S., Dalgliesh, N. P., Foale, M. A. and Hochman, Z. 2012. Farmers use intuition to reinvent analytic decision support for managing seasonal climatic variability. *Agricultural Systems* 106(1), 33–45. doi:10.1016/j.agsy.2011.10.005.

McKinion, J. M., Baker, D. N., Whisler, F. D. and Lambert, J. R. 1989. Application of the GOSSYM/COMAX system to cotton crop management. *Agricultural Systems* 31(1), 55–65. doi:10.1016/0308-521X(89)90012-7.

McSweeney, D., Coughlan, N. E., Cuthbert, R. N., Halton, P. and Ivanov, S. 2019. Micro-sonic sensor technology enables enhanced grass height measurement by a Rising Plate Meter. *Information Processing in Agriculture* 6(2), 279–84. doi:10.1016/j.inpa.2018.08.009.

Melkonian, J., Van Es, H., DeGaetano, A., Joseph, L. and Kosla, R. 2008. ADAPT-N: adaptive nitrogen management for maize using high-resolution climate data and model simulations. *Procedings of the 9th International Conference on Precision Agriculture*.

Mepham, B. 2005. *Bioethics: an Introduction for the Biosciences*. Oxford University Press, Oxford.

Plant Ontology™ Consortium. 2002. The Plant Ontology™ consortium and plant ontologies. *Comparative and Functional Genomics* 3, 137–42.

Rasmussen, N. 2016. From precision agriculture to market manipulation: a new frontier in the legal community. *Minnesota Journal of Law, Science and Technology* 17, 489–516.

Raymundo, R., Asseng, S., Cammarano, D. and Quiroz, R. 2014. Potato, sweet potato, and yam models for climate change: a review. *Field Crops Research* 166, 173–85. doi:10.1016/j.fcr.2014.06.017.

Raymundo, R., Asseng, S., Robertson, R., Petsakos, A., Hoogenboom, G., Quiroz, R., Hareau, G. and Wolf, J. 2018. Climate change impact on global potato production. *European Journal of Agronomy* 100, 87–98. doi:10.1016/j.eja.2017.11.008.

Rijgersberg, H., van Assem, M. and Top, J. 2013. Ontology of units of measure and related concepts. *Semantic Web* 4, 3–13. doi:10.3233/sw-2012-0069.

Roitsch, T., Cabrera-Bosquet, L., Fournier, A., Ghamkhar, K., Jiménez-Berni, J., Pinto, F. and Ober, E. S. 2019. Review: New sensors and data-driven approaches–a path to next generation phenomics. *Plant Science* 282, 2–10. doi:10.1016/j.plantsci.2019.01.011.

Rosenzweig, C., Jones, J. W., Hatfield, J. L., Ruane, A. C., Boote, K. J., Thorburn, P., Antle, J. M., Nelson, G. C., Porter, C., Janssen, S., Asseng, S., Basso, B., Ewert, F., Wallach, D., Baigorria, G. and Winter, J. M. 2013. The Agricultural Model Intercomparison and Improvement Project (AgMIP): protocols and pilot studies. *Agricultural and Forest Meteorology* 170, 166–82. doi:10.1016/j.agrformet.2012.09.011.

Rotz, C. A., Mertens, D. R., Buckmaster, D. R., Allen, M. S. and Harrison, J. H. 1999. A dairy herd model for use in whole farm simulations. *Journal of Dairy Science* 82(12), 2826-40. doi:10.3168/jds.S0022-0302(99)75541-4.

Roudier, P., Hedley, C. B. and Ross, C. W. 2015. Prediction of volumetric soil organic carbon from field-moist intact soil cores. *European Journal of Soil Science* 66(4), 651-60. doi:10.1111/ejss.12259.

Schans, D. A.v.d. and Berg, W.v.d. 2013. *Testen, Validatie en Toepassing van het Veris-sensorplatform Veldanalyse van twee percelen op veenkoloniale grond*. PPO-AGV, Lelystad, p. 47.

Sela, S., van Es, H. M., Moebius-Clune, B. N., Marjerison, R. and Kneubuhler, G. 2018. Dynamic model-based recommendations increase the precision and sustainability of N fertilization in midwestern US maize production. *Computers and Electronics in Agriculture* 153, 256-65. doi:10.1016/j.compag.2018.08.010.

Shannon, D. K., Clay, D. E. and Kitchen, N. R. 2018. *Precision Agriculture Basics*. American Society of Agronomy, Crop Science Society of America, and Soil Science Society of America, Inc., Madison, WI.

Shrestha, R., Arnaud, E., Mauleon, R., Senger, M., Davenport, G. F., Hancock, D., Morrison, N., Bruskiewich, R. and McLaren, G. 2010. Multifunctional crop trait ontology for breeders' data: field book, annotation, data discovery and semantic enrichment of the literature. *AoB Plants* 2010, plq008. doi:10.1093/aobpla/plq008.

Sreelash, K., Buis, S., Sekhar, M., Ruiz, L., Tomer, S. K. and Guérif, M. 2017. Estimation of available water capacity components of two-layered soils using crop model inversion: effect of crop type and water regime. *Journal of Hydrology* 546, 166-78. doi:10.1016/j.jhydrol.2016.12.049.

Steele-Dunne, S. C., McNairn, H., Monsivais-Huertero, A., Judge, J., Liu, P.-W. and Papathanassiou, K. 2017. Radar remote sensing of agricultural canopies: a review. *IEEE Journal of Selected Topics in Applied Earth Observations and Remote Sensing* 10(5), 2249-73. doi:10.1109/JSTARS.2016.2639043.

Tagarakis, A., Mimić, G., Vaessen, H. M., Rodriguez-Moreno, F., van Evert, F. K. and Ćirić, V. 2019. Using a crop growth model to assess within-field variability. ECPA 2019.

Tovihoudji, P. G., Akponikpè, P. B. I., Agbossou, E. K. and Bielders, C. L. 2019. Using the DSSAT model to support decision making regarding fertilizer microdosing for maize production in the sub-humid region of Benin. *Frontiers in Environmental Science* 7. doi:10.3389/fenvs.2019.00013.

Triantafilis, J. and Lesch, S. M. 2005. Mapping clay content variation using electromagnetic induction techniques. *Computers and Electronics in Agriculture* 46(1-3), 203-37. doi:10.1016/j.compag.2004.11.006.

Ulaby, F. T., Allen, C. T., Eger III, G. and Kanemasu, E. 1984. Relating the microwave backscattering coefficient to leaf area index. *Remote Sensing of Environment* 14(1-3), 113-33. doi:10.1016/0034-4257(84)90010-5.

Van der Schans, D., Van Evert, F. K., Malda, J. T. and Dorka-Vona, V. 2012. Sensorgestuurde advisering van stikstof bijbemesting in aardappel: implementatie en integratie. PPO Rapport 520. Praktijkonderzoek Plant & Omgeving (PPO), Wageningen, the Netherlands.

Van der Velde, M. and Nisini, L. 2019. Performance of the MARS-crop yield forecasting system for the European Union: assessing accuracy, in-season, and year-to-year improvements from 1993 to 2015. *Agricultural Systems* 168, 203-12. doi:10.1016/j.agsy.2018.06.009.

Van Diepen, C., Wal, T. and Boogaard, H. 1998. Deterministic crop growth modelling fundamentals and application for regional crop state monitoring and yield forecasting. In: *Phare Multi-country Environment Programme MARS and Environmental Related Applications (MERA) Project: Proceedings 1994-1996 Results Conference*, Bratislava, 10–11 December 1996, pp. 201–27.

Van Evert, F. K., Spaans, E. J. A., Krieger, S. D., Carlis, J. V. and Baker, J. M. 1999a. A database for agroecological research data: I. Data model. *Agronomy Journal* 91(1), 54–62. doi:10.2134/agronj1999.00021962009100010009x.

Van Evert, F. K., Spaans, E. J. A., Krieger, S. D., Carlis, J. V. and Baker, J. M. 1999b. A database for agroecological research data: II. A relational implementation. *Agronomy Journal* 91(1), 62–71. doi:10.2134/agronj1999.00021962009100010010x.

Van Evert, F. K., Van der Schans, D. A., Van Geel, W. C. A., Slabbekoorn, J. J., Booij, R., Jukema, J. N., et al. 2011. Dataset - Droevendaal, Rolde and Colijnsplaat, 1996-2003. Wageningen University and Research Center, Wageningen, the Netherlands. Available at: http://library.wur.nl/WebQuery/wurpubs/409283.

Van Evert, F. K., Booij, R., Jukema, J. N., Ten Berge, H. F. M., Uenk, D., Meurs, E. J. J., van Geel, W. C. A., Wijnholds, K. H. and Slabbekoorn, J. J. 2012. Using crop reflectance to determine sidedress N rate in potato saves N and maintains yield. *European Journal of Agronomy* 43, 58–67. doi:10.1016/j.eja.2012.05.005.

Van Evert, F. K., Meurs, E. J. J., Van der Schans, D., Booij, J. A., Van Geel, W. C. A. and Kempenaar, C. 2016. Using aerial imaging in a large-scale roll-out of N sidedress recommendations for potato. *ASA/CSSA/SSSA Annual Meetings*, Phoenix, AZ, USA.

van Evert, F. K., Fountas, S., Jakovetic, D., Crnojevic, V., Travlos, I. and Kempenaar, C. 2017a. Big Data for weed control and crop protection. *Weed Research* 57(4), 218–33. doi:10.1111/wre.12255.

Van Evert, F. K., Gaitán-Cremaschi, D., Fountas, S. and Kempenaar, C. 2017b. Can precision agriculture increase the profitability and sustainability of the production of potatoes and olives? *Sustainability* 9(10). doi:10.3390/su9101863.

Van Evert, F. K., Baron, F.-J., Booij, J. A., Meurs, B., van Oort, P. A. J. and Kempenaar, C. 2018a. Improving the simulation of crop growth in commercial growers' fields using soil scans, remote sensing imagery and yield monitor data. *2018 ASA and CSSA Meeting*, Baltimore, MD.

Van Evert, F. K., Been, T. H., Booij, J. A., Kempenaar, C., Kessel, G. J. T. and Molendijk, L. P. G. 2018b. Akkerweb: a platform for precision farming data, science, and practice. *14th International Conference on Precision Agriculture*, Montreal, Quebec, Canada.

Wachendorf, M., Fricke, T. and Möckel, T. 2018. Remote sensing as a tool to assess botanical composition, structure, quantity and quality of temperate grasslands. *Grass and Forage Science* 73(1), 1–14. doi:10.1111/gfs.12312.

Wallach, D. 1984. The organization of agronomic experiment data for crop modeling. *Proceedings of the International Symposium at ICRISAT Center*.

White, J. W. and Van Evert, F. K. 2008. Publishing agronomic data. *Agronomy Journal* 100(5), 1396–400. doi:10.2134/agronj2008.0080F.

White, J. W., Hunt, L. A., Boote, K. J., Jones, J. W., Koo, J., Kim, S., Porter, C. H., Wilkens, P. W. and Hoogenboom, G.. 2013. Integrated description of agricultural field experiments and production: a The ICASA Version 2.0 data standards. *Computers and Electronics in Agriculture* 96, 1–12. doi:10.1016/j.compag.2013.04.003.

Wösten, J. H. M. 2012. *BOFEK2012, de nieuwe bodemfysische schematisatie van Nederland*. Alterra Report 2387. Wageningen University and Research Center, Wageningen, the Netherlands.

Zadoks, J. C. 1981. EPIPRE: a disease and pest management system for winter wheat developed in the Netherlands. *EPPO Bulletin* 11(3), 365–9. doi:10.1111/j.1365-2338.1981.tb01945.x.

Zarco-Tejada, P. J., González-Dugo, V. and Berni, J. A. J. 2012. Fluorescence, temperature and narrow-band indices acquired from a UAV platform for water stress detection using a micro-hyperspectral imager and a thermal camera. *Remote Sensing of Environment* 117, 322–37. doi:10.1016/j.rse.2011.10.007.

Zhang, L., Clarke, M. L., Steven, M. D. and Jaggard, K. W. 2011. Spatial patterns of wilting in sugar beet as an indicator for precision irrigation. *Precision Agriculture* 12(2), 296–316. doi:10.1007/s11119-010-9177-2.

Chapter 15

Dealing with uncertainty in crop models

Daniel Wallach, INRA, France

1 Introduction

'Doubt is not a pleasant condition, but certainty is absurd' (Voltaire)

Crop models in general are deterministic. That is, given specific values of inputs (weather, management, soil properties), they always produce the same values for the outputs (yield, soil moisture etc.). However, this determinism is only superficial. In fact, it is clear that the equations for individual processes are uncertain (other equations could reasonably have been chosen), the values of the model parameters are uncertain and the values of at least some of the inputs are often only approximate. If one takes into consideration the alternative possible equations, parameters and values of the inputs, then one obtains a distribution of possible simulated values. By 'model uncertainty' we mean this distribution of simulated results. Uncertainty analysis aims at approximating this distribution. A closely related topic is sensitivity analysis, which aims at quantifying the separate contributions to model uncertainty of each source or combination of sources of uncertainty. Finally, a third topic is prediction[1] uncertainty. Given a simulation result, what can we say about the true value that we are trying to predict? How far is it likely to be from the simulated result?

1 We use here the term 'prediction' in the sense of calculating an output based on known inputs, rather than forecasting the future.

http://dx.doi.org/10.19103/AS.2019.0061.21

This corresponds to a confidence interval in statistics. Prediction uncertainty is not the same as model uncertainty. However, as we shall show, the two are closely related. In this chapter we focus on quantifying model uncertainty and prediction uncertainty and touch only briefly on sensitivity analysis.

Uncertainty is not specific to crop models. It is omnipresent in modeling and concerns modeling of all types, in all fields. Benke et al. (2008), talking of models in hydrology, say:

> 'Uncertainty is inherent in the structure of a predictive model, in the input data, and in the model parameters due to calibration difficulty. The advantage of uncertainty analysis is that it provides methodologies that can add value to conventional risk analysis by providing more information about the outputs of a predictive model, and identifies components of the model where uncertainties can be decreased. For an analyst, model user, or policy maker, uncertainty analysis also has the advantage of providing an error bound and confidence level on the output.'

Onatski and Williams (2002), discussing modeling in economics, say 'Uncertainty is pervasive in economics, and this uncertainty must be faced continually by policy makers. In this paper we propose empirical methods to specify and measure uncertainty associated with economic models, and we study the effects of uncertainty on monetary policy decisions.' In The Guidance Note for Lead Authors of the IPCC Fifth Assessment Report on Consistent Treatment of Uncertainties, Mastrandrea et al. (2010) state:

> 'The AR5 will rely on two metrics for communicating the degree of certainty in key findings: •Confidence in the validity of a finding, based on the type, amount, quality, and consistency of evidence (e.g. mechanistic understanding, theory, data, models, expert judgment) and the degree of agreement. Confidence is expressed qualitatively. •Quantified measures of uncertainty in a finding expressed probabilistically (based on statistical analysis of observations or model results, or expert judgment).'

As the above quotations indicate, uncertainty and sensitivity analysis are very important aspects of modeling. They are necessary for the modeler, in order to evaluate his model, to compare with other models, and to understand the contributions of different sources of uncertainty. They are also essential for the model user, who requires information on how much confidence he should have in the simulated results.

There have been numerous studies devoted to model uncertainty in crop models, but for the most part they focus on a single, or perhaps two, sources of uncertainty. Furthermore, there are serious questions as to the methodologies used to estimate model uncertainty. A major challenge going forward is to improve estimation of model uncertainty, taking into account all major sources of uncertainty. A second challenge is to better integrate estimates of model uncertainty and prediction uncertainty. They are currently usually treated as separate topics, but this ignores the close relationship between them and the information that model uncertainty can provide about prediction uncertainty. A third challenge is to better identify the ways to reduce prediction uncertainty and to better integrate model improvement with estimation of prediction uncertainty so that progress can be quantitatively evaluated.

2 Model uncertainty

2.1 Introduction

When a crop model is used to simulate for a single field, the simulated results are completely determined by the model equations, the values of the model parameters and the values of the inputs used in the model. All of those can be uncertain. Model uncertainty is the distribution of simulated values that results from those uncertainties. Note that model uncertainty is specific to each particular context where the model is used.

Figure 1 illustrates how model uncertainty can be displayed as a probability distribution. A simple summary of a distribution is the mean and the standard deviation of that distribution. The mean of the distribution of simulated values would normally be used as the model prediction, if a single simulated value is required. The standard deviation is the value one would give to summarize uncertainty.

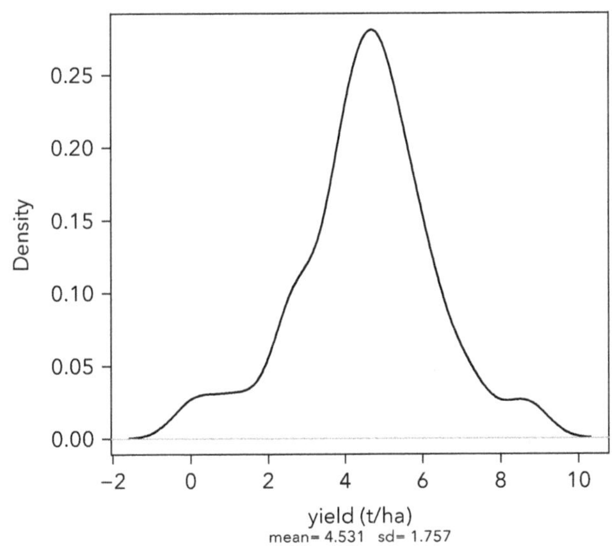

Figure 1 Model uncertainty for yield prediction (artificial example), presented as a probability density.

We use the notation $f(X;\theta)$ to indicate a result simulated by a model, where f represents the equations, X are the inputs and θ are the parameters. When we want to emphasize that a quantity is estimated rather than known exactly, we use the hat notation. Thus $\hat{f}(\hat{X};\hat{\theta})$ is the value simulated by a model, acknowledging that the equations are only approximately known ($\hat{f}$), that input values are only known approximately ($\hat{X}$) and that parameters are also only approximate ($\hat{\theta}$). In the following sections, we first consider each source of uncertainty individually and then discuss overall model uncertainty.

2.2 Model structure uncertainty

By model structure we mean the model equations plus the parameters that are fixed (not estimated by calibration). Crop models consist of coupled equations representing the main processes that affect crop growth and development and soil dynamics. For many or even most of the processes in a crop model, there is more than a single equation, and more than a single value for any fixed parameter, that is consistent with our knowledge about that process. The collection of plausible equations plus fixed parameter values gives the uncertainty in model structure. The fact that multiple crop models exist for all of the major crops is a clear indication that there is uncertainty as to model structure.

Uncertainty due to model structure is probably the hardest source of uncertainty to estimate. Developing a single crop model is a major undertaking; it is unrealistic to suppose that an individual modeling group could produce multiple crop models, representing the range of plausible models. This explains why, until quite recently, structure uncertainty of crop models was largely ignored. This has now changed dramatically, thanks in large part to the Agricultural Modeling Intercomparison and Improvement Project (AgMIP), which proved the usefulness of multi-model ensemble (MME) studies. In these studies, multiple modeling groups, each with their own model, participate in the same simulation exercise. The variability between models is an indication of structure uncertainty. Case study 5.1 illustrates the use of an MME, with 25 different models, to approximate model structure uncertainty.

Model structure uncertainty as estimated by using an MME is a collection of simulated values rather than a continuous distribution function as in Fig. 1. This estimated uncertainty could be displayed as a histogram, a box and whiskers diagram or an empirical cumulative distribution (Fig. 2). Another possibility would be to simply display each of the 25 simulated values, as in case study 5.1. The standard deviation of the simulated values is often used as a simple summary of the uncertainty information.

The use of MMEs has allowed the modeling community to evaluate, in a number of cases, model structure uncertainty, which is generally quite

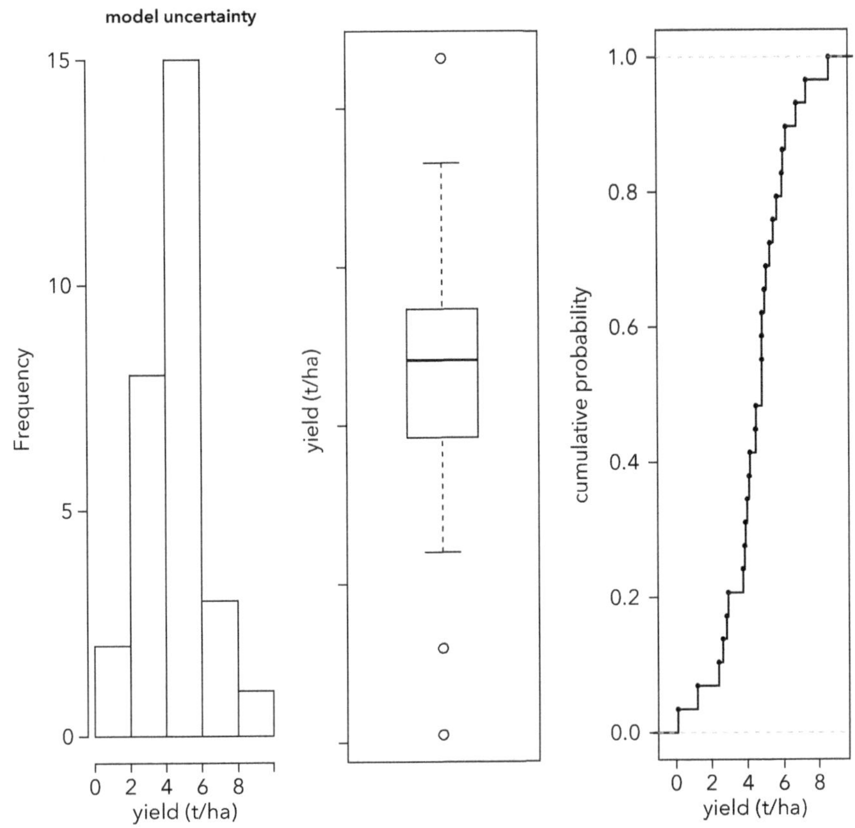

Figure 2 Three ways of representing estimated model structure uncertainty: as a histogram (left), box and whiskers diagram (middle) or as a cumulative distribution function (right). Data represent 25 different models. The mean and standard deviation of the simulated values are respectively 4.53 and 1.76 t/ha. (The simulations are for planting date 9 without supplemental heating, termed environment '9C', of case study 5.1).

large. For example, structural uncertainty, as summarized by the coefficient of variation of the values simulated by the different models, ranges from a low of 23% up to 142% of the mean among the 15 different environments of case study 5.1.

The important underlying assumption in MME studies is that the different models in the study represent a simple random sample from the distribution of plausible models. In practice, however, the MMEs are 'ensembles of opportunity,' signifying that participation is open to any modeling group that wishes to participate, subject only to very minimal conditions. As a result, the notion of random sample is difficult to justify. In the case of climate model ensembles, Knutti (2010) says 'The problem in the case of CMIP3 is that the sample of models is neither random nor systematic.' Wallach et al. (2016b) suggest that the same difficulties apply to crop model MMEs.

One potential problem of this 'all candidates are welcome' approach is that the ensemble might include models that most researchers would not consider acceptable for that application. If the ensemble contains models outside the plausible range, this would unrealistically increase the estimate of structure uncertainty. An ad hoc solution to this problem is to eliminate outliers or to eliminate systematically the models with the smallest and largest simulated values. For example, Asseng et al. (2013) eliminate the lower and upper 10% of simulated values. A different problem, which would tend on the contrary to unrealistically decrease estimated structure uncertainty, is that there may be several related models, for example models derived from the same precursor model. If the simulated values are closely related, this will lead to unrealistically low estimates of model uncertainty, because the variance of simulated values will be smaller than that of a random independent sample. However, it is difficult to quantify model relatedness. It has been found that even the same model, with different parameter values, can give very different simulated results, so relatedness is not just due to similar equations. One suggestion has been to base the measure of model relatedness on the covariance of simulated results (Bishop and Abramowitz, 2013).

When models are used for extrapolation outside the range of conditions where the models were tested, there is the possibility that new unanticipated mechanisms that are not included in any of the current models become important. Some of these mechanisms might be recognized ('known unknowns') while others are unsuspected ('unknown unknowns'). It is important to remember that model structure uncertainty refers to our current knowledge. Thus model structure uncertainty, defined as the distribution of values simulated by plausible models according to current knowledge, does not necessarily include the true response of the system.

If interest is in uncertainty of the full crop model, including growth, development, soil processes etc., then it is hard to sample that uncertainty other than by looking at models developed by different modeling groups. However, if interest is in uncertainty for specific processes in a crop model, it might be possible for a single group to propose a range of models that include the most important options. For example, Ramirez-Villegas et al. (2017) compare two different versions of a regional crop model, where biomass production is either limited just by transpiration or is both transpiration-limited and radiation-limited. Wang et al. (2017) propose four different functions for modeling the effect of temperature on crop phenological development and four other functions for modeling the effect of temperature on biomass growth.

In MME studies, it has often been observed that the median or mean of the simulated values (respectively noted e-median and e-mean) are good predictors, often better than the best individual model. This suggests that predictions should perhaps be based on e-mean or e-median, and this has been done in some cases. It is important then to estimate the structure uncertainty of these predictors.

Assuming that the models of the MME are chosen at random from the distribution of plausible models, the variance of e-mean is the variance of the simulations by plausible models divided by the size of the ensemble:

$$\text{var}\left[(1/n)\sum_{i=1}^{n}\hat{f}_i(X;\theta)\middle|X\right] = (1/n)\text{var}\left[\hat{f}(X;\theta)\middle|X\right] \quad (1)$$

Equation (1) shows that model structure uncertainty of e-mean, as measured by variance (the left-hand side of the equation) is smaller by a factor 1/(ensemble size) than the model uncertainty when one chooses a single model at random. In fact, model uncertainty of e-mean would tend to zero if we could increase the size of the ensemble indefinitely. (Remember however that model uncertainty is not the same as prediction uncertainty, as detailed below.) By $n = 5$, variance of e-mean is only 20% as large as the variance when choosing an individual model at random. Thus averaging over models is an effective way to reduce model uncertainty.

Equation (1) only applies if models are chosen at random for the MME. If for example models are added to the MME in order of past performance (best models first), the behavior of uncertainty with the size of the ensemble can be quite different than for adding models at random (Wallach et al., 2018). For climate models, it has also been suggested that models should not have equal weight in the ensemble, but should be weighted according to their agreement with past data, or to their agreement with the other models in the ensemble, or some combination of the two (Giorgi and Mearns, 2002). This has not yet been tested for crop models. Using weightings would also change the way model uncertainty varies with ensemble size.

2.3 Input uncertainty

The input variables of a crop model are those variables that need to be measured or estimated for each field. Input variables for crop models usually include daily weather variables (maximum and minimum temperature, rainfall, solar radiation etc.), soil characteristics (soil water content at field capacity and permanent wilting point, runoff number etc.), management (crop, variety, sowing date, depth and density, irrigation dates and amounts, fertilizer applications with type, dates and amounts etc.), and initial conditions at sowing or some other date (initial soil water, initial soil mineral N etc.). Many of these variables are difficult to measure and are likely to have appreciable errors. According to Aggarwal (1995): 'Deterministic crop growth models require several inputs relating to crop/variety, soil physical properties, weather and crop management. The input values used could be significantly uncertain due to random and systematic measurement errors and spatial and temporal variation observed in many of these inputs.' Several methods have been used for estimating uncertainty in explanatory variables. For example, in the study by Aggarwal (1995), 'The uncertainty in each input was represented by a statistical distribution of values based on literature review, actual measurement and subjective expert judgement.'

The most important uncertainties in inputs that need to be taken into account will be different depending on the proposed use of the model. Consider the situation where a crop model is intended for aiding management decisions in specific fields. It is important then that the input variables used in the model be as close as possible to the true values for the field in question. Uncertainty can arise from sampling error (one only samples a few points in the field) and from measurement error. For example, consider a crop model that requires as input soil water content at field capacity, averaged over the field. The 'true' value of field capacity is then the true average over the field. If there is spatial variability, then sampling the field at a limited number of points will provide only an approximation to the true average. In addition, measuring field capacity is difficult, so the measured field capacity at a specific point is only approximate.

An example of the effect of input uncertainty is shown in Fig. 6 (Case study 5.2), concerning irrigation management of a vineyard. The figure shows fraction available soil water over time calculated by a vineyard model, and the threshold below which irrigation is recommended. In the example, if uncertainty is ignored, the conclusion is that irrigation is not required. However, when input uncertainty is taken into account, there is a certain probability that irrigation is necessary. In this case uncertainty in the inputs translates into uncertainty in a management decision. If the risk of not irrigating when in fact irrigation is required is deemed too large, a practical conclusion could be that a greater investment in measuring inputs is worthwhile.

A different context is the use of crop models for climate change impact assessment. Here one is not interested in impact for some specific field, but rather for a 'typical' field. In this case, uncertainty in the field characteristics is usually not a major issue. On the other hand, the uncertainty in future climate, which is also an input, is crucial. The common way of estimating uncertainty in future climate is to use multiple climate models. The variability between simulations of the crop model, using different climate models to provide input, is an estimate of the model uncertainty resulting from climate uncertainty. An additional complication is that global climate models need to be downscaled for use at a local level, and several different downscaling techniques have been proposed. The variability between them adds to the uncertainty of future climate (Cammarano et al., 2017).

Climate forecasts depend on the radiative forcing scenario (representative concentration pathway or RCP) that is assumed. One should however be careful as to how to take uncertainty in RCP into account. It has been emphasized that 'The RCPs should not be interpreted as forecasts' (van Vuuren et al., 2011). They are simply alternative possible futures and do not have an associated statistical probability of occurrence. Thus it seems most reasonable to treat each RCP separately and not to include the differences between RCPs in the estimation of uncertainty.

2.4 Parameter uncertainty

Parameters in crop models are typically treated in one of two different ways. Many (usually most) of the model parameters are estimated based on literature results, that is from reports of dedicated experiments, from past model calibrations or from expert opinion that interprets existing information. They are then treated as fixed values. They are not changed when the model is applied to some new context. The remaining parameters are estimated, or re-estimated, by calibration, using data from each new context of interest.

Consider first those parameters obtained from the literature. It has been shown that for many parameters, the values found in the literature have a large variability (Metselaar, 1999). However, it may be very difficult to obtain realistic uncertainty information for each of the model parameters, especially since crop models often have no detailed record of how all the parameters were obtained. A common approach is to consider in detail only a subset of the parameters obtained from the literature. One sets rather large uncertainty bounds on those parameters, in order to guard against over-confidence, and treats the uncertainties as independent as in Tao et al. (2018). This will tend to exaggerate uncertainty in the simulated values, because it does not take into account the fact that in practice models are tested against data, and parameter combinations that give unrealistic results have been eliminated. Acting in the other direction, to reduce uncertainty, is the fact that uncertainty in those parameters not considered in detail is ignored. Overall, there are serious questions as to how to estimate the uncertainty in parameters which are not calibrated.

A more appealing approach may be to consider all the fixed parameters (those not estimated by calibration) as being part of model structure. In that case, estimates of uncertainty in model structure account for both uncertainties in the model equations and in the fixed parameter values.

Consider now the parameters that are estimated by calibration, that is, by fitting the model to field data. These parameters have some distribution as a result of the calibration, and that results in some distribution of simulated values, that is in model uncertainty. Estimation of parameter uncertainty due to calibration can be based on either of the two paradigms of statistical reasoning, frequentist and Bayesian. In both cases, there is extensive literature on estimation of parameter uncertainty.

In a frequentist approach, the uncertainty of the calibrated parameters results from the fact that the data used for calibration are a sample of data. Different data could have been chosen, which would lead to different values for the parameters. This distribution of parameter values is the parameter uncertainty. The statistical literature shows how to estimate this distribution, and in many cases the software for calibration will output the parameter uncertainty.

It is important to keep in mind that standard uncertainty calculations are based on certain assumptions that are often not satisfied for crop models. One basic assumption is that the model is correctly specified, meaning that it has the correct functional dependence on the input variables. Wallach (2011) however argues that crop models are very probably misspecified, because the process equations are only approximate, because they involve fixed parameters which are only approximate and because the coupling of process equations leads to misspecification. In that case the standard formulas for estimating the parameter uncertainty are incorrect. White (1981) proposes corrected estimates, but these have not been used for crop models.

Even if model misspecification is assumed to be slight, there is the problem that the data for crop model calibration often have a complex structure. For example, one might have measurements of several different outputs in the same field (yield, final biomass, LAI etc.) with multiple measurements over time of some of the variables. The model errors for the different measurements in the same field need not be independent. It is known in principle how to correctly estimate the parameters and parameter uncertainty in such cases, for example using generalized least squares, but this is in general not done for crop models. Treating all the model errors as independent when in fact they are correlated will lead in general to an underestimation of uncertainty, since in fact there is less independent data than the number of measurements implies. There is still need for progress in crop model calibration and estimation of parameter uncertainties (Seidel et al., 2018). One promising approach is to explore advantages and disadvantages of various estimation approaches and data structures using simulated data, as in Montesino-San Martin et al. (2018).

The other paradigm in statistics is the Bayesian approach. In this approach, parameters are estimated by combining prior information with the field data. Prior information refers to what is known before taking into account the data used for calibration. Often, the prior information is simply a minimum and maximum possible value for the parameter, based on the literature, and it is assumed that all values between the minimum and maximum are equally likely.

The result of a Bayesian calibration is what is called the posterior distribution, which is directly the estimate of parameter uncertainty. The interpretation here is somewhat different than for the frequentist approach; the uncertainty now represents how much we know about the parameter values, taking into account both the prior information and the information embodied in the field data. There have been several studies where crop model parameters and their uncertainty are estimated using a Bayesian approach (Makowski et al., 2002; Iizumi et al., 2009).

In a Bayesian approach, in principle any number of parameters can be estimated. One does not need to ensure that the data are sufficient to estimate all the calibrated parameters. If the data give information about a parameter, then after calibration the parameter value will be a compromise between the prior information and the value that gives a good fit to the data. If the data provide little or no information about a parameter, then the posterior distribution of the parameter will be close to or the same as the prior distribution. For example if the prior distribution is a uniform probability between a minimum and maximum value, then the posterior distribution will be the same or nearly the same. Furthermore, defining the prior distribution is not really a problem with a Bayesian approach. The prior distribution represents what the user knows about the parameters before invoking the field data, so whatever the user knows, that is the prior information. It might seem then that a Bayesian approach could be used to estimate all the model parameters, and their uncertainties. In fact, this is not done no doubt because of the computational problems in treating a very large number of parameters. It should also be noted that a Bayesian approach, like the frequentist approach, is based on certain assumptions. A simple common assumption is that all model errors are independent and have a normal distribution with the same variance. This is very often not the case, so that Bayesian uncertainties, like frequentist uncertainties, may be suspect.

2.5 Propagating uncertainty from various sources

In the previous sections we have considered separately model uncertainty due to structure uncertainty, due to uncertainty in the values of the model inputs and due to uncertainty in the model parameters. It is of interest to consider the uncertainty due to one particular source, but it is also of interest to estimate total model uncertainty, due to all sources. This is straightforward, once one has estimates of the individual uncertainties.

Suppose that one has a sample of models, assumed representative of the distribution of model structures. Let $\hat{f}_i$ for $i = 1,\ldots,I$ be the sample of model structures. Assume that for each model, one can generate a sample of parameter values for that model, representing the uncertainty in the calibrated parameters. Let $\hat{\theta}_{ik}$, $k = 1,\ldots,K$, be the sample of parameter vectors for model

i. Assume finally that one can generate a sample of possible input values, representing the uncertainty in the inputs. Let $\hat{X}_j$, $j = 1,...,J$ be the sample of approximations to the inputs. Then one calculates $\hat{f}_i(\hat{X}_j;\hat{\theta}_{ik})$ for each model $i = 1,...,I$, for each approximation to the inputs $j = 1,...,J$ and for each parameter vector $k = 1,...,K$. The distribution of $\hat{f}_i(\hat{X}_j;\hat{\theta}_{ik})$ is an estimate of overall model uncertainty.

While there have been multiple studies of model uncertainty due to a single source of uncertainty, there have been relatively few studies that have considered multiple sources of uncertainty. Asseng et al. (2013) considered uncertainty due to model structure and due to future climate, and found that model structure was the more important cause of uncertainty. Wallach et al. (2017) consider structure and parameter uncertainty. Tao et al. (2018) considered uncertainty due to model structure, parameters and climate projections.

2.6 Quantifying the contribution of each source to overall model uncertainty

If overall model uncertainty is estimated as described above, based on uncertainty in multiple sources, then in general one also wants to quantify the contribution of each source. This is the subject of sensitivity analysis, which treats the problem of quantifying the separate contributions to uncertainty. There is important literature on methods of sensitivity analysis (Saltelli et al., 2000).

Here we consider only one specific method of sensitivity analysis, particularly adapted to the case of model uncertainty. We consider three sources of uncertainty, namely model structure (including parameters not estimated by calibration), model inputs (all inputs are treated together) and model calibrated parameters (parameters are specific to each model, but for a given model all parameters are treated together). It is natural to treat all calibrated parameters together, since statistical procedures of calibration will output a joint distribution for the parameters. The input variables are also treated together as a single uncertain vector. In this case, however, one might want to evaluate the separate contributions of different input variables in a second step.

We assume that we have the results of a simulation experiment as described above which has values of $\hat{f}_i(\hat{X}_j;\hat{\theta}_{ik})$ for $i = 1,...,I$, $j = 1,...,J$, $k = 1,...,K$. The total variance of $\hat{f}(\hat{X};\hat{\theta})$ can be written as:

$$\text{var}(\hat{f}(\hat{X};\hat{\theta})) = \sigma_{\hat{f}}^2 + \sigma_{\hat{X}}^2 + \sigma_{\hat{f}\hat{X}}^2 + \sigma_{\hat{\theta}}^2 \quad (2)$$

where $\sigma_{\hat{f}}^2$, $\sigma_{\hat{X}}^2$, $\sigma_{\hat{\theta}}^2$ and $\sigma_{\hat{f}\hat{X}}^2$ are respectively the contributions of structure, input and parameter uncertainty and the contribution of structure–input interaction.

Those individual terms are easily calculated based on the simulation experiment, using standard analysis of variance, as shown below (Wallach et al., 2016a).

Let $m_{ijk} = \hat{f}_i(\hat{X}_j;\hat{\theta}_{ijk})$ be the simulated value for model i, input vector j and parameter vector k. A star indicates an average over the index at that position, so that m_{i*k} for example is the simulated value for structure i, parameter vector k, averaged over all the input values. First calculate the following four sums of squares:

$$SS_f = JK\sum_{i=1}^{I}\left(m_{i**} - m_{***}\right)^2 \tag{3}$$

$$SS_X = IK\sum_{j=1}^{J}\left(m_{*j*} - m_{***}\right)^2 \tag{4}$$

$$SS_{fX} = K\sum_{i=1}^{I}\sum_{j=1}^{J}\left(m_{ij*} - m_{i**} - m_{*j*} + m_{***}\right)^2 \tag{5}$$

$$SS_\theta = \sum_{i=1}^{I}\sum_{j=1}^{J}\sum_{k=1}^{K}\left(m_{ijk} - m_{ij*}\right)^2 \tag{6}$$

Then the contributions to overall variance are respectively:

$$\hat{\sigma}_\theta^2 = SS_\theta / \left[IJ(K-1)\right] \tag{7}$$

$$\hat{\sigma}_{fX}^2 = SS_{fX} / \left[K(I-1)(J-1)\right] - \hat{\sigma}_\theta^2 / K \tag{8}$$

$$\hat{\sigma}_X^2 = SS_X / \left[(J-1)/K\right] - \hat{\sigma}_{fX}^2 / I - \hat{\sigma}_\theta^2 / IK \tag{9}$$

$$\hat{\sigma}_f^2 = SS_f / \left[(I-1)JK\right] - \hat{\sigma}_{fX}^2 / J - \hat{\sigma}_\theta^2 / JK \tag{10}$$

3 Prediction uncertainty

Model developers and even more so model users are particularly interested in how well simulated results compare with what would be observed in reality. This is given by prediction uncertainty, which we now define. Suppose that for a given prediction, the prediction error is $\varepsilon(X) = [y - \hat{f}(\hat{X};\hat{\theta})]$ where y is the value that would be measured. (We assume that the measurements are without error. If there are errors in the measurements, this simply adds on the $\varepsilon(X)$). We imagine doing the prediction for different situations with the same inputs many times. There will then be some distribution of $\varepsilon(X)$. It is this distribution that is prediction uncertainty. Expressed in another way, prediction uncertainty describes, probabilistically, how far simulated values are likely to be from measured values. A good predictor will have $\varepsilon(X)$ that is centered around 0 (unbiased) with small variance, so that we know that the simulated value is close to the value that would be observed (Fig. 3, left). A poor model will have $\varepsilon(X)$ that is biased and /or has large variance; the simulated value will often be far from the value that would be observed (Fig. 3, right).

Model error can be written as the sum of two terms:

$$\begin{aligned}\varepsilon(X) &= \left[y - E(\hat{f}(\hat{X};\hat{\theta})) + E(\hat{f}(\hat{X};\hat{\theta})) - \hat{f}(\hat{X};\hat{\theta})\right] \\ &= \left[y - E(\hat{f}(\hat{X};\hat{\theta}))\right] + \left[E(\hat{f}(\hat{X};\hat{\theta})) - \hat{f}(\hat{X};\hat{\theta})\right] \\ &= \text{error of averaged model} + \text{model variability}\end{aligned} \tag{11}$$

The averaged model is obtained by averaging simulated values over structure, input and parameter uncertainties. This would be the average of the $\hat{f}_i(\hat{X}_j;\hat{\theta}_{ik})$ in the simulation experiment described above. The first term in eq. (11) is the error

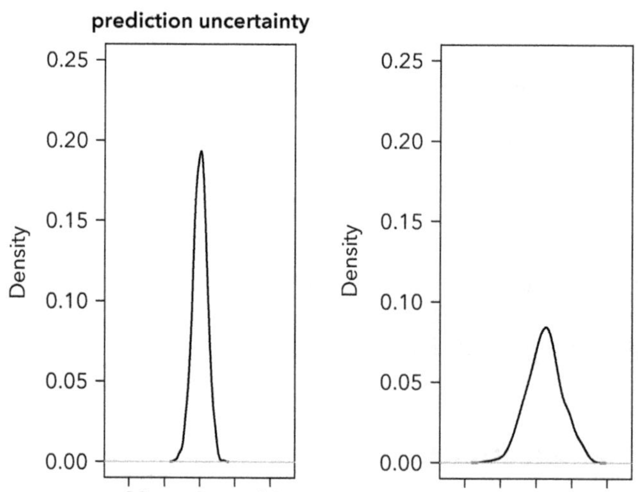

Figure 3 The distribution of model error for a relatively good predictor (left) and a relatively poor predictor (right). For the good predictor, the mean and standard deviation of error are relatively small. For the poor predictor, the mean and standard deviation of error are relatively large.

of the averaged model. The second term is the variability of the simulated values around their average value. The distribution of this term is model uncertainty except that here the mean is 0 (this is model uncertainty, centered at 0).

To summarize $\varepsilon(X)$ we can use the mean of $\varepsilon(X)^2$, called mean squared error of prediction (MSEP). For fixed inputs X, the criterion is:

$$\mathrm{MSEP}(X) = E(\varepsilon(X)^2) = E\left\{\left[\left(y - \hat{f}(\hat{X};\hat{\theta})\right)\right]^2\right\} \tag{12}$$

There are two alternative ways of interpreting eq. (12), depending on what is considered as random and what is considered as fixed (Wallach et al., 2016). One possibility is to treat the simulated values $\hat{f}(\hat{X};\hat{\theta})$ as fixed. That is, we ignore all uncertainties, and just estimate mean squared error for whatever equations, parameter values and estimates of the explanatory variables we have. In this case, the criterion is noted $\mathrm{MSEP}_{\mathrm{fixed}}(X)$. Estimation of $\mathrm{MSEP}_{\mathrm{fixed}}(X)$ is commonly referred to as model evaluation and is obviously of major interest since it concerns prediction error for the specific model to be used for prediction. It is estimated based on a comparison of simulated values with observed data (Wallach et al., 2019). However, it has two major drawbacks. First, because in general there is only limited data, one generally can only estimate an average value and not $\mathrm{MSEP}_{\mathrm{fixed}}(X)$ separately for each set of inputs. Secondly, this approach only estimates total prediction error; it gives no indication of the sources of error.

The alternative is to treat the simulated values in eq. (12) as random, as we did in considering model uncertainty. In this case the criterion is noted $\mathrm{MSEP}_{\mathrm{random}}(X)$, and it represents squared prediction error, averaged over the uncertainties of the model. Using the same expansion as in eq. (11) gives:

$$\begin{aligned}\mathrm{MSEP}_{\mathrm{random}}(X) &= E\left\{\left[y - E(\hat{f}(\hat{X};\hat{\theta})) + E(\hat{f}(\hat{X};\hat{\theta})) - \hat{f}(\hat{X};\hat{\theta})\right]^2\right\} \\ &= E\left\{\left[y - E(\hat{f}(\hat{X};\hat{\theta}))\right]^2\right\} + E\left\{\left[E(\hat{f}(\hat{X};\hat{\theta})) - \hat{f}(\hat{X};\hat{\theta})\right]^2\right\} \\ &= \text{squared error of averaged model} \\ &\quad + \text{model variance}\end{aligned} \tag{13}$$

This shows that $\mathrm{MSEP}_{\mathrm{random}}(X)$ is a sum of two terms. The first term is related to how well the averaged model (i.e. the simulations obtained by averaging over uncertainties) tracks the measurements. The second term is just the summary measure of model uncertainty. We thus see here how model uncertainty and prediction uncertainty are related. Model uncertainty contributes to prediction uncertainty, but in addition there is a term related to how well the average over model uncertainty predicts.

Estimation of the two contributions to $\mathrm{MSEP}_{\mathrm{random}}(X)$ is very different. The model variance term just depends on the model, not on measurements.

We discussed above how this can be estimated, for each input vector X, from a simulation experiment. Estimation of the squared error term is more challenging, because it does require measurements. In general one can only estimate an average value of the squared error, and not values specific to each set of inputs. Nevertheless, $MSEP_{random}(X)$ is different for each X because of the model variance term.

The criterion $MSEP_{random}(X)$ has important advantages compared to $MSEP_{fixed}$. First, $MSEP_{random}(X)$ is specific to each set of inputs X, that is to each prediction problem. This is important, since it indicates to what extent the model is adequate for specific predictions. Secondly, one can estimate the separate contributions of squared error of the averaged simulations, structure uncertainty, input uncertainty, parameter uncertainty and the interaction of structure and input uncertainty to $MSEP_{random}(X)$, and then target for improvement the most important.

3.1 Prediction uncertainty when predicting an average over time

Consider a target population of fields, where for each we want to predict an average over years, perhaps in order to compare management strategies. $MSEP_{random}(X)$ will be smaller for the average over years than for a single year, because there will be some cancellation of errors. We present here an approach for estimating how important this reduction will be.

To estimate how much of the error cancels out when averaging results over years for each field, we need a data set that includes not only multiple fields, but also results for the same field in more than one year. We then define a random effects model for error for field i, year j:

$$y_{ij} - E\left[\hat{f}(\hat{X}_{ij};\hat{\theta})\right] = \varepsilon_{ij}(X) = \mu + \alpha_i + \beta_j + \gamma_{ij} \quad (14)$$

where μ is overall average error (error averaged over all fields and years), α_i is the field effect (how different is field i from the average over fields), β_j is the year effect (how different is year j from the average over years) and γ_{ij} is the field-year interaction. It is assumed that $\alpha_i \sim N(0,\sigma_\alpha^2)$, $\beta_j \sim N(0,\sigma_\beta^2)$, $\gamma_{ij} \sim N(0,\sigma_\gamma^2)$ and that all the random effects are independent. Squared error for predicting an average over n years is then:

$$\text{squared error for average over years} = \mu^2 + \sigma_\alpha^2 + \sigma_\beta^2/n + \sigma_\tau^2/n \quad (15)$$

As the number of years in the average increases, the year and the interaction effects tend to cancel out. On the other hand, the mean error contribution and the field effect contribution do not decrease when averaging over time. One can estimate the values of μ, σ_α^2, σ_β^2 and σ_τ^2, for example using the lmer function of the lme4 package (Lindstrom and Bates, 1990) of the R statistical software package (R Core Team, 2012).

3.2 Prediction uncertainty of averages over models

We have seen that the model uncertainty of e-mean tends to zero as the size of the ensemble increases. However, this does not imply that the prediction uncertainty of e-mean tends to zero. We examine here how MSEP of e-mean varies with ensemble size. The treatment here is very similar to the case of averaging over years.

Once again, we use a random effects model. Here we express model error for model i in environment j using the following random effects model:

$$y_j - \hat{f}_i(\hat{X}_j;\hat{\theta}_i) = \mu + \alpha_i + \beta_j + \gamma_{ij} \tag{16}$$

where y_j is the observed output for environment j, μ is mean error (averaged over all models and environments), and α_i, β_j and γ_{ij} are respectively the model, environment and interaction effects. The usual assumptions are that $\alpha_i \sim N(0,\sigma_\alpha^2)$, $\beta_j \sim N(0,\sigma_\beta^2)$, $\gamma_{ij} \sim N(0,\sigma_\gamma^2)$ and that all the random effects are independent. It can then be shown (Wallach et al., 2018) that MSEP of e-mean based on n independent models is:

$$\text{MSEP}_{\text{e-mean}} = \mu^2 + \sigma_\beta^2 + \sigma_\alpha^2/n + \sigma_\gamma^2/n \tag{17}$$

The last two terms tend to zero as the size of the ensemble increases because model errors tend to cancel. That is the contribution of model uncertainty. The first two terms are the contribution of squared error of the averaged simulations, and they do not decrease as the ensemble size increases. Thus we can make the model uncertainty contribution to squared error tend to zero by having a very large MME, but this does not mean that MSEP tends to zero.

4 Reducing uncertainty

Prediction uncertainty, as quantified by $\text{MSEP}_{\text{random}}(X)$, can be reduced by reducing model uncertainty (the second term on the right-hand side of eq. 13) or by reducing the squared error of the averaged model (the first term on the right-hand side of eq. 13). The approach will be very different for the two contributions.

Consider first the contribution of model uncertainty, which results from structure uncertainty, input uncertainty and parameter uncertainty. Very generally, these contributions can be reduced by obtaining more data. For example, we saw that the variance associated with structure uncertainty of e-mean decreases with the number of models in the MME (eq. (1)). If a very large number of models is included in the MME, this uncertainty will tend to zero.

In general, input uncertainty can be reduced by more, or more accurate, measurements. The increasing availability of sensors offers the opportunity of markedly improving measurement of field characteristics. However, reduced variability usually is associated with increased cost. It is important to quantify the effect of uncertainty in explanatory variables in order to judge whether the increased cost is justified.

Parameter uncertainty can be reduced by obtaining more data for calibration. Current efforts to make crop data more fully available (Antle et al., 2017) could have an important effect on reducing parameter uncertainty.

On the other hand, reducing the squared error of the averaged model requires improving the models so that they better explain the variability in the population of interest. Some insight into what this requires can be obtained by decomposing the squared error of the averaged model as follows:

$$\begin{aligned} E\left\{\left[y - E(\hat{f}(\hat{X};\hat{\theta}))\right]^2\right\} &= E\left\{\left[y - E(y|X) + E(y|X) - E(\hat{f}(\hat{X};\hat{\theta}))\right]^2\right\} \\ &= E\left\{\left(y - E(y|X)\right)^2\right\} + \left[\left(E(y|X) - E(\hat{f}(\hat{X};\hat{\theta})\right)\right]^2 \\ &= \text{variability unexplained by X} \\ &\quad + \text{squared distance from best model} \end{aligned} \tag{18}$$

The first term on the right-hand side is the variability of the output of interest (for example, yield) within the target population when all the inputs X are fixed. This is never exactly zero, because a model never has as inputs every variable that affects a crop. If for example X does not include pest populations, but in fact the level of pest populations does affect the crop, there would be variability unexplained by X. In general, if inputs that have an important effect are not included in the model, then the first term will be large. If on the other hand X does include all the inputs that have a major role in causing differences between fields, the first term will be small.

For a given choice of input variables X, it is known theoretically that the best possible model is E(y|X). The second term in eq. 18 measures how close the actual model, averaged over uncertainties, is to this theoretical best model.

Combining eqs (13) and (18) shows that $MSEP_{random}(X)$ is the sum of three contributions, namely model uncertainty, variability unexplained by model inputs and squared distance of the averaged model from the best model. To these three components correspond three paths for reducing prediction uncertainty. The first contribution is from model uncertainty, and so the first path is to reduce model uncertainty. This can be done by using MMEs to reduce structure uncertainty, by improving or increasing measurements of inputs to reduce input uncertainty, or by obtaining more data for calibration to reduce parameter uncertainty. The second contribution is from variability unexplained by the model input variables. The corresponding path to reducing prediction uncertainty is to add additional input variables to the model. This should be restricted to input variables that are expected to be important in causing differences between fields, since additional input variables usually imply increased model complexity which may increase the third contribution to prediction uncertainty. This third contribution is the difference between the actual model and the best possible model. The corresponding path to reducing prediction uncertainty here is to better understand the soil-plant-atmosphere system and to translate that understanding into equations.

5 Case studies

5.1 MME study of effect of high temperatures on wheat (structure uncertainty)

A large MME study with 25 different models was conducted to simulate the results of the Hot Serial Cereal (HSC) experiment (Ottman et al., 2012; Asseng et al., 2015). The different treatments of this experiment concern a single location and cultivar but multiple planting dates and artificial heat treatments to create a very large gradient of growing season temperatures (Martre, 2018). Since temperature increase is one of the main components of climate change, it is important to evaluate to what extent crop models can simulate the effects of temperature, and what the principle sources of uncertainty are.

Figure 4 shows between-model variability for yield for the different treatments (Wallach et al., 2018). The only uncertainty that is driving the differences between models is structure uncertainty, where model structure includes parameter values.

The distribution of simulated results, that is the model uncertainty, can be summarized by the variance or coefficient of variation (CV). This is specific to each treatment (Table 1). The variability between models (i.e., the model structure uncertainty) is very appreciable, ranging from 23% to 142% of the mean value depending on the treatment. Other MME studies have also found large variability between models.

It is important to understand what the model uncertainty represents here. Suppose that one chooses to use some specific model for prediction. The uncertainty represents the variability in results, depending on which model is chosen. Stated in another way, this is the uncertainty of a single model chosen at random (i.e. all choices have equal probability) among plausible models.

Consider now prediction uncertainty, which is the sum of model variance, just discussed, and squared error of the averaged model (eq. 13). The averaged

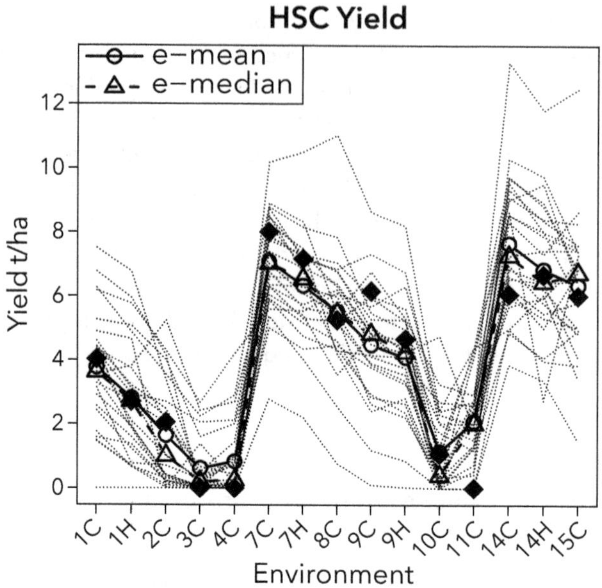

Figure 4 Simulation of yield in the Hot Serial Cereal experiment by 25 different wheat models. Each environment number corresponds to a different sowing date, either without ('C') or with ('H') supplementary heating. Solid diamonds are observed yields. Circles and triangles show respectively the mean and median of simulated results. Values simulated by the same model are connected by thin dotted lines.

model here is just the average over model structures, that is, e-mean. The squared error of e-mean averaged over the 15 environments is:

$$(1/15)\sum_{i=1}^{15}\left(y_i - \frac{\sum_{j=1}^{25} f_j(X_i;\theta_j)}{25}\right)^2 = 0.87(\text{t/ha})^2 \tag{19}$$

For all environments except one, the model variance term is larger than this squared error of e-mean. That is, if one chooses a model at random, it is in general model variance that is making the major contribution to $MSEP_{random}(X)$. To reduce prediction uncertainty, one should focus on reducing model uncertainty.

Rather than using a particular model, one could choose to use the model e-mean for prediction. What are the contributions of squared error of the averaged model and model variance in this case?

Averaging e-mean is the same as averaging over individual models. Thus the squared error of the averaged model is the same for e-mean as for the use

Table 1 The variance and coefficient of variation (CV) of yields simulated by 25 different wheat models for the Hot Serial Cereal Experiment

Environment	Model variance for yield (t/ha)^2	CV of simulated yields (sd of yield/ mean of yield)	Model variance of e-mean = model variance/25
1C	3.29	0.48	0.13
1H	3.49	0.67	0.14
2C	2.74	1.01	0.11
3C	0.76	1.42	0.03
4C	1.27	1.35	0.05
7C	2.54	0.23	0.10
7H	2.69	0.26	0.11
8C	3.64	0.35	0.15
9C	3.47	0.42	0.14
9H	3.13	0.43	0.13
10C	1.64	1.12	0.07
11C	1.69	0.62	0.07
14C	4.65	0.28	0.19
14H	4.72	0.32	0.19
15C	4.47	0.33	0.18

The model variance for e-mean, which is the variance among individual models divided by 25.

of an individual model. From eq. 19, the squared error is 0.87 (t/ha)^2. However, the model variance is now the model variance of the individual models divided by 25 (last column in Table 1). It is now squared error that makes the major contribution to prediction error. Using e-mean has substantially reduced model uncertainty, but does not reduce the squared error contribution to prediction error.

5.2 Irrigation decision support for vineyards (input uncertainty)

The model considered in this study estimates daily fraction transpirable soil water (FTSW) in vineyards in a Mediterranean region, as an aid to irrigation planning (Celette et al., 2010; Roux et al., 2014). A schematic diagram is shown in Fig. 5. The model is used by researchers, engineers of the French Vine and Wine Institute and by advisors (Pellegrino et al., 2006; Ripoche et al., 2011; Gaudin and Gary, 2012). Discussions with users revealed that they frequently rely on approximation procedures to obtain values of model inputs. For example, since use of moisture probes to evaluate soil water at bud-burst (when the model calculations begin) is time-consuming and expensive, initial soil water is often estimated by assuming that available soil water is 0 at the end

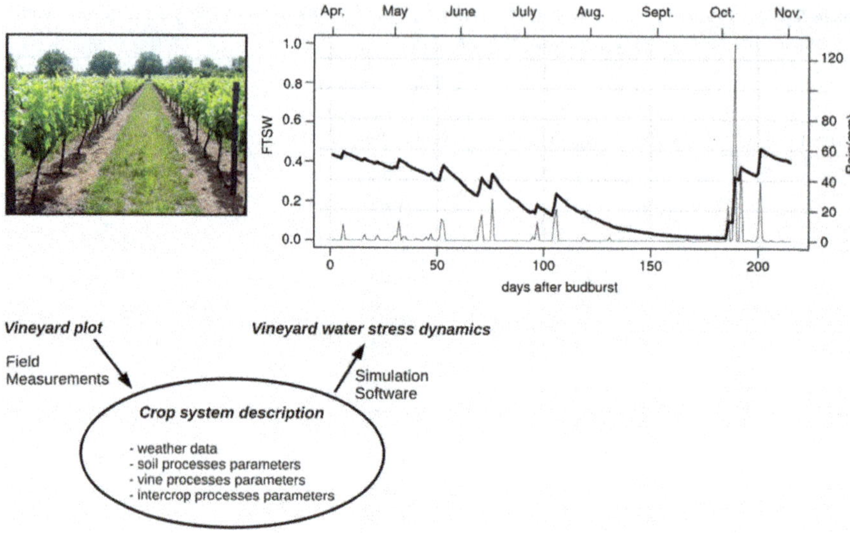

Figure 5 Schematic representation of the vineyard water stress model.

Table 2 Explanatory input variables in the vineyard water stress model, with best method of measurement and approximation method

Input	'Gold standard' (best method of measuring the input)	Approximate method of obtaining input
Daily weather	Weather station at vineyard	Close-by weather station
Curve number	Measure runoff	Estimate from soil group and ground cover
Maximum light interception	Detailed light interception model	Simple relation to plant vigor and height
Maximum transpirable water	Moisture probe after heavy rain and after prolonged drought	Simple relation to soil depth and texture
Fraction transpirable water at bud-burst	Moisture probe at bud-burst	Assume zero at end of summer, use model to estimate at next bud-burst

of summer, and then using the model to estimate soil water at following bud-burst. Those discussions also showed that users are interested in uncertainty information and are aware of the fact that approximating the inputs adds to the uncertainty in the simulations.

The inputs considered in this study are shown in Table 2, together with recommended best approaches to measuring them and simplified approximate approaches that are often used in practice.

For each input, a sample of approximate values was created, for the specific case of an experimental vineyard. For example, for daily weather, there were two weather stations at approximately 15 km from the field studied here. The weather data from the two stations were taken as a sample of size two of possible weather sequences. For approximating maximum transpirable soil water, there is a table that gives the value as a function of soil depth and soil texture. For each cell in the table, that is each combination of depth and texture, there is an amount of maximum water. It was assumed that in fact, the amount could be anywhere up to half way to the values in the adjacent cells. This distribution was sampled to provide a sample of approximate maximum transpirable soil water values. Model uncertainty was then estimated by running the model with input values sampled from the joint distribution of all uncertain inputs.

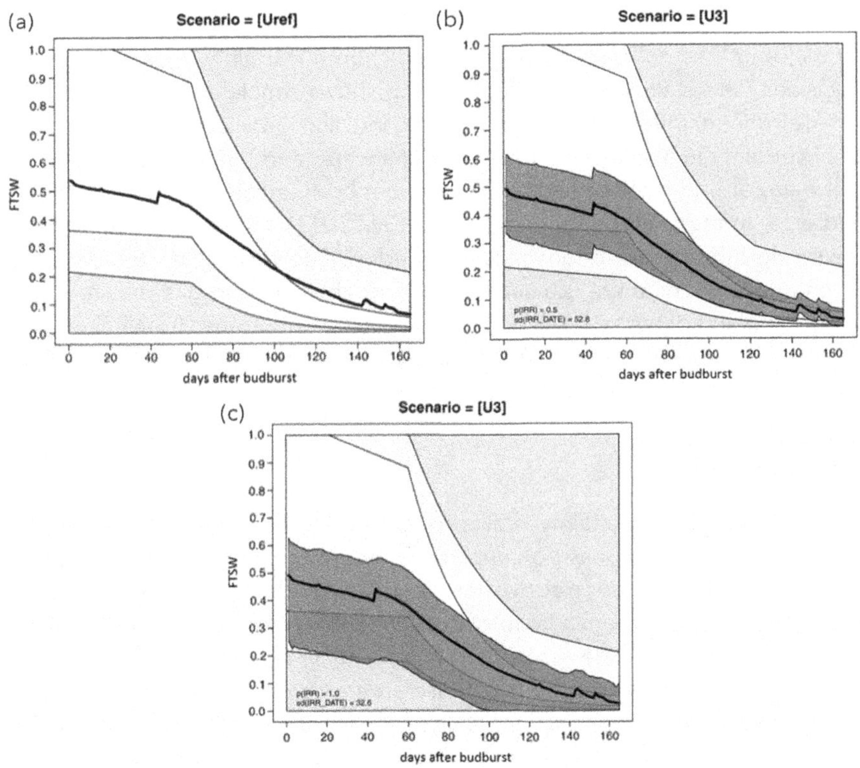

Figure 6 Graphs for irrigation recommendations. The lower limit of the white area is the threshold for irrigation; irrigation is recommended if simulated fraction transpirable soil water (FTSW) goes below that threshold. (a) FTSW ignoring all uncertainties. (b) 90% confidence interval on FTSW taking into account uncertainty in maximum transpirable soil water and in available water at bud-burst. (c) 90% confidence interval on FTSW taking into account errors in inputs as in (b) and also error of averaged model.

For the experimental vineyard, the gold standard values are available for all the inputs except daily weather, and there are in-season measurements of soil moisture. The simulated values using the gold standard inputs were compared to the measured values, and the differences were taken as a sample from the distribution of the error of average simulations (eq. (13)).

Various scenarios were studied, where in each scenario some explanatory variables are estimated, while the others are obtained using the gold standard. For each scenario, a distribution of a fraction of transpirable soil water on each day d, $FTSW(d,\hat{X};\theta)$, was obtained by redoing the following steps 500 times:

1 Draw a value from the distribution of $\hat{X}$ (the uncertain inputs).
2 Run the model to obtain the daily simulated values $FTSW(d,\hat{X};\theta)$.
3 At each daily time step, draw a value b of error of average simulations from its distribution and calculate $FTSW(d,\hat{X};\theta)+b$.

Of major interest to users is how the prediction uncertainty impacts irrigation decisions. This is illustrated in Fig. 6 which shows simulated FTSW over time when all uncertainties are ignored, or taking into account model uncertainty due to uncertainty in maximum transpirable soil water and in fraction transpirable soil water at bud-burst or taking into account both uncertainty in those inputs and error in the averaged model (Roux et al., 2014). If uncertainty is ignored, the results indicate that irrigation is not needed. If however all uncertainties are taken into account, there is a substantial probability that irrigation is required toward the end of the season. Much of this uncertainty comes from uncertainty in the inputs, which could be reduced by adopting the gold standard measurement approach for those variables.

6 Future trends

Evaluating model uncertainty is acknowledged to be an important aspect of modeling, and the emphasis on uncertainty information will no doubt increase in the future. The discussion above has identified problems related to evaluating model uncertainty, in particular related to the choice of models in MMEs and to model calibration and the associated uncertainty in the estimated parameters. We can expect progress in both of these areas. Another likely trend is toward more comprehensive uncertainty analysis, where two or all three sources of uncertainty are considered. Going further, it is important to evaluate prediction uncertainty. Currently this usually ignores model uncertainty, but in the future the estimation of prediction uncertainty may often concern the model with its uncertainty.

We have seen that model and prediction uncertainty depend strongly on what precisely is simulated. Predicting averages in particular can have an important effect in decreasing uncertainty. Also, different methods of estimating input variables have different associated uncertainties. In the future, more attention will be paid to the details of the prediction situation, and uncertainty analysis will be tailored to that situation.

Currently, uncertainty studies require an important effort and so are not done systematically. Software to automate uncertainty analysis for crop models would make uncertainty analysis much more available. Ideally, it could be incorporated into the model software, so that model output would automatically include uncertainty information.

Reducing uncertainty obviously will be a priority in future modeling work, guided by the results of uncertainty and sensitivity analysis which identify the major sources of prediction uncertainty. Since uncertainty can be quite different depending on the prediction situation, the path to reducing uncertainty can also be different for different situations. Thus reducing uncertainty should not be considered as the search for the 'universal best' model, but rather as the search for the best model for each particular situation.

The squared error contribution to prediction uncertainty can be reduced by adding important input variables (for example, based on information from new sensors) and/or by improving the model equations to make them better descriptors of the processes that are modeled. There are no easy guidelines here; progress will require detailed research. However, prediction uncertainty also depends on model uncertainty, and here there are clear recommendations as to how model uncertainty can be reduced. Structure uncertainty can be reduced by using MMEs. Parameter uncertainty can be reduced by obtaining more data for calibration, and by better calibration procedures. Uncertainty in input explanatory variables can be reduced by additional and more accurate measurements of those variables.

7 Conclusion

Model uncertainty measures the distribution in model simulated values due to uncertainty in model structure, in model parameters and in approximations to the model inputs. Uncertainty in model structure can be estimated using MMEs. Statistical calibration can provide estimates of parameter uncertainty.

Prediction uncertainty refers to the distribution of model error. It indicates how far simulated values are likely to be from the true values. It is often prediction uncertainty, rather than model uncertainty that is of major interest to the end user, since decisions are made based on predicted values. Prediction uncertainty can be summarized by the MSEP. This can be thought of in two different ways. First, one can ignore model uncertainty and estimate MSEP for the specific model one has. Secondly, one can take into account model uncertainty. In that case MSEP is the sum of two terms, squared error of the averaged model and model variance, which measures model uncertainty. This shows how model uncertainty and prediction uncertainty are related.

Prediction uncertainty can be reduced through reduction of model uncertainty or reduction of squared error of the average model. Model uncertainty can generally be reduced by averaging over more models, increasing the number of measurements of inputs or increasing the amount of data used for calibration. Reduction of squared error requires improving the model as an approximation to the real world. Two possible paths are through additional input variables or better equations and parameter values.

8 Where to look for further information

The AgMIP Project (http://www.agmip.org/) is a major modeling initiative, open to all, where uncertainty is an important theme. All of the modeling teams in AgMIP are interested in uncertainty, and in addition there is an uncertainty cross-cutting theme.

The book *Working with Dynamic Crop Models* 3rd ed. has a chapter devoted to uncertainty and sensitivity analysis. Several more advanced aspects of uncertainty and sensitivity analysis are covered in other chapters of the book. Specifically for sensitivity analysis, the book by (Saltelli et al., 2000) is an important reference.

All scientific journals that accept modeling studies are likely to include uncertainty and sensitivity studies. A good overview of uncertainty is given in Refsgaard et al. (2007). A compilation of papers on uncertainty specifically in crop and pasture models can be found in a recent special issue of the *European Journal of Agronomy* (Wallach and Thorburn, 2017).

9 References

Aggarwal, P. K. 1995. Uncertainties in crop, soil and weather inputs used in growth models: implications for simulated outputs and their applications. *Agricultural Systems* 48(3), 361–84. doi:10.1016/0308-521X(94)00018-M.

Antle, J. M., Jones, J. W. and Rosenzweig, C. E. 2017. Next generation agricultural system data, models and knowledge products: introduction. *Agricultural Systems* 155, 186–90. doi:10.1016/j.agsy.2016.09.003.

Asseng, S., Ewert, F., Rosenzweig, C., Jones, J. W., Hatfield, J. L., Ruane, A. C., Boote, K. J., Thorburn, P. J., Rötter, R. P., Cammarano, D., Brisson, N., Basso, B., Martre, P., Aggarwal, P. K., Angulo, C., Bertuzzi, P., Biernath, C., Challinor, A. J., Doltra, J., Gayler, S., Goldberg, R., Grant, R., Heng, L., Hooker, J., Hunt, L. A., Ingwersen, J., Izaurralde, R. C., Kersebaum, K. C., Müller, C., Naresh Kumar, S., Nendel, C., O'Leary, G., Olesen, J. E., Osborne, T. M., Palosuo, T., Priesack, E., Ripoche, D., Semenov, M. A., Shcherbak, I., Steduto, P., Stöckle, C., Stratonovitch, P., Streck, T., Supit, I., Tao, F., Travasso, M.,

Waha, K., Wallach, D., White, J. W., Williams, J. R. and Wolf, J. 2013. Uncertainty in simulating wheat yields under climate change. *Nature Climate Change* 3(9), 827–32. doi:10.1038/nclimate1916.

Asseng, S., Ewert, F., Martre, P., Rötter, R. P., Lobell, D. B., Cammarano, D., Kimball, B. A., Ottman, M. J., Wall, G. W., White, J. W., Reynolds, M. P., Alderman, P. D., Prasad, P. V. V., Aggarwal, P. K., Anothai, J., Basso, B., Biernath, C., Challinor, A. J., De Sanctis, G., Doltra, J., Fereres, E., Garcia-Vila, M., Gayler, S., Hoogenboom, G., Hunt, L. A., Izaurralde, R. C., Jabloun, M., Jones, C. D., Kersebaum, K. C., Koehler, A.-K., Müller, C., Naresh Kumar, S., Nendel, C., O'Leary, G., Olesen, J. E., Palosuo, T., Priesack, E., Eyshi Rezaei, E., Ruane, A. C., Semenov, M. A., Shcherbak, I., Stöckle, C., Stratonovitch, P., Streck, T., Supit, I., Tao, F., Thorburn, P. J., Waha, K., Wang, E., Wallach, D., Wolf, J., Zhao, Z. and Zhu, Y. 2015. Rising temperatures reduce global wheat production. *Nature Climate Change* 5(2), 143–7. doi:10.1038/nclimate2470.

Benke, K. K., Lowell, K. E. and Hamilton, A. J. 2008. Parameter uncertainty, sensitivity analysis and prediction error in a water-balance hydrological model. *Mathematical and Computer Modelling* 47(11–12), 1134–49. doi:10.1016/J.MCM.2007.05.017.

Bishop, C. H. and Abramowitz, G. 2013. Climate model dependence and the replicate Earth paradigm. *Climate Dynamics* 41(3–4), 885–900. doi:10.1007/s00382-012-1610-y.

Cammarano, D., Rivington, M., Matthews, K. B., Miller, D. G. and Bellocchi, G. 2017. Implications of climate model biases and downscaling on crop model simulated climate change impacts. *European Journal of Agronomy* 88, 63–75. doi:10.1016/j.eja.2016.05.012.

Celette, F., Ripoche, A. and Gary, C. 2010. WaLIS–a simple model to simulate water partitioning in a crop association: the example of an intercropped vineyard. *Agricultural Water Management* 97(11), 1749–59. doi:10.1016/j.agwat.2010.06.008.

Gaudin, R. and Gary, C. 2012. Model-based evaluation of irrigation needs in Mediterranean vineyards. *Irrigation Science* 30(5), 449–59. doi:10.1007/s00271-012-0349-x.

Giorgi, F. and Mearns, L. O. 2002. Calculation of average, uncertainty range, and reliability of regional climate changes from AOGCM simulations via the 'reliability ensemble averaging' (REA) method. *Journal of Climate* 15(10), 1141–58. doi:10.1175/1520-0442(2002)015.

Iizumi, T., Yokozawa, M. and Nishimori, M. 2009. Parameter estimation and uncertainty analysis of a large-scale crop model for paddy rice: application of a Bayesian approach. *Agricultural and Forest Meteorology* 149(2), 333–48. doi:10.1016/j.agrformet.2008.08.015.

Knutti, R. 2010. The end of model democracy? *Climatic Change* 102(3–4), 395–404. doi:10.1007/s10584-010-9800-2.

Lindstrom, M. L. and Bates, D. M. 1990. Nonlinear mixed effects models for repeated measures data. *Biometrics* 46(3), 673–87. Available at: ftp://www.biostat.wisc.edu/pub/lindstrom/papers/biometrics.1990.pdf.

Makowski, D., Wallach, D. and Tremblay, M. 2002. Using a Bayesian approach to parameter estimation; comparison of the GLUE and MCMC methods. *Agronomie* 22(2), 191–203. doi:10.1051/agro:2002007.

Martre, P., Kimball, B. A., Ottman, M. J., Wall, G. W., White, J. W., Asseng, S., Ewert, F., Cammarano, D., Maiorano, A., Aggarwal, P. K., Anothai, J., Basso, B., Biernath, C., Challinor, A. J., De Sanctis, G., Doltra, J., Dumont, B., Fereres, E., Garcia-Vila, M., Gayler, S., Hoogenboom, G., Hunt, L. A., Izaurralde, R. C., Jabloun, M., Jones, C. D., Kassie, B. T., Kersebaum, K. C., Koehler, A.-K., Müller, C., Kumar, S. N., Liu, B., Lobell,

D. B., Nendel, C., O'Leary, G., Olesen, J. E., Palosuo, T., Priesack, E., Rezaei, E. E., Ripoche, D., Rötter, R. P., Semenov, M. A., Stöckle, C., Stratonovitch, P., Streck, T., Supit, I., Tao, F., Thorburn, P., Waha, K., Wang, E., Wolf, J., Zhao, Z. and Zhu, Y. 2018. The Hot Serial Cereal Experiment for modeling wheat response to temperature: field experiments and AgMIP-Wheat multi-model simulations. *Journal of Agricultural Research* 4, 28–34. doi:10.18174/odjar.v4i0.15830.

Mastrandrea, M. D., Field, C. B., Stocker, T. F., Edenhofer, O., Ebi, K. L., Frame, D. H., Held, H., Kriegler, E., Mach, K. J., Matschoss, P. R., Plattner, P.-J., Yohe, G. W. and Zwiers, F. W. 2010. Guidance note for lead authors of the IPCC fifth assessment report on consistent treatment of uncertainties. Intergovernmental Panel on Climate Change (IPCC). Available at: http://www.ipcc.ch.

Metselaar, K. 1999. Auditing predictive models: a case study in crop growth. WAU Dissertation no. 2570.

Montesino-San Martin, M., Wallach, D., Olesen, J. E., Challinor, A. J., Hoffman, M. P., Koehler, A. K., Rötter, R. P. and Porter, J. R. 2018. Data requirements for crop modelling–applying the learning curve approach to the simulation of winter wheat flowering time under climate change. *European Journal of Agronomy* 95, 33–44. doi:10.1016/j.eja.2018.02.003.

Onatski, A. and Williams, N. 2002. Modeling model uncertainty. Working Paper No. 169.

Ottman, M. J., Kimball, B. A., White, J. W. and Wall, G. W. 2012. Wheat growth response to increased temperature from varied planting dates and supplemental infrared heating. *Agronomy Journal* 104(1), 7–16. doi:10.2134/agronj2011.0212.

Pellegrino, A., Gozé, E., Lebon, E. and Wery, J. 2006. A model-based diagnosis tool to evaluate the water stress experienced by grapevine in field sites. *European Journal of Agronomy* 25(1), 49–59. doi:10.1016/j.eja.2006.03.003.

Ramirez-Villegas, J., Koehler, A.-K. and Challinor, A. J. 2017. Assessing uncertainty and complexity in regional-scale crop model simulations. *European Journal of Agronomy* 88, 84–95. doi:10.1016/J.EJA.2015.11.021.

R Core Team. 2012. *R: A Language and Environment for Statistical Computing*. R Foundation for Statistical Computing, Vienna, Austria. Available at: http://www.r-project.org/.

Refsgaard, J. C., van der Sluijs, J. P., Højberg, A. L. and Vanrolleghem, P. A. 2007. Uncertainty in the environmental modelling process – a framework and guidance. *Environmental Modelling and Software* 22(11), 1543–56. doi:10.1016/j.envsoft.2007.02.004.

Ripoche, A., Rellier, J.-P., Martin-Clouaire, R., Paré, N., Biarnès, A. and Gary, C. 2011. Modelling adaptive management of intercropping in vineyards to satisfy agronomic and environmental performances under Mediterranean climate. *Environmental Modelling and Software* 26(12), 1467–80. doi:10.1016/j.envsoft.2011.08.003.

Roux, S., Delpuech, X., Daudin, G., Brun, F., Wery, J. and Wallach, D. 2014. Providing user oriented uncertainty information with a vineyard model used for irrigation decisions. In: Ahuja, L. R. (Ed.), *Volume 6 of the Advances in Agricultural System Modeling. Practical Applications of Agricultural System Models to Optimize the Use of Limited Water*. ASA.

Saltelli, A., Chan, K. and Scott, E. M. 2000. *Sensitivity Analysis*. Wiley, New York.

Seidel, S. J., Palosuo, T., Thorburn, P. and Wallach, D. 2018. Towards improved calibration of crop models – where are we now and where should we go? *European Journal of Agronomy* 94, 25–35. doi:10.1016/J.EJA.2018.01.006.

Tao, F., Rötter, R. P., Palosuo, T., Gregorio Hernández Díaz-Ambrona, C., Mínguez, M. I., Semenov, M. A., Kersebaum, K. C., Nendel, C., Specka, X., Hoffmann, H., Ewert, F., Dambreville, A., Martre, P., Rodríguez, L., Ruiz-Ramos, M., Gaiser, T., Höhn, J. G., Salo, T., Ferrise, R., Bindi, M., Cammarano, D. and Schulman, A. H. 2018. Contribution of crop model structure, parameters and climate projections to uncertainty in climate change impact assessments. *Global Change Biology* 24(3), 1291–307. doi:10.1111/gcb.14019.

van Vuuren, D. P., Edmonds, J., Kainuma, M., Riahi, K., Thomson, A., Hibbard, K., Hurtt, G. C., Kram, T., Krey, V., Lamarque, J., Masui, T., Meinshausen, M., Nakicenovic, N., Smith, S. J. and Rose, S. K. 2011. The representative concentration pathways: an overview. *Climatic Change* 109(1–2), 5–31. doi:10.1007/s10584-011-0148-z.

Wallach, D. 2011. Crop model calibration: a statistical perspective. *Agronomy Journal* 103(4), 1144–51. doi:10.2134/agronj2010.0432.

Wallach, D. and Thorburn, P. J. 2017. Estimating uncertainty in crop model predictions: current situation and future prospects. *European Journal of Agronomy* 88, A1–7. doi:10.1016/j.eja.2017.06.001.

Wallach, D., Thorburn, P., Asseng, S., Challinor, A. J., Ewert, F., Jones, J. W., Rotter, R. and Ruane, A. 2016a. Estimating model prediction error: should you treat predictions as fixed or random? *Environmental Modelling and Software* 84, 529–39. doi:10.1016/j.envsoft.2016.07.010.

Wallach, D., Mearns, L. O., Ruane, A. C., Rötter, R. P. and Asseng, S. 2016b. Lessons from climate modeling on the design and use of ensembles for crop modeling. *Climatic Change* 139(3–4), 551–64. doi:10.1007/s10584-016-1803-1.

Wallach, D., Nissanka, S. P., Karunaratne, A. S., Weerakoon, W. M. W., Thorburn, P. J., Boote, K. J. and Jones, J. W. 2017. Accounting for both parameter and model structure uncertainty in crop model predictions of phenology: a case study on rice. *European Journal of Agronomy* 88, 53–62. doi:10.1016/j.eja.2016.05.013.

Wallach, D., Martre, P., Liu, B., Asseng, S., Ewert, F., Thorburn, P. J., van Ittersum, M., Aggarwal, P. K., Ahmed, M., Basso, B., Biernath, C., Cammarano, D., Challinor, A. J., De Sanctis, G., Dumont, B., Eyshi Rezaei, E., Fereres, E., Fitzgerald, G. J., Gao, Y., Garcia-Vila, M., Gayler, S., Girousse, C., Hoogenboom, G., Horan, H., Izaurralde, R. C., Jones, C. D., Kassie, B. T., Kersebaum, K. C., Klein, C., Koehler, A. K., Maiorano, A., Minoli, S., Müller, C., Naresh Kumar, S., Nendel, C., O'Leary, G. J., Palosuo, T., Priesack, E., Ripoche, D., Rötter, R. P., Semenov, M. A., Stöckle, C., Stratonovitch, P., Streck, T., Supit, I., Tao, F., Wolf, J. and Zhang, Z. 2018. Multimodel ensembles improve predictions of crop-environment-management interactions. *Global Change Biology* 24(11), 5072–83. doi:10.1111/gcb.14411.

Wallach, D., Makowski, D., Jones, J. W. and Brun, F. 2019. *Working with Dynamic Crop Models: Methods, Tools and Examples for Agriculture and Environment* (3rd edn.). Academic Press, London, UK, 597p.

Wang, E., Martre, P., Zhao, Z., Ewert, F., Maiorano, A., Rötter, R. P., Kimball, B. A., Ottman, M. J., Wall, G. W., White, J. W., Reynolds, M. P., Alderman, P. D., Aggarwal, P. K., Anothai, J., Basso, B., Biernath, C., Cammarano, D., Challinor, A. J., De Sanctis, G., Doltra, J., Dumont, B., Fereres, E., Garcia-Vila, M., Gayler, S., Hoogenboom, G., Hunt, L. A., Izaurralde, R. C., Jabloun, M., Jones, C. D., Kersebaum, K. C., Koehler, A. K., Liu, L., Müller, C., Naresh Kumar, S., Nendel, C., O'Leary, G., Olesen, J. E., Palosuo, T., Priesack, E., Eyshi Rezaei, E., Ripoche, D., Ruane, A. C., Semenov, M. A., Shcherbak, I., Stöckle, C., Stratonovitch, P., Streck, T., Supit, I., Tao, F., Thorburn, P., Waha, K.,

Wallach, D., Wang, Z., Wolf, J., Zhu, Y. and Asseng, S. 2017. The uncertainty of crop yield projections is reduced by improved temperature response functions. *Nature Plants* 3, 17102. doi:10.1038/nplants.2017.102.

White, H. 1981. Consequences and detection of misspecified nonlinear regression models. *Journal of the American Statistical Association* 76(374), 419–33. doi:10.1080/01621459.1981.10477663.

Chapter 16

Crop simulation model inter-comparison and improvement

Senthold Asseng, University of Florida, USA; Pierre Martre, INRA, France; and Frank Ewert, University of Bonn and Leibniz Centre for Agricultural Landscape Research (ZALF), Germany

1 Introduction

Today, there are numerous global challenges for agriculture, including a continuous population growth, demand for more nutritious food, widespread malnutrition and hunger versus obesity, declining soil fertility, depleting water resources for irrigation, optimizing resource use, and increasing costs of fertilizer and pesticide with losses polluting waterways, lakes, and coastlines, and climate change (Ippolito et al., 2015; Smith et al., 2017; West et al., 2014). Crop models have become important tools in agricultural research that can help better understand the complexity of agricultural production at farm, regional, and global scales. They are used for agronomic management, precision agriculture, environmental impacts and externalities, input efficiency, managing climate variability and seasonal forecasting, breeding support, climate change impacts and adaptations, and agricultural policy (Basso et al., 2016; Boote et al., 2013; Chenu et al., 2017; Jeuffroy et al., 2014; Reynolds et al., 2018). The development of crop modeling started in the 1960s by De Wit (1965) and Monteith (1965) and began as computer programs executing mathematical

http://dx.doi.org/10.19103/AS.2019.0061.22

equations that described crop development and growth and its interaction with atmosphere and soil (Jones et al., 2017a). These mathematical equations are simplifications of reality, developed from natural laws and observations, with the aim to represent key features of crop growth and development. Crop models have been developed since then in different research groups around the world, with different scientific focus and data and led to crop models giving widely different simulation results. In the 1990s, researchers started using crop model inter-comparisons in order to better understand the spread of simulation results and the underlying behavior of crop models, as well as their suitability for specific tasks (De Willigen, 1991; Diekkruger et al., 1995; Jamieson et al., 2000; Porter et al., 1993). Some inter-comparisons were also conducted as part of the Global Change and Terrestrial Ecosystems (GCTE) project of the International Geosphere-Biosphere Program (IGBP). Following the initial model inter-comparisons, new networks were developed including COST actions in Europe (https://www.cost.eu/) to assess crop model water and nutrient dynamics (Kersebaum et al., 2007), and larger initiatives such as Modeling European Agriculture with Climate Change for Food Security – MACSUR (https://macsur.eu/), and the Agricultural Model Intercomparison and Improvement Project – AgMIP (www.agmip.org), with emphasis on climate change impact and adaptation, by linking climate, crop, and economic modeling (Rosenzweig et al., 2013). Many of the AgMIP and MACSUR activities have been inter-linked and resulted in a wide range of model inter-comparisons, the identification of model shortcomings, a better understanding of uncertainty, and new approaches to applications for impact assessment and crop model improvements. In this chapter we summarize crop model inter-comparison studies and their contribution to sustainable agriculture and discuss gaps and trends in crop simulation model inter-comparisons and improvements.

2 Crop model inter-comparison

One of the first crop model inter-comparisons was organized through a Collaborative Research Program involving up to 19 crop models that were compared with detailed data for winter wheat (two crops) and sugar beet (one crop) from a site in Germany, including simulated time-series dynamics of soil water, temperature, nitrogen (N), pesticides, evapotranspiration, N uptake, crop growth, leaf area index (LAI), and yield (De Willigen, 1991; Diekkruger et al., 1995). In this model inter-comparison, all models received all measurements for setting up simulations. However, modelers were allowed to choose measurements for calibration and were inconsistent in system processes they simulated. For example, most models simulated water dynamics, growth, and yield; half of them simulated N dynamics; and only three simulated soil pesticide transport. The study showed widely different simulation results.

Individual model shortcomings were identified, and in general the simulation of above-ground biomass was less problematic than below-ground processes (De Willigen, 1991; Diekkruger et al., 1995).

As part of the GCTE project, Porter et al. (1993) compared three wheat crop models with experimental data (three years and up to three sowing dates) under non-limiting growing conditions (sufficient irrigation and N supplied) for a site in New Zealand. Shortcomings of models were indicated for individual models in simulating crop phenology (e.g. the need for one of the models to include photoperiod sensitivity), light interception under a high-light intensity growing environment, and in simulating the rate of leaf senescence and yield components. Jamieson et al. (1998) compared five wheat models with data from a wheat crop experiment performed under a rainout-shelter creating drought treatments in New Zealand. To a varying degree, all of the models simulated a reduction in photosynthetic efficiency due to increasing drought as an important contributor to the reduction in the rate of biomass accumulation. Landau et al. (1998) tested three wheat models and concluded crop models are not able to reproduce a wide range of observed yield data (from 5 to 12 t ha^{-1}) from a large dataset collected across many experimental locations in the UK. While the size of the dataset used was impressive, Jamieson et al. (1999) invalidated their conclusions soon after by pointing to the many unjustified assumptions made by Landau et al. (1998) (e.g. assuming a single soil, single cultivar, no water or N deficit) to fill in missing information on soil characteristics, initial soil water and soil N conditions in each year, and cultivar and crop management (e.g. sowing and N applications). Ewert et al. (2002) compared three wheat crop models with detailed data from CO_2 experiments in the US and Germany and a drought experiment in Spain. They concluded that crop models reproduced wheat growth and yield for different CO_2 and drought treatments, but suggested that crop models performed less well for simulating the combined effect of CO_2 and drought.

These initial crop model inter-comparisons were followed by studies in the European COST action and later MACSUR networks with wheat, including eight models, 49 growing seasons, and eight sites across Europe (Palosuo et al., 2011). Also inter-comparisons were completed with barley, including nine models, 44 growing seasons, also across Europe (Rotter et al., 2012). Inter-comparisons also included eight models, 2 years, and N rate experiments (Salo et al., 2016), with crop rotations; 15 crop models with crop rotations including 301 seasons of ten crop types and a diverse set of treatments (Kollas et al., 2015); and 12 crop models and field data on grain N for seven crops grown in Europe (Yin et al., 2017b). Ehrhardt et al. (2018) analyzed the uncertainty of 24 crop models to production and nitrous oxide (N_2O) emission for wheat, rice, maize, and temperate grasslands. These inter-comparison studies highlighted the need for improvements in N mineralization, light capture, and low temperature effects after emergence (Salo et al., 2016). They also indicated that less-studied crops performed poorer than main crops, such as cereals (Kollas et al., 2015), but showed that crop models correctly simulated the effect of increasing N fertilizer on wheat and barley (Yin et al., 2017a). These studies showed that the use of continuous simulation did not always improve the simulation compared to resetting each single year, suggesting the need for model improvement of crop rotation effects (Yin et al., 2017a). Another study indicated that crop models for rotations performed slightly better than the simulation of single years and single crops (Kollas et al., 2015). Rötter et al. (2011) concluded that due to the wide spread of model simulations across inter-comparison studies, more research in crop model development is needed to improve these critical tools of agricultural science.

MACSUR started in 2012 to create a knowledge hub for modeling of agricultural production. It brought together research in modeling grasslands, livestock, crops, farms, and agricultural trade in order to improve the modeling of climate change impacts on European agriculture. Modeling these impacts can help illustrate to political decision makers how climate will affect regional farming systems and food production in Europe. MACSUR advanced modeling of crop production, grassland production, price development in Europe and produced more than 100 scientific publications, including the most comprehensive crop model inter-comparison on aggregation and scaling (Angulo et al., 2013, 2014; Bakker et al., 2005; Ewert et al., 2011, 2015b; Grosz et al., 2017; Hoffmann et al., 2015, 2016; Kuhnert et al., 2017; van Bussel et al., 2011a,b, 2015, 2016; Volk and Ewert, 2011; Zhao et al., 2015a,b, 2016), which will be discussed in more detail under the section on uncertainty.

AgMIP started in late 2010 with the goal to improve the characterization of world food security under climate change and to enhance adaptation capacity. Protocols were developed to guide coordinated climate, crop modeling, economics, and information technology research activities around the world, along with AgMIP cross cutting themes that address uncertainty, aggregation, and scaling. The development of Representative Agricultural Pathways (RAPs) enabled testing of climate change adaptations in the context of other regional and global trends (Rosenzweig et al., 2013). Model inter-comparisons were organized for wheat, 27 models (Asseng et al., 2013); maize, 23 models (Bassu et al., 2014); rice, 13 models (Li et al., 2015); and potatoes, nine models (Fleisher et al., 2017). These studies followed the same protocols of testing with field data from contrasting global locations, starting with 'blind' simulations where inputs such as soil, weather, and crop management, but no measured output information, like biomass and yields, were supplied. A sensitivity analysis to incremental increases in temperature and atmospheric CO_2 concentrations was also part of the protocol applied in all AgMIP crop inter-comparison studies. Consistently across these studies, temperature changes caused more uncertainty than changes in atmospheric CO_2 concentrations (Makowski et al., 2015), suggesting a need for improving temperature response functions across crop models. These initial crop pilot studies were followed by inter-comparisons with up to 32 models per crop, including field experiments with treatments of temperature for wheat (Asseng et al., 2015), including heat shocks during the grain filling period (Asseng et al., 2019; Liu et al., 2016a), CO_2 in rice (Hasegawa et al., 2017) and wheat (O'Leary et al., 2015), CO_2 by drought in maize (Durand et al., 2018) and temperature by CO_2 by drought in wheat (Asseng et al., 2019), and by assessing model performance for simulating canopy temperature (Webber et al., 2018b). In a study on wheat, Martre et al. (2015) showed that there is little correlation between the model errors for different state variables (e.g. biomass or LAI) and yield. This finding clearly shows that more attention should be paid to the internal consistencies of models and to the compensations for errors (Challinor et al., 2014). A key finding from some of the AgMIP studies included that continuous cropping simulation with wheat (seven models) and maize (five models) pointed to a substantial underestimation of climate change impacts on yield when resetting soils each year in a simulation, a standard practice in most climate change impact studies, in particular, resetting soil organic carbon in models (Basso et al., 2018). A large number of multi-model impact assessments were carried out in AgMIP across locations in Sub-Saharan Africa and Southeast Asia with at least two crop models per crop across various crops and regions (Rosenzweig and Hillel, 2015). In addition, a model inter-comparison and a climate change impact assessment were conducted with seven global gridded models for major crops at a global scale (Rosenzweig et al., 2014). More than 100 scientific publications have been produced in AgMIP. Some of these will be discussed in more detail in the following sections.

3 Crop model uncertainty

There are different sources of crop simulation model uncertainty and include uncertainty of input data, in parameters and in model structure. As crop models are simplified representations of reality with a specific aim, they do include uncertainty in each of their individual simulations. The model structure uncertainty in simulation results is always there, but cannot be determined with a single model simulation. Multi-model ensemble simulations enabled us to quantify structural crop model uncertainty for the first time. In addition, the multi-model ensemble mean or median has been shown to be consistently closer to measurements than any single crop model, offering more accurate simulation results (Martre et al., 2015; Wallach et al., 2018). Clear evidence for the ensemble mean/median being a better predictor (i.e. closer to measurements) than any individual model has been supplied for yield simulations of wheat (Asseng et al., 2013; Palosuo et al., 2011), barley (Rotter et al., 2012; Salo et al.,

2016), maize (Bassu et al., 2014), rice (Li et al., 2015) and potatoes (Fleisher et al., 2017), biomass growth in barley (Salo et al., 2016), and grain N across seven crops (Yin et al., 2017a,b). The same has been observed for other system components, including soil water dynamics, evapotranspiration (ET), N uptake, and LAI for wheat (Martre et al., 2015). Martre et al. (2015) also showed that there is little difference between multi-model ensemble mean and median, and they are interchangeable with larger ensembles (e.g. >10 models). However, the mean of a multi-model ensemble might be more appropriate for small ensembles of five or less models. Recently, Wallach et al. (2018) showed that the multi-model ensemble mean/median is consistently a better predictor than any single crop model across a wide range of growing environments in a number of studies and supplied a statistical explanation for this. They showed analytically that the mean squared error of prediction (MSEP) of an ensemble mean will always be smaller than the MSEP averaged over models.

The possibility to quantify model uncertainty with crop model ensembles now allows to compare uncertainty with other uncertainties from other types of models (e.g. global climate models (GCMs)), which supply the climate projections used as climate input data for crop models in climate change impact studies or when crop model simulations feed into other models as an input. For example, Asseng et al. (2013, 2019) showed that the uncertainty from wheat crop models is larger than the uncertainty from GCMs. This was also confirmed in a barley modeling study (Tao et al., 2018). Model uncertainty is also shown to be larger than parameter uncertainty (Asseng et al., 2013; Tao et al., 2018) and uncertainty of aggregation and scaling, which also depends on the aggregated input and is most pronounced for soil input data as compared to climate input data (Angulo et al., 2013, 2014; Bakker et al., 2005; Ewert et al., 2011, 2015b; Grosz et al., 2017; Hoffmann et al., 2015, 2016; Kuhnert et al., 2017; van Bussel et al., 2011a,b, 2015, 2016; Volk and Ewert, 2011; Zhao et al., 2015a,b, 2016). However, the uncertainty of economic models using crop model simulations as inputs has been shown to be larger than crop model uncertainty (Nelson et al., 2014).

In addition, an often-neglected source of uncertainty is the user of the model. In a study where five crop models and 20 users were arranged in a randomized block design with four replicates, Confalonieri et al. (2016) found that the uncertainty due to the user can be as large as that due to the crop model. This is because of the assumption and rules of pedotransfer functions used by model users to derive the values of soil and cultivar parameters and initial conditions from available field data. For instance, Martre et al. (2015) noted that the uncertainty in soil water and N content in an ensemble of 27 wheat models does not increase over the season; it is thus mainly due to the uncertainty in initial conditions. This points to the need to develop guidelines and rules to estimate crop model parameters and initial conditions from field measurements.

Compared to single model simulations, crop model uncertainty is reduced with crop model ensembles. Crop models in an ensemble with 3-6 models are often sufficient as an ensemble mean to improve model accuracy to be within field experimental measurement errors (Asseng et al., 2013; Bassu et al., 2014).

Wang et al. (2017) compared the temperature functions employed across wheat crop models and showed that the mean of temperature functions placed in models for a process is similar to temperature functions based on the latest available crop physiological data. This supplied the explanation for why the model ensemble mean simulation for temperature impact (Asseng et al., 2015) is a better predictor than individual models in simulating temperature impacts.

Different methods (using point and gridded simulation models, statistical models and observations) were compared for determining temperature impacts on crop yields for wheat and for all four main crops including wheat, rice, maize, and soybean (Zhao et al., 2017). They showed that the ensemble mean of crop simulation scaled up from representative locations, or aggregated from gridded simulations and compared with a statistical model and scaled up from field experiments, all had a similar mean impact, indicating a similar mean and uncertainty range of each of these methods (Zhao et al., 2017) (Fig. 1). Therefore, multi-method ensembles can further improve impact estimate accuracy and reduce uncertainty.

Pirttioja et al. (2015) and Fronzek et al. (2018) used an impact response surface approach to analyze the sensitivity and uncertainty of an ensemble of 26 crop models and identified specific processes and climate factor responses for which crop models will require further improvements. A key avenue for reducing crop model uncertainty is through model improvements (Maiorano et al., 2017) and will be described in the following section.

4 Crop model improvement

Earlier crop model inter-comparisons often indicated the need of general or in some cases specific model improvements, including specific processes or specific crop models that needed improvements. However, earlier studies did not document if, how, and with what effect these proposed improvements were implemented. After modelers had access to a full data set for model improvements, for the first time, Maiorano et al. (2017) showed with another, independent 'blind dataset' (i.e. the crop modelers had no access to the observed growth and yield data before and after their model improvements) that model improvements were effective in

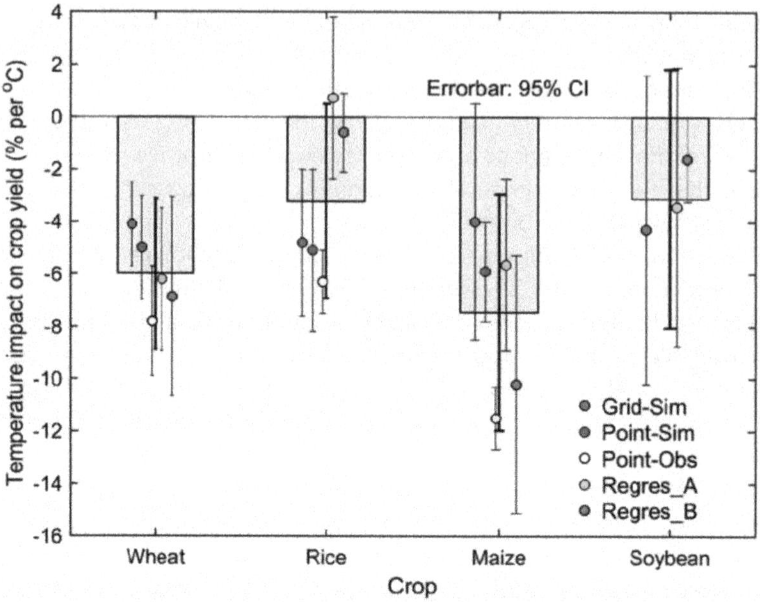

Figure 1 Multi-method estimates of global crop yield changes in response to temperature increase. Impacts on crop yields of a 1°C increase in global temperature with grid-based simulations (Grid-Sim), point-based simulations (Point-Sim), field-warming experiments (Point-Obs), and statistical regressions at the country level (Regres_A) and the global level (Regres_B). Circles are means of estimates from each method or medians for Grid- and Point-Sim. Filled bars are the means of the multi-method ensemble. Error bars show 95% confidence intervals for individual methods (thin gray error bars attached to circles) and the ensemble of methods (thick black error bars attached to bars). Source: adapted from Zhao et al. (2017).

reducing the difference between model simulations and field experimental observation (Fig. 2). They also showed that these model improvements significantly reduced the number of models required in an ensemble to achieve accuracy with the ensemble mean/median to be as low as field experimental errors. One of the key improvements across several models included the introduction of a heat stress function accelerating senescence in wheat. They finally proved that improvements of individual models of an ensemble can significantly reduce uncertainty of a multi-model ensemble (Maiorano et al., 2017).

In another example, after comparing four wheat models (Liu et al., 2016a) and several temperature functions (Liu et al., 2016b) with detailed experimental data, Liu et al. (2017) improved the performance of one model of their previous model ensemble in simulating the effect of post-heading heat stress on biomass growth of winter wheat when compared with field measurements. This was accomplished by including a heat stress function on accelerating

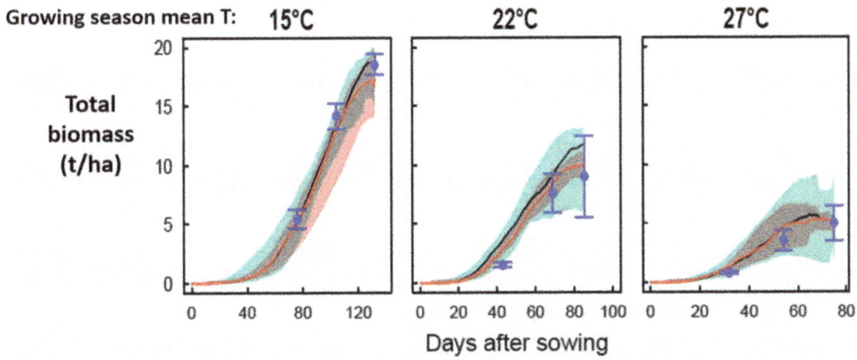

Figure 2 Simulated and measured wheat growth dynamics for total above-ground biomass over days after sowing for mean growing season temperatures 15°C, 22°C, and 27°C. Black lines and blue areas are e-median and the tenth to ninetieth percentile range of the 15 original (unimproved) models, respectively. Red lines and light red areas are e-median and the tenth to nineteenth percentile range of the 15 improved models, respectively. Areas are grey when improved and unimproved ranges overlap. Blue symbols are measured mean ±1 s.d. of three independent replicates. Source: adapted from Maiorano et al. (2017).

senescence with a new cultivar parameter for the sensitivity of biomass growth to heat stress.

Several wheat crop models (9) recently introduced canopy temperature functions as an attempt to improve the complex interactions among air temperature, crop and soil water status, CO_2 concentration, and atmospheric conditions that influence crop canopy temperature, as the temperature of a crop canopy can often vary by several degrees from air temperature (Webber et al., 2017). Using canopy temperature to account for heat stress effects did improve some simulations compared to using only air temperature, but the models that additionally use canopy temperature on other processes did not have better yield simulations. Also, models that simulated yield well under heat stress had varying skills in simulating canopy temperature (Webber et al., 2017). Comparing these models with another detailed field experimental transect dataset suggested that more accurate simulations of heat stress will likely require the use of energy balance approaches that consider atmospheric stability conditions for simulating canopy temperature and its interaction with other processes (Webber et al., 2018b).

Wang et al. (2017) showed the wide range of temperature functions employed in many of the currently used wheat crop models. They developed temperature functions based on the latest available crop physiological data and showed that these functions could improve model performance in comparison to observed data in several models. Wang et al. (2018) used a similar approach to improve the response to temperature in maize crop models.

5 Crop model application

Realizing the enhanced accuracy of multi-model ensemble means/medians, recent crop model applications are now carried out with multi-model ensembles. For example, several recent climate change impact studies used multi-model ensembles for major food crops across the globe (Deryng et al., 2016; Muller et al., 2018; Rosenzweig et al., 2014, 2018; Schauberger et al., 2017; Schleussner et al., 2018), wheat yield (Asseng et al., 2015; Liu et al., 2019), wheat grain quality at global scale (Asseng et al., 2019), and wheat and maize yield across Europe (Webber et al., 2018a). The multi-model ensemble application in Webber et al. (2018a) revealed that for the current genotypes and mix of irrigated and rainfed production, climate change would lead to yield losses for grain maize and gains for winter wheat. Also, on average, heat stress does not increase for either crop in rainfed systems, while drought stress will intensify in many years for maize, with elevated CO_2 offering no yield benefit in these years. Other applications with multi-model ensembles include simulations for cropping systems for a single crop across an individual country, such as Egypt (Asseng et al., 2018) and Mexico (Hernandez-Ochoa et al., 2018), suggesting that negative impacts from increasing heat stress will outweigh possible growth benefits from elevated CO_2. Additional multi-model studies include transects across Europe (Fronzek et al., 2018) and selected global locations for wheat and maize (Hoffmann et al., 2018). Some multi-model ensemble applications were used for designing of barley ideotypes (Tao et al., 2017) and estimating the water foot print (crop productivity per water consumption) across locations in Europe (Kersebaum et al., 2016). Finally, Ruiz-Ramos et al. (2018) analyzed the effect of adaptation options under changed precipitations, temperature, CO_2, and soil type using a response surface approach for an ensemble of 17 crop models. All of these examples took advantage of the improved simulation accuracy of multi-model ensembles compared to single-model applications.

6 Case study: AgMIP-Wheat

The Agricultural Model Intercomparison and Improvement Project for Wheat (AgMIP-Wheat; https://www.agmip.org/wheat/) started shortly after the AgMIP was founded in 2011. The team included wheat crop modelers representing 80–90% of all different wheat crop models in the world, experimentalists supplying various field data, a climate scientist, and a bio-mathematician. Initially, 27 different wheat crop models were tasked to simulate detailed experiments from four contrasting wheat growing locations. Information was supplied to the modelers in different steps of data availability (including blind simulations in which measurements of growth and yields were withheld). The simulated data were collected and analyzed centrally. Comparing simulated with observed data showed that the simulated median or mean grain yield of the multi-model ensemble was consistently a better predictor across these locations (Asseng et al., 2013), which was also shown for other simulated systems variables, including biomass, yield components, leaf area, water use, and soil water contents (Martre et al., 2015). Additional information supplied to the modelers (e.g. biomass and yield) reduced the simulated uncertainty and simulations came closer to observations. Model response to year-to-year variability caused by rainfall and temperature variability had an impact on simulated yields, but the response was often different to simulated climate change scenarios, suggesting that analog years might not be suitable to understand future climate change impacts on crops (Ruane et al., 2016). Differences in evapotranspiration routines caused some model uncertainty (Cammarano et al., 2016) as did the impact of changes in water supply or CO_2 (O'Leary et al., 2015). However, a sensitivity analysis revealed that temperature routines in crop models caused the largest model uncertainty in simulating climate change impacts (Asseng et al., 2013) (Fig. 3).

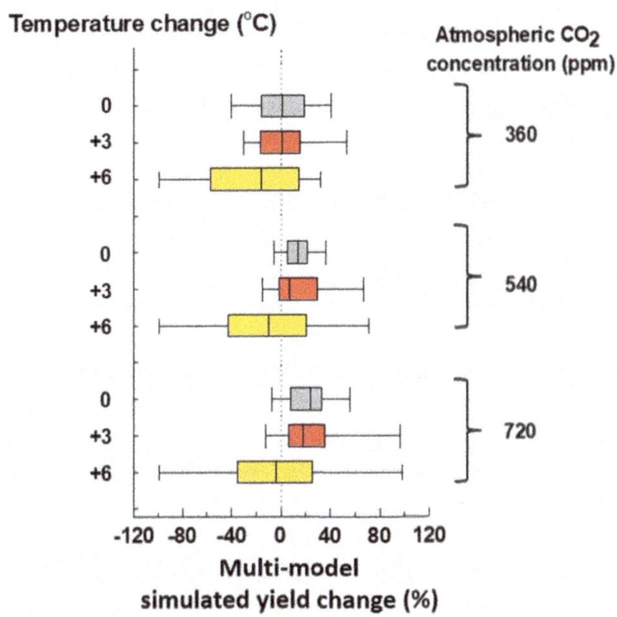

Figure 3 Simulated relative mean (30-year average, 1981-2010) grain yield change for increased temperatures and elevated atmospheric CO_2 concentrations for a location in Australia with 27 wheat crop models. For each box plot, vertical lines represent, from left to right, the tenth percentile, twenty-fifth percentile, median, seventy-fifth percentile, and ninetieth percentile of simulations based on multi-models. Source: adapted from Asseng et al. (2013).

Crop model uncertainty was also shown to be larger than climate model uncertainty (Asseng et al., 2013). Input variables (e.g. soil characteristics) can cause additional model uncertainty and can vary across growing environments (Asseng et al., 2013). The protocols developed for wheat (Asseng et al., 2013) were then applied to other crops such as maize (Bassu et al., 2014), rice (Li et al., 2015), and potatoes (Fleisher et al., 2017) and enabled cross-crop comparisons of sources of uncertainty (Makowski et al., 2015), confirming that simulating temperature change impact is the main cause of model uncertainty for other crops. The crop model inter-comparisons with other crops also confirmed the superior predictor of the ensemble mean/median over any individual model as explained by Wallach et al. (2018). To better understand the temperature routines in the wheat models, these were documented for most wheat models in a Workshop Proceedings (Alderman et al., 2013). Differences in modeling temperature impacts and possible shortcomings in specific models became clear via this documentation. Specific shortcomings in simulating temperature effects on crop growth were also identified in comparison with models and data (Liu et al., 2017; Wang et al., 2017) and led to attempts to improve temperature routines in relation to simulating canopy temperature (Webber et al., 2018b), heat stress effects on canopy senescence (Maiorano et al., 2017), and suggestions of general improvements of temperature response functions for crop phenology and growth (Wang et al., 2017). Additional field data with wheat grown across a wide range of temperatures were then used for testing 30 wheat crop models, and the multi-model ensemble was applied across 30 representative high-rainfall and irrigated locations to estimate the impact of global warming (only temperature change) on global wheat production. This showed that each degree in global temperature increase will result in a 6% yield decline in global wheat production (Asseng et al., 2015). A reduced growing season due to temperature increase was identified as the main cause of yield decline from warming, suggesting that new crops with cultivar characteristics are needed with a delayed flowering combined with an increased grain filling rate to escape heat toward crop maturity, to adapt wheat to global warming (Asseng et al., 2015). In a new simulation experiment, 32 wheat crop models were then tested with detailed field and chamber experimental data on interactions of temperature increase with elevated CO_2 and rainfall change. By working closely with field experimentalists, new evidence from crop physiological breeding field trials showed the existence of real cultivars with delayed flowering linked with an increased grain filling rate. These field experiments were carried out across temperature gradients and showed that such a trait combination can result in yields being higher than current cultivars in warmer growing environments. These crop characteristics were then applied in a global simulation experiment on climate change impacts (considering changes in temperature, rainfall, and atmospheric CO_2

by 2050) across 60 representative locations (also including low rainfall and low input locations). While climate change impacts varied across the globe and are negative in many warm and low input systems near the equator despite a growth stimulus from elevated CO_2, global wheat production was projected not to decline and even increase with the introduction of new cultivars with delayed flowering and an increased grain filling rate. However, simulated grain protein concentration, an important quality and nutritional aspect of wheat, will decline with climate change and require the introduction of additional new crop characteristics to adapt yields and grain protein concentration to global warming (Asseng et al., 2019). A simulation analysis with 1.5°C and 2.0°C temperature increase with the same multi-model ensemble and across the 60 representative locations showed that yield variability will increase, particularly in the frequency in low yields, despite relatively small changes in climate (Liu et al., 2019).

The results from AgMIP-Wheat have been highly cited in the literature influencing and guiding other research (e.g. via protocols). The scientific outcomes from AgMIP-Wheat have also been noticed in media outlets across the world, showing the danger of climate change and how it will impact wheat production and global food security.

7 Conclusion

Model inter-comparison and improvements have shown that many crop models can simulate main crops such as wheat, maize, rice, and potatoes. It has been widely proven now that multi-model ensemble mean or median are better predictors than any individual model across growing environments and are recommended for impact assessments. Multi-model ensembles allow quantifying crop model uncertainty. Crop model inter-comparisons have highlighted areas of model shortcomings and possible model improvements. Model improvements have been documented for simulating temperature impacts and shown to reduce crop model uncertainty.

Using multi-model ensembles has advanced our understanding in assessing climate change impacts on crop yields and global food security.

8 Future trends

There is a clear trend for sharing models, model routines, and data across the crop modeling community. New tools have been developed in AgMIP to support data sharing (Porter et al., 2014), which are likely to be used more across models in the future. However, there is currently no common database to export data to and seamlessly import data from for crop modeling. But, a willingness to share experimental and simulated data is visible through an increased number of data papers in new established data journals, like Open Data Journal for Agricultural Research.

Another trend is using model testing with high-quality, detailed field data for single climate factors of temperature (Asseng et al., 2015; Liu et al., 2016a) and CO_2 (Durand et al., 2018; Hasegawa et al., 2017; O'Leary et al., 2015), and more recently with experimental data with factor interactions and extreme climate events affecting yield (Asseng et al., 2019). Existing crop models usually do not account for nutritional quality apart from a recent study on wheat protein concentration (Asseng et al., 2019), which limits more holistic food security assessments. Dynamic data and modeling approaches are still lacking to address nutritional quality in crop models. Most models for main staple crops account for the interaction of CO_2 with water use and N limitation. However, other nutrients, like phosphorus, are often not considered in crop models, even though P is known to be a main limiting crop nutrient in many soils, particularly across Africa. Efforts are currently underway to consider low input systems with a model inter-comparison approach to better understand the impact of climate change on these systems. Similarly, other factors, like pest and diseases, have their own dependencies on climate drivers and can have large impacts on crop growth and yield when not controlled. These additional factors are currently being developed into and linked in crop models with a model inter-comparison approach.

Other important factors, such as ozone, can damage crop growth and yield and interact with other climate drivers, and these started to receive attention recently (Emberson et al., 2018). The importance of changes in soil processes and their impact on crop growth and yield have been recently highlighted (Basso et al., 2018) and will need to be considered more in future climate impact assessments for cropping systems. There is a recent trend in crop modeling to expand from single-season systems to crop rotations, considering carry-over effects, for example, of soil water, nutrients, and residues (Kollas et al., 2015), from one crop to the next crop and also to consider mixed-cropping systems, including inter-cropping and alley cropping.

There are many crop models for staple crops. However, there are few or no models and data available for most fruit and vegetable crops. There is a recent trend to consider these in the future with crop models in climate impact assessments (Zhao et al., 2018).

Crop models are widely applied in climate impact assessments, and the trend has been to use multi-model ensembles as they improve model accuracy and allow the ability to quantify model uncertainty. However, large gaps still exist in multi-scale application of crop models, including the consistent regional coverage, the consideration of management activities, including technological development to evaluate adaptation options and the propagation of different sources of uncertainties (Ewert et al., 2015a).

New detailed data from phenotyping facilities, collected for the purpose of understanding genetic performance, create opportunities for model development, testing, and improvements and will need to be explored by linking crop modelers with phenotyping experimentalists (Tardieu et al., 2017) and information systems such as the open-source Phenotyping Hybrid Information System (PHIS (Neveu et al., 2019)). New sources of high-resolution temporal and spatial data will become available soon through new technologies on farms (Asseng and Asche, 2019), offering more opportunities and challenges for crop modeling.

There is still a need for more transparency and knowledge exchange in crop model development and applications. There is currently no standard for model algorithms documentation and representation, and many models are not documented at all. Therefore, it remains difficult to analyze and understand the causes of model uncertainty and share model improvements among crop models implemented in different simulation platforms. For model improvements, models should be compared at the process level as shown by Wang et al. (2017). Standards for model algorithms documentation and representation do exist in other communities (Millar et al., 2019) and efforts are currently being made in this direction by the crop modeling community through the Agricultural Model Exchange Initiative (AMEI; https://github.com/AgriculturalModelExchangeInitiative), which has developed a crop modeling meta-language shared between crop simulation platforms named Crop Modeling Meta-Language (Crop2ML). Crop2ML is a simple high-level meta-language based on a declarative architecture of modular model representation with an intermediate modeling language to describe biophysical processes and their transformation to existing platforms. It enables the description and the composition of crop model components independently of the formalism and programming language of the simulation platforms and the exchange of components between them. The semantic representation of the models will greatly facilitate the discovery and reuse of models but also linking models with data through intelligent information systems like PHIS (Brown et al., 2014; Holzworth et al., 2014, 2015; Porter et al., 2014).

9 Where to look for further information

Further general information on model inter-comparison and improvements can be found on the AgMIP (http://www.agmip.org/) and MACSUR (https://macsur.eu/) Websites. A number of reviews have been written recently on crop models, applications, needs, and shortcomings (Basso et al., 2016; Boote et al., 2013; Chenu et al., 2017; Jeuffroy et al., 2014; Reynolds et al., 2018), with specific emphasis on next generation models (Antle et al., 2017; Jones et al., 2017b) and integrated assessment and modeling the linkage of biophysical and economic models (Ewert et al., 2015a; Ruane et al., 2017). A source for high-quality field data for model testing has become the Open Data Journal for Agricultural Research (https://odjar.org/).

10 References

Alderman, P. D., Quilligan, E., Asseng, S., Ewert, F. and Reynolds, M. 2013. *Proceedings of the Workshop on Modeling Wheat Response to High Temperature*, El Batan, Texcoco, Mexico, 19-21 June 2013. CIMMYT, El Batan, Mexico.

Angulo, C., Roetter, R., Trnka, M., Pirttioja, N., Gaiser, T., Hlavinka, P. and Ewert, F. 2013. Characteristic 'fingerprints' of crop model responses data at different spatial resolutions to weather input. *European Journal of Agronomy* 49, 104–14. doi:10.1016/j.eja.2013.04.003.

Angulo, C., Gaiser, T., Rotter, R. P., Børgesen, C. D., Hlavinka, P., Trnka, M. and Ewert, F. 2014. 'Fingerprints' of four crop models as affected by soil input data aggregation. *European Journal of Agronomy* 61, 35–48. doi:10.1016/j.eja.2014.07.005.

Antle, J. M., Basso, B., Conant, R. T., Godfray, H. C. J., Jones, J. W., Herrero, M., Howitt, R. E., Keating, B. A., Munoz-Carpena, R., Rosenzweig, C., Tittonell, P. and Wheeler, T. R. 2017. Towards a new generation of agricultural system data, models and knowledge products: design and improvement. *Agricultural Systems* 155, 255–68. doi:10.1016/j.agsy.2016.10.002.

Asseng, S. and Asche, F. 2019. Future farms without farmers. *Science Robotics* 4(27). doi:10.1126/scirobotics.aaw1875.

Asseng, S., Ewert, F., Rosenzweig, C., Jones, J. W., Hatfield, J. L., Ruane, A. C., Boote, K. J., Thorburn, P. J., Rotter, R. P., Cammarano, D., Brisson, N., Basso, B., Martre, P., Aggarwal, P. K., Angulo, C., Bertuzzi, P., Biernath, C., Challinor, A. J., Doltra, J., Gayler, S., Goldberg, R., Grant, R., Heng, L., Hooker, J., Hunt, L. A., Ingwersen, J., Izaurralde, R. C., Kersebaum, K. C., Mueller, C., Kumar, S. N., Nendel, C., O'Leary, G., Olesen, J. E., Osborne, T. M., Palosuo, T., Priesack, E., Ripoche, D., Semenov, M. A., Shcherbak, I., Steduto, P., Stoeckle, C., Stratonovitch, P., Streck, T., Supit, I., Tao, F., Travasso, M., Waha, K., Wallach, D., White, J. W., Williams, J. R. and Wolf, J. 2013. Uncertainty in simulating wheat yields under climate change. *Nature Climate Change* 3(9), 827–32. doi:10.1038/nclimate1916.

Asseng, S., Ewert, F., Martre, P., Rötter, R. P., Lobell, D. B., Cammarano, D., Kimball, B. A., Ottman, M. J., Wall, G. W., White, J. W., Reynolds, M. P., Alderman, P. D., Prasad, P. V. V., Aggarwal, P. K., Anothai, J., Basso, B., Biernath, C., Challinor, A. J., De Sanctis, G., Doltra, J., Fereres, E., Garcia-Vila, M., Gayler, S., Hoogenboom, G., Hunt, L. A., Izaurralde, R. C., Jabloun, M., Jones, C. D., Kersebaum, K. C., Koehler, A., Müller, C., Naresh Kumar, S., Nendel, C., O'Leary, G., Olesen, J. E., Palosuo, T., Priesack, E., Eyshi Rezaei, E., Ruane, A. C., Semenov, M. A., Shcherbak, I., Stöckle, C., Stratonovitch, P., Streck, T., Supit, I., Tao, F., Thorburn, P. J., Waha, K., Wang, E., Wallach, D., Wolf, J., Zhao, Z. and Zhu, Y. 2015. Rising temperatures reduce global wheat production. *Nature Climate Change* 5(2), 143–7. doi:10.1038/nclimate2470.

Asseng, S., Kheir, A. M. S., Kassie, B. T., Hoogenboom, G., Abdelaal, A. I. N., Haman, D. Z. and Ruane, A. C. 2018. Can Egypt become self-sufficient in wheat? *Environmental Research Letters* 13(9). doi:10.1088/1748-9326/aada50.

Asseng, S., Martre, P., Maiorano, A., Rotter, R. P., O'Leary, G. J., Fitzgerald, G. J., Girousse, C., Motzo, R., Giunta, F., Babar, M. A., Reynolds, M. P., Kheir, A. M. S., Thorburn, P. J., Waha, K., Ruane, A. C., Aggarwal, P. K., Ahmed, M., Balkovic, J., Basso, B., Biernath, C., Bindi, M., Cammarano, D., Challinor, A. J., De Sanctis, G., Dumont, B., Rezaei, E. E., Fereres, E., Ferrise, R., Garcia-Vila, M., Gayler, S., Gao, Y. J., Horan, H., Hoogenboom, G., Izaurralde, R. C., Jabloun, M., Jones, C. D., Kassie, B. T., Kersebaum, K. C., Klein, C., Koehler, A. K., Liu, B., Minoli, S., San Martin, M. M., Muller, C., Kumar, S. N., Nendel, C., Olesen, J. E., Palosuo, T., Porter, J. R., Priesack, E., Ripoche, D., Semenov, M. A., Stockle, C., Stratonovitch, P., Streck, T., Supit, I., Tao, F. L., Van der Velde, M., Wallach, D., Wang, E. L., Webber, H., Wolf, J., Xiao, L. J., Zhang, Z., Zhao, Z. G., Zhu, Y. and Ewert, F. 2019. Climate change impact and adaptation for wheat protein. *Global Change Biology* 25, 155-73. doi: 10.1111/gcb.14481.

Bakker, M. M., Govers, G., Ewert, F., Rounsevell, M. and Jones, R. 2005. Variability in regional wheat yields as a function of climate, soil and economic variables: assessing the risk of confounding. *Agriculture, Ecosystems and Environment* 110(3-4), 195-209. doi:10.1016/j.agee.2005.04.016.

Basso, B., Liu, L. and Ritchie, J. T. 2016. A comprehensive review of the CERES-Wheat, -Maize and -Rice models' performances. *Advances in Agronomy* 136, 27-132. doi::doi:10.1016/bs.agron.2015.11.004.

Basso, B., Dumont, B., Maestrini, B., Shcherbak, I., Robertson, G. P., Porter, J. R., Smith, P., Paustian, K., Grace, P. R., Asseng, S., Bassu, S., Biernath, C., Boote, K. J., Cammarano, D., De Sanctis, G., Durand, J. L., Ewert, F., Gayler, S., Hyndman, D. W., Kent, J., Martre, P., Nendel, C., Priesack, E., Ripoche, D., Ruane, A. C., Sharp, J., Thorburn, P. J., Hatfield, J. L., Jones, J. W. and Rosenzweig, C. 2018. Soil organic carbon and nitrogen feedbacks on crop yields under climate change. *Agricultural and Environmental Letters* 3(1), 5. doi:10.2134/ael2018.05.0026.

Bassu, S., Brisson, N., Durand, J. L., Boote, K., Lizaso, J., Jones, J. W., Rosenzweig, C., Ruane, A. C., Adam, M., Baron, C., Basso, B., Biernath, C., Boogaard, H., Conijn, S., Corbeels, M., Deryng, D., De Sanctis, G., Gayler, S., Grassini, P., Hatfield, J., Hoek, S., Izaurralde, C., Jongschaap, R., Kemanian, A. R., Kersebaum, K. C., Kim, S. H., Kumar, N. S., Makowski, D., Muller, C., Nendel, C., Priesack, E., Pravia, M. V., Sau, F., Shcherbak, I., Tao, F., Teixeira, E., Timlin, D. and Waha, K. 2014. How do various maize crop models vary in their responses to climate change factors? *Global Change Biology* 20(7), 2301-20. doi:10.1111/gcb.12520.

Boote, K. J., Jones, J. W., White, J. W., Asseng, S. and Lizaso, J. I. 2013. Putting mechanisms into crop production models. *Plant, Cell and Environment* 36(9), 1658-72. doi:10.1111/pce.12119.

Brown, H. E., Huth, N. I., Holzworth, D. P., Teixeira, E. I., Zyskowski, R. F., Hargreaves, J. N. G. and Moot, D. J. 2014. Plant Modelling Framework: software for building and running crop models on the APSIM platform. *Environmental Modelling and Software* 62, 385-98. doi:10.1016/j.envsoft.2014.09.005.

Cammarano, D., Rötter, R. P., Asseng, S., Ewert, F., Wallach, D., Martre, P., Hatfield, J. L., Jones, J. W., Rosenzweig, C., Ruane, A. C., Boote, K. J., Thorburn, P. J., Kersebaum, K. C., Aggarwal, P. K., Angulo, C., Basso, B., Bertuzzi, P., Biernath, C., Brisson, N., Challinor, A. J., Doltra, J., Gayler, S., Goldberg, R., Heng, L., Hooker, J. E., Hunt, L. A., Ingwersen, J., Izaurralde, R. C., Müller, C., Kumar, S. N., Nendel, C., O'Leary, G. J., Olesen, J. E., Osborne, T. M., Palosuo, T., Priesack, E., Ripochel, D., Semenov,

M. A., Shcherbak, I., Steduto, P., Stöckle, C. O., Stratonovitch, P., Streck, T., Supit, I., Tao, F., Travasso, M., Waha, K., White, J. W. and Wolf, J. 2016. Uncertainty of wheat water use: simulated patterns and sensitivity to temperature and CO_2. *Field Crops Research* 198, 80–92. doi:10.1016/j.fcr.2016.08.015.

Challinor, A. J., Watson, J., Lobell, D. B., Howden, S. M., Smith, D. R. and Chhetri, N. 2014. A meta-analysis of crop yield under climate change and adaptation. *Nature Climate Change* 4(4), 287–91. doi:10.1038/nclimate2153.

Chenu, K., Porter, J. R., Martre, P., Basso, B., Chapman, S. C., Ewert, F., Bindi, M. and Asseng, S. 2017. Contribution of crop models to adaptation in wheat. *Trends in Plant Science* 22(6), 472–90. doi:10.1016/j.tplants.2017.02.003.

Confalonieri, R., Orlando, F., Paleari, L., Stella, T., Gilardelli, C., Movedi, E., Pagani, V., Cappelli, G., Vertemara, A., Alberti, L., Alberti, P., Atanassiu, S., Bonaiti, M., Cappelletti, G., Ceruti, M., Confalonieri, A., Corgatelli, G., Corti, P., Dell'Oro, M., Ghidoni, A., Lamarta, A., Maghini, A., Mambretti, M., Manchia, A., Massoni, G., Mutti, P., Pariani, S., Pasini, D., Pesenti, A., Pizzamiglio, G., Ravasio, A., Rea, A., Santorsola, D., Serafini, G., Slavazza, M. and Acutis, M. 2016. Uncertainty in crop model predictions: what is the role of users? *Environmental Modelling and Software* 81, 165–73. doi:10.1016/j.envsoft.2016.04.009.

De Willigen, P. 1991. Nitrogen turnover in the soil-crop system – comparison of 14 simulation-models. *Fertilizer Research* 27, 141–9.

De Wit, C. T. 1965. Photosynthesis of leaf canopies. Agric Res Reps Centre agric Publ Docum 663, 57pp

Deryng, D., Elliott, J., Folberth, C., Muller, C., Pugh, T. A. M., Boote, K. J., Conway, D., Ruane, A. C., Gerten, D., Jones, J. W., Khabarov, N., Olin, S., Schapho, S., Schmid, E., Yang, H. and Rosenzweig, C. 2016. Regional disparities in the beneficial effects of rising CO_2 concentrations on crop water productivity. *Nature Climate Change* 6(8), 786–90+. doi:10.1038/nclimate2995.

Diekkruger, B., Sondgerath, D., Kersebaum, K. C. and McVoy, C. W. 1995. Validity of agroecosystem models – a comparison of results of different models applied to the same data set. *Ecological Modelling* 81(1–3), 3–29. doi:10.1016/0304-3800(94)00157-D.

Durand, J. L., Delusca, K., Boote, K., Lizaso, J., Manderscheid, R., Weigel, H. J., Ruane, A. C., Rosenzweig, C., Jones, J., Ahuja, L., Anapalli, S., Basso, B., Baron, C., Bertuzzi, P., Biernath, C., Deryng, D., Ewert, F., Gaiser, T., Gayler, S., Heinlein, F., Kersebaum, K. C., Kim, S. H., Muller, C., Nendel, C., Olioso, A., Priesack, E., Villegas, J. R., Ripoche, D., Rotter, R. P., Seidel, S. I., Srivastava, A., Tao, F., Timlin, D., Twine, T., Wang, E., Webber, H. and Zhao, Z. 2018. How accurately do maize crop models simulate the interactions of atmospheric CO_2 concentration levels with limited water supply on water use and yield? *European Journal of Agronomy* 100, 67–75. doi:10.1016/j.eja.2017.01.002.

Ehrhardt, F., Soussana, J. F., Bellocchi, G., Grace, P., McAuliffe, R., Recous, S., Sandor, R., Smith, P., Snow, V., Migliorati, M. D., Basso, B., Bhatia, A., Brilli, L., Doltra, J., Dorich, C. D., Doro, L., Fitton, N., Giacomini, S. J., Grant, B., Harrison, M. T., Jones, S. K., Kirschbaum, M. U. F., Klumpp, K., Laville, P., Leonard, J., Liebig, M., Lieffering, M., Martin, R., Massad, R. S., Meier, E., Merbold, L., Moore, A. D., Myrgiotis, V., Newton, P., Pattey, E., Rolinski, S., Sharp, J., Smith, W. N., Wu, L. H. and Zhang, Q. 2018. Assessing uncertainties in crop and pasture ensemble model simulations of productivity and N_2O emissions. *Global Change Biology* 24, E603–16. doi:10.1111/gcb.13965.

Emberson, L. D., Pleijel, H., Ainsworth, E. A., van den Berg, M., Ren, W., Osborne, S., Mills, G., Pandey, D., Dentener, F., Buker, P., Ewert, F., Koeble, R. and Van Dingenen, R.

2018. Ozone effects on crops and consideration in crop models. *European Journal of Agronomy* 100, 19–34. doi:10.1016/j.eja.2018.06.002.

Ewert, F., Rodriguez, D., Jamieson, P., Semenov, M. A., Mitchell, R. A. C., Goudriaan, J., Porter, J. R., Kimball, B. A., Pinter, P. J., Manderscheid, R., Weigel, H. J., Fangmeier, A., Fereres, E. and Villalobos, F. 2002. Effects of elevated CO_2 and drought on wheat: testing crop simulation models for different experimental and climatic conditions. *Agriculture, Ecosystems and Environment* 93(1–3), 249–66. doi:10.1016/S0167-8809(01)00352-8.

Ewert, F., van Ittersum, M. K., Heckelei, T., Therond, O., Bezlepkina, I. and Andersen, E. 2011. Scale changes and model linking methods for integrated assessment of agri-environmental systems. *Agriculture, Ecosystems and Environment* 142(1–2), 6–17. doi:10.1016/j.agee.2011.05.016.

Ewert, F., Rötter, R. P., Bindi, M., Webber, H., Trnka, M., Kersebaum, K. C., Olesen, J. E., van Ittersum, M. K., Janssen, S., Rivington, M., Semenov, M. A., Wallach, D., Porter, J. R., Stewart, D., Verhagen, J., Gaiser, T., Palosuo, T., Tao, F., Nendel, C., Roggero, P. P., Bartošová, L. and Asseng, S. 2015a. Crop modelling for integrated assessment of risk to food production from climate change. *Environmental Modelling and Software* 72, 287–303. doi:10.1016/j.envsoft.2014.12.003.

Ewert, F., van Bussel, L. G. J., Zhao, G., Hoffmann, H., Gaiser, T., Specka, X., Nendel, C., Kersebaum, K., Sosa, C., Lewan, E., Yeluripati, J., Kuhnert, M., Tao, F., Rötter, R., Constantin, J., Raynal, H., Wallach, D., Teixeira, E., Grosz, B., Bach, M., Doro, L., Roggero, P. P., Zhao, Z., Wang, E., Kiese, R., Haas, E., Eckersten, H., Trombi, G., Bindi, M., Klein, C., Biernath, C., Heinlein, F., Priesack, E., Cammarano, D., Asseng, S., Elliott, J., Glotter, M., Basso, B., Baigorria, G. A., Romero, C. C. and Moriondo, M. 2015b. Uncertainties in scaling up crop models for large area climate change impact assessments. In: Rosenzweig, C. and Hillel, D. (Eds), *Handbook of Climate Change and Agroecosystems: The Agricultural Model Intercomparison and Improvement Project (AgMIP)* (vol. 4). Imperial College Press, London, pp. 261–77. Chapter 20. doi:10.1142/9781783265640_0010.

Fleisher, D. H., Condori, B., Quiroz, R., Alva, A., Asseng, S., Barreda, C., Bindi, M., Boote, K. J., Ferrise, R., Franke, A. C., Govindakrishnan, P. M., Harahagazwe, D., Hoogenboom, G., Kumar, S. N., Merante, P., Nendel, C., Olesen, J. E., Parker, P. S., Raes, D., Raymundo, R., Ruane, A. C., Stockle, C., Supit, I., Vanuytrecht, E., Wolf, J. and Woli, P. 2017. A potato model intercomparison across varying climates and productivity levels. *Global Change Biology* 23(3), 1258–81. doi:10.1111/gcb.13411.

Fronzek, S., Pirttioja, N., Carter, T. R., Bindi, M., Hoffmann, H., Palosuo, T., Ruiz-Ramos, M., Tao, F. L., Trnka, M., Acutis, M., Asseng, S., Baranowski, P., Basso, B., Bodin, P., Buis, S., Cammarano, D., Deligios, P., Destain, M. F., Dumont, B., Ewert, F., Ferrise, R., Francois, L., Gaiser, T., Hlavinka, P., Jacquemin, I., Kersebaum, K. C., Kollas, C., Krzyszczak, J., Lorite, I. J., Minet, J., Minguez, M. I., Montesino, M., Moriondo, M., Muller, C., Nendel, C., Ozturk, I., Perego, A., Rodriguez, A., Ruane, A. C., Ruget, F., Sanna, M., Semenov, M. A., Slawinski, C., Stratonovitch, P., Supit, I., Waha, K., Wang, E. L., Wu, L. H., Zhao, Z. G. and Rotter, R. P. 2018. Classifying multi-model wheat yield impact response surfaces showing sensitivity to temperature and precipitation change. *Agricultural Systems* 159, 209–24. doi:10.1016/j.agsy.2017.08.004.

Grosz, B., Dechow, R., Gebbert, S., Hoffmann, H., Zhao, G., Constantin, J., Raynal, H., Wallach, D., Coucheney, E., Lewan, E., Eckersten, H., Specka, X., Kersebaum, K. C., Nendel, C., Kuhnert, M., Yeluripati, J., Haas, E., Teixeira, E., Bindi, M., Trombi, G.,

Moriondo, M., Doro, L., Roggero, P. P., Zhao, Z., Wang, E., Tao, F., Roetter, R., Kassie, B., Cammarano, D., Asseng, S., Weihermueller, L., Siebert, S., Gaiser, T. and Ewert, F. 2017. The implication of input data aggregation on up-scaling soil organic carbon changes. *Environmental Modelling and Software* 96, 361–77. doi:10.1016/j.envsoft.2017.06.046.

Hasegawa, T., Li, T., Yin, X., Zhu, Y., Boote, K., Baker, J., Bregaglio, S., Buis, S., Confalonieri, R., Fugice, J., Fumoto, T., Gaydon, D., Kumar, S. N., Lafarge, T., Marcaida Iii, M., Masutomi, Y., Nakagawa, H., Oriol, P., Ruget, F., Singh, U., Tang, L., Tao, F., Wakatsuki, H., Wallach, D., Wang, Y., Wilson, L. T., Yang, L., Yang, Y., Yoshida, H., Zhang, Z. and Zhu, J. 2017. Causes of variation among rice models in yield response to CO_2 examined with Free-Air CO_2 Enrichment and growth chamber experiments. *Scientific Reports* 7(1), 14858. doi:10.1038/s41598-017-13582-y.

Hernandez-Ochoa, I. M., Asseng, S., Kassie, B. T., Xiong, W., Robertson, R., Luz Pequeno, D. N., Sonder, K., Reynolds, M., Babar, M. A., Molero Milan, A. and Hoogenboom, G. 2018. Climate change impact on Mexico wheat production. *Agricultural and Forest Meteorology* 263, 373–87. doi:10.1016/j.agrformet.2018.09.008.

Hoffmann, H., Zhao, G., van Bussel, L., Enders, A., Specka, X., Sosa, C., Yeluripati, J., Tao, F., Constantin, J., Raynal, H., Teixeira, E., Grosz, B., Doro, L., Zhao, Z., Wang, E., Nendel, C., Kersebaum, K., Haas, E., Kiese, R., Klatt, S., Eckersten, H., Vanuytrecht, E., Kuhnert, M., Lewan, E., Rötter, R., Roggero, P., Wallach, D., Cammarano, D., Asseng, S., Krauss, G., Siebert, S., Gaiser, T. and Ewert, F. 2015. Variability of effects of spatial climate data aggregation on regional yield simulation by crop models. *Climate Research* 65, 53–69. doi:10.3354/cr01326.

Hoffmann, H., Zhao, G., Asseng, S., Bindi, M., Biernath, C., Constantin, J., Coucheney, E., Dechow, R., Doro, L., Eckersten, H., Gaiser, T., Grosz, B., Heinlein, F., Kassie, B. T., Kersebaum, K. C., Klein, C., Kuhnert, M., Lewan, E., Moriondo, M., Nendel, C., Priesack, E., Raynal, H., Roggero, P. P., Rotter, R. P., Siebert, S., Specka, X., Tao, F., Teixeira, E., Trombi, G., Wallach, D., Weihermueller, L., Yeluripati, J. and Ewert, F. 2016. Impact of spatial soil and climate input data aggregation on regional yield simulations. *PLoS ONE* 11(4), e0151782. doi:10.1371/journal.pone.0151782.

Hoffmann, M. P., Haakana, M., Asseng, S., Hohn, J. G., Palosuo, T., Ruiz-Ramos, M., Fronzek, S., Ewert, F., Gaiser, T., Kassie, B. T., Paff, K., Rezaei, E. E., Rodriguez, A., Semenov, M., Srivastava, A. K., Stratonovitch, P., Tao, F., Chen, Y. and Rotter, R. P. 2018. How does inter-annual variability of attainable yield affect the magnitude of yield gaps for wheat and maize? An analysis at ten sites. *Agricultural Systems* 159, 199–208. doi:10.1016/j.agsy.2017.03.012.

Holzworth, D. P., Huth, N. I., Devoil, P. G., Zurcher, E. J., Herrmann, N. I., McLean, G., Chenu, K., van Oosterom, E. J., Snow, V., Murphy, C., Moore, A. D., Brown, H., Whish, J. P. M., Verrall, S., Fainges, J., Bell, L. W., Peake, A. S., Poulton, P. L., Hochman, Z., Thorburn, P. J., Gaydon, D. S., Dalgliesh, N. P., Rodriguez, D., Cox, H., Chapman, S., Doherty, A., Teixeira, E., Sharp, J., Cichota, R., Vogeler, I., Li, F. Y., Wang, E. L., Hammer, G. L., Robertson, M. J., Dimes, J. P., Whitbread, A. M., Hunt, J., van Rees, H., McClelland, T., Carberry, P. S., Hargreaves, J. N. G., MacLeod, N., McDonald, C., Harsdorf, J., Wedgwood, S. and Keating, B. A. 2014. APSIM – Evolution towards a new generation of agricultural systems simulation. *Environmental Modelling and Software* 62, 327–50. doi:10.1016/j.envsoft.2014.07.009.

Holzworth, D. P., Snow, V., Janssen, S., Athanasiadis, I. N., Donatelli, M., Hoogenboom, G., White, J. W. and Thorburn, P. 2015. Agricultural production systems modelling and software: current status and future prospects. *Environmental Modelling and Software* 72, 276–86. doi:10.1016/j.envsoft.2014.12.013.

Ippolito, A., Kattwinkel, M., Rasmussen, J. J., Schäfer, R. B., Fornaroli, R. and Liess, M. 2015. Modeling global distribution of agricultural insecticides in surface waters. *Environmental Pollution* 198, 54–60. doi:10.1016/j.envpol.2014.12.016.

Jamieson, P. D., Porter, J. R., Goudriaan, J., Ritchie, J. T., van Keulen, H. and Stol, W. 1998. A comparison of the models AFRCWHEAT2, CERES-wheat, Sirius, SUCROS2 and SWHEAT with measurements from wheat grown under drought. *Field Crops Research* 55(1–2), 23–44. doi:10.1016/S0378-4290(97)00060-9.

Jamieson, P. D., Porter, J. R., Semenov, M. A., Brooks, R. J., Ewert, F. and Ritchie, J. T. 1999. Comments on 'Testing winter wheat simulation models predictions against observed UK grain yields' by Landau, et al. (1998). *Agricultural and Forest Meteorology* 96(1–3), 157–61. doi:10.1016/S0168-1923(99)00060-X.

Jamieson, P. D., Berntsen, J., Ewert, F., Kimball, B. A., Olesen, J. E., Pinter Jr., P. J., Porter, J. R. and Semenov, M. A. 2000. Modelling CO_2 effects on wheat with varying nitrogen supplies. *Agriculture, Ecosystems and Environment* 82(1–3), 27–37. doi:10.1016/S0167-8809(00)00214-0.

Jeuffroy, M. H., Casadebaig, P., Debaeke, P., Loyce, C. and Meynard, J. M. 2014. Agronomic model uses to predict cultivar performance in various environments and cropping systems. A review. *Agronomy for Sustainable Development* 34(1), 121–37. doi:10.1007/s13593-013-0170-9.

Jones, J. W., Antle, J. M., Basso, B., Boote, K. J., Conant, R. T., Foster, I., Godfray, H. C. J., Herrero, M., Howitt, R. E., Janssen, S., Keating, B. A., Munoz-Carpena, R., Porter, C. H., Rosenzweig, C. and Wheeler, T. R. 2017a. Brief history of agricultural systems modeling. *Agricultural Systems* 155, 240–54. doi:10.1016/j.agsy.2016.05.014.

Jones, J. W., Antle, J. M., Basso, B., Boote, K. J., Conant, R. T., Foster, I., Godfray, H. C. J., Herrero, M., Howitt, R. E., Janssen, S., Keating, B. A., Munoz-Carpena, R., Porter, C. H., Rosenzweig, C. and Wheeler, T. R. 2017b. Toward a new generation of agricultural system data, models, and knowledge products: state of agricultural systems science. *Agricultural Systems* 155, 269–88. doi:10.1016/j.agsy.2016.09.021.

Kersebaum, K. C., Hecker, J.-M., Mirschel, M. and Wegenhenkel, M. 2007. Modelling water and nutrient dynamics in soil-crop systems: a comparison of simulation models applied on common data sets. In: Kersebaum, K. C., Hecker, J.-M., Mirschel, M. and Wegenhenkel, M. (Eds), *Modelling Water and Nutrient Dynamics in Soil-Crop Systems*. Springer, Dordrecht.

Kersebaum, K. C., Kroes, J., Gobin, A., Takac, J., Hlavinka, P., Trnka, M., Ventrella, D., Giglio, L., Ferrise, R., Moriondo, M., Dalla Marta, A., Luo, Q. Y., Eitzinger, J., Mirschel, W., Weigel, H. J., Manderscheid, R., Hoffmann, M., Nejedlik, P., Iqbal, M. A. and Hosch, J. 2016. Assessing uncertainties of water footprints using an ensemble of crop growth models on winter wheat. *Water* 8(12). doi:10.3390/w8120571.

Kollas, C., Kersebaum, K. C., Nendel, C., Manevski, K., Muller, C., Palosuo, T., Armas-Herrera, C. M., Beaudoin, N., Bindi, M., Charfeddine, M., Conradt, T., Constantin, J., Eitzinger, J., Ewert, F., Ferrise, R., Gaiser, T., de Cortazar-Atauri, I. Gd, Giglio, L., Hlavinka, P., Hoffmann, H., Hoffmann, M. P., Launay, M., Manderscheid, R., Mary, B., Mirschel, W., Moriondo, M., Olesen, J. E., Ozturk, I., Pacholski, A., Ripoche-Wachter, D., Roggero, P. P., Roncossek, S., Rotter, R. P., Ruget, F., Sharif, B., Trnka, M., Ventrella, D., Waha, K., Wegehenkel, M., Weigel, H. J. and Wu, L. H. 2015. Crop rotation modelling-A European model intercomparison. *European Journal of Agronomy* 70, 98–111. doi:10.1016/j.eja.2015.06.007.

Kuhnert, M., Yeluripati, J., Smith, P., Hoffmann, H., van Oijen, M., Constantin, J., Coucheney, E., Dechow, R., Eckersten, H., Gaiser, T., Grosz, B., Haas, E., Kersebaum, K. C., Kiese, R., Klatt, S., Lewan, E., Nendel, C., Raynal, H., Sosa, C., Specka, X., Teixeira, E., Wang, E. L., Weihermuller, L., Zhao, G., Zhao, Z. G., Ogle, S. and Ewert, F. 2017. Impact analysis of climate data aggregation at different spatial scales on simulated net primary productivity for croplands. *European Journal of Agronomy* 88, 41–52. doi:10.1016/j.eja.2016.06.005.

Landau, S., Mitchell, R. A. C., Barnett, V., Colls, J. J., Craigon, J., Moore, K. L. and Payne, R. W. 1998. Testing winter wheat simulation models' predictions against observed UK grain yields. *Agricultural and Forest Meteorology* 89(2), 85–99. doi:10.1016/S0168-1923(97)00069-5.

Li, T., Hasegawa, T., Yin, X., Zhu, Y., Boote, K., Adam, M., Bregaglio, S., Buis, S., Confalonieri, R., Fumoto, T., Gaydon, D., Marcaida, M., Nakagawa, H., Oriol, P., Ruane, A. C., Ruget, F., Singh, B., Singh, U., Tang, L., Tao, F., Wilkens, P., Yoshida, H., Zhang, Z. and Bouman, B. 2015. Uncertainties in predicting rice yield by current crop models under a wide range of climatic conditions. *Global Change Biology* 21(3), 1328–41. doi:10.1111/gcb.12758.

Liu, B., Asseng, S., Liu, L., Tang, L., Cao, W. and Zhu, Y. 2016a. Testing the responses of four wheat crop models to heat stress at anthesis and grain filling. *Global Change Biology* 22(5), 1890–903. doi:10.1111/gcb.13212.

Liu, B., Liu, L., Asseng, S., Zou, X., Li, J., Cao, W. and Zhu, Y. 2016b. Modelling the effects of heat stress on post-heading durations in wheat: A comparison of temperature response routines. *Agricultural and Forest Meteorology* 222, 45–58. doi:10.1016/j.agrformet.2016.03.006.

Liu, B., Asseng, S., Wang, A. N., Wang, S. H., Tang, L., Cao, W. X., Zhu, Y. and Liu, L. L. 2017. Modelling the effects of post-heading heat stress on biomass growth of winter wheat. *Agricultural and Forest Meteorology* 247, 476–90. doi:10.1016/j.agrformet.2017.08.018.

Liu, B., Martre, P., Ewert, F., Porter, J. R., Challinor, A. J., Müller, C., Ruane, A. C., Waha, K., Thorburn, P. J., Aggarwal, P. K., Ahmed, M., Balkovič, J., Basso, B., Biernatht, C., Bindi, M., Cammarano, D., De Sanctisw, G., Dumont, B., Espadafor, M., Rezaei, E. E., Ferrise, R., Garcia-Vila, M., Gayler, S., Gao, Y., Horan, H., Hoogenboom, G., Izaurralde, R. C., Jones, C. D., Kassie, B. T., Kersebaum, K. C., Klein, C., Koehlerh, A. K., Maiorano, A. and Minoli, S., Montesino San Martine, M., Kumar, S. N., Nendel, C., O'Leary, J. O., Palosuo, T., Priesack, E., Ripoche, D., Rötter, R. P., Semenov, M. A., Stöcklen, C. O., Streck, T., Supit, I., Tao, F., Van der Veldeff, M., Wallach, D., Wang, E., Webber, H., Wolf, J., Xiao, L., Zhang, Z., Zhao, Z., Zhu, Y. and Asseng, S. 2019. Global wheat production with 1.5 and 2.0°C above pre-industrial warming. *Global Change Biology* 25, 1428–44.

Maiorano, A., Martre, P., Asseng, S., Ewert, F., Mueller, C., Rotter, R. P., Ruane, A. C., Semenov, M. A., Wallach, D., Wang, E., Alderman, P. D., Kassie, B. T., Biernath, C., Basso, B., Cammarano, D., Challinor, A. J., Doltra, J., Dumont, B., Rezaei, E. E., Gayler, S., Kersebaum, K. C., Kimball, B. A., Koehler, A.-K., Liu, B., O'Leary, G. J., Olesen, J. E., Ottman, M. J., Priesack, E., Reynolds, M., Stratonovitch, P., Streck, T., Thorburn, P. J., Waha, K., Wall, G. W., White, J. W., Zhao, Z. and Zhu, Y. 2017. Crop model improvement reduces the uncertainty of the response to temperature of multi-model ensembles. *Field Crops Research* 202, 5-20. doi:10.1016/j.fcr.2016.05.001.

Makowski, D., Asseng, S., Ewert, F., Bassu, S., Durand, J. L., Li, T., Martre, P., Adam, M., Aggarwal, P. K., Angulo, C., Baron, C., Basso, B., Bertuzzi, P., Biernath, C., Boogaard, H., Boote, K. J., Bouman, B., Bregaglio, S., Brisson, N., Buis, S., Cammarano, D., Challinor, A. J., Confalonieri, R., Conijn, J. G., Corbeels, M., Deryng, D., De Sanctis, G., Doltra, J., Fumoto, T., Gaydon, D., Gayler, S., Goldberg, R., Grant, R. F., Grassini, P., Hatfield, J. L., Hasegawa, T., Heng, L., Hoek, S., Hooker, J., Hunt, L. A., Ingwersen, J., Izaurralde, R. C., Jongschaap, R. E. E., Jones, J. W., Kemanian, R. A., Kersebaum, K. C., Kim, S. H., Lizaso, J., Marcaida III, M., Mueller, C., Nakagawa, H., Kumar, S. N., Nendel, C., O'Leary, G. J., Olesen, J. E., Oriol, P., Osborne, T. M., Palosuo, T., Pravia, M. V., Priesack, E., Ripoche, D., Rosenzweig, C., Ruane, A. C., Ruget, F., Sau, F., Semenov, M. A., Shcherbak, I., Singh, B., Singh, U., Soo, H. K., Steduto, P., Stoeckle, C., Stratonovitch, P., Streck, T., Supit, I., Tang, L., Tao, F., Teixeira, E. I., Thorburn, P., Timlin, D., Travasso, M., Roetter, R. P., Waha, K., Wallach, D., White, J. W., Wilkens, P., Williams, J. R., Wolf, J., Yin, X., Yoshida, H., Zhang, Z. and Zhu, Y. 2015. A statistical analysis of three ensembles of crop model responses to temperature and CO2 concentration. *Agricultural and Forest Meteorology* 214-215, 483-93. doi:10.1016/j.agrformet.2015.09.013.

Martre, P., Wallach, D., Asseng, S., Ewert, F., Jones, J. W., Rotter, R. P., Boote, K. J., Ruane, A. C., Thorburn, P. J., Cammarano, D., Hatfield, J. L., Rosenzweig, C., Aggarwal, P. K., Angulo, C., Basso, B., Bertuzzi, P., Biernath, C., Brisson, N., Challinor, A. J., Doltra, J., Gayler, S., Goldberg, R., Grant, R. F., Heng, L., Hooker, J., Hunt, L. A., Ingwersen, J., Izaurralde, R. C., Kersebaum, K. C., Mueller, C., Kumar, S. N., Nendel, C., O'Leary, G., Olesen, J. E., Osborne, T. M., Palosuo, T., Priesack, E., Ripoche, D., Semenov, M. A., Shcherbak, I., Steduto, P., Stoeckle, C. O., Stratonovitch, P., Streck, T., Supit, I., Tao, F., Travasso, M., Waha, K., White, J. W. and Wolf, J. 2015. Multimodel ensembles of wheat growth: many models are better than one. *Global Change Biology* 21(2), 911-25. doi:10.1111/gcb.12768.

Millar, A. J., Urquiza, U., Freeman, P. L., Hume, A., Plotkin, G. D., Sorokina, O., Zardilis, A. and Zielinski, T. 2019. Practical steps to digital organism models, from laboratory model species to 'Crops in silico. *Journal of Experimental Botany* 70(9), 2403-18, doi:10.1093/jxb/ery435.

Monteith, J. L. 1965. Light distribution and photosynthesis in field crops. *Annals of Botany* 29(1), 17-37. doi:10.1093/oxfordjournals.aob.a083934.

Muller, C., Elliott, J., Pugh, T. A. M., Ruane, A. C., Ciais, P., Balkovic, J., Deryng, D., Folberth, C., Izaurralde, R. C., Jones, C. D., Khabarov, N., Lawrence, P., Liu, W., Reddy, A. D., Schmid, E. and Wang, X. H. 2018. Global patterns of crop yield stability under additional nutrient and water inputs. *PLoS ONE* 13(6), e0198748. doi:10.1371/journal.pone.0198748.

Nelson, G. C., van der Mensbrugghe, D., Ahammad, H., Blanc, E., Calvin, K., Hasegawa, T., Havlik, P., Heyhoe, E., Kyle, P., Lotze-Campen, H., von Lampe, M., Mason d'Croz, D., van Meijl, H., Müller, C., Reilly, J., Robertson, R., Sands, R. D., Schmitz, C., Tabeau, A., Takahashi, K., Valin, H. and Willenbockel, D. 2014 Agriculture and climate change in global scenarios: why don't the models agree. *Agricultural Economics* 45(1), 85-101. doi:10.1111/agec.12091.

Neveu, P., Tireau, A., Hilgert, N., Negre, V., Mineau-Cesari, J., Brichet, N., Chapuis, R., Sanchez, I., Pommier, C., Charnomordic, B., Tardieu, F. and Cabrera-Bosquet, L.

2019. Dealing with multi-source and multi-scale information in plant phenomics: the ontology-driven Phenotyping Hybrid Information System. *The New Phytologist* 221(1), 588–601. doi:10.1111/nph.15385.

O'Leary, G. J., Christy, B., Nuttall, J., Huth, N., Cammarano, D., Stoeckle, C., Basso, B., Shcherbak, I., Fitzgerald, G., Luo, Q., Farre-Codina, I., Palta, J. and Asseng, S. 2015. Response of wheat growth, grain yield and water use to elevated CO_2 under a Free-Air CO_2 Enrichment (FACE) experiment and modelling in a semi-arid environment. *Global Change Biology* 21(7), 2670–86. doi:10.1111/gcb.12830.

Palosuo, T., Kersebaum, K. C., Angulo, C., Hlavinka, P., Moriondo, M., Olesen, J. E., Patil, R. H., Ruget, F., Rumbaur, C., Takac, J., Trnka, M., Bindi, M., Caldag, B., Ewert, F., Ferrise, R., Mirschel, W., Saylan, L., Siska, B. and Rotter, R. 2011. Simulation of winter wheat yield and its variability in different climates of Europe: A comparison of eight crop growth models. *European Journal of Agronomy* 35(3), 103–14. doi:10.1016/j.eja.2011.05.001.

Pirttioja, N., Carter, T. R., Fronzek, S., Bindi, M., Hoffmann, H., Palosuo, T., Ruiz-Ramos, M., Tao, F., Trnka, M., Acutis, M., Asseng, S., Baranowski, P., Basso, B., Bodin, P., Buis, S., Cammarano, D., Deligios, P., Destain, M. F., Dumont, B., Ewert, F., Ferrise, R., Francois, L., Gaiser, T., Hlavinka, P., Jacquemin, I., Kersebaum, K. C., Kollas, C., Krzyszczak, J., Lorite, I. J., Minet, J., Minguez, M. I., Montesino, M., Moriondo, M., Mueller, C., Nendel, C., Ozturk, I., Perego, A., Rodriguez, A., Ruane, A. C., Ruget, F., Sanna, M., Semenov, M. A., Slawinski, C., Stratonovitch, P., Supit, I., Waha, K., Wang, E., Wu, L., Zhao, Z. and Roetter, R. P. 2015. Temperature and precipitation effects on wheat yield across a European transect: a crop model ensemble analysis using impact response surfaces. *Climate Research* 65, 87–105. doi:10.3354/cr01322.

Porter, J. R., Jamieson, P. D. and Wilson, D. R. 1993. Comparison of the wheat simulation models AFRCWHEAT2, CERES-Wheat and SWHEAT for non-limiting conditions of crop growth. *Field Crops Research* 33(1–2), 131–57. doi:10.1016/0378-4290(93)90098-8.

Porter, C. H., Villalobos, C., Holzworth, D., Nelson, R., White, J. W., Athanasiadis, I. N., Janssen, S., Ripoche, D., Cufi, J., Raes, D., Zhang, M., Knapen, R., Sahajpal, R., Boote, K. and Jones, J. W. 2014. Harmonization and translation of crop modeling data to ensure interoperability. *Environmental Modelling and Software* 62, 495–508. doi:10.1016/j.envsoft.2014.09.004.

Reynolds, M., Kropff, M., Crossa, J., Koo, J., Kruseman, G., Molero-Milan, A., Rutkoski, J., Schulthess, U., Singh, B., Sonder, K., Tonnang, H. and Vadez, V. 2018. Role of modelling in international crop research: overview and some case studies. *Agronomy* 8(12). doi:10.3390/agronomy8120291.

Rosenzweig, C. and Hillel, D. (Eds). 2015. *Handbook of Climate Change and Agroecosystems: the Agricultural Model Intercomparison and Improvement Project (AgMIP) Integrated Crop and Economic Assessments (In Two Parts)*. Imperial College Press, London, UK.

Rosenzweig, C., Jones, J. W., Hatfield, J. L., Ruane, A. C., Boote, K. J., Thorburne, P., Antle, J. M., Nelson, G. C., Porter, C., Janssen, S., Asseng, S., Basso, B., Ewert, F., Wallach, D., Baigorria, G. and Winter, J. M. 2013. The Agricultural Model Intercomparison and Improvement Project (AgMIP): protocols and pilot studies. *Agricultural and Forest Meteorology* 170, 166–82. doi:10.1016/j.agrformet.2012.09.011.

Rosenzweig, C., Elliott, J., Deryng, D., Ruane, A. C., Muller, C., Arneth, A., Boote, K. J., Folberth, C., Glotter, M., Khabarov, N., Neumann, K., Piontek, F., Pugh, T. A. M., Schmid, E., Stehfest, E., Yang, H. and Jones, J. W. 2014. Assessing agricultural risks of climate change in the 21st century in a global gridded crop model intercomparison. *Proceedings of the National Academy of Sciences of the United States of America* 111(9), 3268–73. doi:10.1073/pnas.1222463110.

Rosenzweig, C., Ruane, A. C., Antle, J., Elliott, J., Ashfaq, M., Chatta, A. A., Ewert, F., Folberth, C., Hathie, I., Havlik, P., Hoogenboom, G., Lotze-Campen, H., MacCarthy, D. S., Mason-D'Croz, D., Contreras, E. M., Muller, C., Perez-Dominguez, I., Phillips, M., Porter, C., Raymundo, R. M., Sands, R. D., Schleussner, C. F., Valdivia, R. O., Valin, H. and Wiebe, K. 2018. Coordinating AgMIP data and models across global and regional scales for 1.5°C and 2.0°C assessments. *Philosophical Transactions of the Royal Society, Series A-Mathematical, Physical and Engineering Sciences* 376(2119). doi:10.1098/rsta.2016.0455.

Rötter, R. P., Palosuo, T., Kersebaum, K. C., Angulo, C., Bindi, M., Ewert, F., Ferrise, R., Hlavinka, P., Moriondo, M., Nendel, C., Olesen, J. E., Patil, R. H., Ruget, F., Takac, J. and Trnka, M. 2012. Simulation of spring barley yield in different climatic zones of Northern and Central Europe: A comparison of nine crop models. *Field Crops Research* 133, 23–36. doi:10.1016/j.fcr.2012.03.016.

Rötter, R. P., Carter, T. R., Olesen, J. E. and Porter, J. R. 2011. Crop-climate models need an overhaul. *Nature Climate Change* 1(4), 175–7. doi:10.1038/nclimate1152.

Ruane, A. C., Hudson, N. I., Asseng, S., Camarrano, D., Ewert, F., Martre, P., Boote, K. J., Thorburn, P. J., Aggarwal, P. K., Angulo, C., Basso, B., Bertuzzi, P., Biernath, C., Brisson, N., Challinor, A. J., Doltra, J., Gayler, S., Goldberg, R., Grant, R. F., Heng, L., Hooker, J., Hunt, L. A., Ingwersen, J., Izaurralde, R. C., Kersebaum, K. C., Kumar, S. N., Müller, C., Nendel, C., O'Leary, G., Olesen, J. E., Osborne, T. M., Palosuo, T., Priesack, E., Ripoche, D., Rötter, R. P., Semenov, M. A., Shcherbak, I., Steduto, P., Stöckle, C. O., Stratonovitch, P., Streck, T., Supit, I., Tao, F., Travasso, M., Waha, K., Wallach, D., White, J. W. and Wolf, J. 2016. Multi-wheat-model ensemble responses to interannual climate variability. *Environmental Modelling and Software* 81, 86–101. doi:10.1016/j.envsoft.2016.03.008.

Ruane, A. C., Rosenzweig, C., Asseng, S., Boote, K. J., Elliott, J., Ewert, F., Jones, J. W., Martre, P., McDermid, S. P., Müller, C., Snyder, A. and Thorburn, P. J. 2017. An AgMIP framework for improved agricultural representation in integrated assessment models. *Environmental Research Letters* 12(12). doi:10.1088/1748-9326/aa8da6]: 125003.

Ruiz-Ramos, M., Ferrise, R., Rodriguez, A., Lorite, I. J., Bindi, M., Carter, T. R., Fronzek, S., Palosuo, T., Pirttioja, N., Baranowski, P., Buis, S., Cammarano, D., Chen, Y., Dumont, B., Ewert, F., Gaiser, T., Hlavinka, P., Hoffmann, H., Hohn, J. G., Jurecka, F., Kersebaum, K. C., Krzyszczak, J., Lana, M., Mechiche-Alami, A., Minet, J., Montesino, M., Nendel, C., Porter, J. R., Ruget, F., Semenov, M. A., Steinmetz, Z., Stratonovitch, P., Supit, I., Tao, F., Trnka, M., de Wit, A. and Rotter, R. P. 2018. Adaptation response surfaces for managing wheat under perturbed climate and CO_2 in a Mediterranean environment. *Agricultural Systems* 159, 260–74. doi:10.1016/j.agsy.2017.01.009.

Salo, T. J., Palosuo, T., Kersebaum, K. C., Nendel, C., Angulo, C., Ewert, F., Bindi, M., Calanca, P., Klein, T., Moriondo, M., Ferrise, R., Olesen, J. E., Patil, R. H., Ruget, F., Takac, J., Hlavinka, P., Trnka, M. and Rötter, R. P. 2016. Comparing the performance of 11 crop simulation models in predicting yield response to nitrogen fertilization. *The Journal of Agricultural Science* 154(7), 1218–40. doi:10.1017/S0021859615001124.

Schauberger, B., Archontoulis, S., Arneth, A., Balkovic, J., Ciais, P., Deryng, D., Elliott, J., Folberth, C., Khabarov, N., Muller, C., Pugh, T. A. M., Rolinski, S., Schaphoff, S., Schmid, E., Wang, X. H., Schlenker, W. and Frieler, K. 2017. Consistent negative response of US crops to high temperatures in observations and crop models. *Nature Communications* 8, 13931. doi:10.1038/ncomms13931.

Schleussner, C. F., Deryng, D., Muller, C., Elliott, J., Saeed, F., Folberth, C., Liu, W. F., Wang, X. H., Pugh, T. A. M., Thiery, W., Seneviratne, S. I. and Rogelj, J. 2018. Crop productivity changes in 1.5 degrees C and 2 degrees C worlds under climate sensitivity uncertainty. *Environmental Research Letters* 13(6). doi:10.1088/1748-9326/aab63b.

Smith, M. D., Oglend, A., Kirkpatrick, A. J., Asche, F., Bennear, L. S., Craig, J. K. and Nance, J. M. 2017. Seafood prices reveal impacts of a major ecological disturbance. *Proceedings of the National Academy of Sciences of the United States of America* 114(7), 1512–7. doi:10.1073/pnas.1617948114.

Tao, F. L., Rotter, R. P., Palosuo, T., Diaz-Ambrona, C. G. H., Minguez, M. I., Semenov, M. A., Kersebaum, K. C., Nendel, C., Cammarano, D., Hoffmann, H., Ewert, F., Dambreville, A., Martre, P., Rodriguez, L., Ruiz-Ramos, M., Gaiser, T., Hohn, J. G., Salo, T., Ferrise, R., Bindi, M. and Schulman, A. H. 2017. Designing future barley ideotypes using a crop model ensemble. *European Journal of Agronomy* 82, 144–62. doi:10.1016/j.eja.2016.10.012.

Tao, F. L., Rotter, R. P., Palosuo, T., Diaz-Ambrona, C. G. H., Minguez, M. I., Semenov, M. A., Kersebaum, K. C., Nendel, C., Specka, X., Hoffmann, H., Ewert, F., Dambreville, A., Martre, P., Rodriguez, L., Ruiz-Ramos, M., Gaiser, T., Hohn, J. G., Salo, T., Ferrise, R., Bindi, M., Cammarano, D. and Schulman, A. H. 2018. Contribution of crop model structure, parameters and climate projections to uncertainty in climate change impact assessments. *Global Change Biology* 24(3), 1291–307. doi:10.1111/gcb.14019.

Tardieu, F., Cabrera-Bosquet, L., Pridmore, T. and Bennett, M. 2017. Plant phenomics, From sensors to knowledge. *Current Biology: CB* 27(15), R770–83. doi:10.1016/j.cub.2017.05.055.

van Bussel, L. G. J., Ewert, F. and Leffelaar, P. A. 2011a. Effects of data aggregation on simulations of crop phenology. *Agriculture, Ecosystems and Environment* 142(1–2), 75–84. doi:10.1016/j.agee.2010.03.019.

van Bussel, L. G. J., Muller, C., van Keulen, H., Ewert, F. and Leffelaar, P. A. 2011b. The effect of temporal aggregation of weather input data on crop growth models' results. *Agricultural and Forest Meteorology* 151(5), 607–19. doi:10.1016/j.agrformet.2011.01.007.

van Bussel, L. G. J., Stehfest, E., Siebert, S., Mueller, C. and Ewert, F. 2015. Simulation of the phenological development of wheat and maize at the global scale. *Global Ecology and Biogeography* 24(9), 1018–29. doi:10.1111/geb.12351.

van Bussel, L. G. J., Ewert, F., Zhao, G., Hoffmann, H., Enders, A., Wallach, D., Asseng, S., Baigorria, G. A., Basso, B., Biernath, C., Cammarano, D., Chryssanthacopoulos, J., Constantin, J., Elliott, J., Glotter, M., Heinlein, F., Kersebaum, K. C., Klein, C., Nendel, C., Priesack, E., Raynal, H., Romero, C. C., Rotter, R. P., Specka, X. and Tao, F. L. 2016. Spatial sampling of weather data for regional crop yield simulations. *Agricultural and Forest Meteorology* 220, 101–15. doi:10.1016/j.agrformet.2016.01.014.

Volk, M. and Ewert, F. 2011. Scaling methods in integrated assessment of agricultural systems-State-of-the-art and future directions. *Agriculture, Ecosystems and Environment* 142(1–2), 1–5. doi:10.1016/j.agee.2010.10.014.

Wallach, D., Martre, P., Liu, B., Asseng, S., Ewert, F., Thorburn, P. J., van Ittersum, M., Aggarwal, P. K., Ahmed, M., Basso, B., Biernath, C., Cammarano, D., Challinor, A. J.,

De Sanctis, G., Dumont, B., Eyshi Rezaei, E., Fereres, E., Fitzgerald, G. J., Gao, Y., Garcia-Vila, M., Gayler, S., Girousse, C., Hoogenboom, G., Horan, H., Izaurralde, R. C., Jones, C. D., Kassie, B. T., Kersebaum, K. C., Klein, C., Koehler, A. K., Maiorano, A., Minoli, S., Müller, C., Naresh Kumar, S., Nendel, C., O'Leary, G. J., Palosuo, T., Priesack, E., Ripoche, D., Rötter, R. P., Semenov, M. A., Stöckle, C., Stratonovitch, P., Streck, T., Supit, I., Tao, F., Wolf, J. and Zhang, Z. 2018. Multimodel ensembles improve predictions of crop-environment-management interactions. *Global Change Biology* 24(11), 5072–83. doi:10.1111/gcb.14411.

Wang, E., Martre, P., Zhao, Z. G., Ewert, F., Maiorano, A., Rotter, R. P., Kimball, B. A., Ottman, M. J., Wall, G. W., White, J. W., Reynolds, M. P., Alderman, P. D., Aggarwal, P. K., Anothai, J., Basso, B., Biernath, C., Cammarano, D., Challinor, A. J., De Sanctis, G., Doltra, J., Fereres, E., Garcia-Vila, M., Gayler, S., Hoogenboom, G., Hunt, L. A., Izaurralde, R. C., Jabloun, M., Jones, C. D., Kersebaum, K. C., Koehler, A. K., Liu, L. L., Muller, C., Kumar, S. N., Nendel, C., O'Leary, G., Olesen, J. E., Palosuo, T., Priesack, E., Rezaei, E. E., Ripoche, D., Ruane, A. C., Semenov, M. A., Shcherbak, I., Stockle, C., Stratonovitch, P., Streck, T., Supit, I., Tao, F. L., Thorburn, P., Waha, K., Wallach, D., Wang, Z. M., Wolf, J., Zhu, Y. and Asseng, S. 2017. The uncertainty of crop yield projections is reduced by improved temperature response functions. *Nature Plants* 3. doi:10.1038/nplants.2017.102.

Wang, N., Wang, E. L., Wang, J., Zhang, J. P., Zheng, B. Y., Huang, Y. and Tan, M. X. 2018. Modelling maize phenology, biomass growth and yield under contrasting temperature conditions. *Agricultural and Forest Meteorology* 250–251, 319–29. doi:10.1016/j.agrformet.2018.01.005.

Webber, H., Martre, P., Asseng, S., Kimball, B., White, J., Ottman, M., Wall, G. W., De Sanctis, G., Doltra, J., Grant, R., Kassie, B., Maiorano, A., Olesen, J. E., Ripoche, D., Rezaei, E. E., Semenov, M. A., Stratonovitch, P. and Ewert, F. 2017. Canopy temperature for simulation of heat stress in irrigated wheat in a semi-arid environment: a multi-model comparison. *Field Crops Research* 202, 21–35. doi:10.1016/j.fcr.2015.10.009.

Webber, H., Ewert, F., Olesen, J. E., Muller, C., Fronzek, S., Ruane, A. C., Bourgault, M., Martre, P., Ababaei, B., Bindi, M., Ferrise, R., Finger, R., Fodor, N., Gabaldon-Leal, C., Gaiser, T., Jabloun, M., Kersebaum, K. C., Lizaso, J. I., Lorite, I. J., Manceau, L., Moriondo, M., Nendel, C., Rodriguez, A., Ruiz-Ramos, M., Semenov, M. A., Siebert, S., Stella, T., Stratonovitch, P., Trombi, G. and Wallach, D. 2018a. Diverging importance of drought stress for maize and winter wheat in Europe. *Nature Communications* 9(1), 4249. doi:10.1038/s41467-018-06525-2.

Webber, H., White, J. W., Kimball, B. A., Ewert, F., Asseng, S., Rezaei, E. E., Pinter, P. J., Hatfield, J. L., Reynolds, M. P., Ababaei, B., Bindi, M., Doltra, J., Ferrise, R., Kage, H., Kassie, B. T., Kersebaum, K. C., Luig, A., Olesen, J. E., Semenov, M. A., Stratonovitch, P., Ratjen, A. M., LaMorte, R. L., Leavitt, S. W., Hunsaker, D. J., Wall, G. W. and Martre, P. 2018b. Physical robustness of canopy temperature models for crop heat stress simulation across environments and production conditions. *Field Crops Research* 216, 75–88. doi:10.1016/j.fcr.2017.11.005.

West, P. C., Gerber, J. S., Engstrom, P. M., Mueller, N. D., Brauman, K. A., Carlson, K. M., Cassidy, E. S., Johnston, M., MacDonald, G. K., Ray, D. K. and Siebert, S. 2014. Leverage points for improving global food security and the environment. *Science* 345(6194), 325–8. doi:10.1126/science.1246067.

Yin, X. G., Kersebaum, K. C., Kollas, C., Baby, S., Beaudoin, N., Manevski, K., Palosuo, T., Nendel, C., Wu, L. H., Hoffmann, M., Hoffmann, H., Sharif, B., Armas-Herrera, C. M., Bindi, M., Charfeddine, M., Conradt, T., Constantin, J., Ewert, F., Ferrise, R., Gaiser, T., de Cortazar-Atauri, I. G., Giglio, L., Hlavinka, P., Lana, M., Launay, M., Louarn, G., Manderscheid, R., Mary, B., Mirschel, W. and Moriondo, M., Öztürk, I., Pacholski, A., Ripoche-Wachter, D., Rötter, R. P., Ruget, F., Trnka, M., Ventrella, D., Weigel, H. J. and Olesen, J. E. 2017a. Multi-model uncertainty analysis in predicting grain N for crop rotations in Europe. *European Journal of Agronomy* 84, 152–65. doi:10.1016/j.eja.2016.12.009.

Yin, X., Kersebaum, K. C., Kollas, C., Manevski, K., Baby, S., Beaudoin, N., Öztürk, I., Gaiser, T., Wu, L., Hoffmann, M., Charfeddine, M., Conradt, T., Constantin, J., Ewert, F., de Cortazar-Atauri, I. G., Giglio, L., Hlavinka, P., Hoffmann, H., Launay, M., Louarn, G., Manderscheid, R., Mary, B., Mirschel, W., Nendel, C., Pacholski, A., Palosuo, T., Ripoche-Wachter, D., P. Rötter, R., Ruget, F., Sharif, B., Trnka, M., Ventrella, D., Weigel, H. and E. Olesen, J. 2017b. Performance of process-based models for simulation of grain N in crop rotations across Europe. *Agricultural Systems* 154, 63–77. doi:10.1016/j.agsy.2017.03.005.

Zhao, G., Hoffmann, H., van Bussel, L. G. J., Enders, A., Specka, X., Sosa, C., Yeluripati, J., Tao, F., Constantin, J., Raynal, H., Teixeira, E., Grosz, B., Doro, L., Zhao, Z., Nendel, C., Kiese, R., Eckersten, H., Haas, E., Vanuytrecht, E., Wang, E., Kuhnert, M., Trombi, G., Moriondo, M., Bindi, M., Lewan, E., Bach, M., Kersebaum, K. C., Rötter, R. P., Roggero, P. P., Wallach, D., Cammarano, D., Asseng, S., Krauss, G., Siebert, S., Gaiser, T. and Ewert, F. 2015a. Effect of weather data aggregation on regional crop simulation for different crops, production conditions, and response variables. *Climate Research* 65, 141–57. doi:10.3354/cr01301.

Zhao, G., Siebert, S., Enders, A., Rezaei, E. E., Yan, C. and Ewert, F. 2015b. Demand for multi-scale weather data for regional crop modeling. *Agricultural and Forest Meteorology* 200, 156–71. doi:10.1016/j.agrformet.2014.09.026.

Zhao, G., Hoffmann, H., Yeluripati, J., Xenia, S., Nendel, C., Coucheney, E., Kuhnert, M., Tao, F., Constantin, J., Raynal, H., Teixeira, E., Grosz, B., Doro, L., Kiese, R., Eckersten, H., Haas, E., Cammarano, D., Kassie, B., Moriondo, M., Trombi, G., Bindi, M., Biernath, C., Heinlein, F., Klein, C., Priesack, E., Lewan, E., Kersebaum, K. C., Rötter, R. P., Roggero, P. P., Wallach, D., Asseng, S., Siebert, S., Gaiser, T. and Ewert, F. 2016. Evaluating the precision of eight spatial sampling schemes in estimating regional mean of simulated yields for two crops. *Environmental Modelling and Software* 80, 100–12. doi:10.1016/j.envsoft.2016.02.022.

Zhao, C., Liu, B., Piao, S. L., Wang, X. H., Lobell, D. B., Huang, Y., Huang, M. T., Yao, Y. T., Bassu, S., Ciais, P., Durand, J. L., Elliott, J., Ewert, F., Janssens, I. A., Li, T., Lin, E., Liu, Q., Martre, P., Muller, C., Peng, S. S., Penuelas, J., Ruane, A. C., Wallach, D., Wang, T., Wu, D. H., Liu, Z., Zhu, Y., Zhu, Z. C. and Asseng, S. 2017. Temperature increase reduces global yields of major crops in four independent estimates. *Proceedings of the National Academy of Sciences of the United States of America* 114(35), 9326–31. doi:10.1073/pnas.1701762114.

Zhao, C., Liu, B., Xiao, L., Hoogenboom, G., Boote, K. J., Kassie, B. T., Pavan, W., Shelia, V., Kim, K. S., Hernandez-Ochoa, I. M., Wallach, D., Porter, C. H., Stockle, C. O., Zhu, Y. and Asseng, S. 2018. A SIMPLE crop model. *European Journal of Agronomy* 104, 97–106.

Chapter 17

The future of crop modeling for sustainable agriculture

K. J. Boote, University of Florida, USA

1 Introduction

The early development of dynamic crop and agroecosystem models originated in the 1960s with two important grandfathers of crop modeling: C. T. deWit

http://dx.doi.org/10.19103/AS.2019.0061.23

(1965) and W. G. Duncan et al. (1967), who with other agricultural engineers, soil physicists, and agronomists developed FORTRAN-coded crop models.[1] Those early models were relatively one-dimensional in the sense of simulating one major process, for example, light interception and canopy photosynthesis (deWit, 1965; Duncan et al., 1967) or soil water balance. Crop-specific models subsequently came along for cotton, corn, peanut, and soybean (Duncan, 1972; Duncan et al., 1978; Stapleton et al., 1973), but those models had many hardwired parameters and coefficients in the code (of which some parameters and relationships were truly crop-specific, yet others were generic parameters and coefficients related to processes such as water flow or energy balance for evapotranspiration (ET) computation). The models were not particularly robust to simulate multiple situations and the learning curve continued for another 30 years, during which time agronomists and other scientists were skeptical of the value of crop simulation. By the 1990s, with increased training of new modelers, the models were becoming more advanced and capable of addressing more diverse conditions and were almost totally placed on small desktop and portable computers. By the decade beginning in 2000, another shift occurred, with greater acceptance of crop simulation among the research community. In addition, the interfaces of the crop models had gained the point-and-click functionality of the portable computer windows systems. The next decade beginning about 2010 is when crop modelers truly began to openly share ideas, code, and data, rather than playing the competition game. Of course, competition among modelers continues today, but modelers are collaborating much more now. For a more detailed history of crop simulation, evolution of different models, and their uses over the past 60 years, see Jones et al. (2016).

2 The AgMIP decade of multiple crop model evaluations and improvements

Beginning in 2010, the world community of crop and ecosystem modelers have come together to share models, data, and ideas in the Agricultural Model Intercomparison and Improvement Project (AgMIP) initiated under the leadership of Rosenzweig et al. (2014). This effort was initiated in 2010 at the American Society of Agronomy meetings in Long Beach, California. Initially, the group considered the name 'World Crop Modeling', but soon picked up on the naming convention of other worldwide modeling groups with the letters (MIP), already being used by the global climate modeling community. The AgMIP is a voluntary community of scientists and modelers spanning climate,

1 I had the good fortune to meet both of them and interacted considerably with Professor Duncan at the University of Florida, and Professor Duncan certainly influenced my early career.

crop, and economic disciplines, with the objective of advancing scientific capabilities for assessing climate impacts and adaptation for agriculture, food security, and poverty. It has two-track goals: (1) model intercomparison and improvement and (2) climate change multi-model assessment. AgMIP is not a scientific society nor is it a recipient of major funding. Nevertheless, the group holds conferences at various world sites nearly every year since 2010, and attracts about 200 people per conference. Within the AgMIP umbrella, there are multiple group activities, with each voluntary group acting somewhat independently. There are climate modelers who interact with two types of modeling activities, often culminating in economic analyses. One group, AgMIP Gridded Crop Modeling Initiative, uses ecosystem models that simulate at gridded site/weather scales over the whole world where teams have interest in climate change impact on global and regional production (Elliott et al., 2015). There is another group of modelers who use site-specific crop models that predict climate change impact and adaptation strategies with more specific soil, cultivar, and historical weather for crops such as wheat (Asseng et al., 2013), maize (Bassu et al., 2014), and rice (Li et al., 2015). In addition, AgMIP includes about ten worldwide economic models focused on climate change impact on trade and economic consequences for the world and regions, where global crop yield simulations are based on the weather and climate change scenarios from climate and crop modelers (Muller and Robertson, 2014; Nelson et al., 2014a,b, Rosegrant and IMPACT Development Team, 2012). Most of these groups do not receive any primary funds other than what they as voluntary individuals/local institutions bring to the effort from their own resources or team-oriented grant initiatives. There is no central funding of AgMIP. Within the AgMIP site-specific crop modeling activities which Peter Thorburn and I co-lead and advise (Thorburn et al., 2015; Thorburn et al., 2018), there is AgMIP-wheat (28 models), AgMIP-maize (23 models), AgMIP-rice (16 models), AgMIP-potato (nine models; Fleisher et al., 2017), AgMIP-canola (eight models), AgMIP-sugarcane (five models), AgMIP-grasslands (eight models), AgMIP-maize ET (29 maize models; Kimball et al., 2019), AgMIP low-input agriculture (new, 25 maize models), and AgMIP-soybean (new, ten models). See Asseng et al. (Chapter 16 in this book) for a thorough review of crop model intercomparison activities in AgMIP. These modeling teams have followed several objectives over time: (1) initially they intercompared multiple models for a given crop for simulated response to CO_2, temperature, and water availability (CTW sensitivity analysis) at four or five sentinel sites (in blind and calibrated state, most crop teams have done this so far), (2) in a later step, the rice, maize, and potato modelers conducted important tests/evaluations of model performance against response data from elevated CO_2 sites, (3) similarly, other teams tested against temperature response data from elevated temperature studies (wheat, rice), and (4) during the last 2 years, maize and wheat teams have begun evaluating models for accurate prediction of water loss against ET eddy flux sites (Kimball et al., 2019). The teams have collaborated, shared ideas, code, and data, and have begun making improvements in their individual models. An important issue within the AgMIP community is that this is not to be a competition for 'best' model, but for all groups to learn and make their own improvements. At a later time, there may be selection of a best model or ensemble of models to use for serious projections of agricultural production for Intergovernmental Panel on Climate Change (IPCC) assessment purposes, but this step has not been taken so far. An important finding of the AgMIP crop teams is that the ensemble median or mean of the models is usually a better predictor than any single crop model (Asseng et al., 2013; Bassu et al., 2014; Li et al., 2015; Makowski et al., 2015; Martre et al., 2015). AgMIP crop model intercomparisons have helped discover shortcomings of individual models that led to improvement, and it also quantified the uncertainty among crop models, which is greater for temperature response than for CO_2 response (Asseng et al., Chapter 16 in this book). See chapter of Asseng et al. for a nicely explained case study of intercomparison of multiple wheat models, especially on model improvement for response to elevated temperature stress.

These efforts of AgMIP crop modeling teams have been successful, resulting in improved crop models and improved understanding of where code modifications may be needed and future experiments needed to provide missing knowledge. Concurrently, the sharing of data across modelers is providing a much needed basis for future model improvement and calibration. Techniques of model calibration are being improved and shared. Finally, the crop models are increasingly being accepted as valuable tools for addressing societal and agricultural challenges, such as planning for adapting and mitigating effects of climate change on agricultural production in individual countries and globally. The models are valuable tools for use by individual producers or extension specialists in agriculture.

3 Expanding the role of agroecosystems modeling

Increasingly, agroecosystem models are being accepted as tools with which scientists can integrate, synthesize, and understand missing components in their particular agricultural systems. One of the features of agroecosystem modeling is that the modelers and model users must take a systems approach to consider the important environmental (weather), soils, genetic, and crop management factors at play, that affect the processes of C, water, and N balance for the crop, soil, and atmosphere. There are so many interactions and linkages among these processes, and when existing knowledge is initially employed, the modelers learn that there are often many interactions, linkages, and factors that they have not adequately considered. The modelers (or students) are forced to think about the whole system, rather than the 'reductionist' bit of knowledge that they may have studied as a graduate student. The student-modelers learn to integrate knowledge across disciplines. The crop models are the integrators of all those processes, just as the plant in the field must integrate over all of its processes in its climatic, soil, and management conditions, until producing its final yield.

The future appears to be bright, as universities, companies, and agencies continue to advertise for agroecosystem modeler positions under varying names/titles (digital agronomist, theoretical crop husbandry, nutrient management, ecosystem modeler). The number of trained crop system modelers appears to be short of the demand over the past 5 years and that continues into the present.

The crop models have become tools for multiple uses in classrooms, research laboratories, industry, and government. Boote et al. (1996) introduced potential uses of crop models as research tools, crop system management tools, and policy analysis tools. It appears that many of those suggested uses are coming closer to realization. The paper by Jones et al. (2017) reviews the state of agricultural systems science, with a view toward the uses of agricultural system models for various purposes and at various spatial scales. Some of those use cases and examples are illustrated in this chapter.

4 Crop models in the classroom

Because crop models are integrative tools and can be used to simulate whole agroecosystems, they are great teaching tools. When placed in the hands of students, the students begin to see the big picture of how individual processes they have learned, function together to give an integrated outcome. This is powerful and needed in today's reductionist single-topic education system. With model system software, the students have the chance to understand and interactively simulate the important climatic, management, and genetic factors involved. Several examples of crop models in education are given here. Beginning in the 1980s, at the Agricultural University at Wageningen, crop system modeling was a systematic portion of coursework in which students learned to apply the principles (see book edited by Penning deVries and van Laar, 1982). Graduate courses in crop modeling have been created in the United States in recent years, although relatively more generic systems modeling courses are typical. At the University of Florida, we have conducted a summer semester course on crop modeling for six to eight graduate students per year since 2000, and held informal courses prior to that. This course is now an official course in Agricultural and Biological Engineering graduate course curriculum. Universities throughout the United States and the world are picking up similar courses. At Iowa State University, crop modeling has been an integral part of a distance M.S. program since about 2009 (Archontoulis et al., 2012; Miguez et al., 2012). While not a graduate program *per se*, a week-long course for learning the software of the Decision Support System for Agrotechnology Transfer (DSSAT) has been conducted annually since about 1988, and participants have numbered 50–60 per course since 2004 (Hoogenboom, pers. comm., 2019). The DSSAT crop modeling course targets scientific professionals who see the value of using crop modeling tools in their own work. The participants include young professional scientists and graduate students from around the world (more than half of whom are international). Teaching is assisted by excellent chapters on integrated approaches to crop water balance, N balance, C balance/photosynthesis, and phenology available in crop modeling textbooks (Boote et al., 1998; Tsuji et al., 1998; Wallach et al., 2019) and in chapters written by authors in this book, especially see Chapter 4 by Stockle and Meza; Chapter 3 by Singh and Porter; Chapter 1 by Kim et al.; and Chapter 2 by Evers and Marcelis.

5 Crop modeling for plant breeding and genetics

In the past decade, university scientists and plant breeding companies (DuPont Pioneer and Bayer-Monsanto) have begun to use crop models to evaluate processes and traits that may be used for genetic improvement of yield (Yin et al., 2018). Linking genes and quantitative trait loci (QTLs) to crop models is a new effort by several groups including university and plant breeding companies. Early attempts were made to link specific genes to ecophysiological models to simulate crop growth and yield (Hoogenboom and White, 2003; Hoogenboom et al., 2004; Messina et al., 2006; Boote et al., 2013). This was followed by attempts to link specific QTLs to ecophysiological models to simulate crop performance (Gu et al., 2014; Yin et al., 2018). Some approaches have been modular or 'piece-meal' efforts such as replacement of specific modules affecting rate of leaf appearance (Zhang et al., 2017) or time to flowering (Wallach et al., 2018). In a more holistic approach, DuPont Pioneer scientists have integrated crop growth models with genome-wide prediction (GWAS) to simulate maize performance over multiple environments (Technow et al., 2015; Cooper et al., 2016; Messina et al., 2018). This combination approach for predicting yield was found to be superior to a strictly statistical genomic approach because the crop models provide the mechanisms (and relationships) for dynamic response to climatic factors, for example, life cycle versus temperature and growth process sensitivity to temperature and water deficit, which are aspects that are missing in pure statistical genomic predictions of yield that fail for environments not in statistical 'training' sets. These applications of linking genes and QTLs are preliminary, and will require many years of research and experience to be fully functional, extensive, and useful. It is interesting that genotyping is no longer the limiting factor. Rather, phenotyping of crop traits and performance in multiple environments has now become the most limiting critical factor (Furbank and Tester, 2011), and these authors review use of high technology equipment and sensors to evaluate plant physiological traits. See also Chapter 14 in this book by van Evert, in which phenotyping, data collection, and data handling are reviewed. Even big data phenotyping platforms are relatively limited in ability to cover the number of environments and measurements needed for phenotyping. It is important to connect the crop modelers to the physiologists such as Tardieu et al. (2017) who are doing the phenotyping at a deep process level. Chenu et al. (2009) applied QTLs and environmental effects on leaf area expansion and other organ-level activities, with the goal of evaluating genetic effects on drought tolerance of maize.

Chapter 1 by Kim et al. (in this book) illustrates use of models to evaluate QTLs linked to simulation of life cycle phenology of Arabidopsis that have important implications for similar gene action in agronomic crops. If the phenotyping of the outcome of genes in different environments is adequate and truly works all the way to final yield production, the idea is that one would only need the DNA (QTLs) of a new variety to predict its performance in a new environment without any field phenotyping of that new cultivar.

6 The role of crop models in strategic crop planning and management

The participants in this role include those in university, state extension, insurance, banks, private consultants and companies selling seed, fertilizer, pesticides, machinery, and irrigation services. For this purpose, crop models can be used with long-term weather (30 years or more) for given regions to evaluate strategies and recommendations for optimum N fertilization (rate and timing) or optimum irrigation or optimum sowing dates for a given crop for a given soil, where the decisions are not based on this year's weather, but long-term probabilities. Knowledge of the probability of crop yield and its mean and distribution over multiple seasons can facilitate strategic decisions for bankers and insurance agencies, such as investment in irrigation equipment relative to yield and profit enhancement along with capital and operating cost of irrigation equipment. Considering yearly production over many years of weather will help determine whether irrigation is profitable, considering both current operating cost and long-term amortized investment cost. The same approach is useful for making preseason strategic plans for type of crop, cultivar choice, best sowing dates, and optimum fertilization. Hochman et al. (2009) described a model-based decision support tool called

Yield Prophet® to assist Australian dryland farmers to monitor and manage crops in a variable climate. Yield Prophet® decisions were partially preseason-based on soil moisture left from prior crop as well as assessment of El Nino Southern Oscillation (ENSO) signal projections of drier or wetter season ahead. The DSSAT software with its Seasonal Analysis tool as described by Hoogenboom et al. (in Chapter 7 of this book) has been widely used by a number of scientists to evaluate crop management strategies using long-term weather. For example, Andrea et al. (2018) used the CERES-Maize model to evaluate long-term yield potential of maize for different regions in Brazil with and without irrigation and also comparing to actual yields to determine yield gaps. Heinemann et al. (2016) conducted a similar evaluation of long-term drought impact on dry bean for regions in Brazil by comparison to a crop not limited by water. Battisti et al. (2018) evaluated irrigation, sowing date, and cultivar choices of soybean production to improve yields in Brazil under future climate scenarios. Boote et al. (2008) showed simulated long-term yields over 28 weather years at Patancheru, India, illustrating the probability of grain yield for different cultivars of soybean under rainfed conditions (Fig. 1). Figure 2 shows yield of soybean as affected by sowing date and maturity group under rainfed or irrigated conditions for that site. Figure 3 shows yield of rainfed and irrigated maize as affected by sowing date at Patancheru. Later sowing of maize under irrigation increases yield, takes advantage of cooler fall weather despite the somewhat lower solar radiation. On the other hand, soybean, because of its strong daylength-sensitivity, is reduced in yield with later sowing despite being irrigated (Fig. 2).

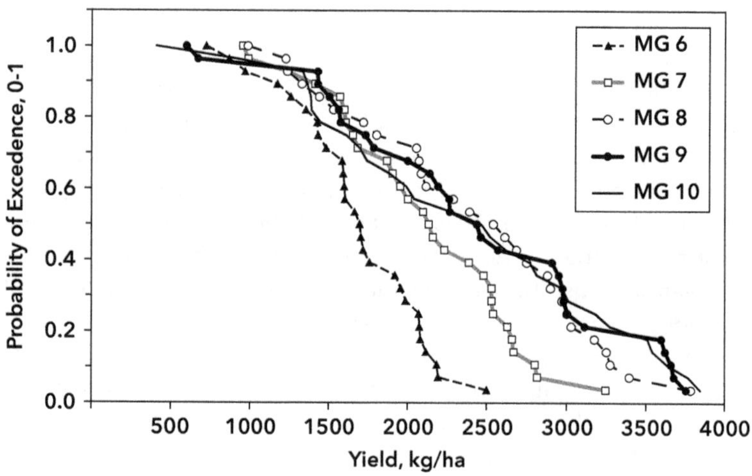

Figure 1 Probability of attaining a given grain yield for five soybean maturity groups under rainfed conditions under early sowing window, for 28 weather years at Patancheru, India.

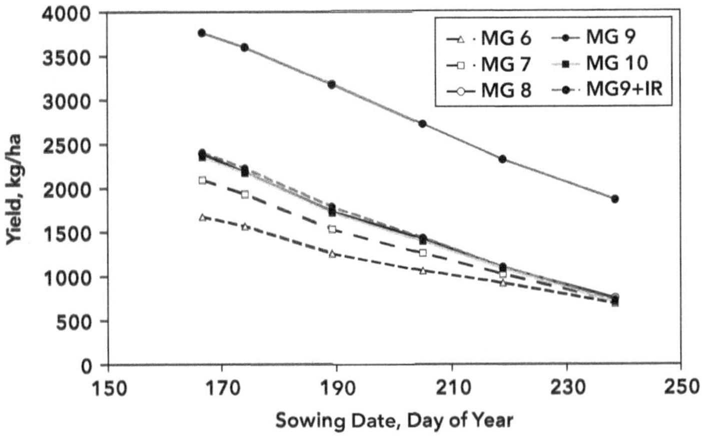

Figure 2 Simulated soybean grain yield response to sowing date for five maturity groups under rainfed and irrigated conditions averaged over 28 weather years at Patancheru, India.

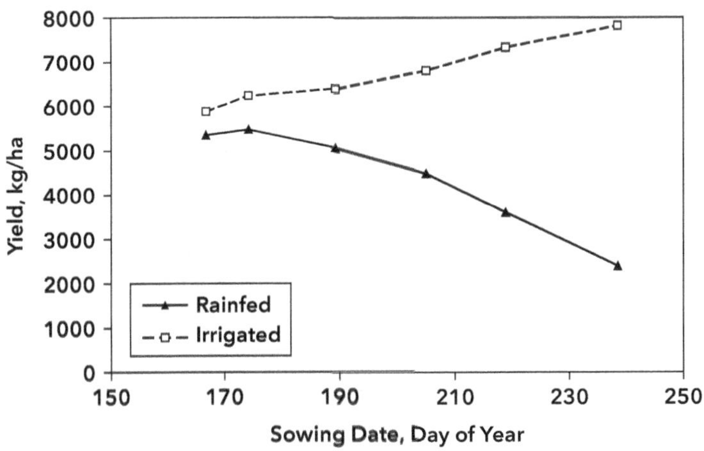

Figure 3 Simulated maize grain yield response to sowing date under rainfed and irrigated conditions, with 120 kg N/ha, averaged over 28 weather years at Patancheru, India.

7 The role of crop models in real-time decision-making by agricultural enterprises

The participants in this role include private consultants, water-management entities, and companies selling advisory services related to pesticides, fertilizer, and irrigation. Insurance and commodity companies may also have interest to foresee yield consequences for crops in larger regions as a result of current weather. For this purpose, crop models are simulated with this year's real weather upto today, along with projected near-term weather, to decide whether to irrigate today, or apply additional fertilizer, control pests, or cut hay. The requirement here is for site-specific known information on soil, crop, cultivar, sowing date, past irrigation, and fertilization, so that the recommendation can apply for a specific field and specific situation. The decisions could include to irrigate today or some future date, to cut hay today or not, to fertilize a given amount today or not, to apply freeze protection, or to apply a fungicide (because of a given growth stage indication or hypothetical pest problem that co-occurs or is predicted as a function of prior weather). Recent weather (rainfall, temperature, and dew duration) can be used to indicate whether a fungicide application is needed. Fraisse et al. (Chapter 11 in this book) illustrate how the AgroClimate decision support system provides weather alerts to Florida strawberry growers to assist them in decisions to spray or not spray for fungal diseases. An example for irrigation decision-making is iCrop, a model-driven decision support tool for optimizing water use in agriculture developed by Kisekka et al. (2016, 2017), in which weather data up till today is available online along with the DSSAT crop models and cultivar information. The consultants/farmers go online to enter minimal information on crop and sowing date, and the software tool gives them a recommendation that irrigation will be required today or in a given number of days if no rainfall occurs. A company/consultant maintaining crop management records could similarly use such simulations to recommend whether additional fertilizer should be applied today, or whether a fungicide should be applied today or tomorrow. In both cases with 'run-out' long-term weather for the remaining season (coming from the historical 30- or 40-year record), the crop yield could be projected for both cases, with or without the added irrigation or N fertilization or pesticide application, to allow an economic decision.

It is possible to consider future 3–6-month weather cycles (wet or dry, warmer or cooler) such as ENSO phases in decision-making that provide both strategic and tactical benefits for agricultural producers. AgroClimate (see Fraisse et al., Chapter 11 in this book) and Yield Prophet® (Hochman et al., 2009) are climate-based decision support systems that consider ENSO phases in decisions on which crop to grow, planting dates to choose, how much fertilizer to apply, and amount of pest control practices to anticipate. These systems are already in operation in the United States and Australia.

Precision agriculture is a highly specialized case where real-time decision recommendations are made for spatial zones in farmer fields, making use of yield history maps and information coming from remote sensing of crop conditions for spatial zones in that field using ground-level equipment such as green-seeker for N deficiency or satellite or UAV drones to determine crop heat stress via infrared sensors. Decisions of such real-time applications can be assisted by 'big data' handling features for storing spatial information on weather, soils, cultivar (and its traits), crop management conditions, and spatial crop yield records on the web. That way, the model user only needs to enter minimal information to select the field, run the model, and make a new decision with updated weather (the weather should be automatically updated by the system software). Then the applications of irrigation and N fertilizer are made by precision irrigation or fertilization equipment that can handle spatial variation of application. Providing the 30-year long-term weather and simulated yields for that crop for those spatial zones provides knowledge of long-term probabilities of yield distribution to provide perspective on this year's decision-making.

8 The role of crop models in government policy: predicting famine risk

Government, commodity firms, and various nongovernmental organizations (NGOs) are interested in projecting global food production for the coming year, including the risk of production and famine for specific regions. Commodity firms and various government agencies wish to project global production with an eye to the effect on commodity pricing and commodity availability for the coming year either for the world or their individual country. NGOs have the same interest, but with a focus on production in specific countries with the goal of evaluating risk of famine. Both crop-specific models and global ecosystem models can be used for these purposes. A good example of this is the CraftTool developed by Shelia et al. (2019) which is used with the DSSAT models (see Chapter 7 by Hoogenboom et al.) for regional agricultural forecasting by CCAFS (Climate Change, Agriculture and Food Security) program. These crop and ecosystem models require relatively generic inputs of soils, cultivars, sowing dates, and fertilization amounts for regions. For example, sowing dates and fertilization records are derived from current or past government survey information. An important feature is that these models use current real-time weather data for the regions and have no-concern for long-term climate scenarios. The models are simulated with real weather to project yield reduction for specific regions and the information is used to forecast food shortage in given regions when compared to prior production and consumer demand. Generally, the weather risk is primarily to address effects of regional droughts with less concern for total world production. The CraftTool and similar tools may also include additional climatic signal information, such as sea surface temperature (SST) in the future forecast of weather and effects on yield for a given region.

9 The role of crop models in government policy: predicting and mitigating the impacts of climate change

The modeling participants in this role include those in university, insurance, city, government, and various national and international agencies who wish to evaluate the impact of climate change on agricultural production and food security as well as strategies for adapting to climate change or mitigating climate change effects. Because the strategic decisions by these agencies are desired for larger regions and whole countries, the modeling exercise is less requiring of highly site-specific soil and crop management information. Long-term historical weather data is needed, however, to allow simulating probabilities of interannual yield distribution and mean for given regions, to provide evaluation of risk and uncertainty. Climate change impact assessments are being conducted with global gridded models with gridded soil and weather information, for the CO_2, temperature, and rainfall levels projected to occur under future climate change pathways predicted by multiple climate models (Muller and Robertson, 2014; Ruane et al., 2018). Inputs for those models are provided by various 'data products', because the soils and cultivars can be more general for regions, and the sowing dates and fertilization records are derived from government survey information. Unfortunately, the global gridded crop models may inadequately account for N fertility as well as increasing yield trends associated with genetic yield potential and improved production technology. They also fail to account for slow degradation of soils associated with warming. Simulated outputs are used to project the general degree of increase or decrease in crop yield responses to temperature, CO_2, and water availability. Simulations by Singh et al. (2013) demonstrated the benefits of heat tolerance and drought tolerance (genetic attributes) of groundnut to sustain or enhance productivity under future climate scenarios.

10 The role of crop models in assessing ecosystem services and resource use

Ecosystem services and sustainable intensification are discussed in Chapter 13 by Hatfield et al., Chapter 12 by Nendel and Zander, and Chapter 7 by Hoogenboom et al. of this book. They describe issues involving water and nutrient flow (surface or subsurface) from agriculture, water requirements (irrigation demand), erosion, C sequestration, and vegetative cover/biodiversity of the landscape. While most agroecosystem models can predict irrigation demand, deep drainage, N leaching, and soil C sequestration, only a small set can also simulate erosion, water runoff, and greenhouse gas emission. An example of resolving multiple conflicting goals in the hands of land managers (stewards) is the use of agroecosystem models to evaluate best strategies for fertilization or irrigation of a given crop with the goal of optimizing economic production, while minimizing N leaching and minimizing irrigation required. Nitrogen fertilization is crucial to sustaining high yields; however, application of more N than the crop takes up leads to

N leaching with water that passes through the soil and below the root zone. One of the problems is that rainfall varies considerably, thus causing highly variable N leaching, especially on sandy soil. From a producer viewpoint, N application should be sufficient to provide economic yields over the long term. This can be evaluated with simulations of different N fertilization rates using long-term weather records and applying economic analysis. Short-term analysis can also be used to the farmer's benefit, because model simulations may indicate that early N applied is now completely leached out by heavy rains and the farmer needs to apply additional side-dress N fertilizer.

From an environmental concern, the amount of N leached can be simulated with the long-term weather to suggest strategies to minimize N leaching over the long term, or understand dynamics of N leaching in the short term of this season. For N leaching, it is important to simulate 365 days per year and over all subsequent fallow or cover-crop phases because the nitrate N left in the soil at the end of an N-fertilized crop cycle is subject to leaching after the crop is harvested. Basso, Martinez-Feria, and Dumont in Chapter 8 of this book emphasized the importance of legacy effects, meaning carry-over N from the prior crops which is influenced by weather after the cycle of the fertilized crop ends and by cover-crop choices. Management strategies to minimize N leaching in off season include cover crops that take up that remaining N and re-cycle it into organic N. These two simulations to optimize economic yield or minimize N leaching have a natural conflict, as there have typically been no costs assessed against the amount of N leached, although government regulations limiting N fertilization are increasingly being applied in European farming environments and in some US urban environments. It may be possible to include a hypothetical cost assessment against N leached into the groundwater or against residual inorganic N above a certain threshold that is left in soil after crop harvest. Agroecosystem models offer an attractive way to estimate these thresholds, as opposed to extensive measurements in every field for such residual N values.

11 The role of crop models in assessing long-term soil fertility and sustainability

While the title of this book and chapter is aimed at sustainable agriculture, the term 'sustainability' can have multiple meanings. Goals of sustainability could be to sustain forests, biodiversity, and soil 'health' while ignoring human food shortages, or alternatively to feed the world population regardless of losses in biodiversity and soil health. These various goals of sustainability are often at odds, as shown by the following example. One goal can be toward sustaining farmer profitability in the near term with less regard to what may happen over a century of climate change. An alternate valuable target can be to sustain the farmer and his descendants (indeed our global population) by truly maintaining the soil fertility and soil organic C for decades/centuries so that the farmer and his descendants can continue to produce profitably from the same land area.

A major shortcoming in past efforts is that most crop model simulations of climate change impact are set up to do single-year simulations (always re-initiated annually) and therefore they ignore year-to-year carry-over of soil C, soil water, crop residue, soil inorganic N, and soil P pools. Singh and Porter in their chapter illustrated how soil C can decline over the long term even under ambient conditions. These are the legacy effects that Basso et al. discussed in Chapter 8 of this book. Future scenario simulations in 're-set' mode basically ignore the long-term consequences of elevated temperature (with climate change) that cause depletion of soil C which in turn causes reduced productivity (Basso et al., 2018). For example, no-till farming practices (no-till with substantial residue return) can result in long-term sustainability of soil organic C, whereas conventional tillage and lack of residue return can provide profitability in the short term, but cause non-sustainable agriculture in the long term.

Failure to apply fertilizer is a factor that can contribute to non-sustainable agriculture (from economic, food-provision, and soil C viewpoints). With less fertilizer (N, P, K, etc.) application to the crop, there is not only less food production, but there is less residue return to sustain soil C. Agriculture in much of the African continent involves production of maize, sorghum, and millet under production practices in which no N or P fertilizer is applied and nearly all the crop-produced surface residues are not returned (they are commonly removed and the remainder is allowed to be grazed by animals). Under this situation, the soil C gradually declines (as shown by Singh and Porter, see Chapter 3). The longer-term effect of loss of soil C is a decline in soil water-holding capacity and N-mineralization rate, both of which cause yields to decline over time (Fig. 4) as shown with simulations of maize over 25 years for the Patancheru site where the CENTURY soil C module in DSSAT was used. With no N fertilizer applied, the mean rainfed yield was 1437 kg ha^{-1} and declined 0.86% per year, while total soil C (mass) declined 0.42% per year. By contrast, application of 100 kg N ha^{-1} (split applied) per crop increased rainfed yield to 5012 kg ha^{-1} with no yield decline over time. N fertilization also increased simulated biomass production which returned more residue which acted to slow the rate of soil C depletion. The practical economic factor limiting this practice of residue return is that farmers place a high value on the residue to feed the livestock, hence there is conflict between short- and long-term goals. While fertilization can be a significant offset factor, it requires cash outlay, another limiting factor for farmers in developing countries.

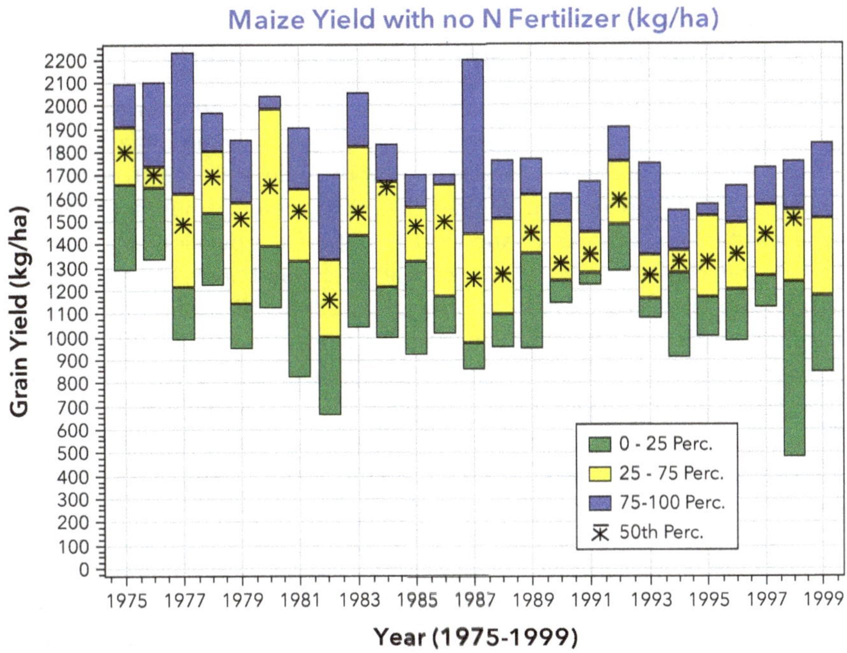

Figure 4 Maize yield decline with no N fertilization over time (25 years in sequence, repeated ten times with generated weather) at Patancheru, India, simulated with the DSSAT-CERES-maize model using the CENTURY soil C module.

12 The impact of machine learning and big data on agroecosystem models

Over the past decade, the agricultural community, industry, and government agencies have increasingly collected and stored large-scale extensive data from farmers' yield monitors, from satellite sensing of crop production and crop management, and industry-assisted recording of fertilization and agrichemical applications on farmer fields. These data can be input into crop models or used to calibrate the models, thus allowing more precise management decisions. These data can be accessed by machine learning tools and data-discovery tools to enhance the model function by calibrating parameters (optimization function) or to assist model code improvement (a type of artificial intelligence function). We will see more of this in the future, at individual farm-scale with precision management, and at large regions for industry/government to project production. See Kersebaum et al. (2015) concerning standards for analysis and classification of data used for model calibration. See Chapter 14 in this book by van Evert that describes data for developing, testing, and applying crop and farm models.

13 Connecting agroecosystem models to other components of farming: livestock and whole-farm models

Farming and agricultural production involves many components, and economic decision-making is crucial to link all farm enterprises including off-farm income. Antle in his chapter in this book and in prior work (Antle et al., 2014) describe the combined use of crop and household economic models to simulate intervention scenarios to improve economic and food security of a population of farmers in a given region, considering all of their capital assets, land basis, livestock, human labor for the farm, and off-farm income. Laing et al. in Chapter 6 of this book described three whole-farm models used for scenario analysis in Australia that consider these same components, with emphasis on connections to livestock enterprises. They also highlighted the need to consult frequently with farmers in the development and use of such whole-farm models. Descheemaeker and Bell in Chapter 9 of this book give a good example of a mixed-farm enterprise system, in which ruminant animals are dependent on pastures and hay fields and other animals are dependent on grain production on the farm. To simulate the total farm production requires models for each crop commodity as well as livestock models of the required types (ruminants, grain-fed beef, swine, chicken) and manure handling modules, all to address the different biomass C, energy, and nutrient flows among components. To be realistic over multiple years, the soil-crop-livestock models for these systems must be run in sequence/carry-over mode to address the effects of nutrient return in manure and consequences of removal or nonremoval of crop stover on soil fertility and soil C dynamics. To be sustainable over multiple years (for the farmers) also requires multiyear simulations with livestock and their dependence on annual climatic variation in crop productivity. An important aspect of these integrated farm-type simulations is the need for appropriate time coupling (daily or less frequent) of the different component models (animal size, stocking rate, sale of animals, periodic crop production, manure handling). Because of the complexity of the total system, the types of crop models are typically relatively simple, although outputs of forage and grain quality are needed to create sufficient interactions with the livestock productivity models. And finally the bottom line is economics and food security, which requires a whole-farm economic enterprise of the type that Antle et al. (2014) developed. Descheemaeker and Bell in Chapter 9 provide an example of economic outcomes (winners vs. losers) where the strategy of adding the legume mucuna forage crop to the agronomic package (70% for animal fodder, and 30% to soil fertility improvement) provided a successful economic outcome.

14 What is new in modeling agroecosystems?

FSP models: Chapter 2 by Evers and Marcelis describes important advances of functional-structural plant (FSP) models. The ability to consider

3D architecture will facilitate simulating competition among plants of different types (crop-weed, or intercropped plants, or even plants of the same species that vary in time of emergence). The FSP models nicely address competition for light, and have potentials to consider root competition for uptake of water and nutrients. FSP approaches will increase, both as standalone models and as components of existing crop models. In my view, FSP models have the potential to replace or add to 'object-oriented' modeling approaches because successive phytomeres/organs in FSP models can have their own photosynthesis, growth, and life cycle dynamics that can vary dynamically.

QTLs and genes in crop models: In the past one or two decades, there have been efforts to link molecular genetics directly to crop simulation models, thus bypassing the empirical constructs currently called cultivar coefficients. Unfortunately, a promised chapter on this topic for this book was not delivered, so readers are referred to my brief review in Section 5.

Accounting for pest and disease losses: Fresh efforts have recently been initiated to account for yield losses caused by insect pests and diseases (see Savary et al., 2018, and Chapter 5 in this book by Fernandes et al.). Until now, most crop models have failed to account for yield losses caused by insect pests and diseases, something that I attribute to several factors: (1) there are too many individual pests to attempt to simulate, (2) there is difficulty linking existing models of two different organisms that may be coded in different programming languages, and (3) there is less natural connection of the disciplines of pathology and entomology with engineers and agronomists. Hatfield et al. (Chapter 13 in this book) review insects, diseases, weeds, and pollinators as a part of the ecosystem, including a discussion of yield losses to pest damage. Fernandes et al. (Chapter 5 in this book) give good examples of linking diseases and insects to the CROPGRO model that use new programming techniques to enable daily coupling points (message passing of shared state variables) between two mechanistic models (the crop model and the pest model). This maintains the good mechanism and function of the pest models and allows multiple pest modules to share the same coupling points. This feature allows 're-use' of the same coupling points and methods for information transfer. It allows linkage to relatively simple mechanistic pest models written in different programming languages such as proposed by Donatelli et al. (2017) and used by Savary and Willocquet (2014) and Savary et al. (2015). The same approaches for coupling points could be followed to facilitate daily communication between forage crop models and livestock productivity models (especially under grazing) where animal stocking rate and consumption interact with forage 'on-offer'.

15 What improvements are needed in crop models?

Need to better account for soil fertility: Crop models need to be improved for their ability to simulate the effects of soil fertility and soil chemistry (pH, N, P, K, micronutrients). In recent years, soil fertility as a science has gone out of fashion in favor of fancy canopy-green indices sensed by remote means, and other high technology devices. However, that technology does not improve understanding or ability to model. We as an agroecosystem modeling community lack adequate understanding of mineralization of N and P from soil C pools and availability of various nutrients from different soils under different temperature and soil water conditions, thus the rate of N release and availability of nutrients from soils is not adequately simulated. This limitation also applies to sufficient understanding of soil C accumulation/sequestration, which is a critical component for climate change mitigation. Effects of soil pH should be easy to model, but in fact this has been ignored for years in crop models, especially considering effects of pH and aluminum saturation for deeper soil layers which inhibit root growth and affect root access to soil water, especially in tropical soils. A limited number of agroecosystem models such as APEX (Gassman et al., 2010), DSSAT (Dzotsi et al., 2010), and APSIM (Keating et al., 2003) consider effects of P fertility (see Chapter 3 in this book by Singh and Porter), but those models are not well tested, and simulating P fertility effects requires good soil tests for regions of application, thus those models are restrictive in use. A similar deficiency exists for modeling potassium (K) fertility effects with only a few models identified by Singh and Porter in Chapter 3, including DSSAT (Singh et al., 2007) and APSIM (Scanlan et al., 2015). The present modules for uptake of P and K have not advanced since those proposed by Barber (1984, 1995) in the 1970s and 1980s, and in fact, the Barber style of mechanistic modeling of P and K uptake is barely present in existing crop models. The most recent attempts toward a mechanistic approach to nutrient uptake modeling were de Willigen and van Noordwijk (see their 1987 book). Those scientists have retired and their code is likely no longer available.

Improved root growth simulation: Root growth is another area of crop modeling that is stuck in a time warp, with little measurable improvement in 25 years (Jones et al., 1991). While there have been some highly mechanistic studies of root type and root angle with genetic emphasis, those studies and their root 'models' do not seem to connect to practical crop models. Present root growth modules in crop models are relatively simple, because of lack of data and/or lack of emphasis on the part of the modelers. Root growth is important, because water and nutrient uptake in the crop models depends on the simulated shape and depth of rooting (at least for accessing water and nutrients). Root growth in crop models is also limited because of poor characterization of soils with depth, despite many efforts of various worldwide soil databases to provide data. Those databases are mostly insufficient for good characterization of root growth. Finally, a reason for poor characterization of rooting and root growth patterns is the lack of data on root growth patterns as affected by different crops and soil characteristics. Different species have different root growth patterns and may have different tolerances to low pH and high Al saturation of soils.

Account for lateral water flows (run-on, runoff) and vertical upflux: Future work with agroecosystem models includes the need to better account for the spatial aspects of water flows, both surface run-on as well as runoff and subsurface water flows and water tables, as discussed by Nendel and Zander in Chapter 12, and also Stockle and Meza in Chapter 4 of this book. This is needed to improve accuracy for spatial simulations of yield and management recommendations in precision agriculture applications. A critical need for water balance in spatial and precision agriculture is to be able to simulate spatially the surface and subsurface water flows (run-on and runoff). Most models totally lack these features, and rather consider each land unit as having no impact from adjacent land units. There are hydrological models such as SWAT (Arnold and Fohrer, 2005; Neitsch et al., 2011) that address this, but the crop component of SWAT is limited in capacity. There have been some attempts to link hydrological models to inform more mechanistic crop models and this effort needs to be enhanced.

Improved simulation of ET: Despite the critical importance of water for plant production, crop models are not sufficiently accurate in simulating ET (Kimball et al., 2019), partly related to problems of simulating soil water availability. The AgMIP-maize model effort revealed that maize models were highly variable in their simulation of crop ET (Bassu et al., 2014; Kimball et al., 2019), and also uncertain in the degree of reduction in transpiration under elevated CO_2 (Bassu et al., 2014). Furthermore, Durand et al. (2018) reported that maize models tended to underestimate yield enhancement under the combination of drought and elevated CO_2, a feature I believe is associated with insufficient 'model' reduction in transpiration under elevated CO_2. Measured reduction in ET of maize and sorghum was shown to be 18% with a doubling of CO_2 from 350 to 700 ppm (Allen Jr. et al., 2011), a target not achieved by most maize models in the paper of Bassu et al. (2014). While these examples are provided for maize, similar issues may apply for ET simulation and soil water balance for C-3 crops such as wheat.

Better parameterization for elevated temperature effects: In my opinion, many crop models are poorly parameterized for elevated temperature effects, because supraoptimal temperatures reduce reproductive growth and yield (Boote et al., 2005). There are two reasons for this inadequacy of the models: (1) for many crops, there is limited data for documenting elevated temperature effects because of insufficient research and because the elevated temperatures anticipated under future climate change are above the present-day growth temperatures and hence poorly researched, and (2) there is inertia and insufficient funding for crop model improvement, even where there may be existing data sufficient to parameterize the models for elevated temperature effects. There have been recent efforts to improve models for elevated temperature effects on reproductive growth associated with low pollen viability and poor grain-set (see Boote et al., 2018 for issues and improvements in some DSSAT models). Also, see efforts by the AgMIP-wheat group (Asseng et al., 2015), which evaluated 29 wheat crop models against spring wheat experiments conducted under a range of sowing dates and infrared heating (Reynolds et al., 1994; Ottman et al., 2012), which has resulted in modifications and improvements of many wheat crop models (Wang et al., 2017). The AgMIP-rice team evaluated 16 rice crop models against elevated temperature data with the finding that nearly two-thirds of the models initially failed to reduce yield sufficiently at elevated temperature, although the modelers are now making modifications (Boote et al., 2019, unpublished). Most maize crop models, until recently, have had inadequate coding for elevated temperature effects, for the added reason of few elevated temperature experiments, because maize plants are tall and not easily grown in elevated temperature growth facilities. Grain yield, grain number, and grain size of maize were evaluated in field and greenhouse treatments at cool versus elevated temperatures (Cicchino et al., 2010; Rattalino-Edreira et al., 2011; Lizaso et al., 2018). Elevated temperature (38–40°C daytime) during the anthesis period (compared to other phases) caused the largest yield reductions, with fewer grains formed because of lower pollen viability (Lizaso et al., 2018); however, the female component/heat stress on foliage was also important as fresh 'cool treatment' pollen did not completely compensate (Lizaso et al., 2018; Rattalino-Edreira et al., 2011). Maize grain size was modestly reduced by continued heat stress. The IXIM-Maize model (Lizaso et al., 2017) and SIMPLACE (Gabaldón-Leal et al., 2016) are examples of models recently modified to account for such heat stress effects.

Another feature is that most crop models lack the ability to predict foliage temperature. Normally, models are parameterized with data of experiments conducted under air temperature (and more typically under moderate humidity environments). But for arid regions with very large vapor pressure deficits associated with high air temperature, crop models tend to predict too much reduction in yield with high temperature and too much acceleration of crop life cycle. The problem occurs because foliage temperature is much lower than air temperature under large VPD and high air temperature, thus the functions derived under moderate humidity and air temperature do not apply well. The solution is not simple because modules predictive of foliage temperature are not very robust at present, and more critically, the crop models are all parameterized for temperature responses based on air temperature experiments (e.g. the foliage temperature is not often measured). Modelers are beginning to address this issue, but this requires models that simulate energy balance more mechanistically and which solve for foliage temperature (Webber et al., 2018).

Crop rotation effects: The crop rotation effects for residue, water, and N carry-over, as well as breaking disease, insect, and nematode pest effects, have not been adequately considered in crop simulations to-date (Basso et al., Chapter 8 in this book and Fernandes et al., Chapter 5 in this book). There are two issues: (1) many agroecosystem models have the capability to simulate crop rotations, yet most of the work is done with annual re-set and ignorance of carry-over effects (despite being easy to switch to continuous sequence), (2) however, the crop rotation break-cycle effects on disease, insect, and nematode effects are not adequately addressed and few models have even attempted this. The break-cycle effects on pests are not simple or easy to solve mechanistically, and may require more empirical region-specific solutions (yield offset factors).

Ozone effects are mostly missing: Most present crop models lack the ability to simulate ozone effects on growth and production, despite rising ozone levels throughout the world becoming an increasing yield detriment. Recent work on this has been initiated by a few modeling groups (Emberson et al., 2018). The approach followed by these first modelers should be to make generic modules that function with other crop models.

Climate effects on crop quality: Recent papers have documented climate change impacts on quality of cereal grains and vegetables (Taub et al., 2008; Broberg et al., 2017; Nuttall et al., 2017; Zhu et al., 2018). Crop models are poorly coded and insufficiently parameterized to simulate those quality factors. While many crop models predict protein composition of grains as part of overall crop N balance, other quality factors are generally ignored (Nuttall et al., 2017). There have been some attempts for fruits and horticultural crops, but not much for agronomic crops although Senthold Asseng (pers. comm., 2019) is leading an effort for predicting yield and quality of agronomic and horticultural crops. Heuvelink et al. (2004) described the state of modeling product quality for horticulture commodities. Bertin et al. (2006) reviewed the challenges of modeling fruit quality for peach and tomato ecophysiological models.

Need models for fruits, vegetables, forages: There are inadequate efforts for developing simulation models for lesser known commodities including fruits, vegetables, and forages. This is partly caused by the minor status of the vegetable commodities, but also by the lack of connection of the modeling community to horticulturalists and forage scientists. Some progress is being made. There are models for some vegetables in DSSAT (tomato, green bean, sweet corn, green pepper, cabbage, potato), often developed on existing DSSAT templates. Another solution is to create a very simple generic model that works for many of the lesser commodities (Zhao et al., 2019), and this model was calibrated to the same primary six vegetable data sets that already exist in DSSAT. Zhao et al. (2019) additionally simulated carrot and banana. But in my opinion, there is risk in creating simple models if they are not robust for diverse environments. See the list of grassland, pasture, and rangeland models reviewed by Descheemaeker and Bell (Chapter 9 in this book). The DSSAT system recently incorporated perennial forage models for brachiaria and cynodon (Pequeno et al., 2014, 2018), and alfalfa (Malik et al., 2018).

16 Where to look for further information

For a more detailed history of crop simulation over multiple decades, see Jones et al. (2016). Also, see Jones et al. (2017) for a review titled 'state of agricultural systems science'.

For more information on crop modeling efforts in the AgMIP Project, see http://www.agmip.org/. See Chapter 16 by Asseng et al. in this book for their review of AgMIP projects aimed at intercomparison of multiple crop models.

For crop modeling textbooks, see Tsuji et al. (1998) 'Understanding Options for Agricultural Production', and Wallach et al. (2019) 'Working with Dynamic Crop Models'.

Professional societies that host crop modeling activities in their conferences and professional journals include the American Society of Agronomy (https://www.agronomy.org), Crop Science Society of America (https://www.crops.org), and Soil Science Society of America (https://www.soils.org) which publish Agronomy Journal, Crop Science, and Soil Science Society of America Journal. These three societies jointly host a conference every year (go to: acsmeetings@sciencesocieties.org). The European Society for Agronomy (european-agronomy.org) hosts conferences that include crop modeling and publish the European Journal of Agronomy. Agrosystems modeling is also covered in the conferences and the Transactions of the American Society of Agricultural and Biological Engineers (https://www.asabe.org). In addition to the above journals, Field Crops Research, Agricultural Systems, Global Change Biology, and Agricultural and Forest Meteorology are important publication outlets for crop modeling papers.

17 References

Allen Jr., L. H., Kakani, V. G., Vu, J. C. V. and Boote, K. J. 2011. Elevated CO_2 increases water use efficiency by sustaining photosynthesis of water-limited maize and sorghum. *J. Plant Physiol.* 168(16), 1909–18. doi:10.1016/j.jplph.2011.05.005.

Andrea, M. C., Boote, K. J., Sentelhas, P. C. and Romanelli, T. L. 2018. Variability and limitations of maize production in Brazil: potential yield, water-limited yield and yield gaps. *Agric. Syst.* 165, 264–73. doi:10.1016/j.agsy.2018.07.004.

Antle, J. M., Stoorvogel, J. J. and Valdivia, R. O. 2014. New parsimonious simulation methods and tools to assess future food and environmental security of farm populations. *Philos. Trans. R. Soc. Lond. B Biol. Sci.* 369(1639), 20120280. doi:10.1098/rstb.2012.0280.

Archontoulis, S. V., Miguez, F. E., Moore, K. J., Loynachan, T. and Knapp, A. 2012. Incorporating farming system modeling approaches into agronomy MS program. *ASA, CSSA and SSSA 2012 International Annual Meeting*, 21–24 October, Cincinnati, OH.

Arnold, J. G. and Fohrer, N. 2005. SWAT2000: current capabilities and research opportunities in applied watershed modelling. *Hydrol. Process.* 19(3), 563–72. doi:10.1002/hyp.5611.

Asseng, S., Ewert, F., Rosenzweig, C., Jones, J. W., Hatfield, J. L., Ruane, A. C., Boote, K. J., Thorburn, P. J., Rötter, R. P., Cammarano, D., Brisson, N., Basso, B., Martre, P., Aggarwal, P. K., Angulo, C., Bertuzzi, P., Biernath, C., Challinor, A. J., Doltra, J., Gayler, S., Goldberg, R., Grant, R., Heng, L., Hooker, J., Hunt, L. A., Ingwersen, J., Izaurralde, R. C., Kersebaum, K. C., Müller, C., Kumar, S. N., Nendel, C., O'Leary, G., Olesen, J. E., Osborne, T. M., Palosuo, T., Priesack, E., Ripoche, D., Semenov, M. A., Shcherbak, I., Steduto, P., Stöckle, C., Stratonovitch, P., Streck, T., Supit, I., Tao, F., Travasso, M., Waha, K., Wallach, D., White, J. W., Williams, J. R. and Wolf, J. 2013. Uncertainties in simulating wheat yields under climate change. *Nat. Clim. Change* 3, 827–32.

Asseng, S., Ewert, F., Martre, P., Rötter, R. P., Lobell, D. B., Cammarano, D., Kimball, B. A., Ottman, M. J., Wall, G. W., White, J. W., Reynolds, M. P., Alderman, P. D., Prasad, P. V. V., Aggarwal, P. K., Anothai, J., Basso, B., Biernath, C., Challinor, A. J., De Sanctis, G., Doltra, J., Fereres, E., Garcia-Vila, M., Gayler, S., Hoogenboom, G., Hunt, L. A., Izaurralde, R. C., Jabloun, M., Jones, C. D., Kersebaum, K. C., Koehler, A., Müller, C., Naresh Kumar, S., Nendel, C., O'Leary, G., Olesen, J. E., Palosuo, T., Priesack, E., Eyshi Rezaei, E., Ruane, A. C., Semenov, M. A., Shcherbak, I., Stöckle, C., Stratonovitch, P., Streck, T., Supit, I., Tao, F., Thorburn, P. J., Waha, K., Wang, E., Wallach, D., Wolf, J., Zhao, Z. and Zhu, Y. 2015. Rising temperatures reduce global wheat production. *Nat. Clim. Change* 5(2), 143–7. doi:10.1038/nclimate2470.

Barber, S. A. 1984. *Soil Nutrient Bioavailability: a Mechanistic Approach*. John Wiley, New York, 398pp.

Barber, S. A. 1995. *Soil Nutrient Bioavailability: Mechanistic Approach* (2nd edn.). Wiley, New York, 398pp.

Basso, B., Dumont, B., Maestrini, B., Shcherbak, I., Robertson, G. P., Porter, J. R., Smith, P., Paustian, K., Grace, P. R., Asseng, S., Bassu, S., Biernath, C., Boote, K. J., Cammarano, D., De Sanctis, G., Durand, J.-L., Ewert, F., Gayler, S., Hyndman, D. W., Kent, J., Martre, P., Nendel, C., Priesack, E., Ripoche, D., Ruane, A. C., Sharp, J., Thorburn, P. J., Hatfield, J. L., Jones, J. W. and Rosenzweig, C. 2018. Soil organic carbon and nitrogen feedbacks on crop yields under climate change. *Ael* 3(1). doi:10.2134/ael2018.05.0026.

Bassu, S., Brisson, N., Durand, J. L., Boote, K., Lizaso, J., Jones, J. W., Rosenzweig, C., Ruane, A. C., Adam, M., Baron, C., Basso, B., Biernath, C., Boogaard, H., Conijn, S., Corbeels, M., Deryng, D., DeSanctis, G., Gayler, S., Grassini, P., Hatfield, J., Hoek, S., Izaurralde, C., Jongschaap, R., Kemanian, A. R., Kersebaum, K. C., Kim, S. H., Kumar, N. S., Makowski, D., Mueller, C., Nendel, C., Priesack, E., Pravia, M. V., Sau, F., Shcherbak, I., Tao, F., Teixeira, E., Timlin, D. and Waha, K. 2014. How do various maize crop models vary in their responses to climate change factors? *Glob. Change Biol.* 20(7), 2301–20. doi:10.1111/gcb.12520.

Battisti, R., Sentelhas, P. C., Parker, P. S., Nendel, C., Câmara, G. M. D. S., Farias, J. R. B. and Basso, C. J. 2018. Assessment of crop-management strategies to improve soybean resilience to climate change in Southern Brazil. *Crop Pasture Sci.* 69(2), doi:10.1071/CP17293.

Bertin, N., Bussières, P. and Génard, M. 2006. Ecophysiological models of fruit quality: a challenge for peach and tomato. *Acta Hortic.* 718(718), 633–46. doi:10.17660/ActaHortic.2006.718.74.

Boote, K. J., Jones, J. W. and Pickering, N. B. 1996. Potential uses and limitations of crop models. *Agron. J.* 88(5), 704–16. doi:10.2134/agronj1996.00021962008800050005x.

Boote, K. J., Jones, J. W., Hoogenboom, G. and Pickering, N. B. 1998. The CROPGRO model for grain legumes. In: Tsuji, G. Y., Hoogenboom, G. and Thornton, P. K. (Eds), *Understanding Options for Agricultural Production*. Kluwer Academic Publishers, Dordrecht, pp. 99–128.

Boote, K. J., Allen, L. H., Prasad, P. V. V., Baker, J. T., Gesch, R. W., Snyder, A. M., Pan, D. and Thomas, J. M. G. 2005. Elevated temperature and CO_2 impacts on pollination, reproductive growth, and yield of several globally important crops. *J. Agric. Meteorol.* 60, 469–74.

Boote, K. J., Jones, J. W. and Hoogenboom, G. 2008. Crop simulation models as tools for agro-advisories for weather and disease effects on production. *J. Agrometeorol.* 10, 9–17.

Boote, K. J., Jones, J. W., White, J. W., Asseng, S. and Lizaso, J. I. 2013. Putting mechanisms into crop production models. *Plant Cell Environ.* 36(9), 1658–72. doi:10.1111/pce.12119.

Boote, K. J., Prasad, V., Allen Jr., L. H., Singh, P. and Jones, J. W. 2018. Modeling sensitivity of grain yield to elevated temperature in the DSSAT crop models for peanut, soybean, dry bean, chickpea, sorghum, and millet. *Eur. J. Agron.* 100, 99–109. doi:10.1016/j.eja.2017.09.002.

Broberg, M. C., Högy, P. and Pleijel, H. 2017. CO_2 - induced changes in wheat grain composition: meta-analysis and response functions. *Agronomy* 7(2), 32. doi:10.3390/agronomy7020032.

Chenu, K., Chapman, S. C., Tardieu, F., McLean, G., Welcker, C. and Hammer, G. L. 2009. Simulating the yield impacts of organ-level quantitative trait loci associated with drought response in maize: a 'gene-to-phenotype' modeling approach. *Genetics* 183(4), 1507–23. doi:10.1534/genetics.109.105429.

Cicchino, M., Rattalino-Edreira, J. I. R., Uribelarrea, M. and Otegui, M. E. 2010. Heat stress in field-grown maize: response of physiological determinants of grain yield. *Crop Sci.* 50(4), 1438–48. doi:10.2135/cropsci2009.10.0574.

Cooper, M., Technow, F., Messina, C., Ghu, C. and Totir, L. R. 2016. Use of crop growth models with whole-genome prediction: application to a maize multi-environment trial. *Crop Sci.* 56, 2141–56. doi:10.2135/cropsci2015.08.0512.

De Willigen, P. and Van Noordwijk, M. 1987. Roots, plant production and nutrient use efficiency. PhD thesis. Agricultural University, PUDOC, Wageningen, the Netherlands, 282pp.

DeWit, C. T. 1965. Photosynthesis of leaf canopies. Agr. Res. Rep. 663. PUDOC, Wageningen, the Netherlands, 57pp.

Donatelli, M., Magarey, R. D., Bregaglio, S., Willocquet, L., Whish, J. P. M. and Savary, S. 2017. Modelling the impacts of pests and diseases on agricultural systems. *Agric. Syst.* 155, 213–24. doi:10.1016/j.agsy.2017.01.019.

Duncan, W. G. 1972. SIMCOT: a simulation of cotton growth and yield. In: Murphy, C. M. (Ed.), *Proceedings of a Workshop for Modeling Tree Growth*. Duke University, Durham, NC, pp. 115-8.

Duncan, W. G., Loomis, R. S., Williams, W. A. and Hanau, R. 1967. A model for simulating photosynthesis in plan communities. *Hilgardia* 38(4), 181-205. doi:10.3733/hilg.v38n04p181.

Duncan, W. G., McCloud, D. E., McGraw, R. L. and Boote, K. J. 1978. Physiological aspects of peanut yield improvement. *Crop Sci.* 18(6), 1015-20. doi:10.2135/cropsci1978.0011183X001800060028x.

Durand, J.-L., Delusca, K., Boote, K., Lizaso, J., Manderscheid, R., Weigel, H. J., Ruane, A. C., Rosenzweig, C., Jones, J., Ahuja, L., Anapalli, S., Basso, B., Baron, C., Bertuzzi, P., Biernath, C., Deryng, D., Ewert, F., Gaiser, T., Gayler, S., Heinlein, F., Kersebaum, K. C., Kim, S.-H., Muller, C., Nendel, C., Olioso, A., Priesack, E., Villegas, J. R., Ripoche, D., Rotter, R. P., Seidel, S. I., Srivastava, A., Tao, F., Timlin, D., Twine, T., Wang, E., Webber, H. and Zhao, Z. 2018. How accurately do maize crop models simulate the interactions of atmospheric CO_2 concentration levels with limited water supply on water use and yield? *Eur. J. Agron.* 100, 67-75. doi:10.1016/j.eja.2017.01.002.

Dzotsi, K. A., Jones, J. W., Adiku, S. G. K., Naab, J. B., Singh, U., Porter, C. H. and Gijsman, A. J. 2010. Modeling soil and plant phosphorus within DSSAT. *Ecol. Modell.* 221(23), 2839-49. doi:10.1016/j.ecolmodel.2010.08.023.

Elliott, J., Müller, C., Deryng, D., Chryssanthacopoulos, J., Boote, K. J., Büchner, M., Foster, I., Glotter, M., Heinke, J., Iizumi, T., Izaurralde, R. C., Mueller, N. D., Ray, D. K., Rosenzweig, C., Ruane, A. C. and Sheffield, J. 2015. The global gridded crop model intercomparison: data and modeling protocols for phase 1 (v1.0). *Geosci. Model Dev.* 8, 261-77.

Emberson, L. D., Pleijel, H., Ainsworth, E. A., van den Berg, M., Ren, W., Osborne, S., Mills, G., Pandey, D., Dentener, F., Büker, P., Ewert, F., Koeble, R. and Van Dingenen, R. 2018. Ozone effects on crops and consideration in crop models. *Eur. J. Agron.* 100, 19-34. doi:10.1016/j.eja.2018.06.002.

Fleisher, D. H., Condori, B., Quiroz, R., Alva, A., Asseng, S., Barreda, C., Bindi, M., Boote, K. J., Ferrise, R., Franke, A. C., Govindakrishna, P. M., Harahagazwe, D., Hoogenboom, G., Kumar, S. N., Merante, P., Nendel, C., Olesen, J. E., Parker, P. S., Raes, D., Raymudo, R., Ruane, A. C., Stockle, C., Supit, I., Vanuytrecht, E., Wolf, J. and Woli, P. 2017. A potato model intercomparison across varying climates and productivity levels. *Glob. Change Biol.* 23(3), 1258-81. doi:10.1111/gcb.13411.

Furbank, R. T. and Tester, M. 2011. Phenomics-technologies to relieve the phenotyping bottleneck. *Trends Plant Sci.* 16(12), 635-44. doi:10.1016/j.tplants.2011.09.005.

Gabaldón-Leal, C., Webber, H., Otegui, M. E., Slafer, G. A., Ordóñez, R. A., Gaiser, T., Lorite, I. J., Ruiz-Ramos, M. and Ewert, F. 2016. Modelling the impact of heat stress on maize yield formation. *Field Crops Res.* 198, 226-37. doi:10.1016/j.fcr.2016.08.013.

Gassman, P. W., Williams, J. R., Wang, X., Saleh, A., Osei, E., Hauck, L. M., Izaurralde, R. C. and Flowers, J. D. 2010. The Agricultural Policy/Environmental Extender (Apex) Model: an emerging tool for landscape and watershed environmental analyses. *Trans. ASABE* 53, 711-40.

Gu, J., Yin, X., Zhang, C., Wang, H. and Struik, P. C. 2014. Linking ecophysiological modelling with quantitative genetics to support marker-assisted crop design for improved yields of rice (*Oryza sativa*) under drought stress. *Ann. Bot.* 114(3), 499-511. doi:10.1093/aob/mcu127.

Heinemann, A. B., Ramirez-Villegas, J., Livio, T., Souza, P. O., Didonet, A. D., di Stefano, J. G., Boote, K. J. and Jarvis, A. 2016. Drought impact on rainfed common bean production areas in Brazil. *Agric. For. Met.* 225, 1-18.

Hochman, Z., van Rees, H., Carberry, P. S., Hunt, J. R., McCown, R. L., Gartmann, A., Holzworth, D., van Rees, S., Dalgliesh, N. P., Long, W., Peake, A. S., Poulton, P. L. and McClelland, T. 2009. Re-inventing model-based decision support with Australian dryland farmers. Yield Prophet® helps farmers monitor and manage crops in a variable climate. *Crop Pasture Sci.* 60(11), 1057-70. doi:10.1071/CP09020.

Hoogenboom, G. and White, J. W. 2003. Improving physiological assumptions of simulation models by using gene-based approaches. *Agron. J.* 95(1), 82-9. doi:10.2134/agronj2003.0082.

Hoogenboom, G., White, J. W. and Messina, C. D. 2004. From genome to crop: integration through simulation modeling. *Field Crop Res.* 90(1), 145-63. doi:10.1016/j.fcr.2004.07.014.

Heuvelink, E., Tijskens, P. and Kang, M. Z. 2004. Modelling product quality in horticulture: an overview. In: *Proc. Intl. WS on Models Plant Growth &. Contr. Prod. Qual. in Hort. Prod.* Available at: https://core.ac.uk/download/pdf/29287934.pdf (Accessed on 3 August 2019).

Jones, C. A., Bland, W. L., Ritchie, J. T. and Williams, J. R. 1991. Simulation of root growth. In: Hanks, J. and Ritchie, J. T. (Eds), *Modelling Plant and Soil Systems*. Amer. Soc. Agron. no. 31. ASA-CSSA-SSSA, Madison, WI, pp. 91-123.

Jones, J. W., Antle, J. M., Basso, B. O., Boote, K. J., Conant, R. T., Foster, I., Godfray, H. C. J., Herrero, M., Howitt, R. E., Janssen, S., Keating, B. A., Munoz-Carpena, R., Porter, C. H., Rosenzweig, C. and Wheeler, T. R. 2016. Brief history of agricultural systems modeling. *Agric. Syst.* 155, 240-54. doi:10.1016/j.agsy.2016.05.014.

Jones, J. W., Antle, J. M., Basso, B., Boote, K. J., Conant, R. T., Foster, I., Godfray, H. C. J., Herrero, M., Howitt, R. E., Janssen, S., Keating, B. A., Munoz-Carpena, R., Porter, C. H., Rosenzweig, C. and Wheeler, T. R. 2017. Toward a new generation of agricultural system data, models, and knowledge products: state of agricultural systems science. *Agric. Syst.* 155, 269-88. doi:10.1016/j.agsy.2016.09.021.

Keating, B. A., Carberry, P. S., Hammer, G. L., Probert, M. E., Robertson, M. J., Holzworth, D., Huth, N. I., Hargreaves, J. N. G., Meinke, H., Hochman, Z., McLean, G., Verburg, K., Snow, V., Dimes, J. P., Silburn, M., Wang, E., Brown, S., Bristow, K. L., Asseng, S., Chapman, S., McCown, R. L., Freebairn, D. M. and Smith, C. J. 2003. An overview of APSIM, a model designed for farming systems simulation. *Eur. J. Agron.* 18(3-4), 267-88. doi:10.1016/S1161-0301(02)00108-9.

Kersebaum, K. C., Boote, K. J., Jorgenson, J. S., Nendel, C., Bindi, M., Frühauf, C., Gaiser, T., Hoogenboom, G., Kollas, C., Olesen, J. E., Rotter, R. P., Ruget, F., Thorburn, P. J., Trnka, M. and Wegehenkel, M. 2015. Analysis and classification of data sets for calibration and validation of agro-ecosystem models. *Environ. Modell. Softw.* 72, 402-17. doi:10.1016/j.envsoft.2015.05.009.

Kimball, B. A., Boote, K. J., Hatfield, J. L., Ahuja, L. R., Stockle, C., Archontoulis, S., Baron, C., Basso, B., Bertuzzi, P., Constantin, J., Deryng, D., Dumont, B., Durand, J.-L., Ewert, F., Gaiser, T., Gayler, S., Hoffmann, M. P., Jiang, Q., Kim, S.-H., Lizaso, J., Moulin, S., Nendel, C., Parker, P., Palosuo, T., Priesack, E., Qi, Z., Srivastava, A., Stella, T., Tao, F., Thorp, K. R., Timlin, D., Twine, T. E., Webber, H., Willaume, M. and Williams, K.

2019. Simulation of maize evapotranspiration: an inter-comparison among 29 maize models. *Ag. For. Met.* 271, 264–84. doi:10.1016/j.agrformet.2019.02.037.

Kisekka, I., Aguilar, J., Rogers, D. and Klocke, N. 2016. Assessing deficit irrigation strategies for corn using simulation. *Trans. ASABE* 59(1), 303–17. doi:10.13031/trans.59.11206.

Kisekka, I., Arabi, M., Sung, K. J., Gowda, P., Prasad, P. V., Aguilar, J. and Rogers, D. 2017. iCrop a model-driven decision support tool for optimizing water use in agriculture. *Abstract for the 2017 ASABE Annual Internal Meeting in Spokane*, Washington, 16–19 July 2017.

Li, T., Hasegawa, T., Yin, X., Zhu, Y., Boote, K., Adam, M., Bregaglio, S., Buis, S., Confalonieri, R., Fumoto, T., Gaydon, D., Marcaida III, M., Nakagawa, H., Oriol, P., Ruane, A. C., Ruget, F., Singh, B., Singh, U., Tang, L., Tao, F., Wilkens, P., Yoshida, H., Zhang, Z. and Bouman, B. 2015. Uncertainties in predicting rice yield by current crop models under a wide range of climatic conditions. *Glob. Change Biol.* 21(3), 1328–41. doi:10.1111/gcb.12758.

Lizaso, J. I., Ruiz-Ramos, M., Rodriguez, L., Gabaldon-Leal, C., Oliveira, J. A., Lorite, I. J., Rodriguez, A., Maddonni, G. A. and Otegui, M. E. 2017. Modeling the response of maize phenology, kernel set, and yield components to heat stress and heat shock with CSM-IXIM. *Field Crops Res.* 214, 239–54. doi:10.1016/j.fcr.2017.09.019.

Lizaso, J. I., Ruiz-Ramos, M., Rodriguez, L., Gabaldon-Leal, C., Oliveira, J. A., Lorite, I. J., Sanchez, D., Garcia, E. and Rodriguez, A. 2018. Impact of high temperatures in maize: phenology and yield components. *Field Crops Res.* 216, 129–40. doi:10.1016/j.fcr.2017.11.013.

Makowski, D., Asseng, S., Ewert, F., Bassu, S., Durand, J. L., Martre, P., Adam, M., Aggarwal, P. K., Angulo, C., Baron, C., Basso, B., Bertuzzi, P., Biernath, C., Boogaard, H., Boote, K. J., Bouman, B., Bregaglio, S., Brisson, N., Buis, S., Cammarano, D., Challinor, A. J., Confalonieri, R., Conijn, J. G., Corbeels, M., Deryng, D., De Sanctis, G., Doltra, J., Fumoto, T., Gaydon, D., Gayler, S., Goldberg, R., Grant, R., Grassini, P., Hatfield, J., Hasegawa, T., Heng, L., Hoek, S., Hooker, J., Hunt, L., Ingwersen, J., Izaurralde, C., Jongschaap, R., Jones, J., Kemanian, A., Kersebaum, C., Kim, S., Lizaso, J. I., Marcaida, M., Müller, C., Nakagawa, H., Naresh Kumar, S., Nendel, C., O'leary, G., Olesen, J.E., Oriol, T., Osborne, T., Palosuo, T., Pravia, M., Priesack, E., Ripoche, D., Rosenzweig, C., Ruane, A., Ruget, F., Sau, F., Semenov, M., Shcherbak, I., Singh, B, Singh, Z., Soo, H., Steduto, P., Stöckle, C., Stratonovitch, P., Streck, T., Supit, I., Tang, L., Tao, F., Teixeira, E., Thorburn, P., Timlin, D., Travasso, M., Rotter, R. P., Waha, K., Wallach, D., White, J., Wilkens, P., Williams, J., Wolf, J., Yin, X., Yoshida, H., Zhang, Z. and Zhu, Y. 2015. A statistical analysis of three ensembles of crop model responses to temperature and CO_2 concentration. *Ag. For. Met.* 214–215, 483–93.

Malik, W., Boote, K. J., Hoogenboom, G., Cavero, J. and Dechmi, F. 2018. Adapting the CROPGRO model to simulate alfalfa growth and yield. *Agron. J.* 110, 1777–90.

Martre, P., Wallach, D., Asseng, S., Ewert, F., Jones, J. W., Rötter, R. P., Boote, K. J., Ruane, A. C., Thorburn, P. J., Cammarano, D., Hatfield, J. L., Rosenzweig, C., Aggarwal, P. K., Angulo, C., Basso, B., Bertuzzi, P., Biernath, C., Brisson, N., Challinor, A. J., Doltra, J., Gayler, S., Goldberg, R., Grant, R. F., Heng, L., Hooker, J., Hunt, L. A., Ingwersen, J. C., Izaurralde, R. C., Kersebaum, K. C., Müller, C., Kumar, S. N., Nendel, C., O'Leary, G. J., Olesen, J. E., Osborne, T. M., Palosuo, T., Priesack, E., Ripoche, D., Semenov, M. A., Shcherbak, I., Steduto, P., Stöckle, C. O., Stratonovitch, P., Streck, T., Supit, I., Tao, F., Travasso, M., Waha, K., White, J. W. and Wolf, J. 2015. Multimodel ensembles of wheat growth: many models are better than one. *Glob. Change Biol.* 21(2), 911–25 doi:10.1111/gcb.12768.

Messina, C. D., Jones, J. W., Boote, K. J. and Vallejos, C. E. 2006. A gene-based model to simulate soybean development and yield responses to environment. *Crop Sci.* 46(1), 456–66. doi:10.2135/cropsci2005.04-0372.

Messina, C. D., Technow, F., Tang, T., Totir, R., Gho, C. and Cooper, M. 2018. Leveraging biological insight and environmental variation to improve phenotypic prediction: integrating crop growth models (CGM) with whole genome prediction (WGP). *Eur. J. Agron.* 100, 151–62. doi:10.1016/j.eja.2018.01.007.

Miguez, F. E., Moore, K. J. and Archontoulis, S. V. 2012. A modeling approach to unifying concepts taught in a professional MS agronomy Curriculum. *NACTA J.* 56, 12. Abstract # 0247.

Muller, C. and Robertson, R. D. 2014. Projecting future crop productivity for global economic modeling. *Agric. Econ.* 45(1), 37–50. doi:10.1111/agec.12088.

Neitsch, S. L., Arnold, J. G., Kiniry, J. R. and Williams, J. R. 2011. Soil and water assessment tool: theoretical documentation version 2009. Texas A & M, AgriLife Research & Extension, TR-406.

Nelson, G. C., Valin, H., Sands, R. D., Havlík, P., Ahammad, H., Deryng, D., Elliott, J., Fujimori, S., Hasegawa, T., Heyhoe, E., Kyle, P., Von Lampe, M., Lotze-Campen, H., Mason d'Croz, D., van Meijl, H., van der Mensbrugghe, D., Müller, C., Popp, A., Robertson, R., Robinson, S., Schmid, E., Schmitz, C., Tabeau, A. and Willenbockel, D. 2014a. Climate change effects on agriculture: economic responses to biophysical shocks. *Proc. Natl. Acad. Sci. U.S.A.* 111(9), 3274–9. doi:10.1073/pnas.1222465110.

Nelson, G. C., van der Mensbrugghe, D., Ahammad, H., Blanc, E., Calvin, K., Hasegawa, T., Havlik, P., Heyhoe, E., Kyle, P., Lotze-Campen, H., von Lampe, M., Mason d'Croz, D., van Meijl, H., Muller, C., Reilly, J., Robertson, R., Sands, R. D., Schmitz, C., Tabeau, A., Takahashi, K., Valin, H. and Willenbockel, D. 2014b. Agriculture and climate change in global scenarios: why don't the models agree. *Agric. Econ.* 45, 85–101.

Nuttall, J. G., O'Leary, G. J., Panozzo, J. F., Walker, C. K., Barlow, K. M. and Fitzgerald, G. J. 2017. Models of grain quality in wheat – a review. *Field Crops Res.* 202, 136–45. doi:10.1016/j.fcr.2015.12.011.

Ottman, M. J., Kimball, B. A., White, J. W. and Wall, G. W. 2012. Wheat growth response to increased temperature from varied planting dates and supplemental infrared heating. *Agron. J.* 104(1), 7–16. doi:10.2134/agronj2011.0212.

Penning deVries, F. W. T. and van Laar, H. H. (Eds). 1982. *Simulation of Plant Growth and Crop Production*. PUDOC, Wageningen, The Netherlands.

Pequeno, D. N. L., Pedreira, C. G. S. and Boote, K. J. 2014. Simulating forage production of Marandu palisade grass (*Brachiaria brizantha*) with the CROPGRO-Perennial Forage model. *Crop Pasture Sci.* 65(12), 1335–48. doi:10.1071/CP14058.

Pequeno, D. N. L., Pedreira, C. G. S., Boote, K. J., Alderman, P. D. and Faria, A. F. G. 2018. Species-genotypic parameters of the CROPGRO Perennial Forage Model: implications for comparison of three tropical pasture grasses. *Grass Forage Sci.* 73(2), 440–55. doi:10.1111/gfs.12329.

Rattalino-Edreira, J. I., Budakli Carpici, E., Sammarro, D. and Otegui, M. E. 2011. Heat stress effects around flowering on kernel set of temperate and tropical maize hybrids. *Field Crop. Res.* 123(2), 62–73. doi:10.1016/j.fcr.2011.04.015.

Reynolds, M. P., Balota, M., Delgado, M. I. B., Amani, I. and Fischer, R. A. 1994. Physiological and morphological traits associated with spring wheat yield under hot, irrigated conditions. *Functional Plant Biol.* 21(6), 717–30. doi:10.1071/PP9940717.

Rosegrant, M. W. and IMPACT Development Team. 2012. International Model for Policy Analysis of Agricultural Commodities and Trade (IMPACT) model description. Available at: http://www.ifpri.org/publication/international-model-policy-analysis-agriculturalcommodities-and-trade-impact-1 (accessed on June 2013).

Rosenzweig, C., Elliott, J., Deryng, D., Ruane, A. C., Arneth, A., Boote, K. J., Folberth, C., Glotter, M., Müller, C., Khabarov, N., Neumann, K., Piontek, F., Pugh, T., Schmid, E., Stehfest, E., Yang, H. and Jones, J. W. 2014. Assessing agricultural risks of climate change in the 21st century in a global gridded crop model intercomparison. *Proc. Natl. Acad. Sci.* 111, 3268–73. doi:10.1073/pnas.1222463110.

Ruane, A. C., Antle, J., Elliott, J., Folberth, C., Hoogenboom, G., Mason-D'Croz, D., Müller, C., Porter, C., Phillips, M., Raymundo, R., Sands, R., Valdivia, R., White, J., Wiebe, K. and Rosenzweig, C. 2018. Biophysical and economic implications for agriculture of+1.5 degrees and+2.0 degrees C global warming using AgMIP coordinated global and regional assessments. *Clim. Res.* 76(1), 17–39. doi:10.3354/cr01520.

Savary, S. and Willocquet, L. 2014. Simulation modeling in botanical epidemiology and crop loss analysis. *APSnet Education Center The Plant Health Instructor*. Available at: https://www.apsnet.org/edcenter/advanced/topics/BotanicalEpidemiology/Pages/ModelingYieldLossesDuetoPests.aspx.

Savary, S., Stetkiewicz, S., Brun, F. and Willocquet, L. 2015. Modelling and mapping potential epidemics of wheat diseases-examples on leaf rust and Septoria tritici blotch using EPIWHEAT. *Eur. J. Plant Pathol.* 142(4), 771–90. doi:10.1007/s10658-015-0650-7.

Savary, S., Nelson, A. D., Djurle, A., Esker, P. D., Sparks, A., Amorim, L., Bergamin Filho, A., Caffi, T., Castilla, N., Garrett, K., McRoberts, N., Rossi, V., Yuen, J. and Willocquet, L. 2018. Concepts, approaches, and avenues for modelling crop health and crop losses. *Eur. J. Agron.* 100, 4–18. doi:10.1016/j.eja.2018.04.003.

Scanlan, C. A., Bell, R. W. and Brennan, R. F. 2015. Simulating wheat growth response to potassium availability under field conditions in sandy soils. I. Model development. *Field Crop. Res.* 178, 125–34. doi:10.1016/j.fcr.2015.03.019.

Shelia, V., Hansen, J., Sharda, V., Porter, C., Aggarwal, P., Wilkerson, C. J. and Hoogenboom, G. 2019. A multi-scale and multi-model gridded framework for forecasting crop production, risk analysis, and climate change impact studies. *Environ. Modell. Softw.* 115, 144–54. doi:10.1016/j.envsoft.2019.02.006.

Singh, U., Wilkens, P. W., Gopi, R. and Potter, C. 2007. Development and application of a potash decision support system. *2007 Annual Meetings Abstracts*, ASA/CSSA/SSSA.

Singh, P., Nedumaran, S., Ntare, B. R., Boote, K. J., Singh, N. P., Srinivas, K. and Bantilan, M. C. S. 2013. Potential benefits of drought and heat tolerance in groundnut for adaptation to climate change in India and West Africa. *Mitig. Adapt. Strateg. Glob. Change* 19(5), 509–29 (doi:10.1007/s11027-012-9446-7).

Stapleton, H. N., Buxton, D. R., Watson, F. L., Notling, D. J. and Baker, D. N. 1973. COTTON: a computer simulation of cotton growth. Univ. of Arizona, Agric. Exp. Sta. Tech. Bull. 206.

Tardieu, F., Cabrera-Bosquet, L., Pridmore, T. and Bennett, M. 2017. Plant phenomics, from sensors to knowledge. *Curr. Biol.* 27(15), R770–83. doi:10.1016/j.cub.2017.05.055.

Taub, D. R., Miller, B. and Allen, H. 2008. Effects of elevated CO_2 on the protein concentration of food crops: a meta-analysis. *Glob. Change Biol.* 14(3), 565–75. doi:10.1111/j.1365-2486.2007.01511.x.

Technow, F., Messina, C. D., Totir, L. R. and Cooper, M. 2015. Integrating crop growth models with whole genome prediction through approximate bayesian computation. *PLoS ONE* 10(6), e0130855. doi:10.1371/journal.pone.0130855.

Thorburn, P., Boote, K., Hargreaves, J., Poulton, P. and Jones, J. 2015. Crop systems modeling in AgMIP: a new protocol-driven approach for regional integrated assessments. In: Hillel, D. and Rosenzweig, C. (Eds), *Handbook of Climate Change and Agroecosystems: Agricultural Model Intercomparison and Improvement Project Integrated Crop and Economic Assessments (Vol. 3 and 4). ICP Series on Climate Change Impacts, Adaptation, and Mitigation*. Imperial College Press.

Thorburn, P. J., Boote, K. J., Nendel, C., Rötter, R. P. and Ewert, F. 2018. Recent advances in crop modelling to support sustainable agricultural production and food security under global change. *Eur. J. Agron.* 100, 1–3. doi:10.1016/j.eja.2018.10.009.

Tsuji, G. Y., Hoogenboom, G. and Thornton, P. K. (Eds). 1998. *Understanding Options for Agricultural Production*. Kluwer Academic Publishers, Dordrecht.

Wallach, D., Hwang, C., Correll, M. J., Jones, J. W., Boote, K. J., Hoogenboom, G., Gezan, S., Bhakta, M. and Vallejos, C. E. 2018. A dynamic model with QTL covariables for predicting flowering time of common bean (*Phaseolus vulgaris*) genotypes. *Eur. J. Agron.* 101, 200–9. doi:10.1016/j.eja.2018.10.003.

Wallach, D., Makowski, D. and Brun, F. 2019. *Working with Dynamic Crop Models: Methods, Tools and Examples for Agriculture and Environment*. Academic Press, London, UK.

Wang, E. L., Martre, P., Zhao, Z. G., Ewert, F., Maiorano, A., Rotter, R. P., Kimball, B. A., Ottman, M. J., Wall, G. W., White, J. W., Reynolds, M. P., Alderman, P. D., Aggarwal, P. K., Anothai, J., Basso, B., Biernath, C., Cammarano, D., Challinor, A. J., De Sanctis, G., Doltra, J., Dumont, B., Fereres, E., GarciaVila, M., Gayler, S., Hoogenboom, G., Hunt, L. A., Izaurralde, R. C., Jabloun, M., Jones, C. D., Kersebaum, K. C., Koehler, A. K., Liu, L. L., Muller, C., Kumar, S. N., Nendel, C., O'Leary, G., Olesen, J. E., Palosuo, T., Priesack, E., Rezaei, E. E., Ripoche, D., Ruane, A. C., Semenov, M. A., Shcherbak, I., Stockle, C., Stratonovitch, P., Streck, T., Supit, I., Tao, F. L., Thorburn, P., Waha, K., Wallach, D., Wang, Z. M., Wolf, J., Zhu, Y. and Asseng, S. 2017. The uncertainty of crop yield projections is reduced by improved temperature response functions. *Nat. Plants* 3(10), 833-. doi:10.1038/s41477-017-0032-6.

Webber, H., White, J. W., Kimball, B. A., Ewert, F., Asseng, S., Rezaei, E. E., Pinter Jr., P. J., Hatfield, J. L., Reynolds, M. P., Ababei, B., Bindi, M., Doltra, J., Ferrise, R., Kage, H., Kassie, B. T., Kersebaum, K.-C., Luig, A., Olesen, J. E., Semenov, M. A., Stratonovitch, P., Ratjen, A. M., LaMorte, R. L., Leavitt, S. W., Hunsaker, D. J., Wall, G. W. and Matre, P. 2018. Physical robustness of canopy temperature models for crop heat stress simulation across environments and production conditions. *Field Crops Res.* 216, 75–88. doi:10.1016/j.fcr.2017.11.005.

Yin, X., Liden, C. G. and Struik, P. C. 2018. Bringing genetics and biochemistry to crop modelling, and vice versa. *Eur. J. Agron.* 100, 132–40. doi:10.1016/j.eja.2018.02.005.

Zhang, L., Gezan, S. A., Eduardo Vallejos, C., Jones, J. W., Boote, K. J., Clavijo-Michelangeli, J. A., Bhakta, M., Osorno, J. M., Rao, I., Beebe, S., Roman-Paoli, E., Gonzalez, A., Beaver, J., Ricaurte, J., Colbert, R. and Correll, M. J. 2017. Development of a QTL-environment-based predictive model for node addition rate in common bean. *Theor. Appl. Genet.* 130(5), 1065–79. doi:10.1007/s00122-017-2871-y.

Zhao, C. B., Liu, B., Xiao, L., Hoogenboom, G., Boote, K. J., Kassie, B. T., Pavan, W., Shelia, V., Soo-Kim, K. S., Hernandez-Ochoa, I. M., Wallach, D., Porter, C. H., Stockle, C. O., Zhu, Y. and Asseng, S. 2019. A SIMPLE crop model. *Europ. J. Agron.* 104, 97–106. doi:10.1016/j.eja.2019.01.009.

Zhu, C., Kobayashi, K., Loladze, I., Zhu, J., Jiang, Q., Xu, X., Liu, G., Seneweera, S., Ebi, K. L., Drewnowski, A., Fukagawa, N. K. and Ziska, L. H. 2018. Carbon dioxide (CO_2) levels this century will alter the protein, micronutrients, and vitamin content of rice grains with potential health consequences for the poorest rice-dependent countries. *Sci. Adv.* 4(5), eaaq1012. doi:10.1126/sciadv.aaq1012.

Index